MUSCLE ENERGY AND "BEYOND" TECHNIQUE

VOLUME III

INTEGRATIVE MANUAL THERAPY

FOR BIOMECHANICS: APPLICATION OF MUSCLE ENERGY AND "BEYOND" TECHNIQUE

Treatment of the Spine, Ribs, and Extremities

Sharon (Weiselfish) Giammatteo, Ph.D. P.T., I.M.P., C.
Thomas Giammatteo, D.C., P.T., I.M.P., C.
Photographs by John Giammatteo

North Atlantic Books
Berkeley, California

Published by
North Atlantic Books
P.O. Box 12327
Berkeley, California 94712

Cover and book design by Andrea DuFlon
Production by Jan Camp
Photography by John Giammatteo
Printed in the United States of America

Integrative Manual Therapy for Biodynamics is sponsored by the Society for the Study of Native Arts and Sciences, a nonprofit educational corporation whose goals are to develop an educational and crosscultural perspective linking various scientific, social, and artistic fields; to nurture a holistic view of arts, sciences, humanities, and healing; and to publish and distribute literature on the relationship of mind, body, and nature.

Library of Congress Cataloging-in-Publication Data

Weiselfish-Giammatteo, Sharon.
Integrative manual therapy for biomechanics: application of muscle energy and "beyond" technique: treatment of the spine, ribs, and extremities / by Sharon Weiselfish Giammatteo and Thomas Giammatteo.
p. ; cm.
"Integrative manual therapy, Volume III"—TOC p. vii
Includes bibliographical references (p. 335) and index.
ISBN 1-55643-435-9 (hardcover)
1. Manipulation (Therapeutics) 2. Sacroiliac joint.
3. Joints—Diseases—Patients—Rehabilitation. I. Giammatteo, Thomas. II. Title.
[DNLM: 1. Joint Diseases—therapy. 2. Biomechanics. 3. Muscular Diseases—therapy. 4. Musculoskeletal Manipulations—methods. 5. Physical Therapy Techniques—methods. 6. Spinal Diseases—therapy. WE 304 W427i 2003]
RD736.M25W45 2003
615.8'2—dc21 CIP
2003008121

1 2 3 4 5 6 7 8 9 DATA 07 06 05 04 03

DEDICATION

To our children, Nimrod, Ayelet, and Amir, whose love we cherish.

TABLE OF CONTENTS

FOREWORD

The journey in the field of biomechanics is never-ending. The concepts and techniques presented in this book are new, yet there are foundation blocks placed by many original thinkers, scientists and philosophers, contributed to over the past several hundred years.

Medical doctors, chiropractic physicians, physical therapists, and manual therapist have developed skills in order to affect joint mobility. This science has expanded to include articular balance as well as the concept of intra-articular space. Quantum physics has contributed theories of quanta and energy which fill in the space and maintain the distance between joint surfaces. Noetic sciences which investigate emotional energies are looking into the effect of energy on joint spaces.

This book builds upon the contributions of many clinical and theoretical scientists. Because the authors are physical therapists, including manual therapists, the information in this text fits in the realm of rehabilitation. The author has graduate and doctoral studies and research in the field of neurologic sciences, so the text offers clinical skills relevant to the client with neurologic disorders. The author has chiropractic and acupuncture education and training, thus much of this information is presented in different modes for the benefit of all disciplines.

The first book was published in 1994: *Manual Therapy with Muscle Energy Technique for the Pelvis, Sacrum, Cervical, Thoracic and Lumbar Spine.* During the following ten years, more information has been added to the body of knowledge relevant to the health and movement of joints. This book includes new and additional information. The text presents techniques and skills for the rib joints, and also for the peripheral joints of the arms and legs. The presentation is clear and concise, with graphics and photographs for easier and self-directed learning.

The graphics have been created with computer technology of the highest quality. The author combined his knowledge and skill from health care and computer technology to provide these exceptional learning tools. The photographs are contributions that are easy to understand. With the text and graphics and photographs, practitioners who are new to the field of biomechanics will feel comfortable using the techniques in the clinic.

Pre and post tests and measurements are unique. Assessment of posture is common in physical therapy and rehabilitation. Yet the interpretation of postural findings regarding the meaning of biomechanical dysfunction is contributed to by the author. The postural analysis will make the clinical decision making process easy. The author suggests a more dynamic pre- and post-testing procedure. In such a manner, the therapist will be able to see before and after results.

The philosophies in the text expand into theories of disc derangement, organ dysfunction, and muscle spasm: how do these and other somatic problems cause biomechanical dysfunction of the spine, rib cage, and extremities.

The 3-planar, dynamic approach to correction of joint dysfunction is a remarkable method to achieve pain free movement.

Sharon (Weiselfish) Giammatteo
Thomas Giammatteo

INTRODUCTION

Pain is an entity that has meaning for all. My own pain, which was disabling in its severity, was the driving force toward my rehabilitation and recovery process. I studied and was treated with Muscle Energy Technique. After two decades of research and considerable modifications and additions to the original techniques, I developed a process: Muscle Energy and "Beyond" Technique for the pelvis, sacrum, and spinal joints. I discovered that my low back pain, my sciatica, and my intermittent headaches had disappeared. I went on to discover that you never know how good you can feel until you get there.

With an attitude toward positive growth and continued recovery, I wanted to learn, as I still do, just how good I can feel. I was introduced to the many other wonderful manual therapy approaches: fascial release, Strain and Counterstrain Technique, cranial therapies, visceral manipulation, manual lymph drainage, and so much more. I realized pain could be a process to teach us, to educate us for our further discovery about the enormous joys of pain-free living.

In this book, the information regarding the sacrum is my most significant contribution. The mobility of the sacroiliac joints and the lumbosacral junction appears to be the key to total systems health for many patients. The approach described in this book can restore joint mobility to the sacral joints. Only when complete joint mobility is restored to the sacrum is realization of normal ranges of spinal motion possible.

Pelvic joints need to be treated in order to disimpact the sacrum. I strongly believe that the majority of spinal pain and disability experienced in the world is because of sacral disorders. Other causes contribute to spinal pain and disability, yet clinical research indicates that these other causes are not accessible to therapy when the sacral joints are stuck. The majority of spinal surgery occurs at L5/S1. Health care professionals need to restore mobility to this joint in order to reduce worldwide pain and suffering. When sacrum mobility is restored with Muscle Energy and "Beyond" Technique, over 70 percent of normal spinal motion throughout the cervical, thoracic, and lumbar spine is attained. Lumbosacral junction movement is attained, and discogenic pathology is eliminated.

The effect on healing attained by restoration of sacral mobility will allow the practitioner and the patient to experience hope. It is an indication for treatment.

Sharon (Weiselfish) Giammatteo

ACKNOWLEDGMENTS

The authors give thanks to the professionals, practitioners, clients, family, and friends who helped them attain this final product. They give thanks to all those who contributed: teaching, mentoring, facilitating throughout three decades of clinical investigation into solutions for the lumbosacral junction, spinal pain, and disability.

Most of the concepts presented in this book were developed during years of investigation. Only a small percentage of the understanding and techniques presented in this book is similar to that in previous books. Yet the philosophy of a direct, 3-planar approach for treatment of biomechanical dysfunction has found its place in the millennium.

Acknowledgments for this book include those who supported research in this field and who contributed to new philosophies, greater insight, and contemporary twentieth-century understanding. Thanks for these contributions are extended to Marion Eagan, Carol Gordon, Jay Kain, clients, and practitioners.

The authors extend their appreciation to Margaret Loomer, whose contribution with skills in word processing are invaluable.

The authors wish to thank many persons for their support, contributions, and encouragement:

John Giammatteo, photographer, whose professional contributions make this an exceptional learning tool.

Ayelet Giammatteo Weiselfish, the exceptional editor whose dedication to perfection made this book an experience for the learner.

Our colleagues at Regional Physical Therapy, Center of Integrative Manual Therapy and Dialogues in Contemporary Rehabilitation for their clinical research and their passion and mission for learning.

Ron Rosetti, president of Northeast Seminars, who supported the process of continuing education to get this work to the public.

Our clients, whose pursuit of pain-free movement and function is the reason for this body of knowledge.

Many, many thanks to colleagues who contributed their joy of learning, creative materials, expertise in clinical research studies, and clinical practice results at Regional Physical Therapy and Dialogues in Contemporary Rehabilitation.

For those above mentioned, and for those who are not mentioned, we remind you of our respect and appreciation for your caring, excellence in practice, and outcomes.

Sharon (Weiselfish) Giammatteo
Tom Giammatteo

CHAPTER 1

INTEGRATIVE MANUAL THERAPY

Integrative Manual Therapy

Integrative Manual Therapy is a process of assessment, diagnosis, and treatment. In 1983, the term was first used by the author to embrace a new type of rehabilitation medicine that included, the traditional functional rehabilitation approaches presented in physical therapy, occupational therapy, speech therapy, vocational therapy and psychotherapy. The term also embraced the concept of structural rehabilitation.

Structural rehabilitation plus functional rehabilitation equals Integrative Manual Therapy. This has not changed since 1983. Yet the concept has expanded. Many disciplines are empowered to utilize this work. Currently Integrative Manual Therapy is practiced by allopathic physicians, osteopathic physicians, chiropractic physicians, homeopathic physicians, and dentists. Physical therapists, occupational therapists, speech therapists, psychotherapists, counselors and social workers, wellness and body workers, massage therapists, nurses, dental hygienists and more incorporate all aspects of Integrative Manual Therapy into their work.

The phrase "Structure dictates function; function influences structure" is not new. There are many metaphors that use this concept.

Software and Hardware: A System

Functional rehabilitation can be considered the software of Integrative Manual Therapy. Structural rehabilitation can be considered the hardware of Integrative Manual Therapy. Functional rehabilitation will contribute the fundamental components of a person's program: walking, talking, listening, and all aspects of daily living. Structural rehabilitation will contribute the essence of the body.

Consider the computer: boards with chips. Consider each board as a "body system." There are many body systems, including bones, joints, muscles, connective tissue, organs, nerves, skin, glands, energy. If each board represents a system, then each "chip" on the board is an integral element of that system. For example, on the board that is the system of joints, one chip is the elbow joint; one chip is the hip joint; and so on.

Structural rehabilitation will investigate each chip of each system. Integrative Diagnostics permits investigation via mobility testing as well as motility testing. Motility is the biologic (circadian) rhythm of that system. Techniques are used to correct the dysfunction discovered in the chip.

Structural or Functional Rehabilitation

Functional rehabilitation is performed within disciplines. In the new millennium, there is little indication (yet) that functional rehabilitation will become transdisciplinary.

Physical therapists restore the function of movement. Usually there is a discrimination process: is the movement abnormal or normal? If the movement is normal, is the problem one of balance, coordination, strength, endurance, proprioception, or some other?

Occupational therapists also restore movement function. Typically, unless they are in the field of neuropediatric rehabilitation, the focus is often on upper extremity movement. This discipline has also encompassed the field of sensory integration: restoration of exteroception as well as proprioception. Occupational therapists have also embraced the field of psychotherapy as it relates to vocational rehabilitation.

Psychotherapists focus on the restoration of emotional and social reintegration. Their realm

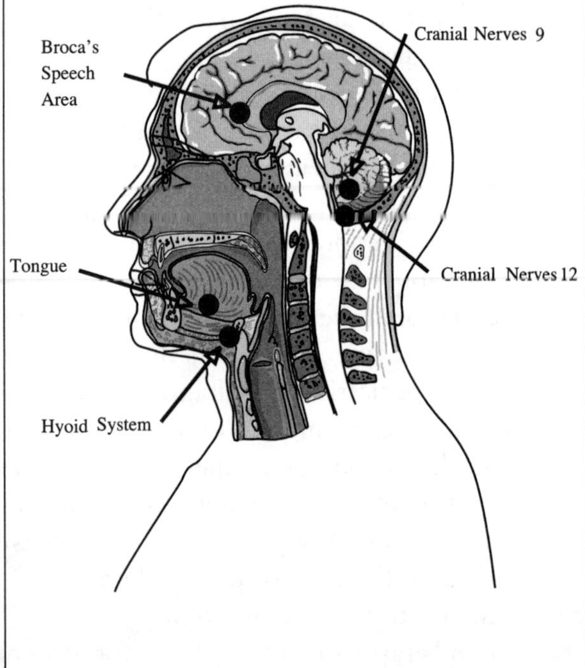

of discipline can include pediatrics, geriatrics, neurology, psychiatry, and more.

Speech therapists intend to improve speech. Drooling, speech impediments, swallowing dysfunction, and other aspects of communication are encompassed within this discipline. Within the realm of speech therapy, there is an easy differentiation between functional rehabilitation and structural rehabilitation. The speech pathologist focuses on function: respiration, phonation, and articulation. The practitioner's goal is to improve speech and swallowing. Exercises to attain this objective are used. Facilitation, breathing, strengthening, and inhibition as well as other tools are utilized. The intent of the speech therapist is to improve speech *depending upon the potential of the client.* The traditional speech pathologist *has no intention of changing the potential for speech.* How is structural rehabilitation used for the purpose of changing the potential for speech? Structure must be directly addressed. The cells, fibers, tissues, and organs of the system of speech must be improved. The pathoanatomy must change so that the functional potential can increase.

Structural rehabilitation to improve the potential for speech includes assessment, diagnosis, and treatment of the anatomical structures of speech. This means that the following areas must be addressed with Integrative Manual Therapy: (1) Broca's speech area in the brain; (2) cranial nerves #9 (glossopharyngeal) and #12 (hypoglossal); (3) the tissues and structures of the hyoid (tongue) system.

How can these areas be treated? An Integrated Systems Approach for assessment, diagnosis, and treatment of Broca's speech area, cranial nerves #9 and #12, and the joints and soft tissues of the hyoid system is recommended. What is the Integrated Systems Approach?

The Integrated Systems Approach

For every type of cell there needs to be a cell-oriented technique. The technique to correct a joint problem will not affect the muscle directly. The technique to correct a skeletal muscle problem will not affect the heart or any other organ directly. The technique to correct an organ problem will not affect the nervous tissue directly. And so on. There are many approaches, *many* techniques that are designed to facilitate healing. Each technique is adapted to the characteristics of each system. Muscle Energy and "Beyond" Technique corrects *joint problems*. Other techniques, developed for other systems, may correct organ problems, muscle problems, connective tissue problems, nerve problems, skin problems, and so on.

CHAPTER 2

DECREASING HYPERTONICITY WITH MANUAL THERAPY

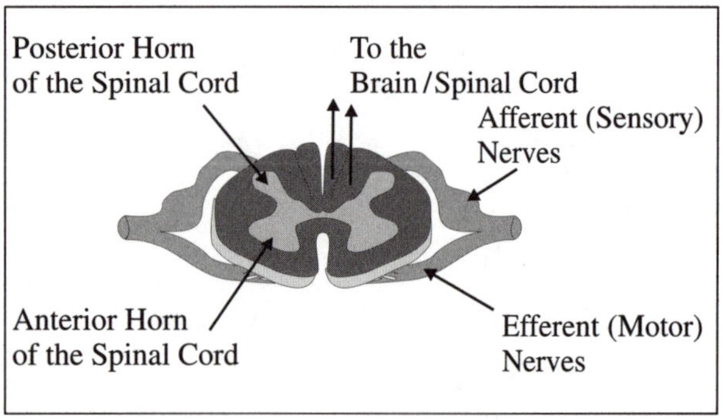

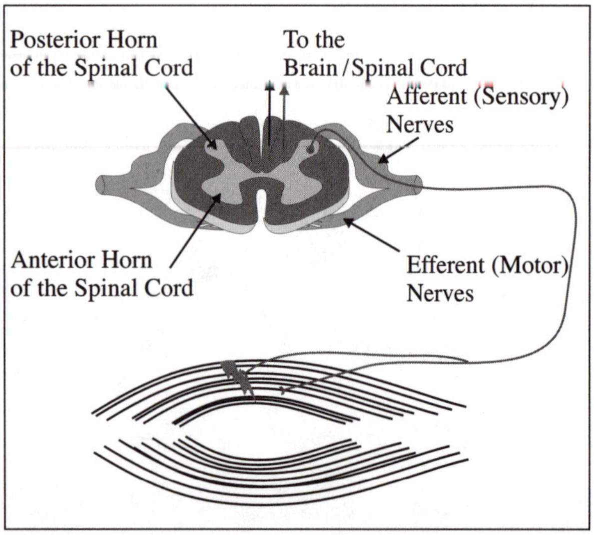

There has been a great deal of new understanding regarding protective muscle spasm during the past decade. Clinical practice of Integrative Manual Therapy (see Chapter One) has provided evidence regarding all aspects of muscle tone: low tone; excessive tone; spasticity.

This book is meant to correct biomechanical dysfunction: the dysfunction of joints. There is always the presence of excessive tone when there is joint dysfunction. This excessive tone (or hypertonicity) can contribute to biomechanical dysfunction. Often, the joint dysfunction will not correct without treatment of muscle spasm (hypertonicity).

When there is biomechanical dysfunction, there is *afferent* (sensory) feedback from the tissues surrounding the joint. Because there is a strain at the joint, the capsule, ligaments, tendons, and muscles will all be affected. Receptors will perceive the tension of the joint strain pattern. These receptors will transmit the information via the afferent nerves.

The afferent nerves carry the information about the tissue damage into the spinal cord via the posterior horn. From the posterior horn, the information about the strain patterns of the joints and surrounding tissues will be distributed throughout the central nervous system. Some of the information is passed along the nerve fibers into the brain. The information reaches all regions of the brain.

Some of the information is passed along to the anterior horn of the spinal cord.

At the anterior horn of the spinal cord, two "programs" occur. Program one: there is a synapse to the alpha nerve. The alpha nerve innervates the muscle fiber, the sarcomere.

Program two: There is a synapse to the gamma nerve. The gamma nerve innervates the muscle spindle.

How is this information passed along? The information about tissue dysfunction is passed along the nerve fibers as excessive, high-frequency discharge.

What is "discharge" along the length of a nerve fiber? The nerve fiber always has activity: Impulses are always traveling along all nerve fibers, at a certain frequency. When the impulse reaches an area of the nerve fiber, there is an action potential.

When the action potential passes that area of the nerve fiber, the nerve returns to its resting state, to a resting potential.

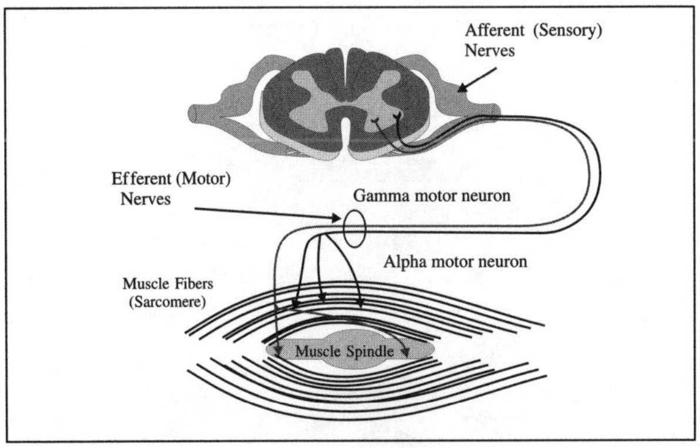

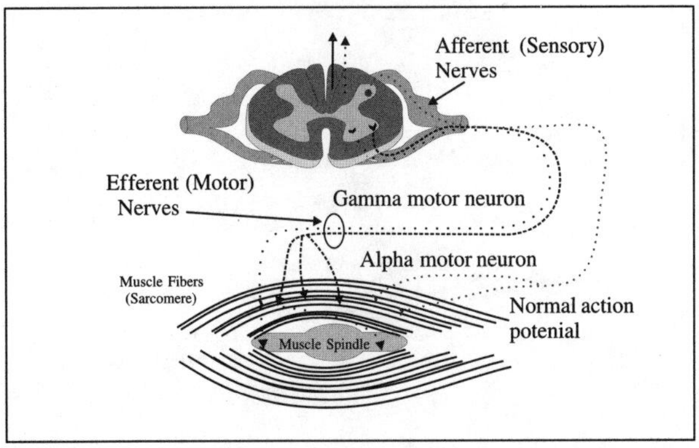

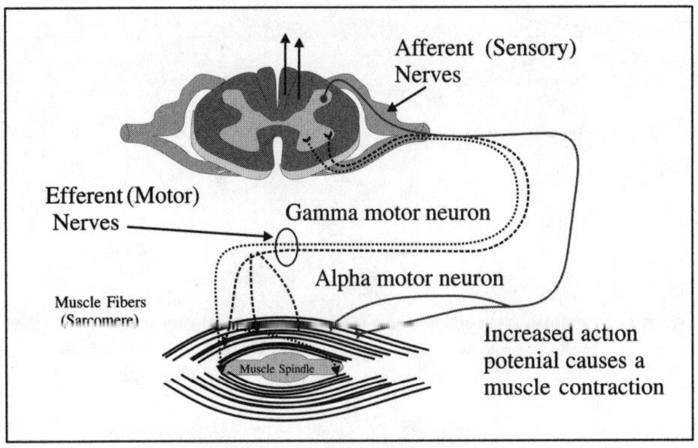

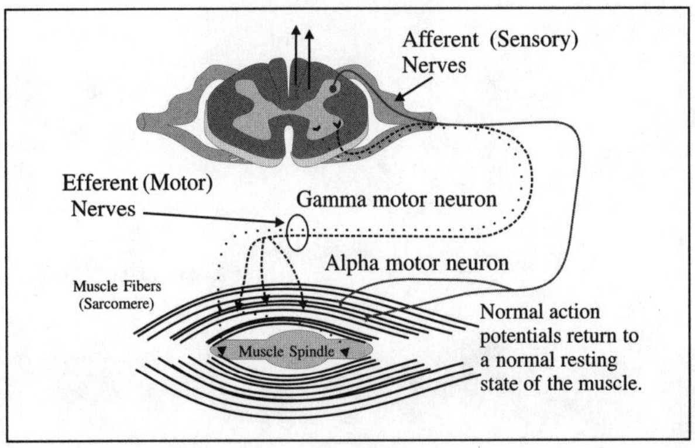

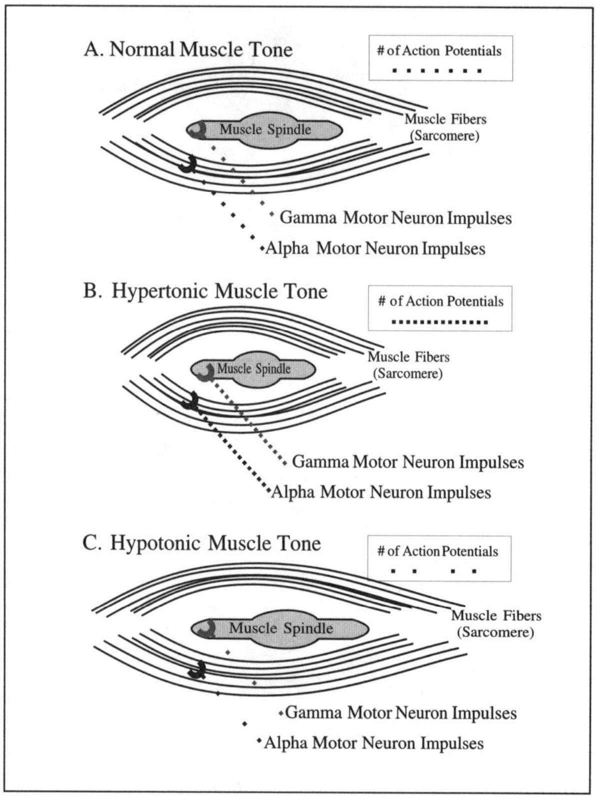

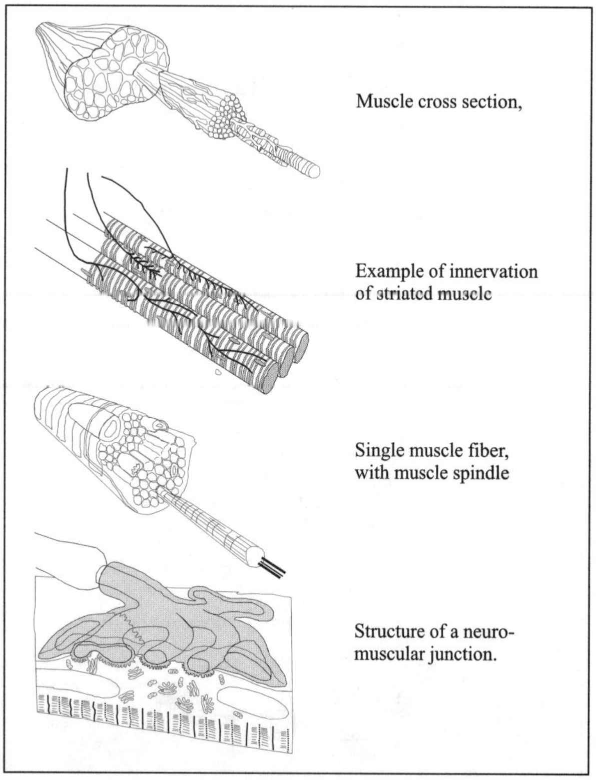

When too many impulses pass along the nerve fiber, there is excessive, high-frequency discharge of that nerve.

When there is a strain pattern at a joint, the receptors transmit that information via the sensory nerve fibers as excessive, high-frequency discharge. In other words, too many impulses are sent from the receptors along the sensory nerves into the spinal cord. These impulses are sent throughout the central nervous system and are passed along to the anterior horn of the spinal cord. This leads to an excessive amount of impulses crossing the synapses to the alpha nerves and to the gamma nerves. These impulses travel along the length of the alpha nerves and along the length of the gamma nerves.

What are the consequences of excessive, high-frequency discharge? Two are most evident:

1. Too many impulses affect the sarcomere.
2. Too many impulses affect the muscle spindle.

What occurs when too many impulses affect the sarcomere? Every time an impulse crosses the motor end plate (where the synapse occurs), calcium spills out of the sarcoplasmic reticulum. The muscle fiber then contracts.

When the muscle fiber contracts, there is a shortening of the muscle fiber. The muscle fiber "pulls" on the bone. The bone "moves" (shifts). Then the muscle fiber goes into a resting state. It can relax and elongate after the contraction. This occurs with normal muscle tone.

When too many impulses cross the motor end plate, too much calcium spills out of the sarcoplasmic reticulum. Too many contractions of the muscle fiber occurs (excessive). The muscle fiber contractions occur too quickly (high frequency). This is called tetany. In other words, the muscle fiber is maintained in a state of contraction. This means that the muscle fiber cannot relax and elongate. This is muscle spasm.

What occurs when too many impulses affect the muscle spindle? The muscle spindle is our protection. When the muscle spindle is stretched, the protection that occurs is a muscle contraction. Why? If we have a strain pattern at a joint, we cannot use the joint to its full capacity. If we do, we can further hurt the joint. The body must limit the amount of motion that is performed at that joint. The muscle spindle that is receiving too many impulses is more sensitive than the muscle spindle which is receiving normal amounts of impulses. How will a sensitive muscle spindle respond?

When the muscle spindle is sensitive because it is receiving too many impulses, the muscle spindle will respond to a lower level stretch. A lower level stretch means a lower amplitude stretch, less forceful stretch, and lower velocity stretch. When the muscle spindle responds to stretch, there is a reaction. The reaction is a muscle contraction. The muscle that belongs to the muscle spindle will contract.

If the muscle spindle and the spindle's nerve innervation are healthy, the muscle contraction (the reaction of the spindle being stretched) will normally occur when there is *end range motion.* In the case of a sensitive muscle spindle, the muscle contraction will occur at *inner ranges of motion* (for severe cases) and *mid ranges of motion* (for moderate cases).

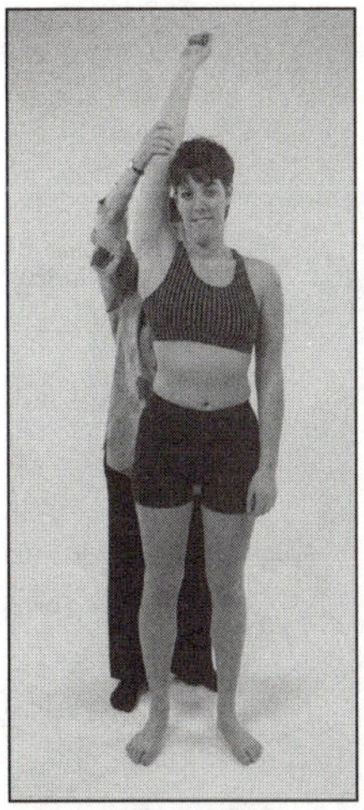

End range motion

Mid range motion

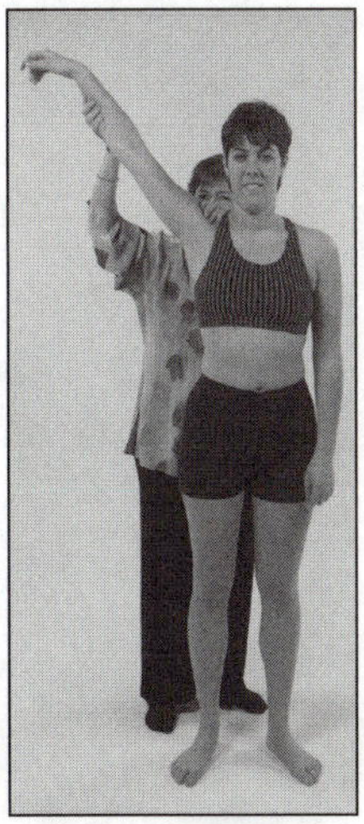

Inner range motion

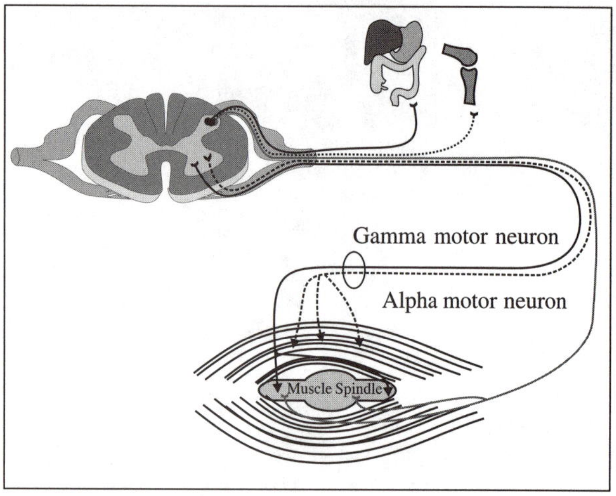

There is a lot of evidence from clinical practitioners that this elevated sensitivity of the muscle spindle is responsible for protection. This protection provided by the muscle spindle is a muscle contraction that occurs when motion is too great for the muscle spindle, and therefore the joint, to tolerate!

What is protective muscle spasm?

1. Muscle spasm is sustained sarcomere fiber shortening. The sarcomere cannot relax and elongate to a normal resting length. This is secondary to alpha nerve excessive, high-frequency discharge.
2. Protective muscle spasm is the hypersensitive muscle spindle causing a muscle contraction to protect the joint whenever there is a strain pattern of that joint. This is secondary to gamma nerve excessive, high-frequency discharge.

The purpose of Muscle Energy and "Beyond" Technique is to correct strain patterns in the joints. When the strain patterns in the joints are corrected, muscle spasm may be reduced. If the muscle spasm is severe and chronic, it may have become a primary problem. This means that when there was no joint problem, there was no muscle spasm. Then a strain pattern occurred at the joint. There was excessive, high-frequency discharge along the afferent nerves. Then there was excessive, high-frequency discharge along the length of the alpha and gamma nerves. This resulted in a secondary problem: muscle spasm. When the secondary problem is chronic and severe, it may transform into a primary problem. If the muscle spasm transforms into a primary problem, the Integrated Systems Approach must change. The techniques for joints are different from the techniques for muscle. (See Chapter One.)

Techniques for muscles can be reviewed in books/videos/seminars on the subject of Strain and Counterstrain Technique (Lawrence Jones, D.O.) and Advanced Strain and Counterstrain Technique (Thomas Giammatteo, D.C., P.T. and

Sharon (Weiselfish) Giammatteo, Ph.D., P.T.).

The joint that has problems will have receptors associated with a specific spinal segment. For example, if the glenohumeral joint has problems, the associated spinal segment is C5. C5 (the fifth cervical vertebra) supplies innervation to the dermatome, myotome, and scleratome of the shoulder girdle region. This means, if there is a strain pattern of the glenohumeral joint, the following problems (secondary problems) could occur:

1. Pain in the shoulder girdle region
2. Muscle spasm of the muscles innervated by C5
3. Limitations of motion of the shoulder girdle and shoulder joint
4. Fascial problems of the connective tissue, especially the specialized clavipectoral fascia (of the anterior shoulder girdle)
5. Disc dysfunction
6. Nerve tension of the C5 nerve roots, cervical plexus, and brachial plexus (which are supplied by C5 nerves)
7. Superior lung and dome of the lung problems
8. Tongue problems (leading to speech and swallowing difficulties)

If the joint is a more complex joint, the innervation may be supplied from more than one spinal segment. For example, the sacroiliac joint (as found on cadaver dissection) is innervated by any or all combinations between L1 and S2. This means that one person may have the sacroiliac joint innervated by L1, L2, L3, L4, and L5, and another person may have the sacroiliac joint innervated by L4, L5, and S1. Multiple possibilities cause many problems for diagnostics. If the sacroiliac joint has a strain pattern, receptors from the joint and surrounding tissue will transmit excessive, high-frequency discharge along the afferents into the posterior horns of all the spinal vertebrae that contribute innervation to the dysfunctional joint. Then there will be excessive, high-frequency discharge along all of the alpha

Patient is 36 years old with complaint of low back pain

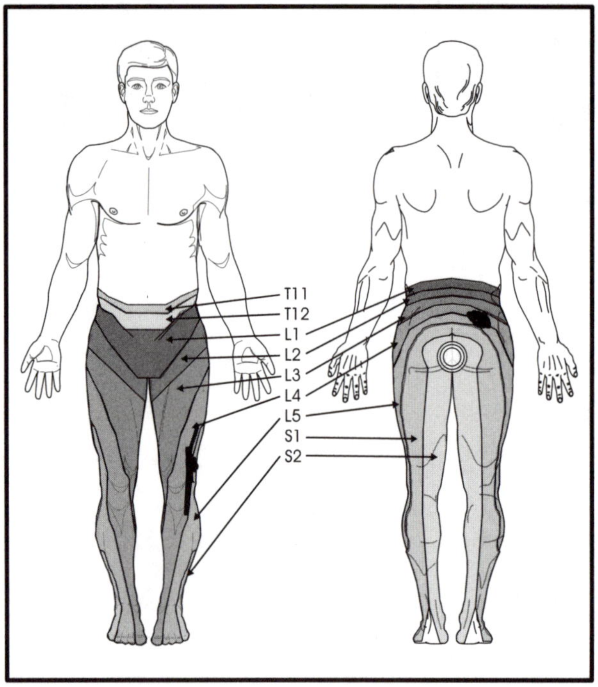

Patient is 35 years old with complaint of low back pain

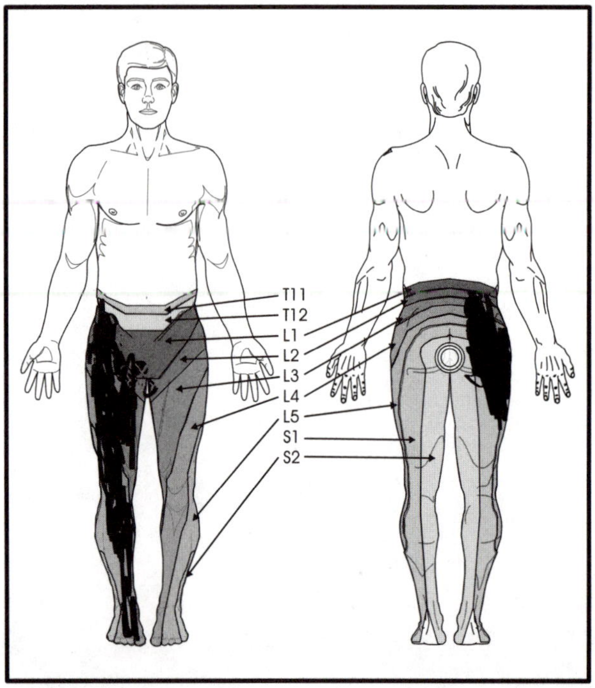

as well as all of the gamma nerves, which exit from all of those segments. This means that all the muscles innervated by the alpha nerves can go into muscle spasm and all the muscle spindles can be hypersensitive, causing "protective modes" in all these muscles.

Summary: The person whose sacroiliac joint is innervated by only L4 and L5 may feel pain in dermatomes L4 and L5, with muscle spasm of muscles innervated by those segments. Yet for the person whose sacroiliac joint is innervated by L1, L2, L3, L4, L5, S1, and S2 may feel a lot more pain, and a lot more protective muscle spasm.

CHAPTER 3

POSTURE

Posture or Posturing?

Whenever there are significant postural deviations, the practitioner must consider the following:

1. Posture is correctable with facilitated manual therapy. This means that there are postural deviations that occur directly because of joint and soft tissue dysfunction.
2. Posturing is not habit. Posturing can be the result of emotional stress, mental/thought activity, or spiritual practice. Posturing requires more than manual therapy.

Postural Deviations

Posture is a reflection of dysfunction. Postural deviations are an indication for treatment with Muscle Energy and "Beyond" Technique. Certain postures reflect specific biomechanical dysfunction. The information in this chapter will provide shortcuts: Years of clinical research are condensed in this chapter.

The following systems contribute most often to postural deviations:

1. Biomechanics
2. Muscle
3. Connective tissue
4. Circulation (especially arterial)

Joint (biomechanical) dysfunction will cause postural deviations on all three planes: sagittal, coronal, and transverse. The most significant changes in posture will occur when biomechanics of the pelvis, sacrum, and lumbosacral junction are corrected. This means that marked shifts in flexion and extension, side bending and rotation occur when joint dysfunction of the iliosacral joints (pelvis), accessory joints (pelvis), sacroiliac joints (sacrum), and lumbosacral junction are corrected. The most significant changes in posture will occur in the following body regions:

1. Neck
2. Thoracic spine
3. Lumbar spine

In other words, correction of biomechanical dysfunction will change the posture of the neck, trunk, and low back. When skeletal muscle is treated, posture of the arms and legs changes. When connective tissue is treated, posture of the arms and legs changes. When circulation (especially arterial) is treated, total body posture changes. The body protects arteries; posture will shift dramatically in order to provide protection to a compromised blood vessel.

Evaluation with a Neutral Base of Support

Posture is evaluated on three planes: sagittal, coronal, and transverse. Have the patient stand with a neutral base of support during all posture evaluations:

1. The feet are acetabular distance apart.
2. There is approximately 15 degrees of equal external rotation of both feet.
3. The feet are evenly positioned. It is important that one foot not be in front of the other.
4. The knees should be equally flexed/extended. If there is recurvatum of one knee, it should be maintained in neutral to equal the posture of the other knee.

When the client stands with an asymmetric base of support, the true underlying physical postural dysfunction cannot be observed. While standing with a symmetric base of support, true postural deviations of the spine are evident. If a person has the opportunity to stand in an asymmetric position, he or she will do so. In this stance, the body will diffuse the problems in order to decrease the compromising of the tissues and structures in dysfunction. In other words, people stand asymmetrically in order to decrease discomfort.

Sagittal Plane Posture

Static Postural Evaluation

Sagittal plane posture is assessed from a side view. Sagittal plane movements are flexion and extension. As long as there is fair standing balance, sagittal plane posture can be assessed. The therapist images an imaginary plumb line, which touches the apex of what would be a normal thoracic kyphosis. This plumb line should also touch the occiput and the buttocks. The distance from this plumb line to the apex of the lumbar lordosis is approximately 5 to 6 cm in healthy sagittal plane posture. The distance of this plumb line to the apex of the cervical lordosis should be approximately 5 to 6 cm.

When sagittal plane posture is viewed from the side, the following may be noted:

1. Tension of the anterior cervical muscles, with problems affecting tongue, larynx, and pharynx
2. Compromise at sternum, where there is a depression of that bone posterior
3. Compromise at the respiratory abdominal diaphragm
4. Compromise of the abdominal cavity tissue

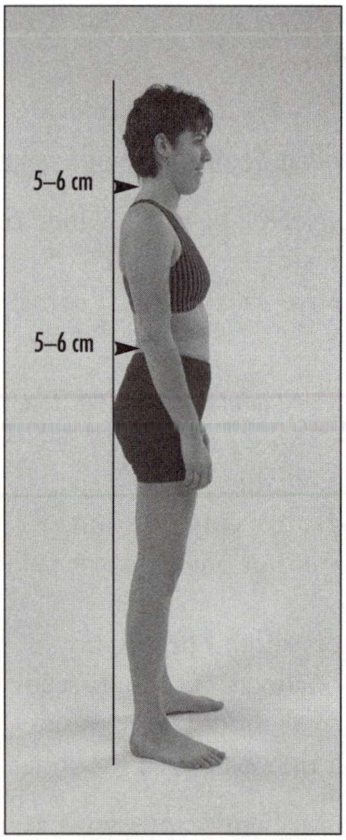

Normal sagittal plane posture.

Typical Static Postural Dysfunction of the Spine

1. Forward head and neck posture: This reflects a loss of cervical spine lordosis; the cervical vertebrae are "stuck." One or more of the vertebrae of the neck will be stuck: flexed, anterior and/or lateral shear.
2. Loss of thoracic kyphosis: Loss of thoracic kyphosis is caused by thoracic vertebrae that are stuck extended or anterior and/or laterally sheared. The primary reason for decreased thoracic kyphosis will *always* be secondary to internal rib cage soft tissue dysfunction. Specifically: pleura, lung tissue, heart or circulatory vessels, pericardium, mediastinum, and other.

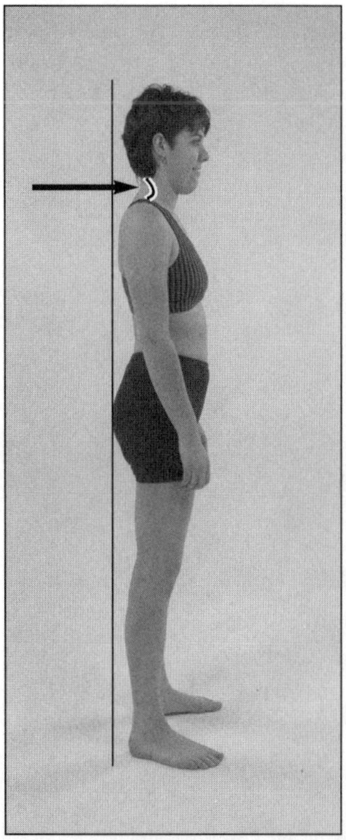

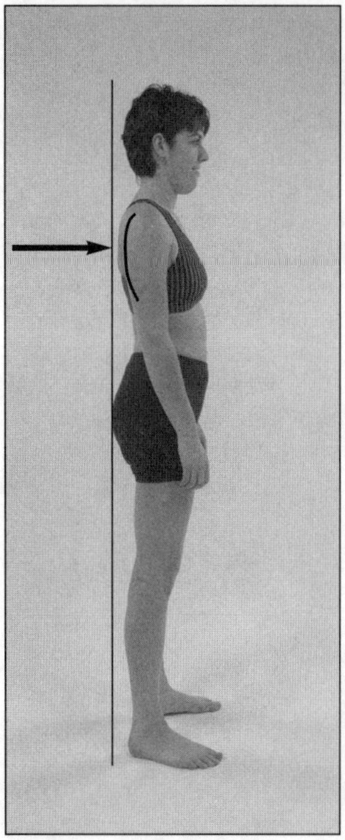

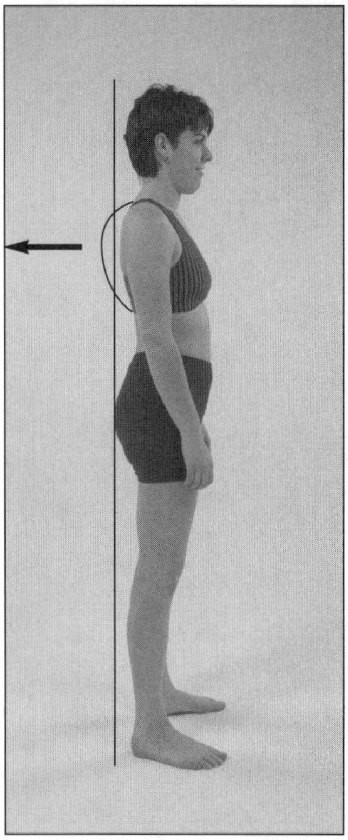

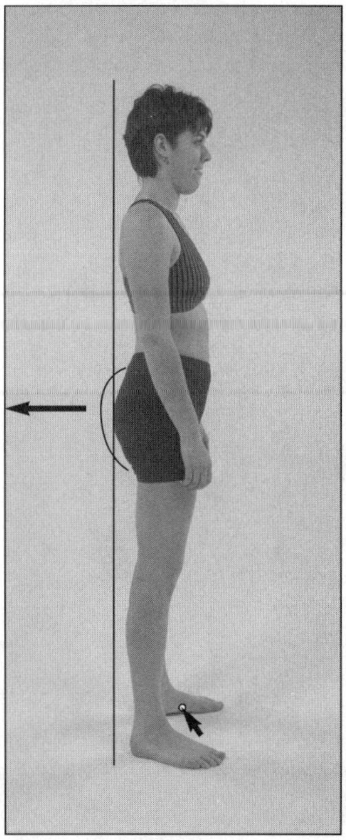

3. Excessive thoracic kyphosis: An excessive thoracic kyphosis is due to thoracic vertebrae that are stuck flexed. Often, there is spinal cord and/or nerve root tissue tension as well as osteopenia/osteoporosis of those thoracic vertebrae. Often, the kyphosis occurs to protect the internal organs, especially the heart. The kyphosis is excessive because the person simply cannot stand up straight. This may also be a protective mode. For example, if there is dysfunction of the femoral artery wall, when the person stands up straight there is increased tension on the femoral artery. The body will protect itself from increased tension and pressure on arteries that are in stress.
4. Posterior buttocks: When the buttocks are posterior (posterior to the imaginary plumb line) there is always dysfunction of the sacroiliac joints and the lumbosacral junction. Therefore, when biomechanics are correct at the sacroiliac joints and at the lumbosacral junction, the posterior buttocks phenomenon will disappear!

   Often there is an appearance of excessive lumbar lordosis, rather than posterior buttocks. In fact, loss of lumbar lordosis is common; rarely is there real excessive lumbar lordosis. The appearance of excessive lumbar lordosis is always because of the presentation of posterior buttocks.

5. Loss of lumbar lordosis: When there is loss of lumbar lordosis (which can be viewed on x-ray), the lumbar vertebrae are stuck flexed. Often, there is also a component of an anterior and/or lateral shear of those lumbar vertebrae, which contributes to the loss of lumbar lordosis.
6. Protruding chin: Whenever the chin is protruding, there is hyperextension of occiput on atlas. There should always be space between occiput and atlas, with free anterior and posterior glides of occiput on atlas. Often, the suboccipital space becomes compressed, causing tension on the occipital region, inferior brain stem, and other neurovascular tissues. Headaches are common.

   There are typical causes that result in compromise of the suboccipital space (craniocervical junction), and those causes require therapy. They include: muscle spasm of the suboccipital musculature; foramen pressures of the foramen magnum, jugular foramen, and mastoid foramen; tension of the tentorium cerebellum; sacral biomechanical dysfunction; arterial insufficiency, including circle of willis, basilar artery, and spinal artery; venous compromise, including the great vein of Galen; dural membrane tension, including transverse sinus and straight sinus.

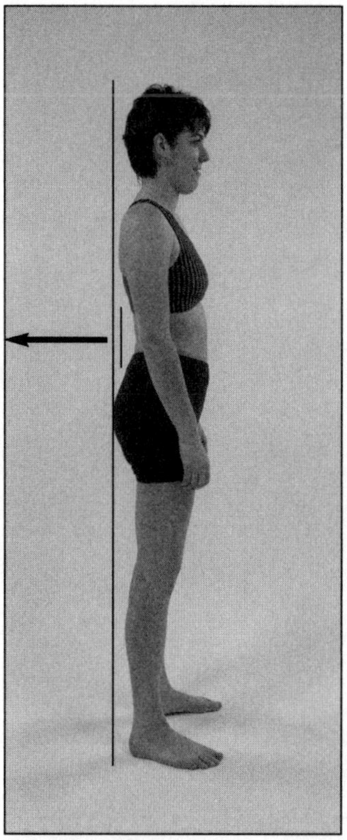

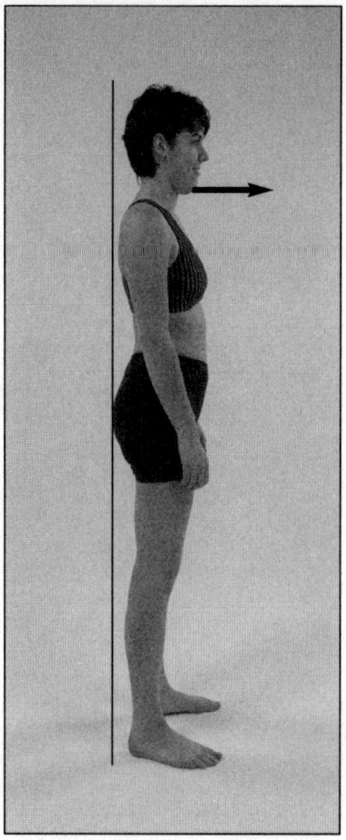

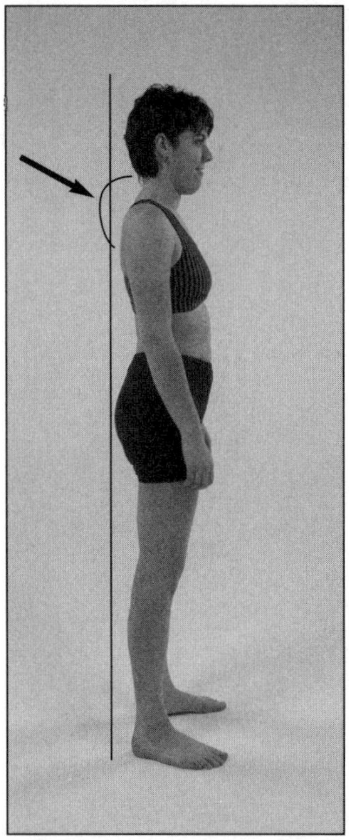

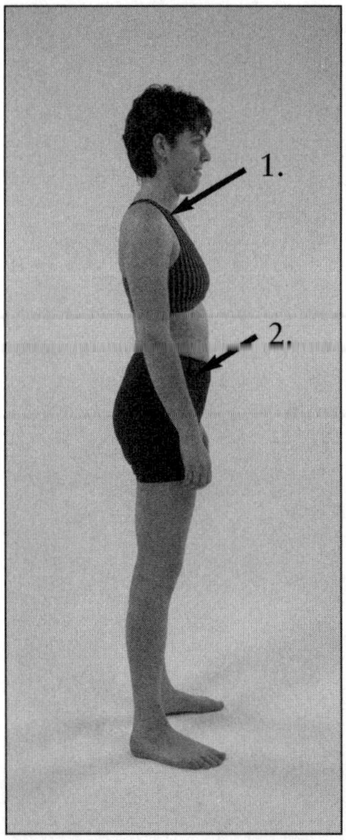

7. Dowager's hump: The Dowager's Hump is a dysfunction of the cervicothoracic junction. Typical problems associated with and contributing to this postural deviation are: flexed C7 and/or anterior shear; flexed T1; compressed C7/T1 interspace, with compromise of the C7/T1 disc; nerve root impingement of C7 nerve roots; cervical plexus neural tissue tension; compromise of the spinal artery; compromise of dural mobility; spinal cord fibrosis at the level of the cervicothoracic spine; biomechanical dysfunction of sacrum.

Typical Static Postural Dysfunction of the Extremities

1. Anterior shoulders: Anterior shoulders are most often caused by muscle spasm of the pectoralis minor and the subscapularis.
2. Flexed hips: More often, flexed hips are secondary to sacroiliac and lumbosacral junction dysfunction (biomechanics). Additional common causes include: arterial and venous compromise of the femoral artery and vein, and protective muscle spasm of the iliacus and anterior flexors of L5.

3. Recurvatum of the knees: This postural deviation is most often caused by ligamentous laxity of the articular and intra-articular ligaments of the knee as well as muscle spasm of the quadriceps. Muscle spasm of the gastrocnemius and biomechanical dysfunction is a common contributor. Also, limitation of dorsiflexion is a common cause. When the above mentioned dysfunctions are corrected, the recurvatum should be eliminated.
4. Posterior tibial glide: A very common problem is limitation of dorsiflexion. This appears to be a world-wide problem. When there is dorsiflexion, tibia can glide anterior over talus. When there is a limitation of dorsiflexion, tibia is stuck in posterior glide on talus. This must be corrected for optimal biomechanical correction.

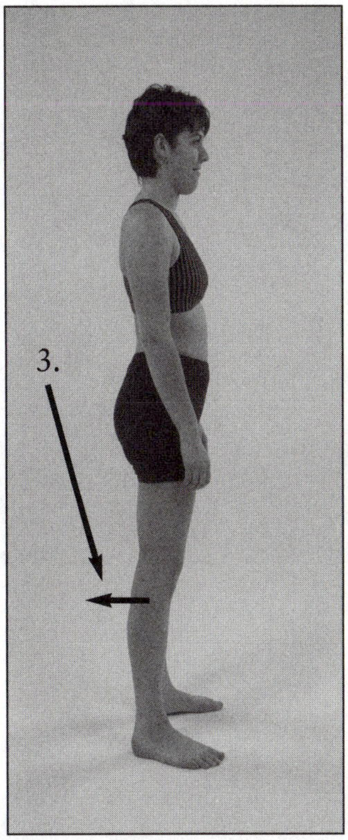

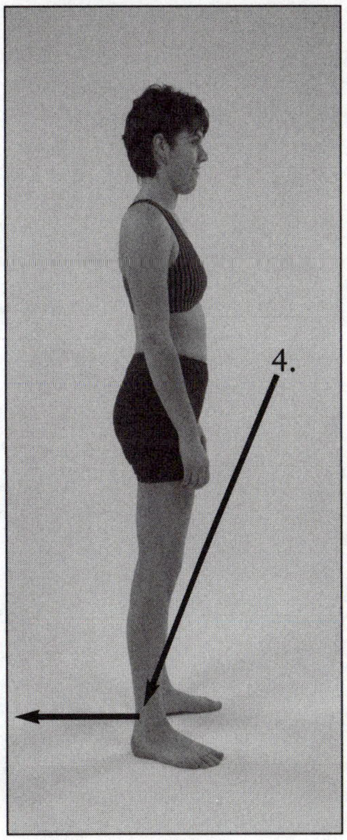

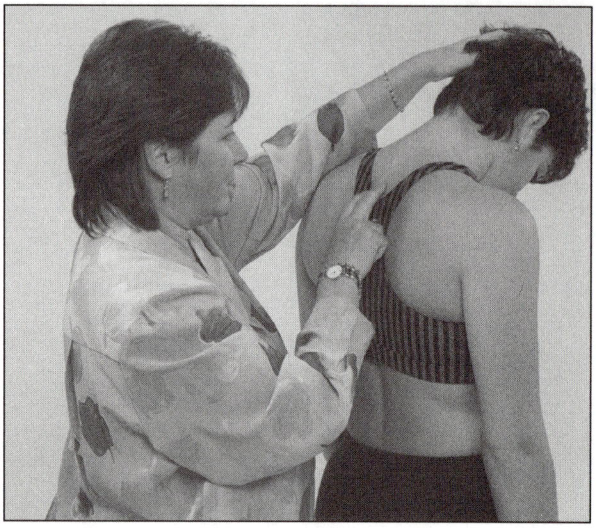

Dynamic Postural Evaluation

Usually, practitioners do not correctly assess ranges of motion. Flexion must be anterior displacement of the body. Extension must be posterior displacement of the body.

Flexion is attained "down the kinetic chain" as follows:

1. Tuck the chin.
2. Flex the neck down the kinetic chain, maintaining chin tuck.
3. Continue to flex the thoracic spine. Do not allow the buttocks to protrude in a posterior direction.
4. Continue to flex the lumbar spine. Do not allow the buttocks to protrude in a posterior direction.

Per definition, during extension, the body part extending must be totally posterior to midline. Per definition, during flexion, the body part flexing must be completely anterior to midline. This means, if during cervical extension any part of the anterior neck is more anterior to midline than when in neutral resting position, "trick" or compensatory motions are occurring. The extension is not true extension. Rather, anterior shears of the cervical vertebrae are being elicited in an attempt to extend in a physiologic manner.

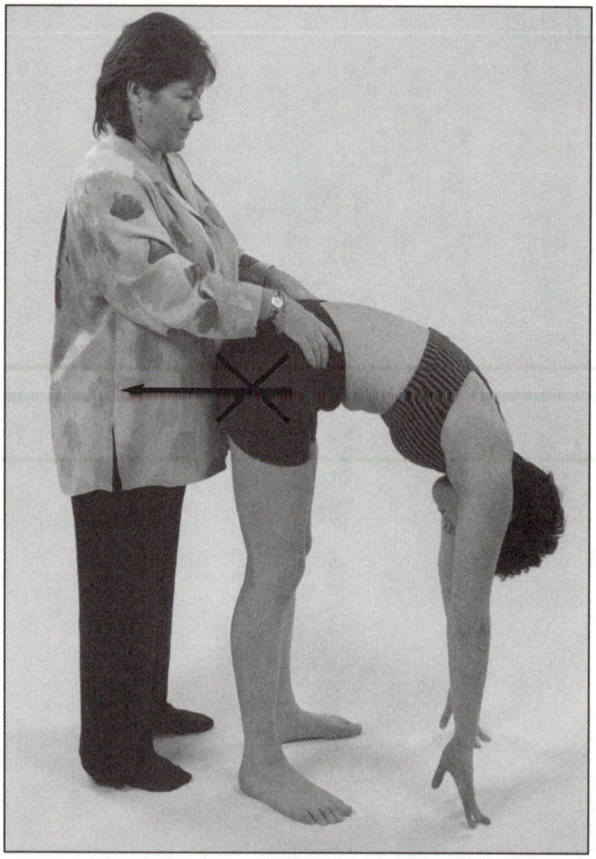

Extension is attained down the kinetic chain as follows:

1. Attain axial neutral of the head and neck.
2. Extend down the kinetic chain, maintaining axial neutral of the head and neck. Do not allow the chin to protrude.
3. Extend further down the kinetic chain, one segment at a time, through each thoracic segment.
4. Extend down the kinetic chain, one lumbar segment at a time. Each lumbar segment will contribute minimally to extension until L5 is contacted.
5. Focus on the lumbosacral junction. Perceive the motion of L5 on S1. Perceive how the sacral base (S1) glides anterior, while L5 extends in a posterior direction. Do not allow "cheating" at the abdominal cavity and hips. The abdominal cavity and hips should not move in an anterior direction.

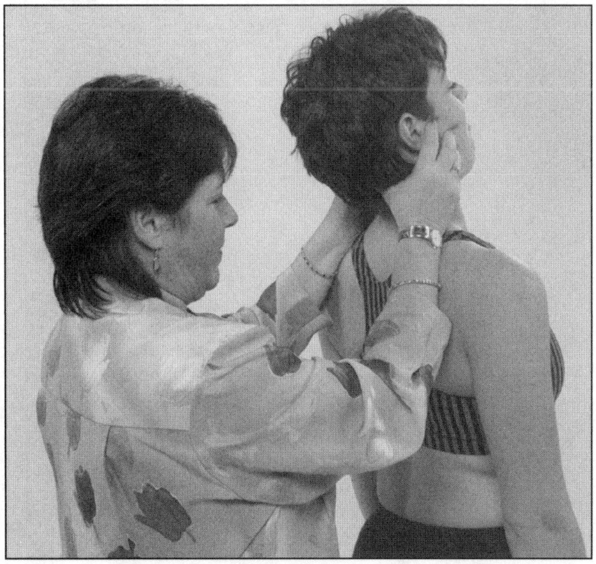

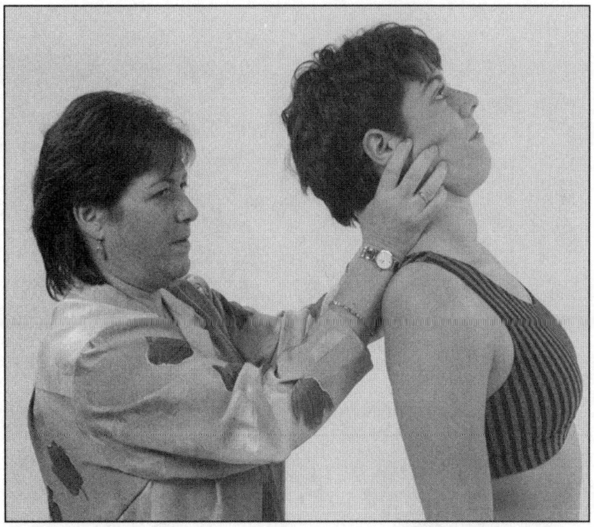

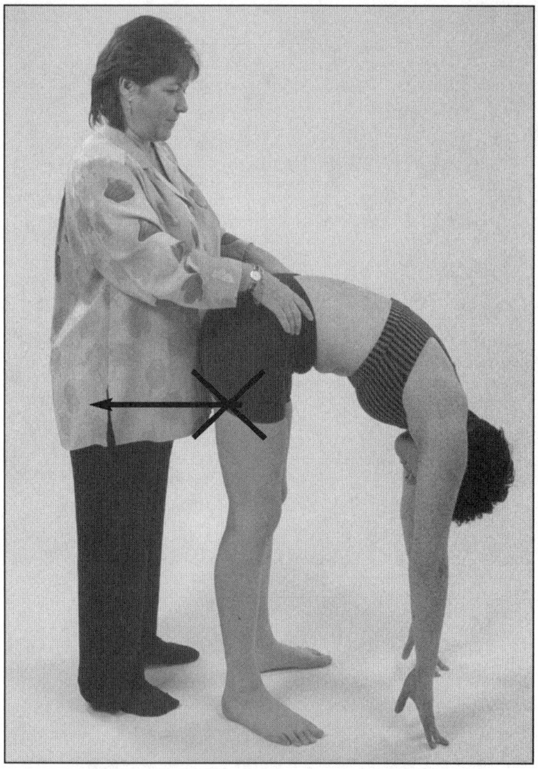

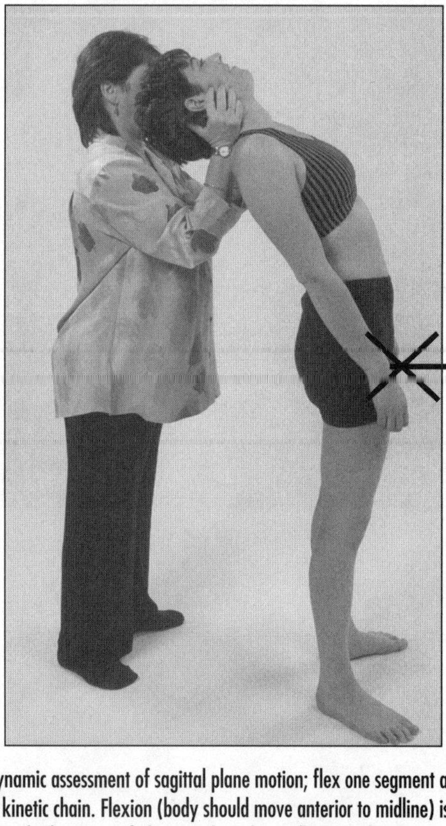

Correct dynamic assessment of sagittal plane motion; flex one segment at a time down the kinetic chain. Flexion (body should move anterior to midline) is attained by posterior displacement of tibia on talus (plantar flexion). This is not spinal flexion.

Typical Deviations of Motion

Movement must be assessed as unidirectional and uniplanar motion. This means that when there are strain patterns at joints, movement is restricted. The body will attempt to compensate for these restrictions. How will these compensations occur? The body will deviate during movement into other planes of motion. For example, if pure spinal flexion is limited, the body will add rotation and/or side bending in an attempt to gain more range. Inhibit all compensations.

During spinal flexion:

1. Inhibit posterior displacement of buttocks.
2. Inhibit rotation and / or side bending of spine.
3. Maintain chin tuck.

During spinal extension:

1. Inhibit anterior displacement of abdomen and hips.
2. Inhibit rotation and / or side bending of spine.

Coronal Plane Posture

Static Postural Evaluation

Coronal plane posture can be evaluated in the standing position. A posterior view and an anterior view are appropriate and will typically provide the same information. If the patient cannot stand, then coronal plane posture in a sitting, supine, or prone position is possible.

Coronal plane postural dysfunction is assessed by comparing right and left discrepancies in elevation (superior height) of anatomical landmarks. The following anatomical landmarks are assessed:

Posterior view:

1. Gluteal folds
2. PSIS
3. Iliac crests
4. Twelfth ribs, lateral aspects
5. Elbow levels, with flexed elbows
6. Space between lateral wall of rib cage and ilium
7. Inferior angle of scapula
8. Shoulder girdles
9. Lateral "shift" of neck
10. Ears

Anterior view:

1. Eyes
2. Ears
3. Shoulder girdles
4. Nipple line

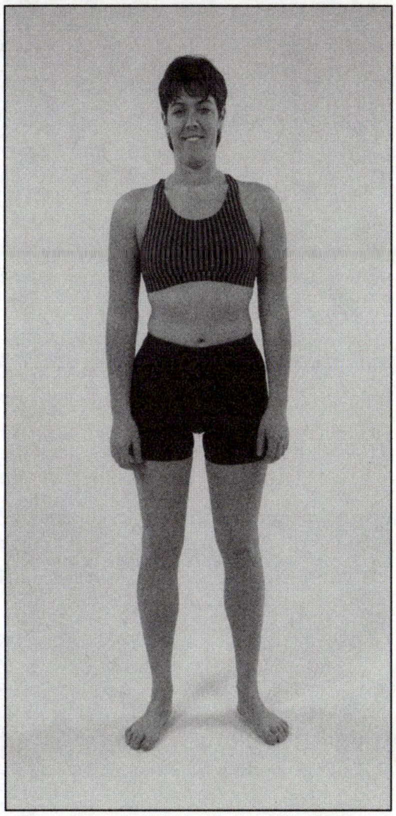

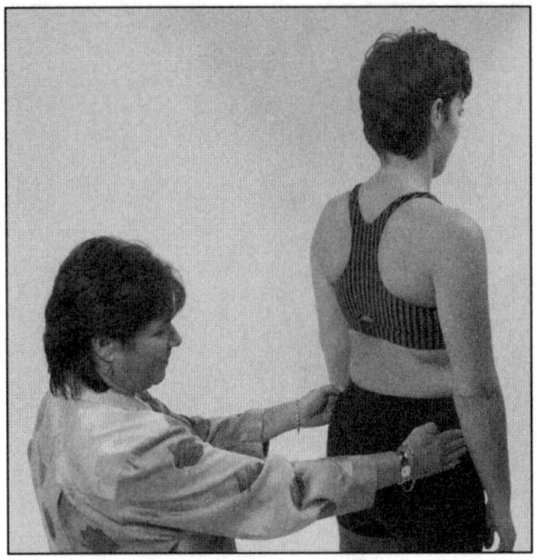

Extremities:

1. Lateral displacement or compression of the greater trochanters
2. Lateral glide of the tibial plateau at the knee joint
3. Pronation or supination of the hind foot, mid foot, forefoot
4. Cubital angle

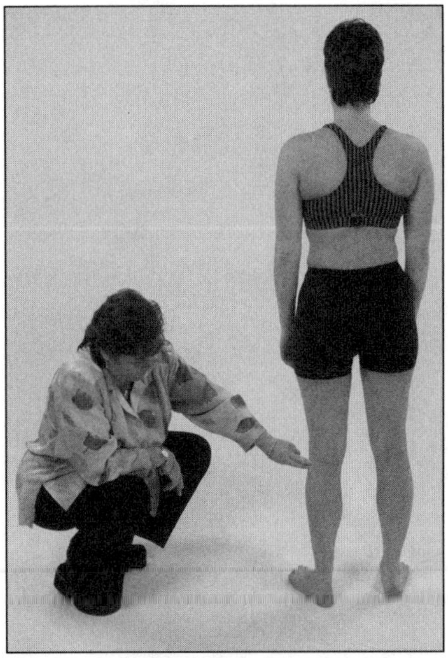

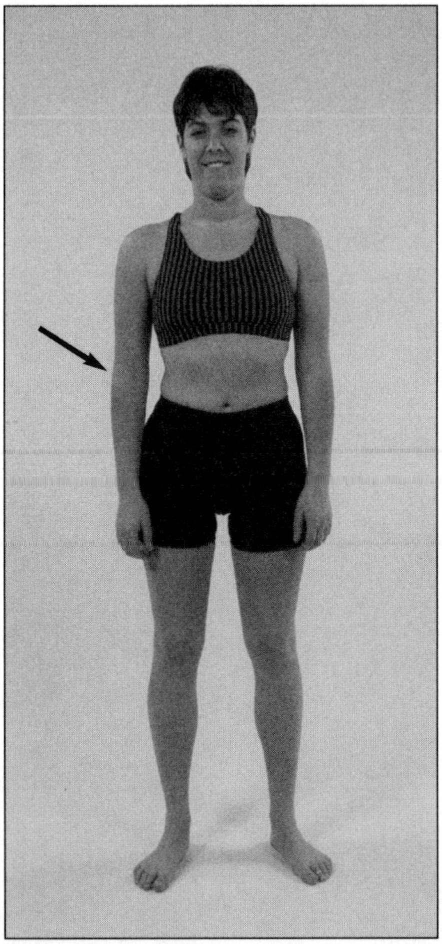

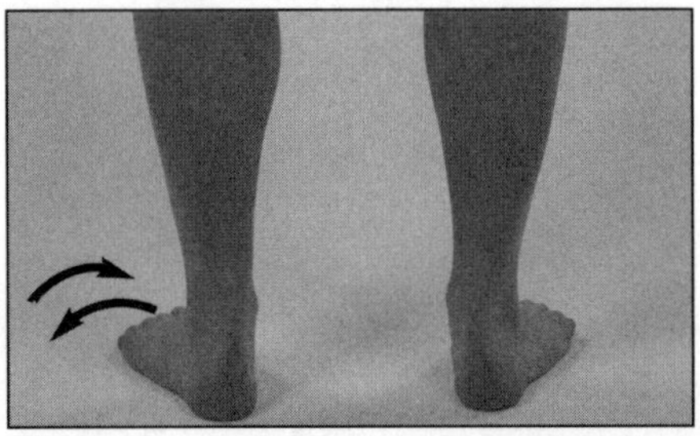

Coronal plane dysfunction of the spine is scoliosis.

Coronal plane dysfunction of the arms and legs is reflected in the following postural deviations:

1. Elevated shoulder girdle
2. Increased cubital angle
3. Ulnar deviation of the wrist
4. Adduction of the hip and upper leg
5. Lateral glide of the proximal tibial head
6. Supination or pronation of the foot and ankle

Typical Deviations of Motion

Static postural deviations can be interpreted to indicate physiologic limitations in ranges of motion. This means: Assess posture. Consider: If the postural deviation is causing a limitation of motion, what motion will be found to be limited? For example, if there is a right pelvic obliquity, then there is also a right side bending of the lumbar spine. There may be full right side bending of the lumbar spine, but there cannot be full left lumbar side bending. Another example: If there is a left shoulder girdle obliquity, there will often be a compensatory deviation of the cervical spine to the right, with limitations in left side bending of the cervical spine. (According to Fryette's third law of biomechanics, postural dysfunction will correlate with limitations in ranges of motion.) Side bending will always be limited to one side.

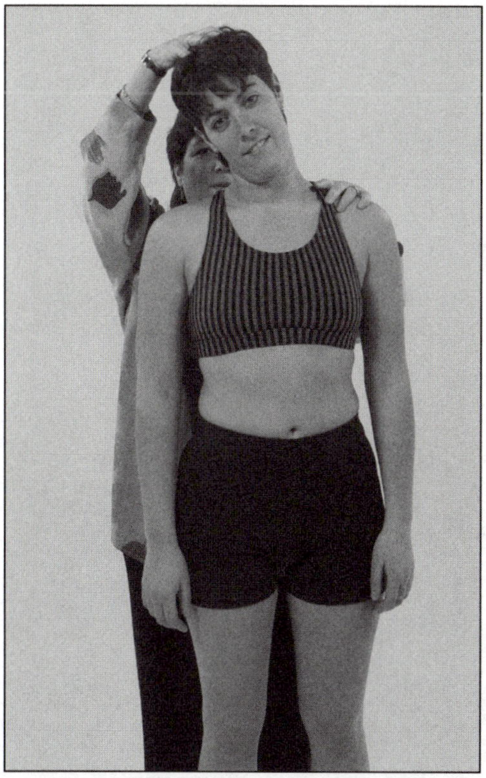

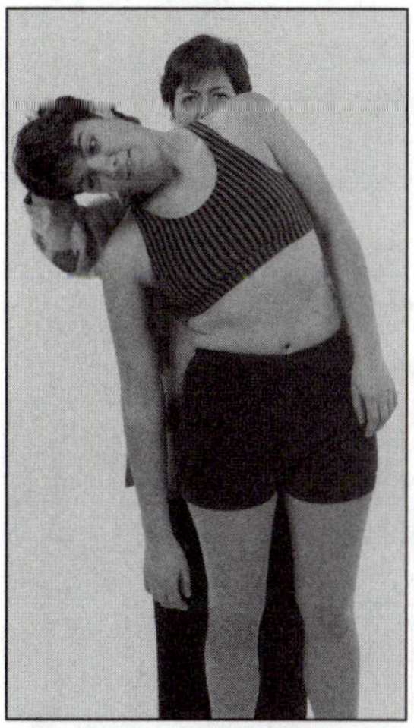

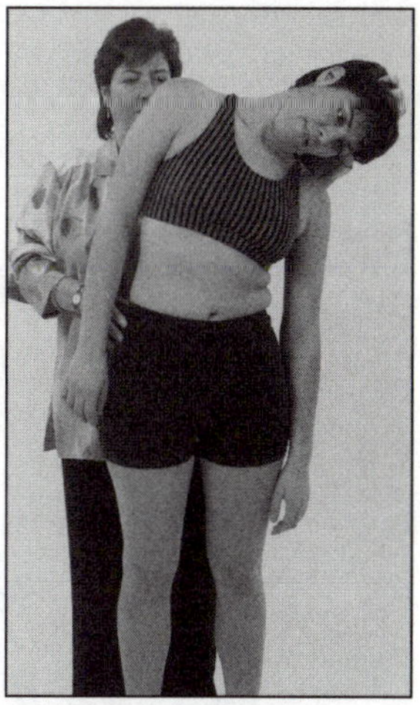

[Above, right] Coronal plane movement: Left side bending. Note: Any compensatory (trick) right lateral shear of pelvis. [Above, left] Note: Rib restrictions vs. spinal coronal plane limitations in motion. Note: Elevation of ipsilateral shoulder girdle and rotation of neck to compensate for true limitations in cervical side bending.

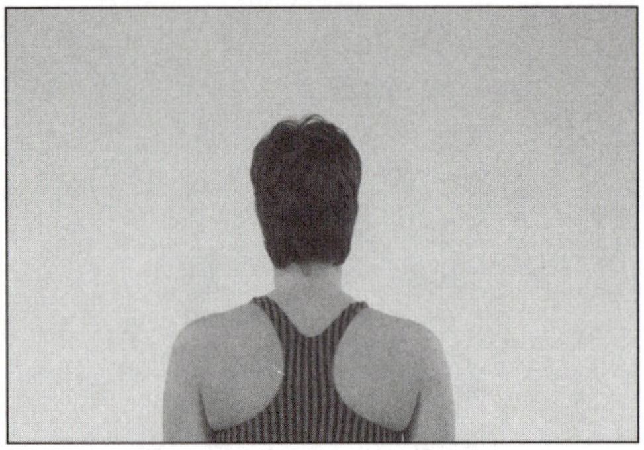

Transverse Plane Posture

Static Postural Evaluation

Transverse plane postural dysfunction represents rotation dysfunction of the spine. Transverse plane dysfunction of the spine can be assessed by palpation of the *posterior prominence* of the paravertebral soft tissue and transverse processes. The therapist's fingers are lightly placed on both sides of the spine over the paravertebral muscles. The hand that is more prominent posteriorly reflects rotation of those vertebrae to that side. The total spine including cervical, thoracic, lumbar, and sacral segments can be evaluated for transverse plane postural asymmetry.

For example, if the right transverse processes are prominent posteriorly from T4 through T8, this represents a rotation dysfunction of the vertebrae of T4 through T8 to the right with a limitation in left rotation of those segments.

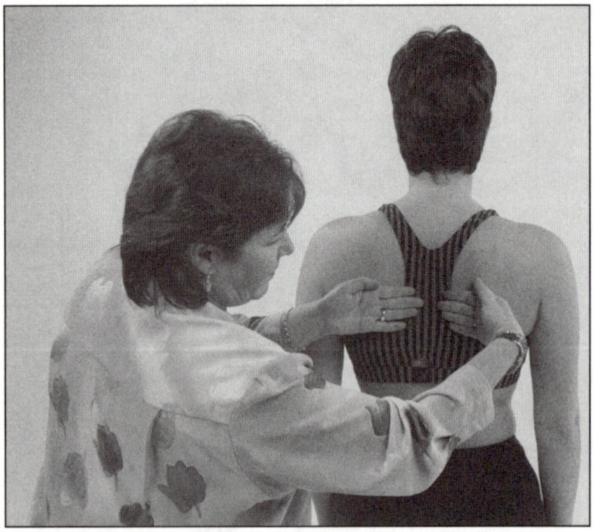

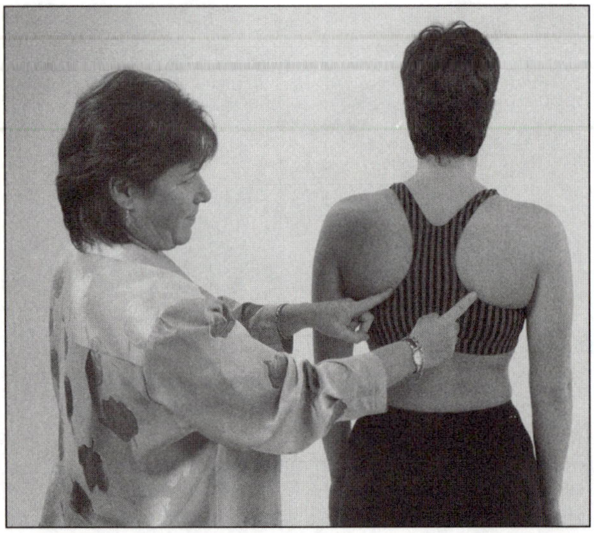

If the lumbar transverse processes and buttocks on the left are prominent posteriorly, this represents a transverse plane rotation dysfunction of L1 through sacrum to the left, with a limitation in right rotation of the lumbosacral spine.

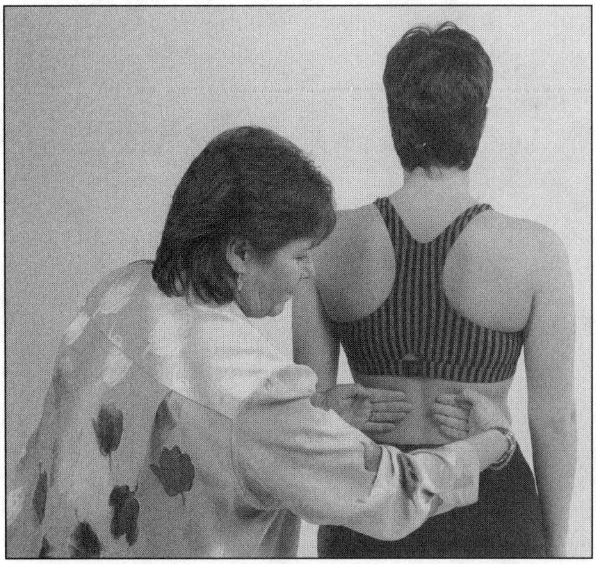

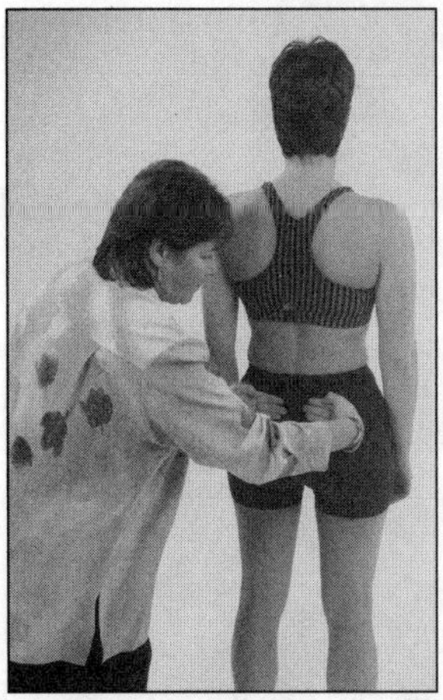

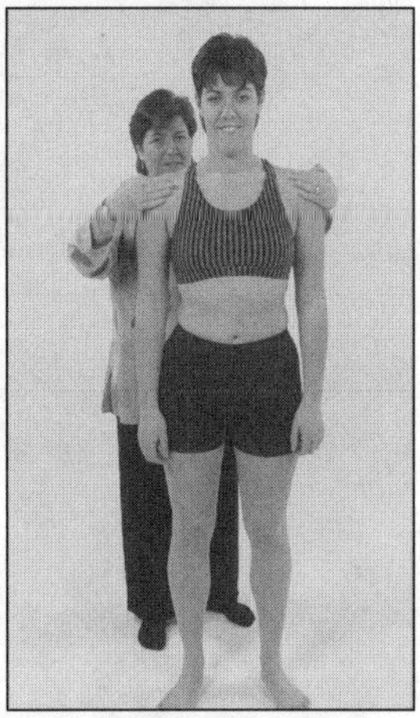

Observe posterior prominence of paravertebral soft tissue as a reflection of rotation dysfunction to the side of the prominence.
Note: Observe only paravertebral prominence, not rib cage or shoulder girdle, as indication of spinal transverse plane dysfunction.
Observe transverse plane dysfunction of shoulder girdles. Note: If the right shoulder girdle is protracted, there will be a limitation in right horizontal abduction and shoulder girdle retraction.

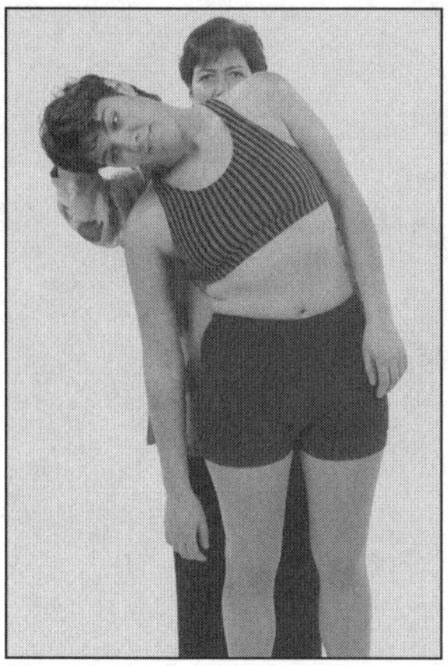

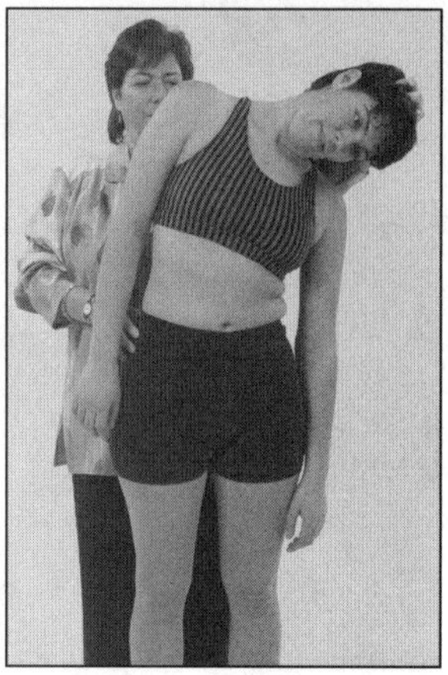

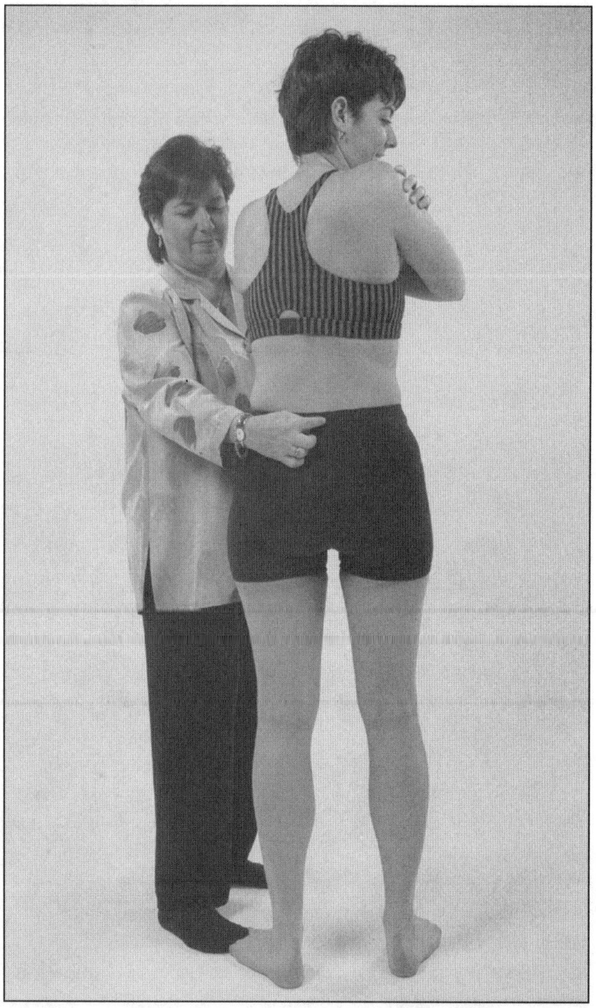

Transverse plane dynamic movement: Right rotation of the spine.
Note: Side bending of neck to compensate for true limitations in cervical rotation (trick movement).

Typical Transverse Plane Deviations of Motion

If there is a static postural deviation on a transverse plane, the therapist can anticipate finding a physiologic limitation of motion in rotation to the opposite side. Note: Fryette's first law of biomechanics states that in neutral, rotation and side bending of the thoracic and lumbar spine occur to opposite sides. For example, if T4 through T8 are stuck rotated to the right, there will be a limitation in left rotation of T4 through T8. If on postural assessment T4 through T8 are stuck in right rotation, they are also stuck in left side bending. These segments, from T4 through T8, will be limited in right side bending as well as left rotation.

There will be correlating information between posterior and anterior views during assessment of transverse plane dysfunction. If T4 through T8 is prominent posteriorly on the right, the chest at this level will be prominent anteriorly on the left. Differentiation of transverse plane dysfunction of the spine from rib cage and shoulder girdle dysfunction is important.

Typical Deviations of Motion of the Extremities

1. Protracted shoulder girdles at rest; actively protracting shoulder girdles during movement
2. Increasing ulnar deviation with forearm pronation
3. Adducting hip and thigh (rotation that occurs at proximal femur)
4. Abducting tibia-which is truly internal or external rotation of the tibia

Posture as an Indication for Treatment with Muscle Energy and "Beyond" Technique

Use postural deviations for indication of correction of biomechanics. Always consider the spine:

Sagittal plane: Posterior buttocks: Treat sacrum, treat Type II's of L5. Loss of lumbar lordosis: Treat Flexed Type II's of L1 through L5. Loss of thoracic kyphosis: Treat Extended Type II's of thoracic spine. Excessive kyphosis of the thoracic spine: Treat flexed Type II's of the thoracic spine. Forward head and neck: Treat Flexed Type II's of the cervical spine.

Coronal plane: Pelvic obliquity: Treat pubic symphysis and iliosacral joints. Pelvic shear: treat sacrum; treat Type II's of L5. "C" or "S" scoliosis curve: Treat Type I's of the thoracic and lumbar spine. Shoulder girdle obliquity: Treat pubic symphysis and iliosacral joints. Lateral shift of the neck: Treat sacrum.

Transverse plane: Posterior prominence of the paravertebral musculature of the spine: Treat sacrum.

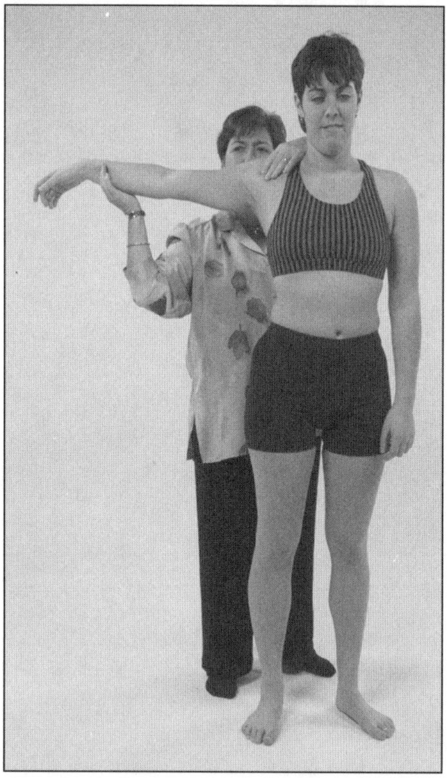

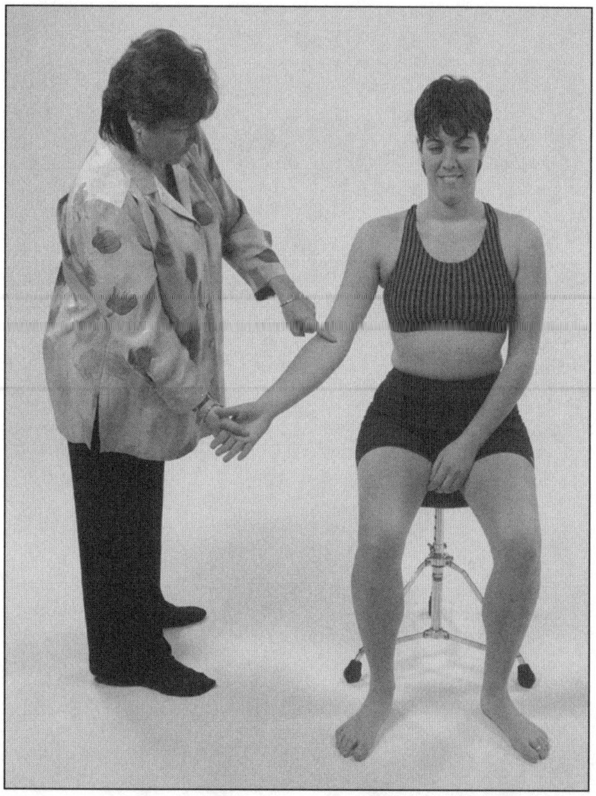

Static and Dynamic Postural Evaluation: A Summary

The extremities can be evaluated for static postural deviations and deviations of motion on the sagittal, coronal, and transverse planes. For example, a protracted shoulder girdle is a static postural deviation on a transverse plane. Because the right shoulder girdle is protracted, there may be full horizontal adduction, but there cannot be full horizontal abduction because that shoulder girdle cannot retract. This would be the deviation of motion on a transverse plane.

Upper Extremity: Compensatory Movements

In order to attain apparent full physiologic ranges of motion, the body is capable of performing involuntary compensatory (trick) movements to successfully achieve the desired target goal of movement. If there is a protracted shoulder girdle, it is not possible to attain full horizontal abduction, or full extension, of the shoulder joint on that side. So, in order to attain this apparent physiologic range of motion, the humeral head will typically deviate from midline during movement in a compensatory attempt to attain the movement goal. The typical compensatory deviation of the humeral head is a tendency to sublux anteriorly within the glenoid fossa.

There are typical compensatory movements discovered during upper extremity motion testing.

For example, if there is a limitation in elbow extension and forearm supination, the proximal radial head will tend to shear anteriorly; the proximal ulnar head will abduct, causing a medial gap; the distal ulnar head will shear anteriorly; the distal radial head will shear posteriorly; the proximal carpal row will shear anteriorly; the proximal first metacarpal head will deviate anteriorly and distally.

For normal pain-free movement, good articular balance without deviation or shears must be maintained throughout the physiologic ranges of motion. The true limitations in joint mobility will then be noticeable. These compensatory deviations from midline during movement further exacerbate a hypomobile and painful condition.

Hypomobility and Hypermobility

In the field of manual therapy, *hypermobility* is not typically an issue to be addressed. True hypermobility with ligamentous laxity (in cases such as pregnancy and rheumatoid arthritis) is not one of the common causes of dysfunction treated with manual therapy. For example, if there is a glenohumeral joint dislocation or subluxation, this joint is *not* hypermobile in Integrative Manual Therapy concepts. The humeral head is hypermobile on one plane, in one direction only. In all other planes and in all other directions of movement, this humeral head is hypomobile. All physiologic ranges of motion are limited; the majority of accessory movements are limited. Therefore, a dislocated glenohumeral joint is a hypomobile problem, and manual therapy will address the dysfunction in the context of the presenting hypomobility. The result of therapy should be normal articular balance with normal joint mobility of the glenohumeral joint throughout full physiologic ranges of motion, once the humeral head is repositioned in the glenoid fossa.

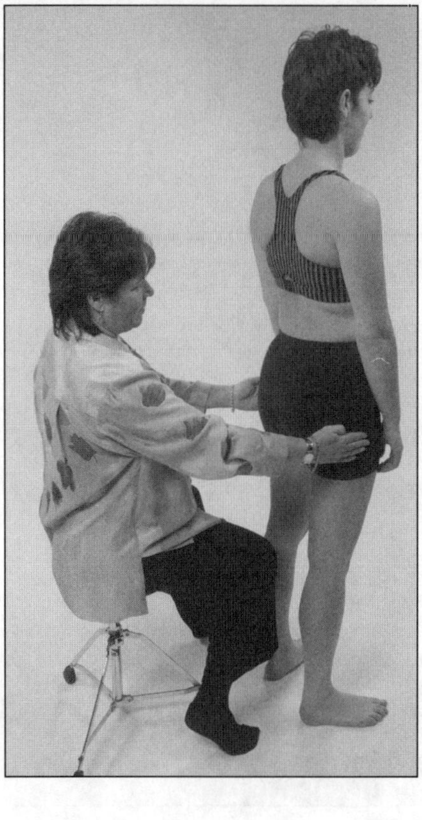

Lower Extremity: Compensatory Movements

Typically, hip joint dysfunction includes an approximated, adducted, internally rotated and inferiorly displaced femoral head within the acetabulum. In standing and in the recumbent position, this approximation of hip joint surfaces may be palpated by placing the hands over the greater trochanter.

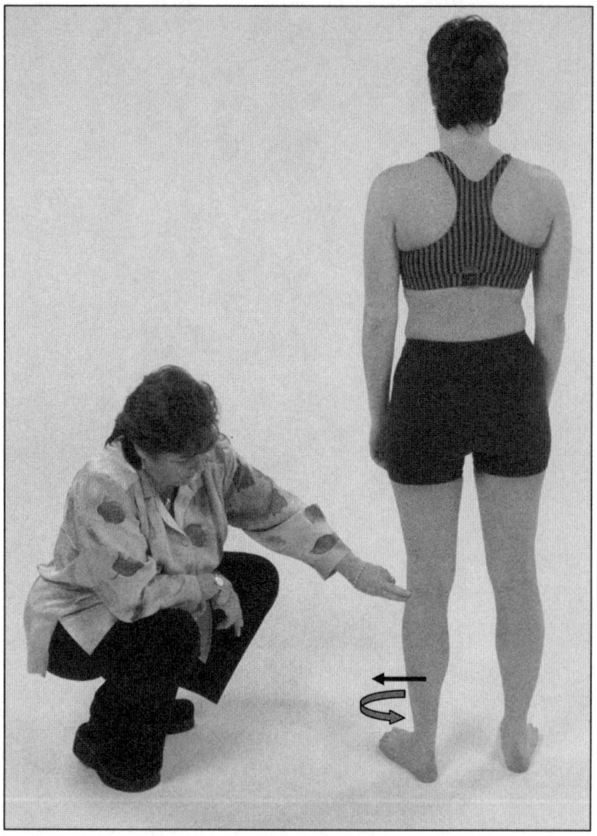

Knee joint dysfunction typically presents with an externally rotated and laterally sheared proximal tibial articular surface.

Ankle joint dysfunction, which includes the tibiotalar and subtalar joints, typically presents with a posterior shear of tibia on talus, an inferior lateral malleolus, an inverted calcaneus and intra-articular adhesions of the talocalcaneal joint. Supinated (high arch) and pronated (flat foot) postural deviations of the feet are common.

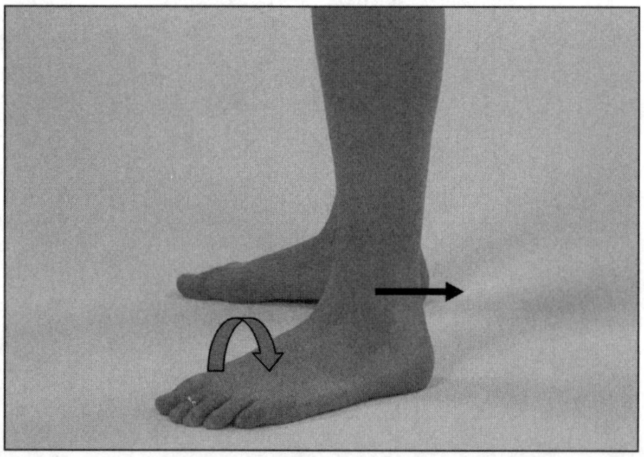

Dynamic Postural Evaluation: Specificity of Assessment

When evaluating movement, it is necessary to slowly and gently move the extremities passively, or assisted actively, through the ranges of motion, maintaining a neutral position throughout (without deviation from midline: pure unidirectional, uniplanar movement). This process is called *dynamic postural evaluation.* It is essential to maintain articular balance throughout the movement. If the limb is not allowed to deviate from midline during the movement, the limitations of motion will be obvious. With static postural dysfunction on any one plane, it is impossible to have full accessory and physiologic ranges of motion of the joint(s) involved. If the patient quickly and grossly moves the body part through an apparent full range of motion without maintaining articular balance, the compensatory (trick) movements will not be noted. Typical compensatory movements of the extremities were outlined previously in this chapter.

When the articular balance is maintained throughout the physiologic range of motion, the limitations are more easily identified. If there is dysfunction but the patient appears to attain the movement goal, this occurs by using compensatory movements. Much energy exertion is required by the patient to complete the movement when compensatory movement replaces normal joint accessory and physiologic function.

Postural Dysfunction and Articular Balance

Whenever there is postural dysfunction on a sagittal, coronal, or transverse plane, there will be a compromise of articular balance of the joints in that region. Articular balance is the normal neutral relationship of two articular surfaces, at rest and throughout a complete physiologic range of motion. For example, with a protracted shoulder girdle, the articular balance will be compromised in all the joints that contribute to the shoulder girdle anatomy: the scapulothoracic joint, the sternocostal joint, the costovertebral joints, the glenohumeral joint, the acromioclavicular joint, and the sternoclavicular joint.

Articulations in the body do not occur solely between bony joint surfaces. Soft tissue articulates and glides on bony structures. Soft tissue also articulates and glides on soft tissue. The observant therapist can assess and palpate postural dysfunction that is not limited to bony joint surfaces, but also soft tissue compromise. All movement should be full and friction-free, without compensatory deviations!

Articular Balance and Accessory Movement

There is a compromise of articular balance present with postural dysfunction that correlates with a limitation in accessory movements and joint play at those joints. For example, in a protracted shoulder girdle, there will be hypomobility discovered on mobility testing of accessory joint movements of all joints that contribute to the shoulder girdle: costovertebral joints; sternocostal joints; scapulothoracic joint, glenohumeral joint, acromioclavicular joint, and sternoclavicular joint.

Accessory Joint Movement and Physiologic Ranges of Motion

Whenever there is hypomobility of the accessory movements and joint play in a joint, there must be correlating limitations in physiologic ranges of motion. There may also be hypertonicity (protective muscle spasm) of the musculature surrounding that joint. There may be fascial dysfunction of the connective tissue surrounding that joint. Yet, if the body receives a command from the brain cortex to move in spite of the hypomobility, the person will attempt to move and to reach the movement goal. In order to attain the movement goal, the body will overcome the obstacle presented by limitations of motion. Instead of utilizing the required range of motion at the joint to achieve the movement goal, the body will involuntarily compromise the accessory movements of the joint with trick movements.

For example, a tennis player with a protracted right shoulder girdle must attain full shoulder extension. In order to attain this movement goal, the humeral head will sublux anteriorly during the movement to allow a compromise, to reach the target end movement which requires more than the existing range of true physiologic extension. This "'pseudo" extension is not within the normal healthy constraints of joint movement. The accessory movements at the joint surface are now pathological. As a result of this compensation, the glenohumeral joint capsule will become dysfunctional, possibly developing fascial dysfunction of the anterior joint capsule and tightening of the posterior joint capsule. But if the range of motion is not evaluated while maintaining normal articular balance (between the glenoid fossa and the humeral head), and instead gross apparent physiologic extension is measured while allowing excessive and pathologic anterior glide of the humeral head, then the practitioner will not discover the total dysfunctional pattern that requires structural rehabilitation and functional reeducation and retraining.

Postural dysfunction indicates articular imbalance, joint hypomobility, and limitations in physiological ranges of motion. Manual therapy techniques can be performed at areas of postural dysfunction in order to improve articular balance, and increase joint mobility and ranges of physiologic motions. Postural deviations on all planes are an indication for treatment.

Observation with the "Preferred Eye"

The practitioner should emphasize using his or her *preferred eye* (previously known as the "dominant eye") during evaluation for postural dysfunction, in order to attain accurate perception of anatomical landmarks. Otherwise, it is difficult to accurately perceive mild discrepancies in anatomical landmarks.

Self-evaluation for eye preference:

Step 1. Hold the hands at shoulder level approximately 18 inches from the body.

Step 2. Form a circle with both index fingers and thumbs.

Step 3. Choose a point (landmark) in the room. With both eyes open, look at that point through the circle formed by the thumbs and index fingers.

Step 4. Close only the right eye; then close only the left eye.

Result: The eye that perceives the same location of the landmark as the two eyes together is the preferred eye. (When significant unilateral or bilateral visual discrepancy is present, vision perception can be corrected with Integrative Manual Therapy; a combination of Advanced Strain and Counterstrain, Cranial Therapy, Neural Tissue Tension, and vision training with a neurooptometrist/behavioral optometrist.)

Visual Assessment:
Wide-Angle Lens to Zoom Lens

Step 1. Observe the patient's gross posture on sagittal, coronal, and transverse planes.

Step 2. Then proceed to focus on specific localized regions of postural dysfunction:
Phase 1: Wide-angle lens view of gross postural deviations
Phase 2: Zoom lens view of articular postural deviations

Philosophy and Hypothesis

Limitations of physiologic ranges of motion should correlate with static postural deviations. Postural deviations indicate the body's potential for dynamic movement. This means that postural deviations correlate with deviations of motion and, therefore, limitations of motion.

There are mild, moderate, and severe neuromusculoskeletal dysfunctions. Severe postural deviations correlate with severe limitations in ranges of motion with severe imbalance of articular surfaces. Pain and/or compensation patterns will be observed in inner ranges of motion. Moderate postural deviations correlate with moderate limitations in physiologic ranges of motion. Pain and/or compensation patterns will not be observed until mid ranges of motion. Mild postural deviations will correlate with only mild limitations in ranges of motion. Pain and/or compensation patterns will only be observed in outer ranges of motion.

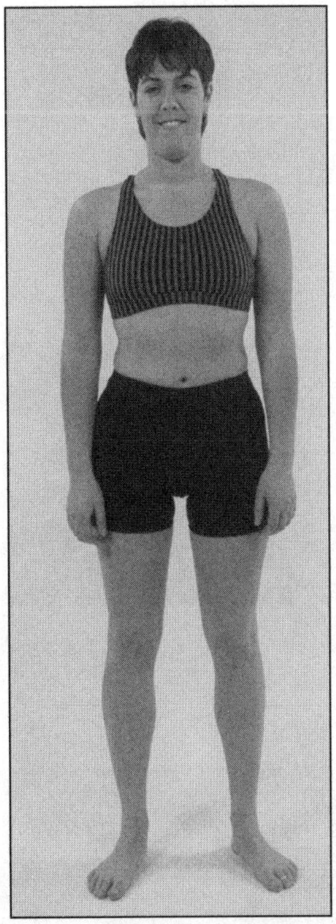

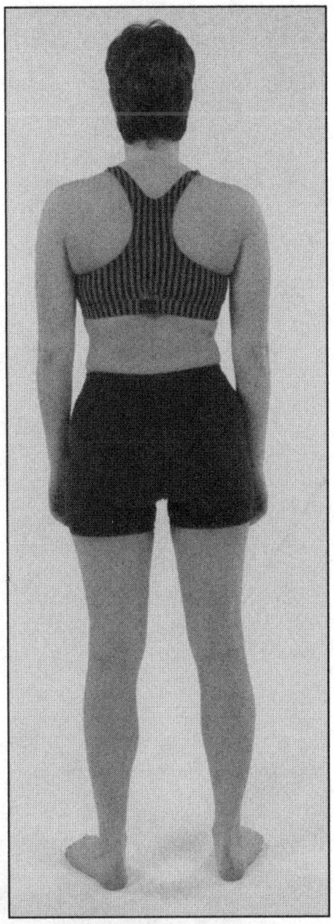

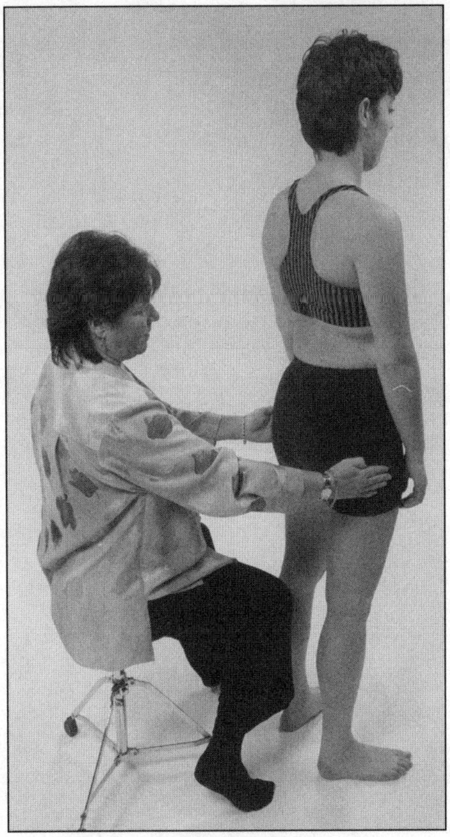

Pictorial Documentation of Postural Deviations

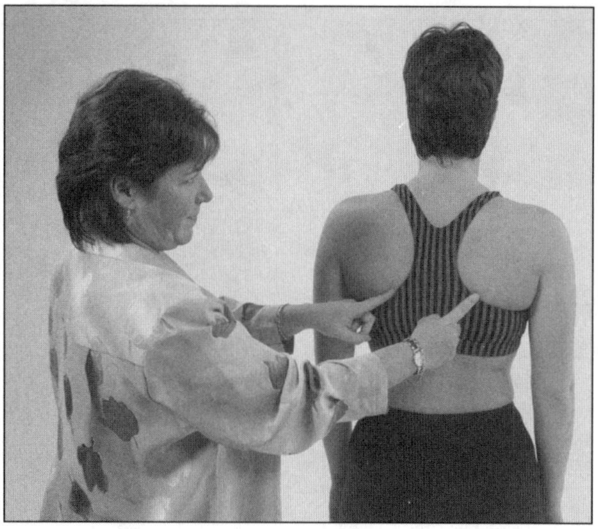

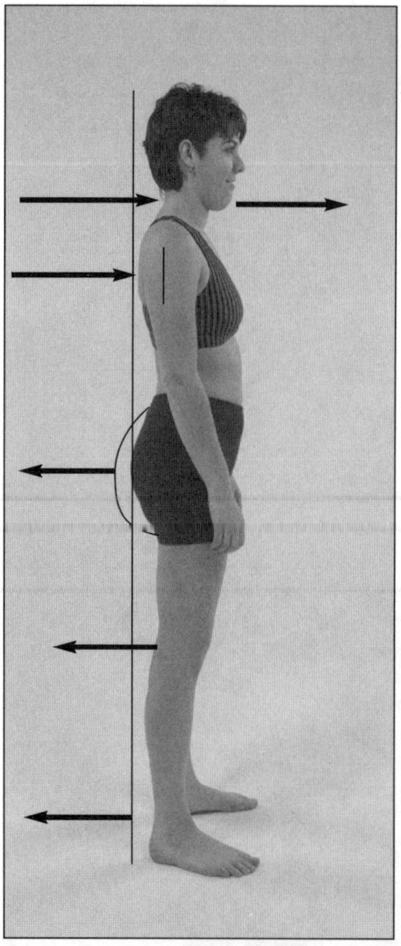

Sagittal

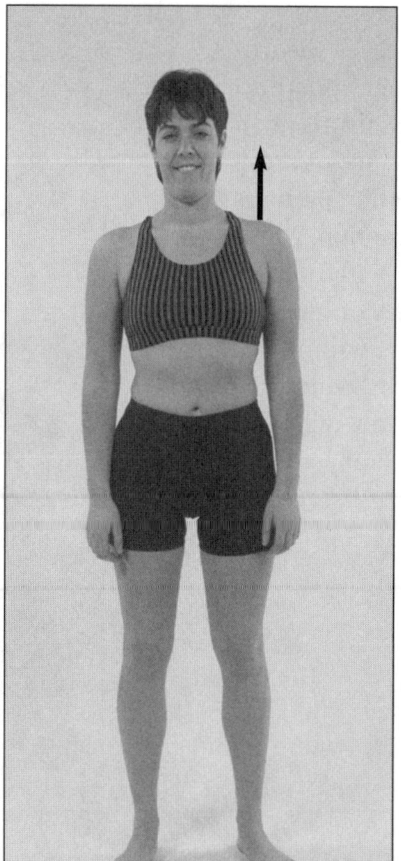

Coronal

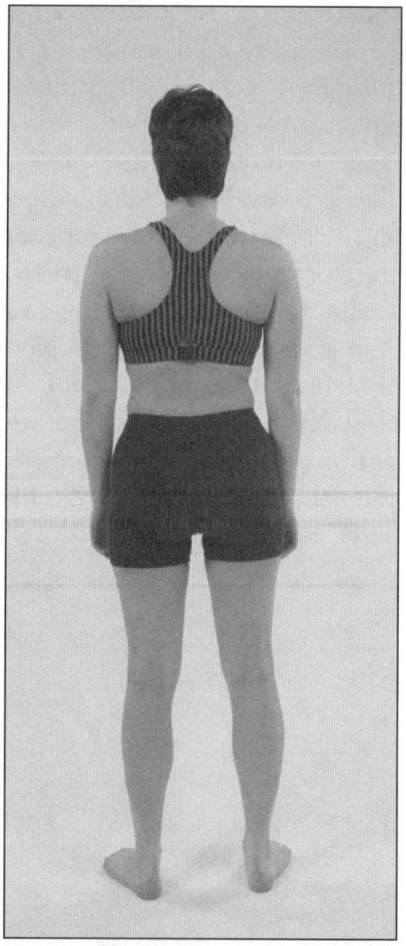

Transverse

Normal Ranges of Spinal Motion

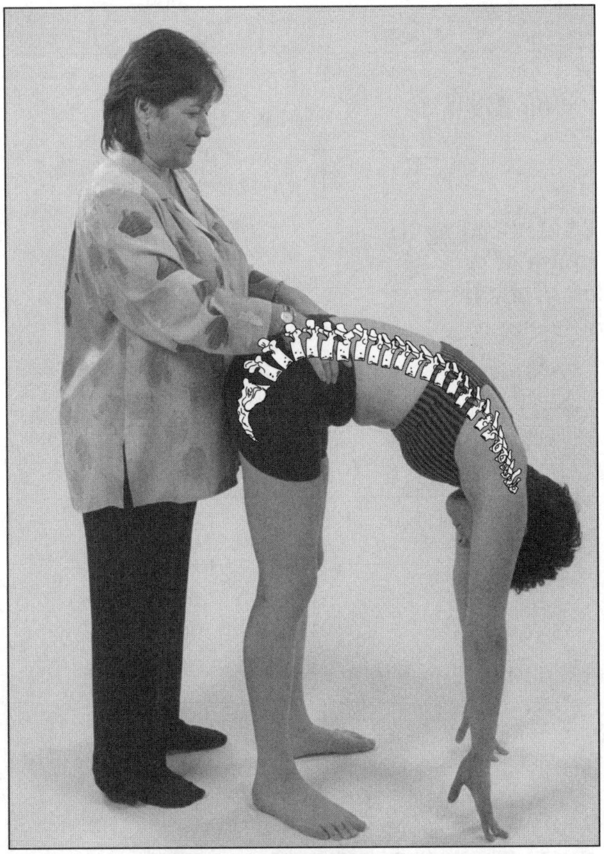

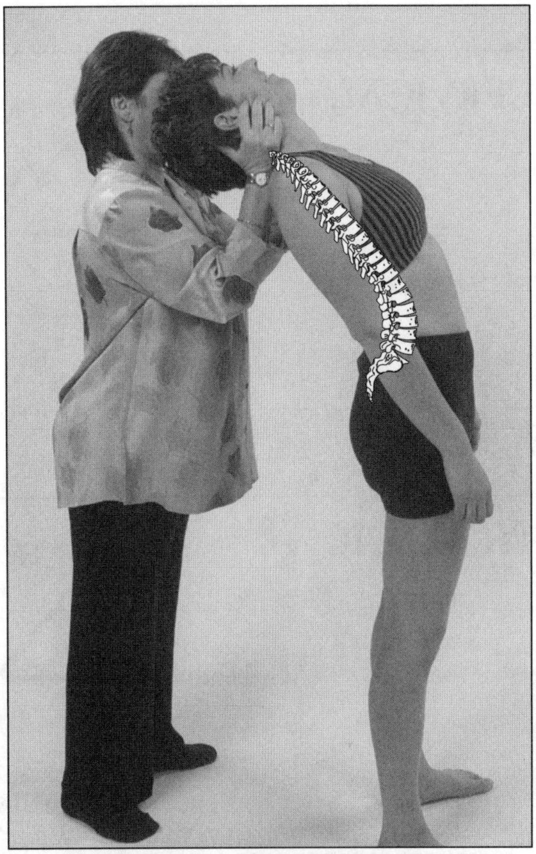

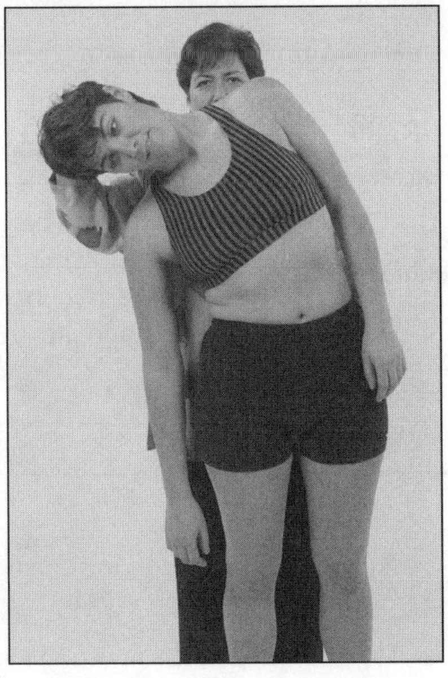

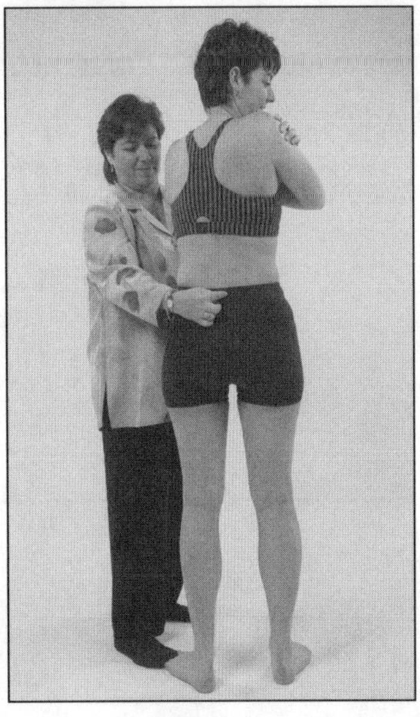

Documentation of Ranges of Spinal Motion and Limitations of Spinal Motion

CERVICAL

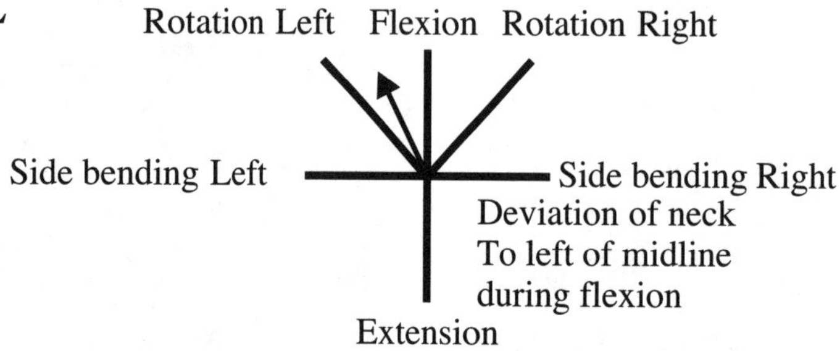

Note: Observe pure segmental movement without excessive chin protrusion or anterior shear of C7 during extension

THORACIC

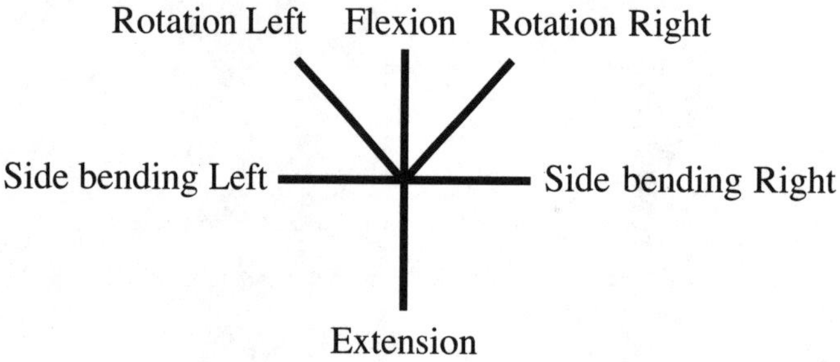

Note: Observe segmental movement

LUMBAR

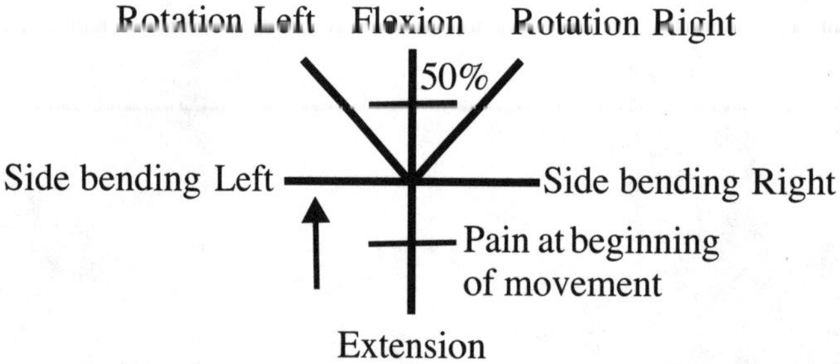

Note: Hip drop test flex one knee in standing. This will result in side bending up the Kinetic chain to the opposite side

Normal Ranges of Motion of Extremities

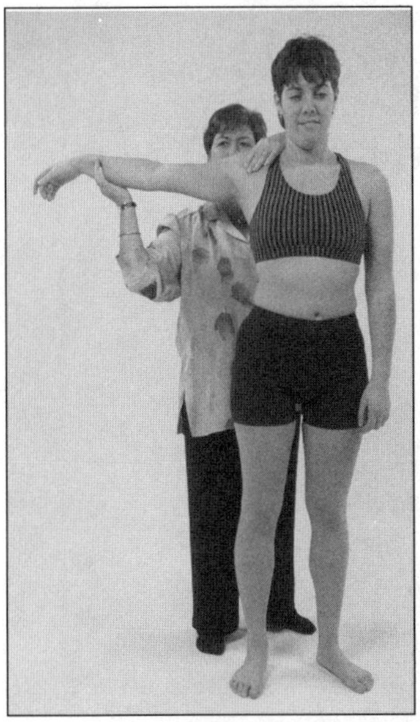

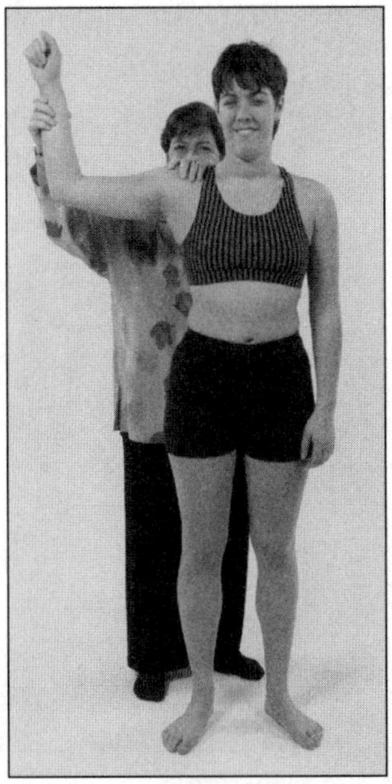

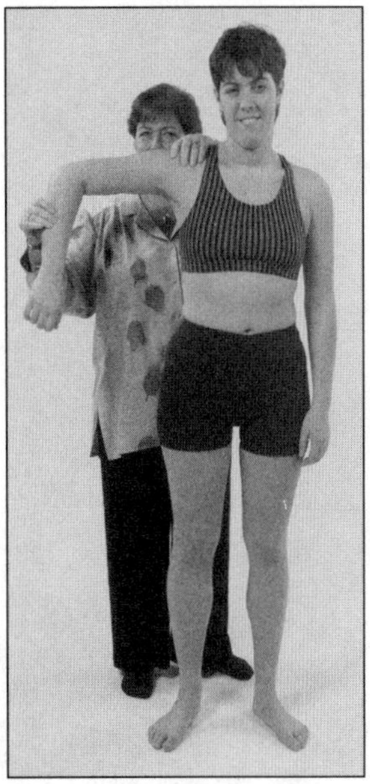

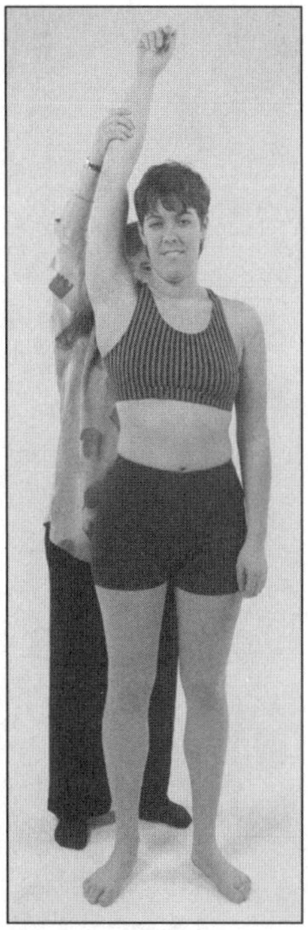

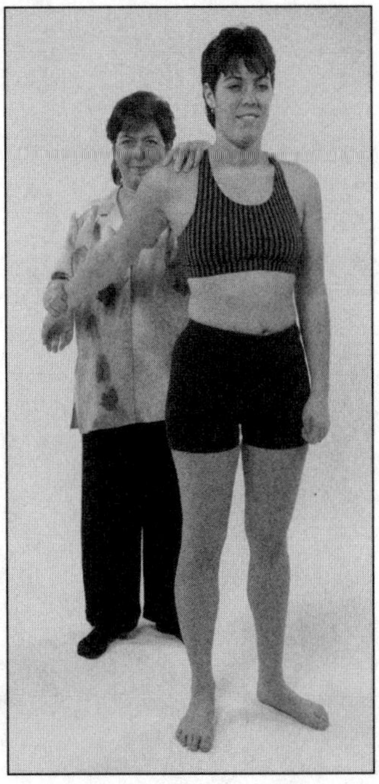

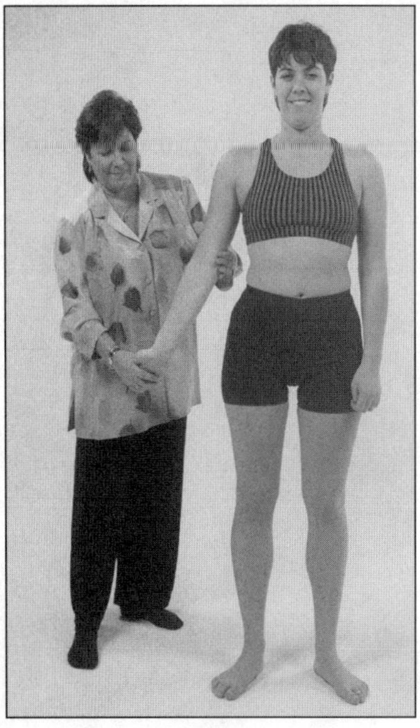

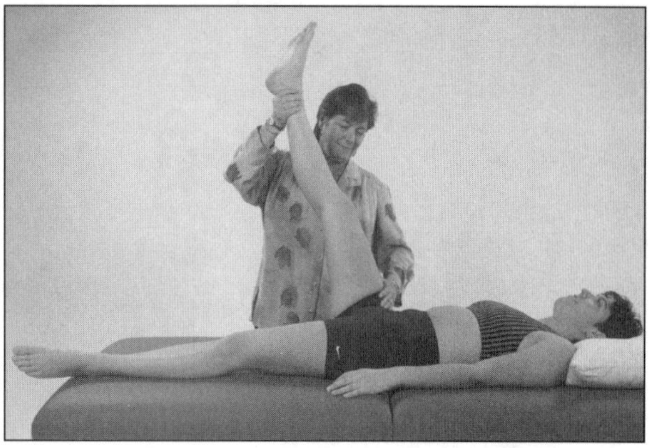

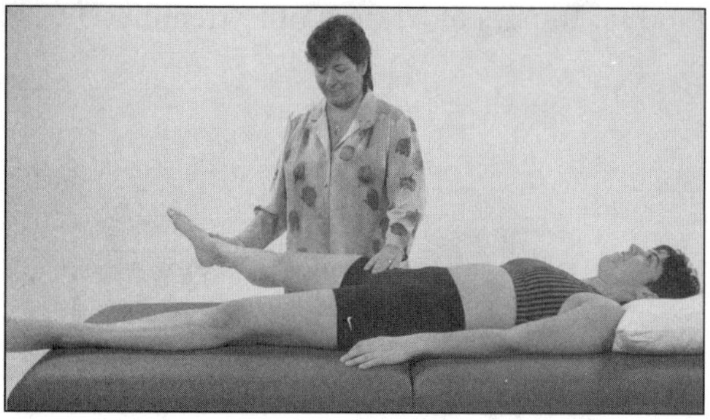

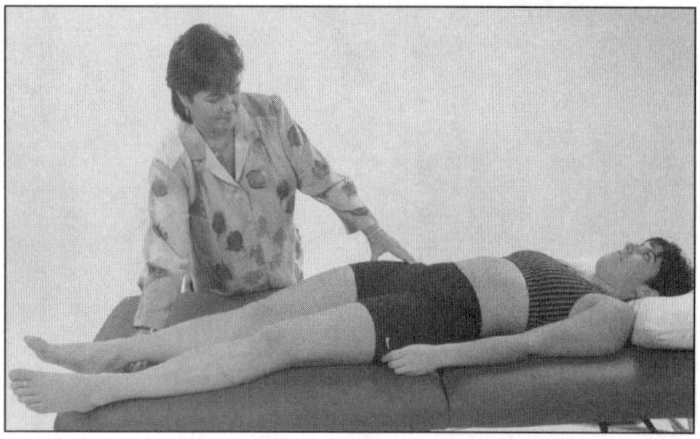

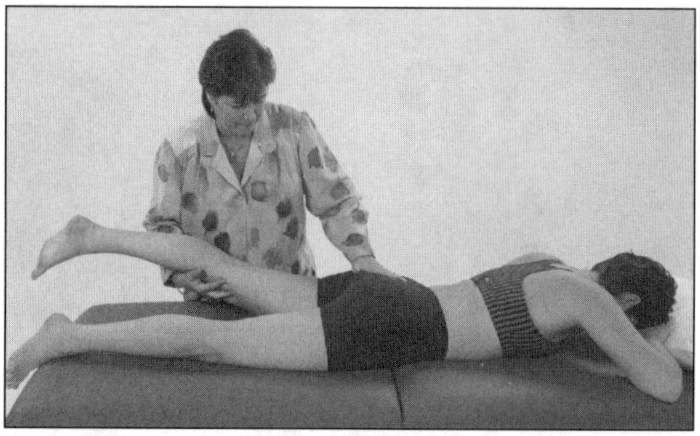

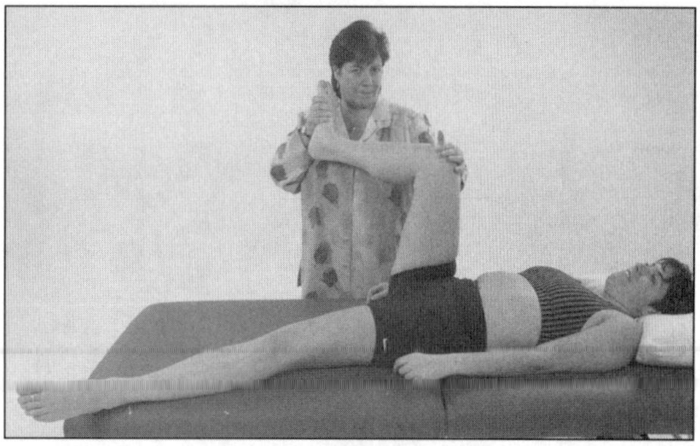

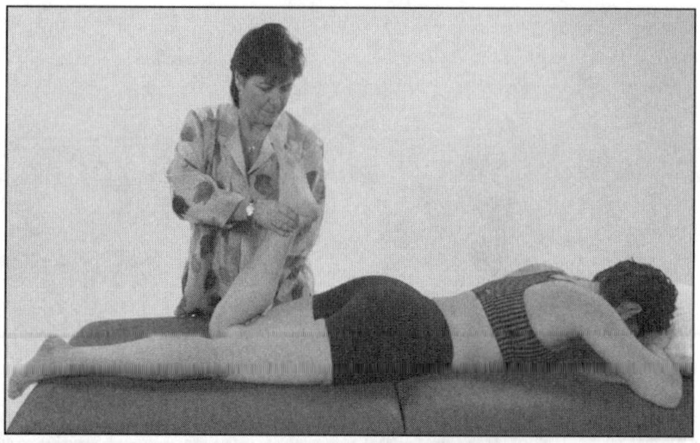

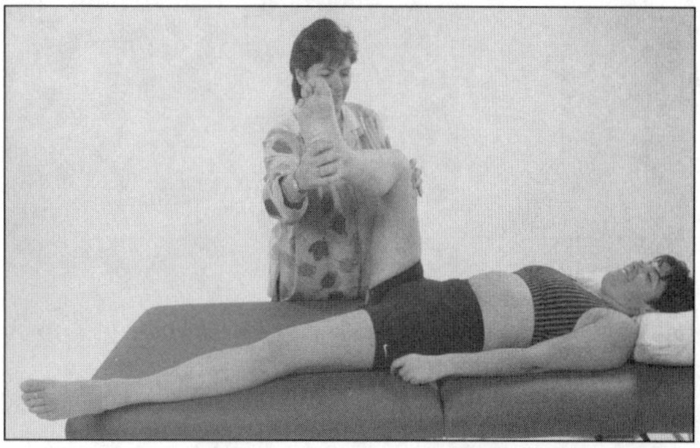

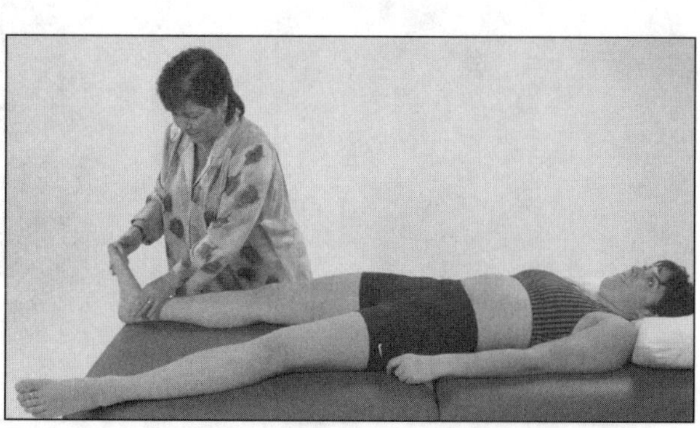

CHAPTER 4

THE MUSCLE BARRIER

The Muscle Barrier
is
the threshold of the muscle spindle
of the stretched muscle fiber(s)
to stimulus of stretch.

The term *muscle barrier* is often used historically in the literature of Muscle Energy Technique, but also in manual medicine literature. The concept is important. When stretching *beyond* a barrier is performed, there is resistance. When stretching *up to* a barrier is performed, there is no resistance. The resistance comes *after* the barrier. The barrier comes *before* the resistance.

The concept is confusing. When the word barrier is used, immediately we think about resistance, pushing or pulling against a resistant force. In traditional forms of body work, stretching or even elongation of tissues up to a barrier brings up a thought of force.

Yet, in Muscle Energy terms, barrier does not elicit the thought of force or resistance. There is no force; there is no resistance. The barrier is when there is very subtle tissue tension so few sarcomeres are resisting, that no force is perceived, and no resistance is perceived.

This aspect of tissue tension, which develops secondary to a minimal response of a few sarcomeres, is intrinsic to Muscle Energy philosophy.

Scenario of a Muscle Barrier

The muscle spindle is innervated by the gamma nerve. The muscle fiber, the sarcomere, is innervated by the alpha nerve.

When there is excessive, high-frequency discharge as described in Chapter Two, there is hyperinnervation. Excessive, high-frequency discharge along the gamma nerve is called *gamma gain.* Too many impulses reach the target. If there is excessive, high-frequency discharge along the gamma nerve fiber, the muscle spindle, which is the target, will be hyperinnervated.

If the muscle spindle is hyperinnervated due to gamma gain, the threshold to stretch of that muscle spindle is lowered. When a muscle spindle is not hyperinnervated, the muscle spindles should not be stimulated with passive movements throughout the normal range of motion. There should be a reasonable stretch on each muscle before the muscle spindle (the stretch receptor) is activated. Consider the term passive movement. The term passive movement assumes that there is no muscle contraction. The muscle is passive. In passive movement, there should not be any assistance or any resistance by the muscle fibers. This means that during passive movement the muscle is not contributing to the movement, nor is the muscle resisting the movement.

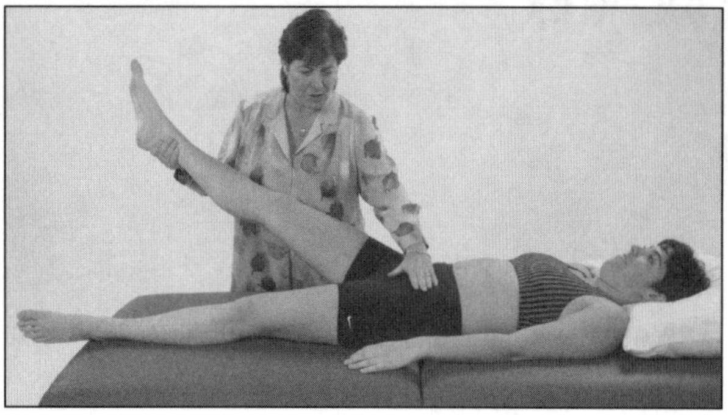

If the muscle spindle is hyperinnervated, during passive stretch the muscle spindle will react, and a stretch reflex arc will be activated. Afferent impulses from the spindle will go into the dorsal root of the horn, and many of these impulses will continue along to the anterior horn. The afferent neuron will synapse with the alpha motor neuron in the anterior horn. The impulses will be passed along the neuron, and the muscle fiber will contract.

Example: Hamstring Muscle Barrier

Picture the following scenario: A straight leg raise is performed passively. Initially the movement is passive: There is no assistance by the hip flexors and no resistance by the hamstrings. But the muscle barrier is finally reached. The degree of straight leg raise that finally exerts too much of a stretch on the muscle spindle of at least one hamstring muscle fiber. The threshold to stretch of the muscle spindle of this first hamstring muscle fiber is met. The muscle spindle begins to fire. The impulses are passed along the afferent neuron, into the spinal cord via the posterior horn, and over to the anterior horn. The impulses cross the synapse and continue along the axon of the alpha motor neuron to the hamstring muscle fiber. That hamstring muscle fiber contracts. The straight leg raise is now no longer passive; it is no longer resistance-free. There is resistance from the muscle contraction of the hamstring muscle fiber. If the therapist's hand palpates the ASIS, motion will be perceived due to the pull of the hamstring's contracted fiber. Often the barrier is palpated by the hand on the opposite ASIS (the hand on the same ASIS would be distracted by the leg movement) before it is felt by the hand holding the leg.

Definitions

The barrier is the place in the passive range of motion when the first muscle fiber contracts and resists the passive movement. The *interbarrier zone* is a few degrees before the muscle barrier is engaged (a few degrees before the first muscle fiber goes into a state of contraction because the muscle spindle is stretched). The *neutral position* is the loose-packed, relaxed position where the soft tissues surrounding the joints are most relaxed, and the joint surfaces are the least congruent. This is the midpoint in the normal range of motion. The *pathological neutral position* is the new, pathological loose-packed position. With neuromusculoskeletal dysfunction, there is a shift in the neutral because of adaptation of the soft tissues, due to protective muscle spasm and fascial dysfunction. This new, pathological neutral position is the midpoint in the new, reduced existing range of motion.

Analysis and Interpretation of Postural Dysfunction

With static postural dysfunction, the body part deviates toward the protective muscle spasm and spasticity due to contracted and shortened muscle fibers. The range of motion will be limited in the opposite direction. In the case of dynamic postural dysfunction, consider the muscle in spasm as the agonist. The range of motion is limited in the opposite direction, opposite the agonist's direction of primary movement, because of the resistance by the agonist. For example, if the biceps (the agonist) are in spasm, then elbow extension is limited because the biceps will resist the pull of the triceps (antagonist).

Application of the Muscle Barrier Concept

Prior to correcting biomechanical dysfunction of the pelvic joints, the therapist treats the muscle barriers of the long muscles of the legs. If there is hypertonicity in one or more of the long muscles that insert on the pelvis, they will exert a force on the bony pelvis that can cause articular imbalance at the pelvic joints. The long muscles to be treated include: hip flexors, hip extensors, hip adductors, hip abductors, hip internal rotators, hip external rotators, knee flexors, and knee extensors.

Hypertonicity (protective muscle spasm or spasticity) in a muscle will inhibit range of motion in the opposite direction. For example, hamstrings (hip extensors) spasm will inhibit straight leg raise (hip flexion). In addition, hamstrings (knee flexors) spasm will inhibit knee extension. Right hip adductors spasm will inhibit right hip abduction and right lumbar side bending. Right scalenes spasm will inhibit left cervical side bending. Iliacus (hip flexors) spasm will inhibit hip extension and lumbar extension. Spasm of the upper left rib depressor muscles will inhibit left shoulder abduction.

When performing muscle barrier testing, be sure to maintain a neutral position of the body part during passive range of motion. Remember, passive movement assumes there is no muscle contraction. This means there is no assistance and no resistance. This means the muscle is not contributing to the movement, nor is the muscle resisting the movement. For example, when testing the hamstrings (hip extensors and knee flexors) for hypertonicity: Patient is in supine position. Straight leg raise, keeping the hip in neutral (no rotation, abduction, adduction). Lift the leg slowly and passively. As soon as there is any resistance or any increased tension in the leg, or there is any subtle movement of the pelvis, you have reached the threshold to stretch of at least one muscle spindle of one hamstring fiber when assessing muscle barriers, the barrier can be palpated at the ipsilateral or contralateral PSIS.

Another example is when testing combined iliopsoas and quadriceps muscle barriers (hip flexors and knee extensors) for hypertonicity. Patient is in prone position. Maintain the hip and knee in neutral. Passively bend the knee. Do not allow the hip and pelvis to shift. As soon as there is any resistance (i.e., the quadriceps and iliopsoas resisting the stretch of prone knee bending) or there is any subtle movement of the pelvis, you have reached the threshold of at least one muscle spindle of the hip flexors/knee extensors fibers to stretch.

Energy Expenditure Due to the Resistance of a Muscle in Protective Muscle Spasm

Observation: Test assisted active prone knee bending. Observe all postural deviations of the leg. Maintain the hip and pelvis at neutral. Do not allow foot/ankle/knee to deviate from neutral/midline during movement. The therapist may need to passively maintain subtalar neutral during movement. Ask the patient to actively bend the knee. Observe any difficulty in initiating and continuing the movement when the foot and leg are forced to maintain neutral/midline posture. Movement requires good articular balance.
Correlation: With postural asymmetry there are limitations in normal ranges of motion.
Correlation: With protective muscle spasm, there are limitations in normal ranges of motion.

Muscle Energy and "Beyond" Techniques

The following protocol should be followed when treating pathological leg muscle barriers:

1. Position the joint at the interbarrier zone on three planes.
2. The patient controls the position of the joint.
3. There is active contraction by the patient.
4. There is a specific direction to the contraction.
5. There is a distinct counterforce by the therapist.
6. The intensity of the contraction is controlled.
7. The resistance to the contraction is small.
8. The contraction is isometric.

When testing with Muscle Energy and "Beyond" Techniques, a normalization of position of joint surfaces is attained. The result is improved articular balance, decreased hypertonicity of the surrounding musculature, and increased joint mobility and range of motion.

Treatment of the Muscle Barriers of the Legs for Symmetry of Muscle Tone

Each leg muscle is evaluated for its muscle barrier. Right and left symmetry are compared. The muscle barrier evident earlier in the range of motion indicates greater hypertonicity of that muscle than the muscle of the other leg. The muscle with the greatest hypertonicity is treated. With decreased tone there will be increased symmetry of muscle pull on the pelvis. Focus on asymmetry of the leg barriers as a cause of pelvic imbalance.

Treatment of Long Muscle Barriers with Muscle Energy and "Beyond" Technique

1. Start in the resting position.
2. Maintain a neutral position.
3. Move the leg passively (no assistance or resistance) into the range of motion.
4. Rest hand lightly on the pelvis at the ASIS to palpate the movement.
5. Find the barrier.
6. The barrier is reflected by the appearance of tissue tension (resistance to passive leg movement) and movement of the pelvis.
7. Retreat a few degrees to the interbarrier zone.
8. Perform a contract/relax for 6 to 10 seconds.
9. Wait for complete relaxation.
10. Move toward the second new barrier further into the range of motion. Repeat this process three to five times. There is no need for maximal recruitment during the contraction. The contraction should occur at the interbarrier zone. Contractions may utilize reciprocal or autogenic inhibition by antagonist or agonist contraction.

Treatment of Pathological Leg Muscle Barriers

1. **Hamstring barriers:**
   Movement: Supine; straight leg raise
   (Attain specificity with hip and knee flexion and rotation)

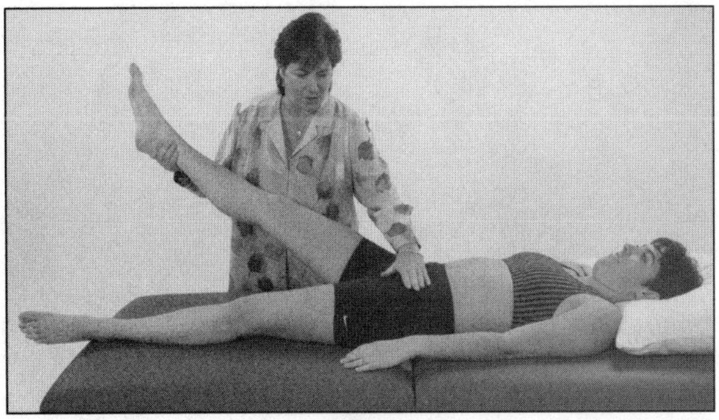

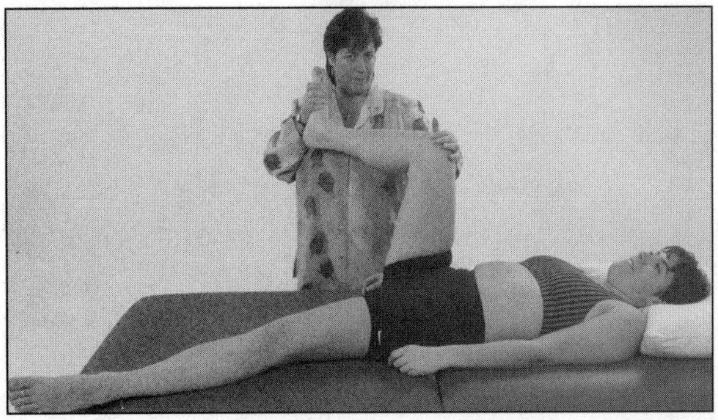

2. **Abductor barriers:**
   Movement: Supine; adduction

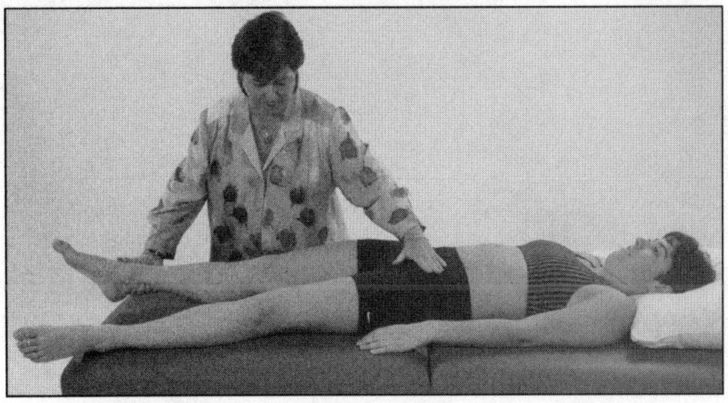

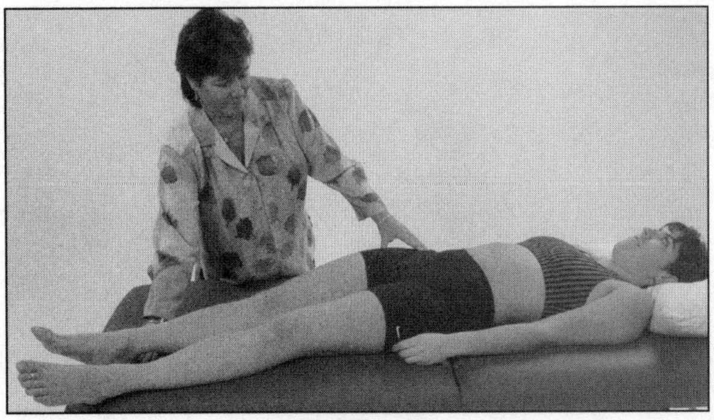

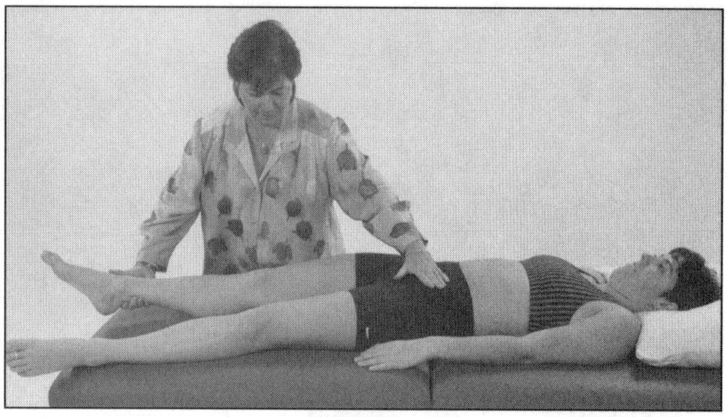

3. **Adductor barriers:**
   Movement: Supine; abduction

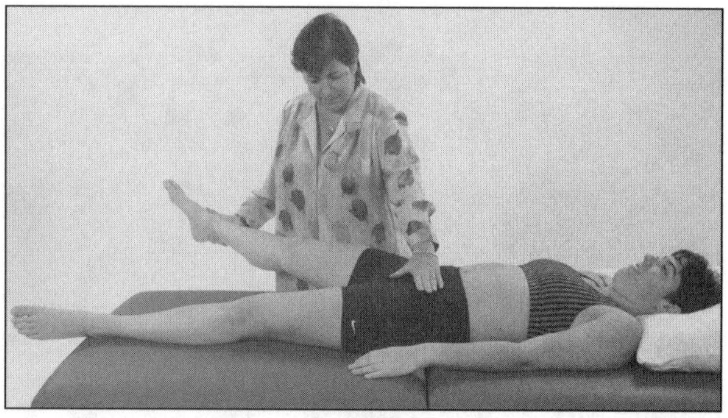

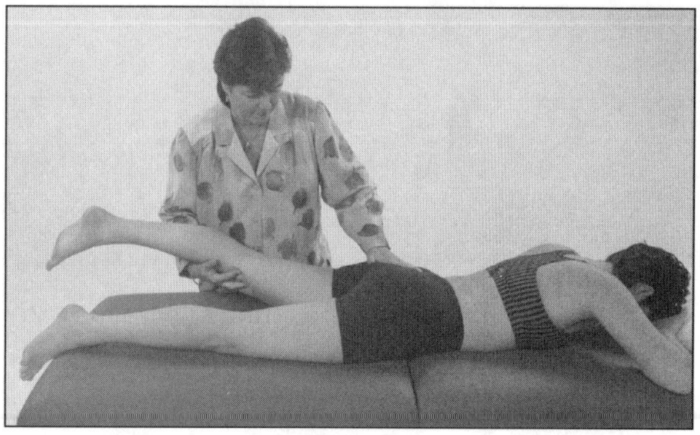

4. **Hip Flexor barriers** (Iliopsoas):
   Movement: Prone knee bending (Attain specificity by using the Thomas position)

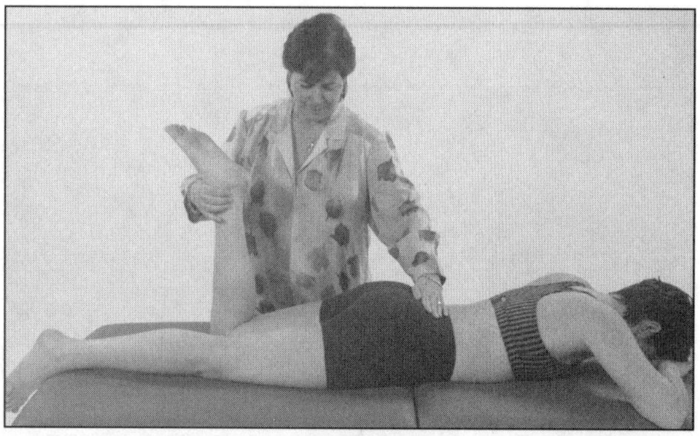

5. **Knee Extensor barriers** (Rectus femoris):
   Movement: Prone knee bending (Attain specificity by using the Thomas position)

   Note: Prone knee bending tests barriers for hip flexors and knee extensors. (To differentiate between these barriers, use the Thomas position: Supine, buttocks at end of table, contralateral knee held up to chest in hip and knee flexion, ipsilateral tested leg off edge of bed.)

6. **Internal Rotator barriers:**
   Movement: Prone; external rotation

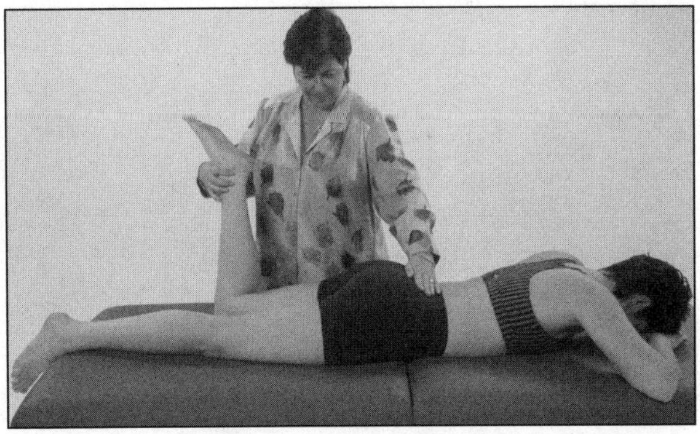

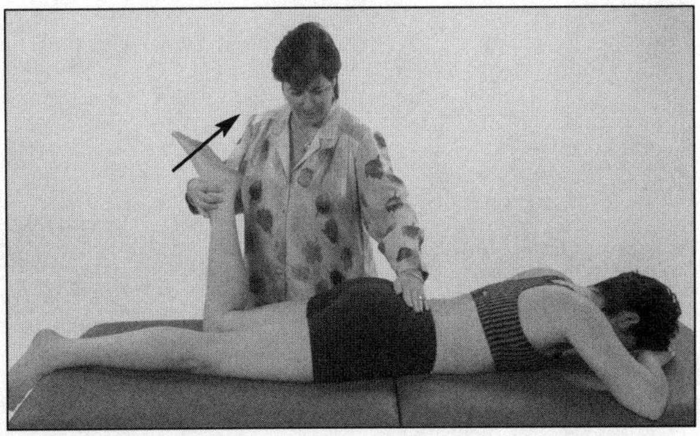

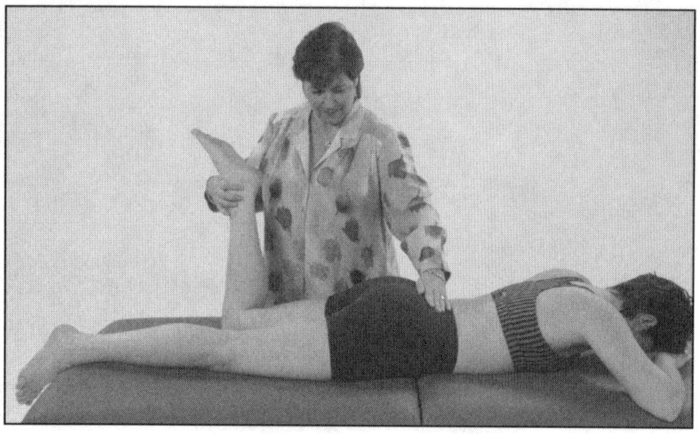

7. **External Rotator barriers:**
   Movement: Prone; internal rotation

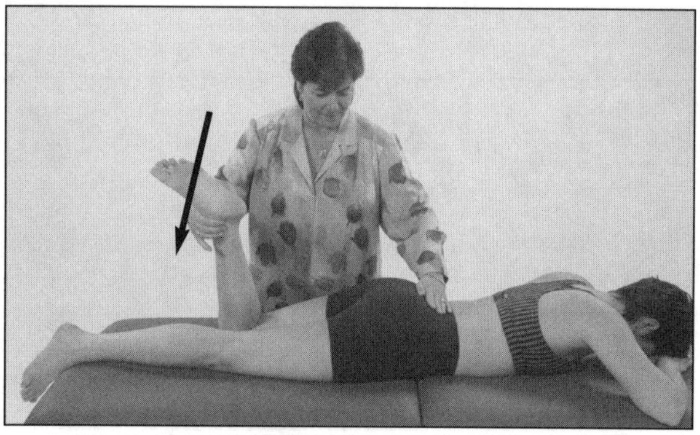

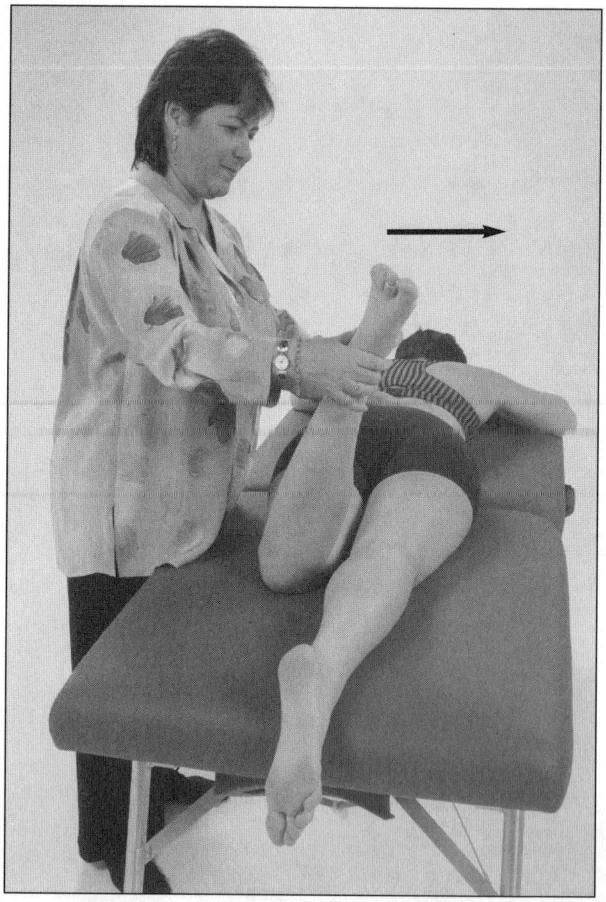

CHAPTER 5

THE ACCESSORY SACROILIAC JOINTS: HYPOMOBILITY AND TREATMENT

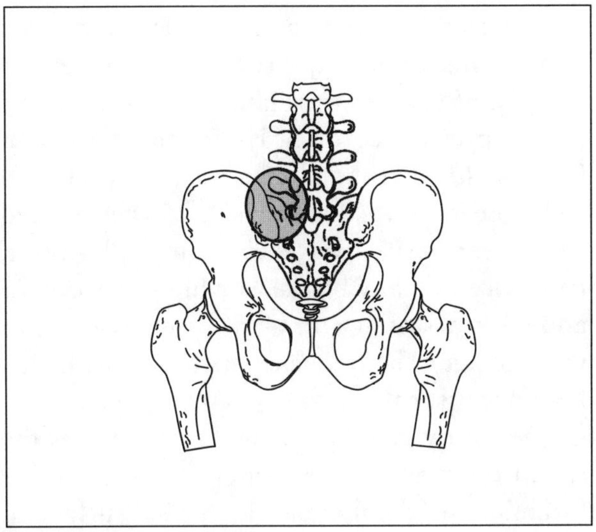

The reader should refer to Chapter Seven for important information relevant to the accessory sacroiliac joints.

Accessory or "axial" sacroiliac joints are the cause of iliosacral joint and sacroiliac joint hypomobility. These joints are situated between the posterior surface of the sacrum at the most superior and lateral aspect and the anterior surface of the PSIS (posterior superior iliac spine).

If there is hypomobility of the accessory sacroiliac joints, or merely limitations of mobility between the PSIS and sacrum, the ilium has compromised mobility. Often, a posterior rotation cannot be successfully treated, and ambulation is affected. This can cause recurrent biomechanical dysfunction of the iliosacral joints. The most significant clinical issue: The sacrum cannot be mobilized and lumbosacral junction mobility cannot be attained.

Everyone should have normal mobility of the PSIS on sacrum. The accessory sacroiliac joints are not present in all people. If the therapist is skilled with the assessment technique of Myofascial Mapping, when there is biomechanical dysfunction of these joints, there will be positive mapping over the PSIS.

The following excerpt is from an article, "The Sacroiliac Joint: A Critical Review," written by Joan M. Walker, Ph.D., P.T., a professor and director of the School of Physiotherapy at Dalhousie University, Nova Scotia, Canada, published in *The Journal of Physical Therapy* 72(12), 1992.

"Accessory" or "Axial" Sacroiliac Joints

A number of investigators have reported the presence of "supernumerary articular facets" or "axial" or accessory sacroiliac joints (ASIJs), which may contribute to, or be responsible for, SIJ dysfunction. These joints have been observed in 8 percent to 40 percent of samples, sometimes two or three at one SIJ. Trotter's comprehensive study of 958 pelvic articulations demonstrated a higher prevalence of ASIJs in males than in females (40 percent and 21 percent, respectively) and more in whites than in blacks (50 percent and 21 percent, respectively). Her sample, however, had twice as many blacks as whites between 15 and 44 years of age and half as many over 55 years of age. The frequency increased with age, but no statistical analysis was reported.

Accessory joints are located at the level of the sacral crest at the first and second posterior foramina, on the ilium at the medial surface of the posterior superior iliac spine (PSIS), and on the tuberosity. Trotter's data did not support Seligmann's theory (cited by Trotter) that prevalence of ASIJs was related to the distance between PSISs.

Hadley described a "superficial" joint at the level of the S2 posterior foramina between the PSIS and the lateral crest of the sacrum and a "deep" articulation at the level of the S1 posterior foramina. The joint plane was variable, at any angle up to 60 degrees relative to the auricular surface.

Bakland and Hansen are the only investigators to distinguish between axial and accessory joints. They described, from a small sample of 27 anatomical specimens, the axial joint as extracapsular and occurring regularly in contrast to accessory joints, which were stated by citing other authors, as "true synovial joints" occurring sporadically. Bakland and Hansen's axial joint appears similarly located to Hadley's deep ASIJ, slightly caudal to the S2 posterior foramina with the concavity on the sacral surface. Histological study of 22 specimens showed most surfaces to be fibrocartilage, thus supporting Schunke's and Trotter's observations; however, Trotter did not report any histological study. The axial joint was considered a syndesmosis on the basis of the surface composition and presence of loose connective tissue in the joint space.

Ehara, et al. termed the ASIJ a "common variant" and concurred with others in its location, that the shape of the joint surfaces was variable but that the surfaces were generally congruent. Hadley observed more pathological changes (arthritis, ankylosis) in his sample than did Ehara, et al. Although some investigators may have failed to detect these joints, others, using CT scans, make no mention of either axial or accessory joints. Dijkstra, et al., in a discussion of analyzing radiographic films, stressed the need for detailed knowledge of the configuration of the SIJ, including the so-called axial joint, which may encourage radiologists to note the presence or absence of ASIJs.

I found no reports of the presence of either axial or accessory joints in fetuses or children, giving no support to the statement by Ehara, et al. that the ASIJ is present at birth. The higher frequencies observed in adult samples with increasing age supports the theory that the ASIJ may be acquired as a result of the stress of weight bearing. Stewart commented that the ASIJ was more frequent in great apes than in humans. Great apes tend to be quadrupedal and lack hip extension. Could the ASIJ be more frequent in long-term wheelchair users or older humans who tend to have some degree of hip flexion contracture, sit for long periods, and often have low back pain? Given the observed frequency of these joints and their demonstrated pathology, it is reasonable that, when present, these joints could be a source of localized low back pain. No reports of the innervation of the ASIJ were located; pain referral patterns are therefore unknown. Whether ASIJ play or motion can be detected clinically is also unknown. Because the ASIJ is shown to be superficially located, it is possible that play may be detected by experienced manual therapists. Careful examination of radiographs for presence of such joints appears to be indicated. When present, it seems possible that motion at an ASIJ could be mistaken for SIJ motion and treatment intervention may be mistakenly applied.

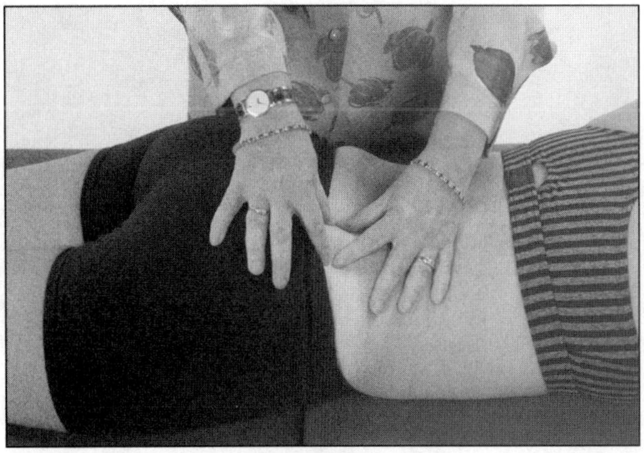

Mobility Testing of the Accessory Sacroiliac Joints

1. The patient is prone.
2. The PSIS is gripped (pinched).
3. Mobility testing is performed in all directions:

- Superior
- Inferior
- Lateral
- Medial
- All oblique directions

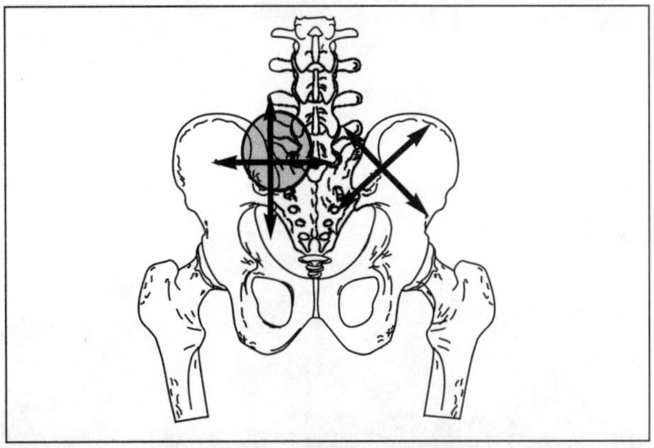

Clinical research indicates that normal mobility is approximately 0.5 to 1.0 cm excursion of the PSIS in each direction.

If mobility testing indicates limited glide manipulate with high velocity, low amplitude, low force. Do not compress the PSIS onto sacrum; this inhibits the mobilization.

Mobilization of the Accessory Sacroiliac Joints

The PSIS must be fully mobile in all directions to attain normal biomechanical function of the iliosacral joints, the sacroiliac joints, and the lumbosacral junction.

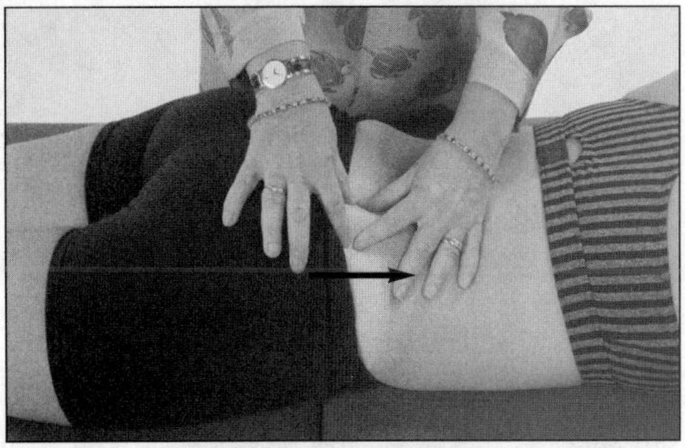

1. **Hypomobile superior glide**
   Treatment: The therapist stands on the ipsilateral side, facing the patient's head. Thenar or hypothenar eminences push the PSIS into full existing superior glide with overpressure. Manipulate to attain full superior glide.
2. **Hypomobile inferior glide**
   Treatment: The therapist stands on the ipsilateral side, facing the patient's feet. Thenar or hypothenar eminence push the PSIS into full existing inferior glide with overpressure. Manipulate to attain full inferior glide. (Note: This is the most difficult direction to manipulate.)

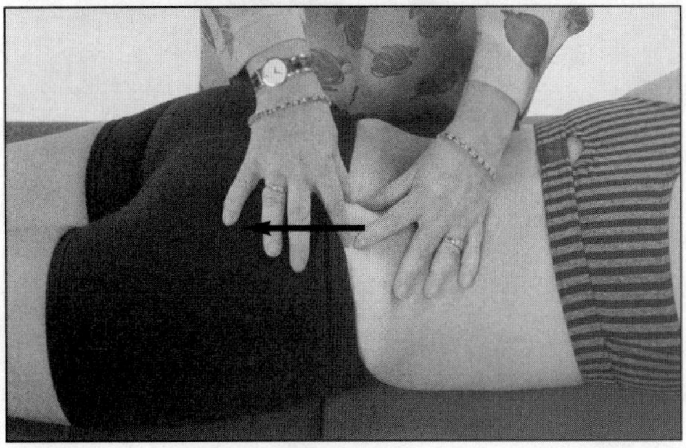

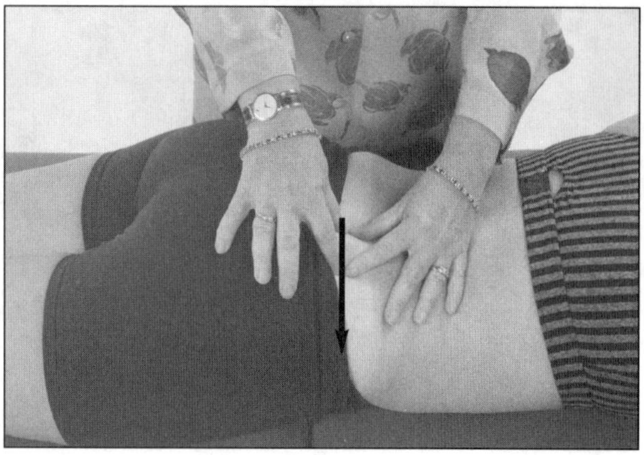

3. **Hypomobile lateral glide**
   Treatment: The therapist stands on the contralateral side. Thenar or hypothenar eminences push the PSIS into full existing lateral glide with overpressure. Manipulate to attain full lateral glide.

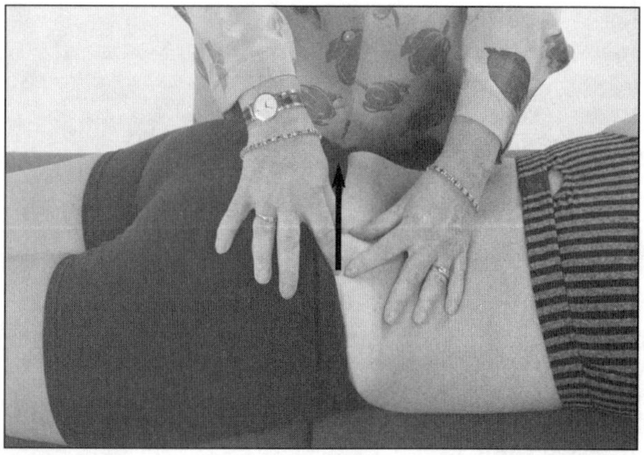

4. **Hypomobile medial glide**
   Treatment: The therapist stands on the ipsilateral side. Thenar or hypothenar eminence push the PSIS into full existing medial glide with overpressure. Manipulate to attain full medial glide.

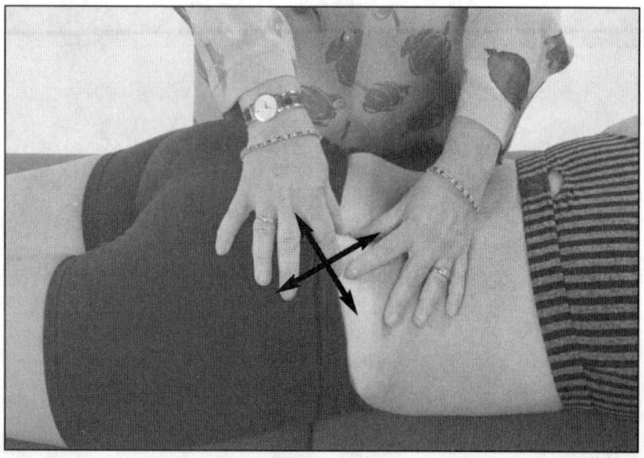

5. **Hypomobile oblique glides**
   Treatment: The therapist tests all oblique directions of mobility/glides of the PSIS. Manipulate in the directions of hypomobility.

Decompress the Accessory Sacroiliac Joints with Integrative Manual Therapy

Manipulation of the accessory sacroiliac joints can be facilitated by decompressing and distracting the joint surfaces utilizing various Integrative Manual Therapy techniques. Strain and Counterstrain Technique (Jones) can be performed first, treating the Tender Points of the posterior pelvis and sacrum, especially the HSO Tender Point. Following this, the therapist can perform an Articular Fascial Release ((Weiselfish) Giammatteo) of the posterior iliosacral joint to facilitate mobilization of the accessory sacroiliac joints.

Sequence of Treatment

Step 1: The accessory sacroiliac joints
Step 2: The pelvic joints: The pubic symphysis
Step 3: The pelvic joints: The iliosacral joints
Step 4: The sacrum: The sacroiliac joints and the lumbosacral junction

CHAPTER 6

MOBILITY TESTS

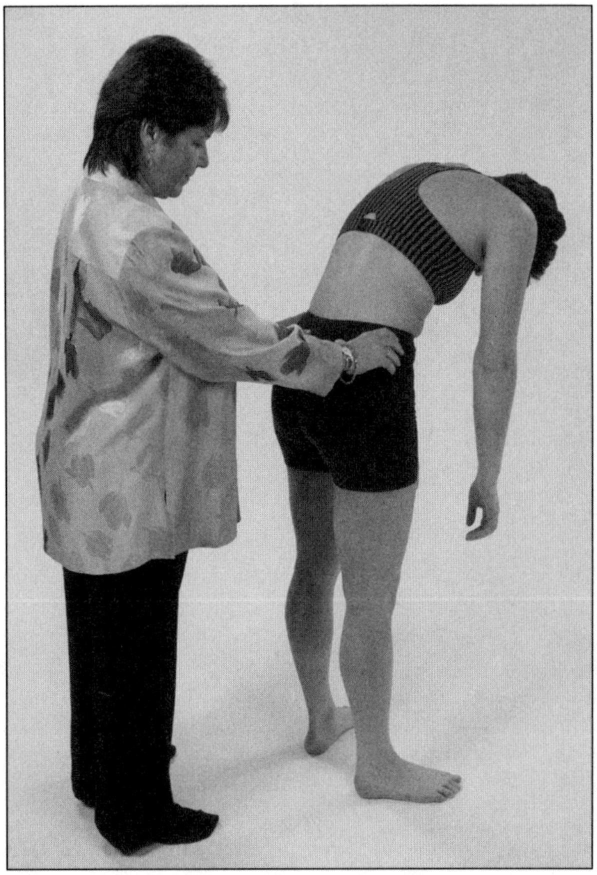

A test is as valid as the specificity of that test. A test is more specific as the number of tissues and structures compromised during that test is reduced. Mobility tests to the pelvic joints will determine which side, right or left, is more hypomobile. The side of the greater hypomobility is treated first.

Standing Flexion Test (Standing Forward Bending Test)

This test is used for pelvic joints (pubic symphysis and iliosacral joints). The test has poor specificity because multiple tissues and structures are compromised. Test muscle barriers of the long muscles of the legs first, especially for asymmetric pull of the hamstrings, which may be influencing the results of this test.

1. The patient is standing with legs acetabular distance apart and feet equally rotated (approximately 15 to 20 degrees external rotation).
2. The patient's knees are equally extended. There should not be more flexion or recurvatum allowed on one side.
3. The therapist places the pads of the thumbs just inferior to the PSIS bilaterally.
4. Ask the patient to bend forward (flex down the spinal kinetic chain, with eyes closed) as far as possible, one vertebral segment at a time, maintaining the chin against the chest.

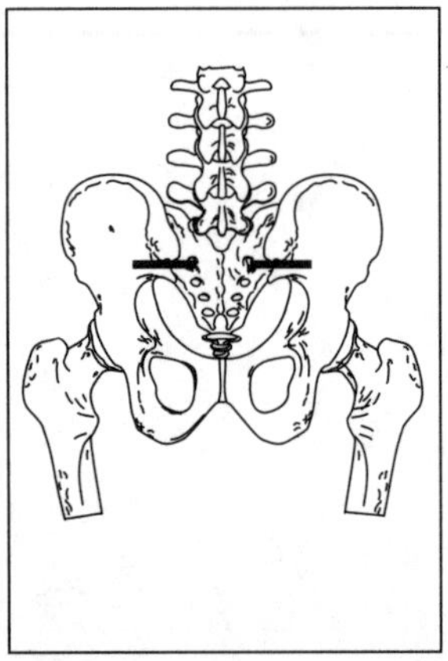

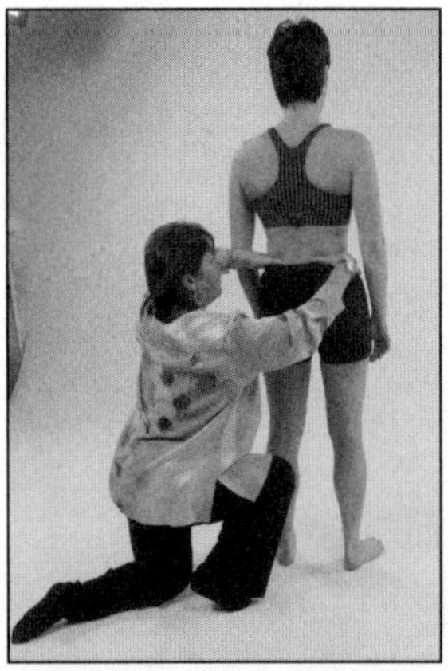

Results and Interpretation: One PSIS moves cephalad more than the other at the end range of forward bending. If the right side moves more cephalad, then there is a right positive standing flexion test. This signifies that the right pelvic joint is the more hypomobile side.

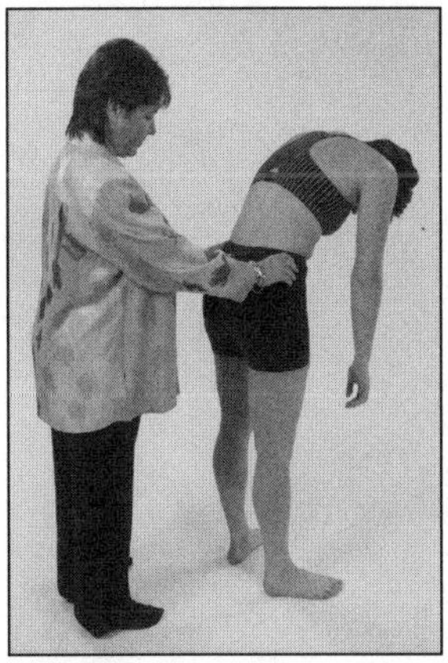

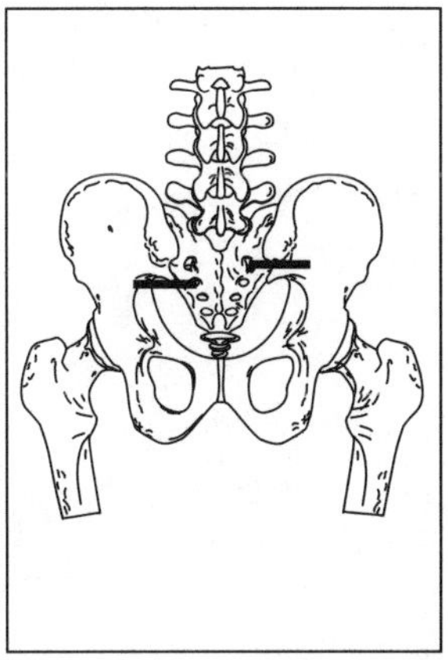

Rationale: A healthy joint has a certain amount of joint play. During standing, when flexing down the spine, the hamstrings maintain a posterior/inferior pull on the ischial tuberosities. When the sacrum begins to extend, the ilia remain posterior because of the hamstrings' pull. As the sacrum extends, there is a disassociation between sacrum and ilium at their joint surfaces bilaterally. This disassociation of joint surfaces continues until the end of normal joint play, unless there is dysfunction. If there is dysfunction, the joint play will be compromised. The disassociation of the joint surfaces of ilium and sacrum will be limited accordingly. The disassociation of ilia and sacrum during spinal flexion continues until the hypomobility due to dysfunction limits the joint play between the two bones. At that time, the hypomobile ilium will be pulled superior/anterior by the sacrum because there is limited joint play and inhibited disassociation between ilium and sacrum. The superior PSIS on the side of the ilium with the greater hypomobility will reflect this limited disassociation of articular surfaces.

Note: This is not a good test for the neurologic patient, because it requires normal balance and equilibrium.

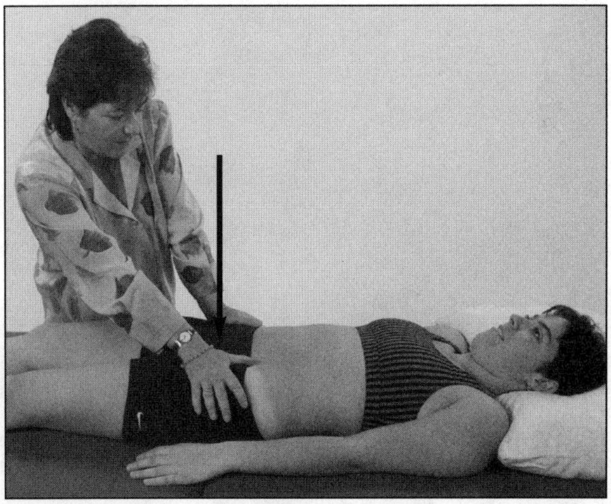

Squish Test (Mobility Test or SI and IS Joint Mobility)

1. The patient is supine. Push the right ilium to be tested for hypomobility in a posteromedial direction.
2. Place the other hand lightly over the opposite left ASIS.
3. Push the right ilium slowly and steadily. (Think of the law of inertial mass in physics.) If the joint is mobile, the right ilium will be displaced posteromedial. If the joint is hypomobile, the force from the push will cause the opposite left ASIS to be displaced in an anterior direction.
4. Palpate over the opposite left ASIS and discern how much force was utilized on the right ilium (which was pushed

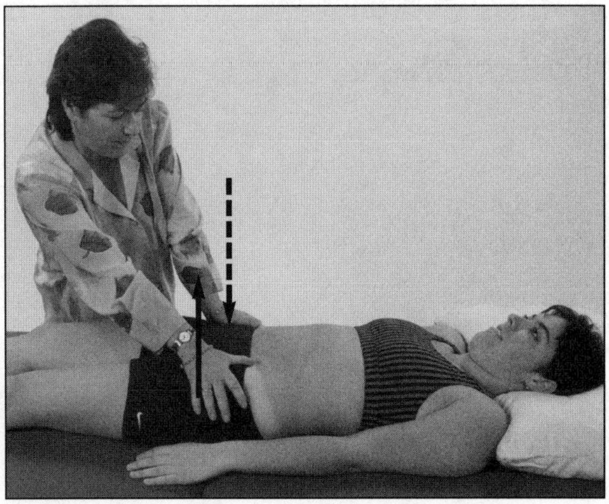

posteromedial), during how much time (seconds), until the opposite left ASIS moved anteriorly.
5. Compare the mobility of the opposite left side.

Results and interpretation: The ilium that used less force and less time of posteromedial push until the opposite ASIS was displaced anteriorly is the hypomobile ilium.

The Longitudinal Distraction and Compression Test

This test is used to diagnose upslips and downslips. The asymmetry of the position of the ischial tuberosities is often mild and difficult to distinguish. Often there are bilateral upslips and occasionally bilateral downslips, which do not present with asymmetrically positioned ischial tuberosities. Occasionally bilateral downslips, or bilateral upslips where one side is more significant than the other side, can confuse interpretation and the differential diagnostic process.

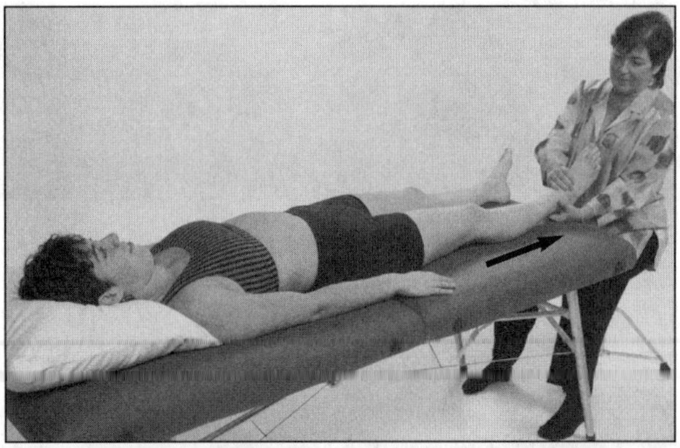

1. The patient is prone. The therapist is sitting at the foot of the treatment table, holding the foot/ankle.

2. **Phase A—Longitudinal Distraction Test.** The therapist places an inferior longitudinal distraction force on the leg from the foot/ankle. The force and palpation is subtle: First feel a distraction of the knee; then feel a distraction at the hip joint; then focus on the mobility of the caudal glide of the ilium with the distraction force. The therapist can distract each leg separately or both legs together.

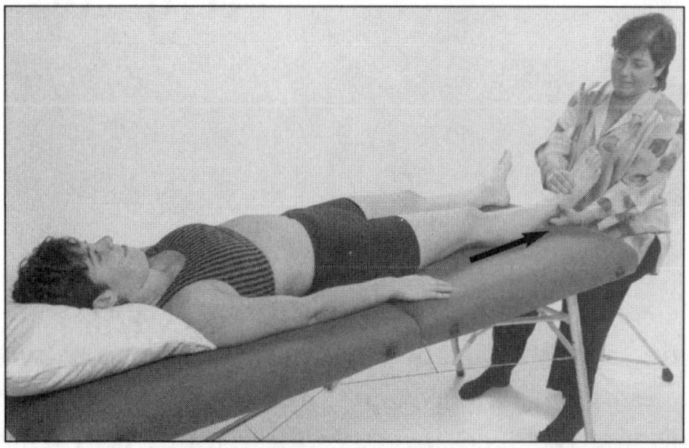

Results and Interpretation: The therapist palpates and discerns the mobility of caudal glide of each ilium. If there is a restriction in caudal glide, there is an upslip.

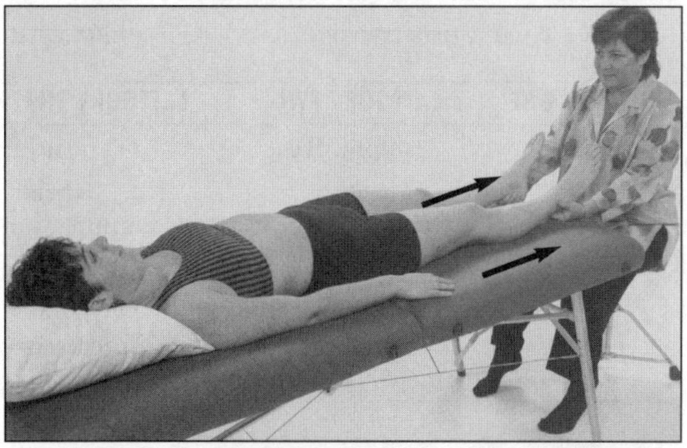

3. **Phase B—Longitudinal Compression Test.** The therapist places a superior longitudinal compression force on the leg from the foot/ankle. The force and palpation is subtle: First feel the compression of the knee joint; then feel a compression of the hip joint; then focus on the mobility of the superior glide of the ilium with the compression force. The therapist can compress each leg separately or both legs together.

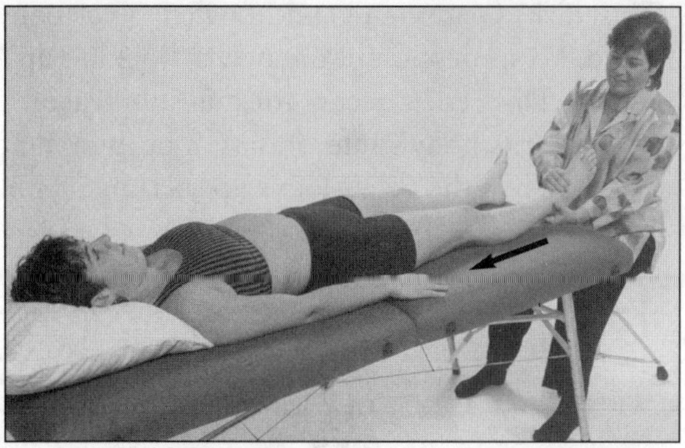

Results and Interpretation: The therapist palpates and discerns the mobility of superior glide of each ilium. If there is a restriction in superior glide, there is a downslip.

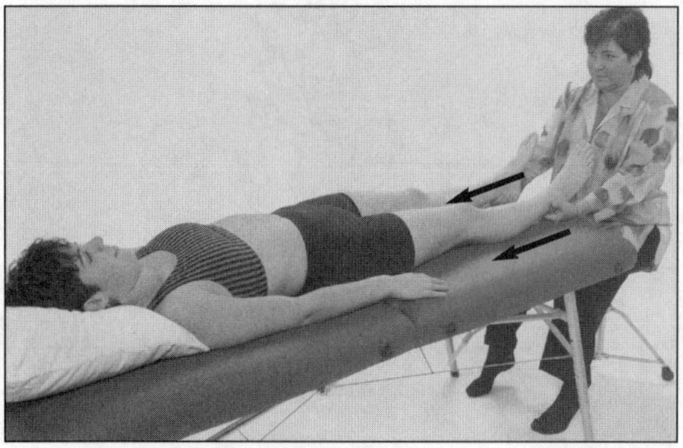

Note: Another way to test superior and caudal glide of the ilium is in prone. Grasp the medial aspect of the iliac crest and the ischial tuberosity. Push ilium up (superior glide); push ilium down (inferior glide).

CHAPTER 7

PELVIC COMPRESSION REFLEX

There are three categories of low back pain clients. Category one is the client population without a pelvic compression reflex. Category two clients have a pelvic compression reflex without a positive accessory sacroiliac joint. Category three clients have a pelvic compression reflex together with a positive accessory sacroiliac joint.

CATEGORY ONE	CATEGORY TWO	CATEGORY THREE
(-) negative	(+) positive	(+) positive
pelvic compression reflex	pelvic compression reflex	pelvic compression reflex
(-) negative	(-) negative	(+) positive
accessory sacroiliac joint	accessory sacroiliac joint	accessory sacroiliac joint

The pelvic compression reflex was noted by (Weiselfish) Giammatteo and Lowen. They confirmed a compression of the pelvis "locking up" sacrum. This reflex occurs for multiple reasons, including the body's interpretation of imminent danger. When the pelvis locks up sacrum, spinal motion is inhibited on all planes: sagittal, coronal, and transverse. The most important reason for reflexogenic inhibition of spinal motion occurs on the sagittal plane. When the pelvis locks up sacrum, spinal flexion and spinal extension are inhibited.

"Why is it necessary for the body to inhibit spinal flexion and spinal extension? Why is there a requirement for a pelvic compression reflex for the purpose of inhibition of spinal flexion and spinal extension? Why would the body compromise function, tolerate low back pain and disability, and undermine functional activities of daily living for the sake of spinal motion inhibition? The answer: when spinal motion is dangerous for one's health."

How does the pelvic compression reflex work? The PSIS (posterior superior iliac spine) "sticks to sacrum" at the superior aspect of the posterior and lateral sacrum. This area of sacrum is superior to the sacroiliac joint and has been found to articulate with ilium in approximately 35 percent of cadavers. This articulation has been called the *accessory sacroiliac joint* or the *axial sacroiliac joint* (Walker, J. M. *The Journal of Physical Therapy* 72(12), 1992.). Although this joint is mentioned rarely in medical literature, it is fairly well documented in the literature of basic biologic sciences.

A diagnostic technique called Myofascial Mapping, documented by (Weiselfish) Giammatteo in 1984, facilitates diagnostics for accessory sacroiliac joints. Myofascial Mapping is a hands-on skill for localizing areas of pathoanatomy in the body. When Myofascial Mapping is positive at the PSIS, there is hypomobility of the PSIS. This is called a *positive accessory sacroiliac joint*.

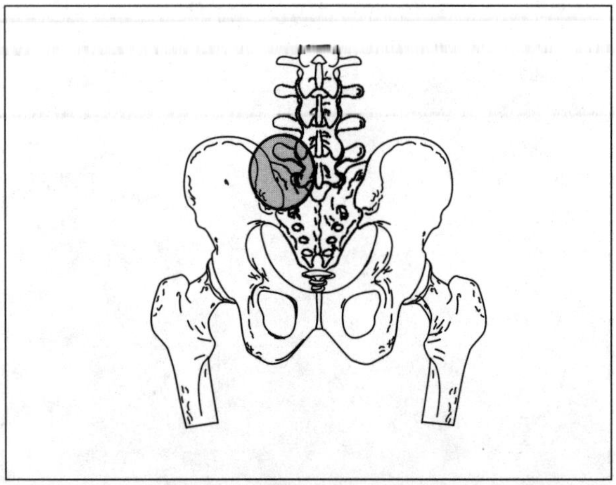

Location of the accessory sacroiliac joint

The positive accessory sacroiliac joint can be mobilized. Normal motion of the PSIS is 0.5 to 1.0 cm excursion (amplitude) in all of the following mobility tests: superior glide, inferior glide, medial glide, lateral glide, oblique glides. When there is a positive accessory sacroiliac joint the glides are lacking in mobility, i.e., mobility tests are positive for hypomobility in one or more directions. Mobility tests may be positive (i.e., hypomobile) during medial glide, lateral glide, inferior glide, superior glide, or oblique glides. (See Chapter Six.)

When there is a positive accessory sacroiliac joint the ilium cannot disassociate from the sacrum during spinal flexion and spinal extension, thus inhibiting spinal flexion and extension. During normal motion, the ilium rotates in a posterior direction while spinal flexion occurs; the ilium rotates in an anterior direction while spinal extension occurs.

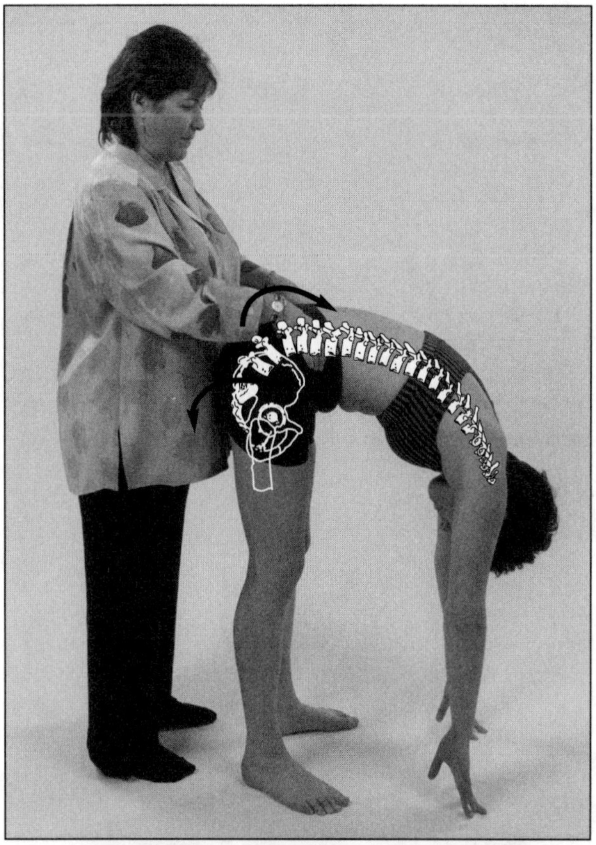

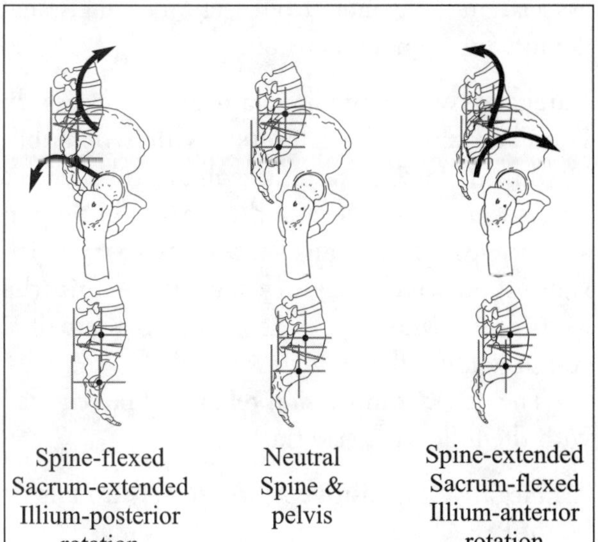

Posterior rotation of ilium during spinal flexion; anterior rotation of ilium during spinal extension.

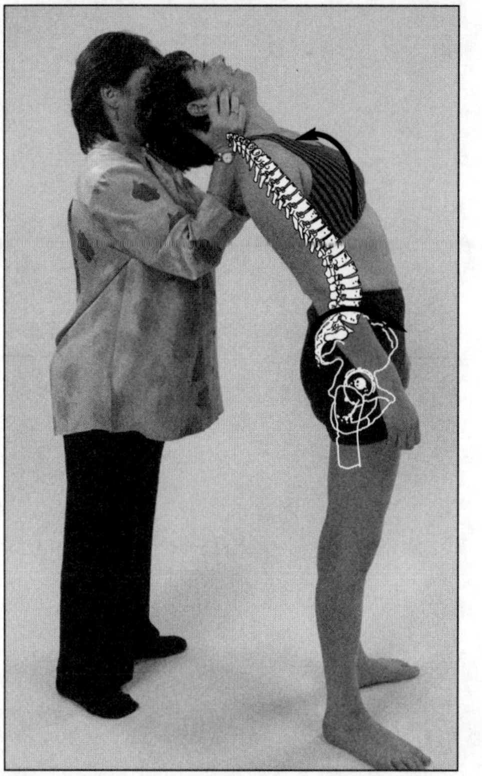

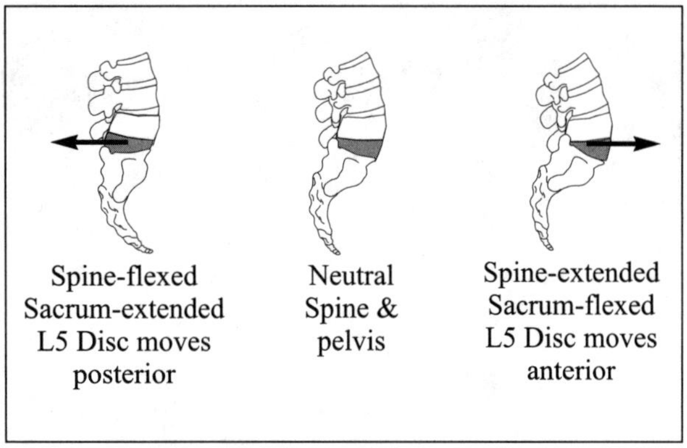

Sacral extension with L5 flexion and posterior L5 disc displacement.

When the PSIS is stuck to sacrum, which occurs when there is a positive accessory sacroiliac joint, the sacral base (S1) is pulled into posterior glide, which is sacral extension. Sacrum moves in reciprocal motion with L5. Therefore, when sacrum is in extension, L5 is in flexion. When L5 is flexed, there is a posterior displacement of the L5 disc. This means that when there is a positive accessory sacroiliac joint, there may be positive discopathy of the L5/S1 disc. Certainly, dysfunctional hydrostatic pressures will influence the L5 disc to be displaced in a posterior direction.

The PSIS is stuck to sacrum when there is a positive accessory sacroiliac joint. The evidence of a positive accessory sacroiliac joint is reflected with positive Myofascial Mapping at the PSIS. The PSIS can be mobilized. When the PSIS is successfully mobilized, there is no longer positive Myofascial Mapping at the PSIS. This indicates that the positive accessory sacroiliac joint is now negative.

When the accessory sacroiliac joint is negative, there should be good mobility of PSIS on sacrum. The glides should have 0.5 to 1.0 cm movement in all directions. There should be disassociation of the ilium from sacrum during spinal flexion and spinal extension.

Category two clients have a negative accessory sacroiliac joint, but still present with hypomobility of the PSIS. As a result, the ilium still does not disassociate from sacrum during spinal flexion and extension. The cause of a stuck PSIS, which is not a positive accessory sacroiliac joint (the Myofascial Mapping is negative), is the pelvic compression reflex.

The pelvic compression reflex will be reflected with the following criteria:

1. Negative or positive Myofascial Mapping of PSIS
2. Positive mobility testing of the PSIS (hypomobility of one or more of the following glides: superior glide, inferior glide, medial glide, lateral glide, oblique glides)

3. Painful PSIS with pressure palpation when there is a positive accessory sacroiliac joint; no pain on pressure palpation of PSIS when there is a negative accessory sacroiliac joint
4. Limitation of disassociation of ilium from sacrum during spinal flexion and extension: No posterior rotation of ilium during spinal flexion; no anterior rotation of ilium during spinal extension
5. Local listening (Barral) that is positive at the PSIS

The pelvic compression reflex appears to be a higher level brain stem reflex, rather than a spinal level reflex. (Example of a spinal level reflex: Protective muscle spasm from a myotatic reflex arc hyperactivity. Examples of a brain stem level reflex: ATNR (asymmetric tonic neck reflex), flexor withdrawal reflex.) Brain stem level reflexes occur when there is a significant body problem that requires protection. When spinal flexion or extension is "dangerous for one's health," there will be a pelvic compression reflex. The pelvic compression reflex will inhibit spinal flexion and spinal extension so that the body is not in danger.

There are recognized body problems that elicit the pelvis compression reflex. These body problems can pose a danger. The danger is more significant than the L5/S1 disc problems, which occur secondary to a sustained pelvic compression reflex.

There are many vascular problems, especially arterial, that elicit this brain stem protective reflex. These vascular problems may be stenosis, occlusion, dysbiosis (infection, toxicity), disruption of membrane (excessive permeability of a membrane wall). These vascular problems (which elicit the pelvis compression reflex) may be located at the aorta (at the aortic arch and superior to the aortic biforcation), iliac artery, spinal artery, circle of Willis, middle cerebral artery, or vertebral artery.

Another problem that elicits the pelvis compression reflex is a bone bruise. The *bone bruise* has been defined by the author as a dysfunction that may affect cortex, trabeculae, or marrow. There has been no documentation on this dysfunction until recently. The bone bruise was defined as a microtrabecular fracture in the journal *Physical Therapy Case Reports,* Volume 2. This author describes the problem in clinical terms: hairline or greater bone cortex lesion, or osteomalacia. The bone bruise occurs often in the following regions of the axial skeleton, eliciting the pelvis compression reflex: lateral to the PSIS, on the lateral border of sacrum, acetabulum, occiput, sphenoid, spinal vertebrae.

Another condition that may elicit a pelvic compression reflex is described in the medical literature as "leaky gut syndrome." When there is excessive permeability of an organ wall, dysbiosis may occur. *Dysbiosis* is when the normal flora of the environment are overwhelmed by pathologic microorganisms, causing a toxicity (sepsis, infection, inflammation) in the region. For example, there can be a leaky gut syndrome of the sigmoid colon. The sigmoid colon could leak toxic waste, which should be contained within the organ wall. When this waste is leaked into the environment, there can be a local inflammation and/or infection of the pelvic cavity. The sigmoid colon is situated anterior to L5 and sacrum. When there is spinal extension, tension is placed upon the sigmoid colon. Spinal extension could place excessive tension upon an already compromised organ wall membrane. The spinal extension must be inhibited. The brain stem recognizes the requirement: Inhibit spinal extension to protect an already dysfunctional sigmoid colon. The pelvis compression reflex is elicited.

Returning to the three categories of low back clients listed above, there is an understanding that the lumbosacral junction is paramount in dysfunctional process. One hundred percent of the American population will have spinal pain within their lifetime. Low back pain incidence is 80 percent. More than 85 percent of spinal surgery occurs at the lumbosacral region. What are we not doing? Why are we not affecting the incidence of lumbosacral junction dysfunction?

Investigation into the problems of L5/S1 always ends at the same point of reference: hypomobility. The hypomobility of the lumbosacral junction is reflected in the following manners and more:

1. There are limitations of spinal flexion, spinal extension, spinal rotations and spinal side bending.
2. The client cannot touch the floor (touching the floor indicates good lumbosacral flexion).
3. There is less than 40 degrees of lumbosacral extension.
4. There is limited posterior glide of sacral base during lumbosacral flexion.
5. There is limited anterior glide of sacral base during lumbosacral extension.

When there is a pelvic compression reflex, there is always a problem of lumbosacral junction hypomobility. Relatively easy techniques will not be successful; for example, functional rehabilitation, which might include stretching, strengthening, movement reeducation, sensory reeducation, respiratory rehabilitation for increased endurance, non-weight-bearing exercises, weight-bearing exercises, and more. Structural rehabilitation to restore joint mobility at the lumbosacral junction will not be successful, including mobilization, manipulation, muscle energy technique, Muscle Energy and "Beyond" Technique, Strain and Counterstrain for the low back, Myofascial Release for the low back, Visceral Mobilization and Visceral Manipulation, Neural Tissue Tension Technique and mobilization of the nervous tissue, Cranial Therapy and Craniosacral Therapy, etc. When there is a pelvic compression reflex, the practitioner must discover and correct the cause: the reason the body is in a protective mode, the reason the body is inhibiting motion.

What are common causes of pelvic compression reflex? They may be vascular, neural, bone, or visceral. Typically, causes of pelvic compression reflex do not include muscle or fascia. The muscle will be in protective muscle spasm in order to protect. The fascia will thicken (hypertrophy) and the fibers will multiply (hyperplasia) in order to contribute to the protective mode. Under those conditions, the hypertrophy and hyperplasia become a primary problem, which increases the requirements for protection.

Low back pain is caused by skeletal (axial) joint dysfunction. The primary successful approach for treatment of low back pain and disability is correction of biomechanical dysfunction with Muscle Energy and "Beyond" Technique. This approach, developed by the author after studying the foundation osteopathic approach developed by Fred Mitchell, Sr., D.O., called Muscle Energy Technique, can help the client attain full pain-free function. The guidelines and criteria for success include:

1. Elimination of low back pain
2. Full ranges of spinal motion, including the ability to touch the floor without a posterior deviation of buttocks (i.e., standing against the wall and flexing down the spinal kinetic chain until the fingers touch the floor, without loss of equilibrium and balance); the ability to extend the lumbosacral junction to 45 degrees (without an anterior deviation of chest and abdomen); at least 70 percent of all cervical, thoracic, and lumbar ranges of motion (the ranges of motion of the cervical, thoracic, and lumbar spine are dependent upon biomechanical joint integrity of the pelvic joints and sacrum up to 70 degrees)

Essential joint dysfunctions that contribute directly to low back pain and disability of pelvis and sacrum include: upslip of the ilium; downslip of the ilium; positive accessory sacroiliac joint; flexed sacrum (bilateral or unilateral); extended sacrum (bilateral or unilateral); anterior rotation of sacrum on a vertical axis; posterior rotation of sacrum on a vertical axis; anterior sacral torsion (a Type II dysfunction of S1); posterior sacral torsion (a Type II dysfunction of S1); ascended sacrum; descended sacrum; flexed, rotated, and side bent L5 (a Type II dysfunction); extended, rotated, and side bent L5 (a Type II dysfunction); Type III of S1 (an anterior or lateral shear/subluxation); Type III of L5 (an anterior or lateral shear/subluxation).

When there is a pelvic compression reflex and the above described joint dysfunctions of the pelvis and sacrum are present, the pelvis will not correct with biomechanical treatment. When there is no pelvis compression reflex, Muscle Energy and "Beyond" Techniques (attained in DCR seminars METSP and advMET) can successfully correct these joint dysfunctions. When Muscle Energy and "Beyond" Techniques are applied, low back pain and disability can be eliminated.

The problem with therapeutic intervention for low back pain arises when Muscle Energy and "Beyond" Techniques cannot be applied because there is a positive accessory sacroiliac joint or there is a pelvic compression reflex. When there is a negative accessory sacroiliac joint and there is no pelvic compression reflex, Muscle Energy and "Beyond" Techniques can be applied successfully.

1. If there is a negative accessory sacroiliac joint and there is no pelvic compression reflex, then effective and efficient results can be attained with Muscle Energy and "Beyond" Techniques.
2. If there is a positive accessory sacroiliac joint and there is no pelvic compression reflex, then effective and efficient results will occur only when the accessory sacroiliac joint is corrected.
3. If there is a positive accessory sacroiliac joint and a pelvic compression reflex, the pelvic compression reflex will not be evident until the accessory sacroiliac joint is mobilized and corrected.
4. If there is a pelvic compression reflex, there is a reason or cause for protection. This reason must be discerned. The practitioner must correct the cause or causes for protection so that the pelvic compression reflex will be eliminated. Once the pelvis compression reflex is eliminated, good results will be attained with Muscle Energy and "Beyond" Techniques.

CHAPTER 8

INTERPRETATION: ANATOMICAL LANDMARKS PLUS RESULTS OF MOBILITY TESTING

ANATOMICAL LANDMARKS	MOBILITY TEST	DIAGNOSIS	
1. The right superior pubic ramus is more superior than the left superior pubic ramus.	The right pelvic joints are more hypomobile (standing flexion test and squish test).	Right elevated pubes	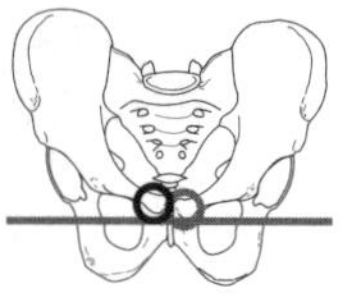
2. The right superior pubic ramus is more superior than the left superior pubic ramus.	The left pelvic joints are more hypomobile (standing flexion test and squish test).	Left descended pubes	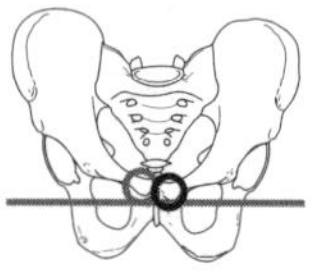
3. The right ASIS is more medial than the left ASIS.	The right pelvic joints are more hypomobile (standing flexion test and squish test).	Right inflare	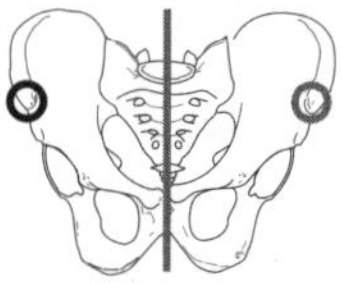
4. The right ASIS is more medial than the left ASIS. (The left ASIS is more lateral than the right ASIS.)	The left pelvic joints are more hypomobile (standing flexion test and squish test).	Left outflare	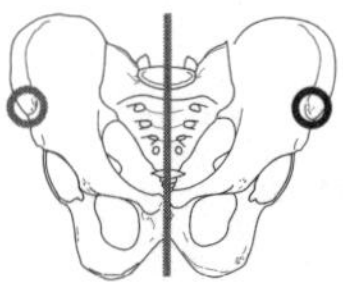
5. The right ischial tuberosity is more superior than the left ischial tuberosity.	The right pelvic joints are more hypomobile (standing flexion test and squish test). The right longitudinal distraction test is limited, but the compression test is normal. (The left compression and distraction tests are normal.)	Right upslip	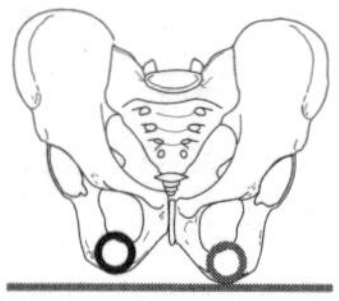
6. The right ischial tuberosity is more superior than the left ischial tuberosity.	The left pelvic joints are more hypomobile (standing flexion test and squish test). Also, the left longitudinal compression test is limited, but the distraction test is normal. (The right compression and distraction tests are normal.)	Left downslip	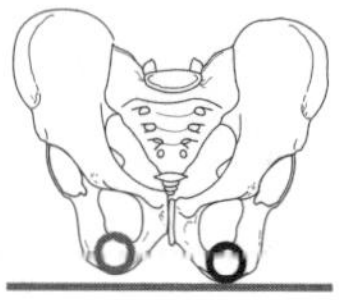
7. The right ASIS is more superior than the left ASIS, and the right PSIS is more inferior than the left PSIS.	The right pelvic joints are more hypomobile (standing flexion test and squish test).	Right posterior rotation (innominate)	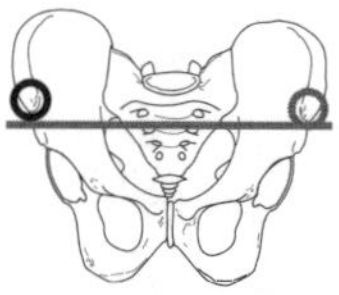
8. The right ASIS is more superior than the left ASIS, and the right PSIS is more inferior than the left PSIS.	The left pelvic joints are more hypomobile (standing flexion test and squish test).	Left anterior rotation (innominate)	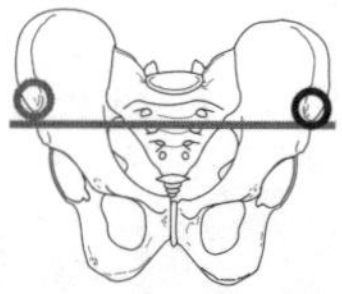

Pelvic Joint Biomechanical Dysfunctions

1. Right/left elevated pubes
2. Right/left descended pubes
3. Right/left inflare
4. Right/left outflare
5. Right/left upslip
6. Right/left downslip
7. Right/left anterior rotation (innominate)
8. Right/left posterior rotation (innominate)

Differential Diagnosis of Biomechanical Dysfunction of Pelvic Joints

Anatomical Landmarks for Differential Diagnosis of Pubic and Iliosacral Joint Dysfunction

Four anatomic landmarks for diagnosis of pelvic biomechanical dysfunction:

1. Superior pubic ramus
2. ASIS
3. PSIS
4. Ischial tuberosity

Five observations of anatomic landmarks for differential diagnosis of pelvic biomechanical dysfunction:

1. Which superior pubic ramus is more superior?
2. Which ASIS is more medial?
3. Which ASIS is more superior?
4. Which PSIS is more superior?
5. Which ischial tuberosity is more superior?

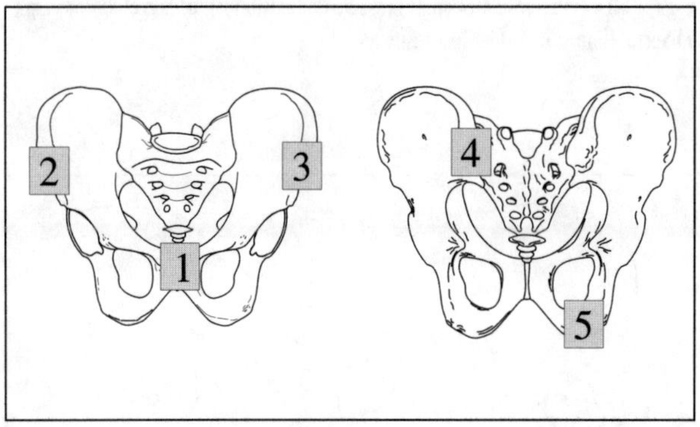

CHAPTER 9

MUSCLE ENERGY AND "BEYOND" TECHNIQUE—THE PELVIC JOINTS

Pelvic Biomechanics: Anatomical Landmarks for Differential Diagnosis of Iliosacral and Sacroiliac Joint Dysfunction

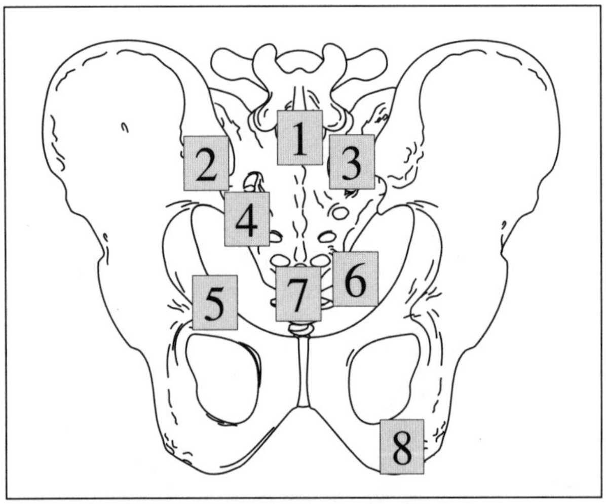

Posterior View: 1) L5/S1 interspace; 2) PSIS; 3) Depth of Sacral Sulcus; 4) Median Crest; 5) ILA (interior lateral angle) (caudal and posterior); 6) Sacrotuberous Lig.; 7) Sacral Hiatus; 8) Ischial Tuberosity

POSTERIOR VIEW

1. L5/S1 interspace
2. PSIS
3. Depth of Sacral Sulcus
4. Median Crest
5. Sacrotuberous Ligament
6. ILA (inferior lateral angle) (caudal and posterior)
7. Sacral Hiatus
8. Ischial Tuberosity

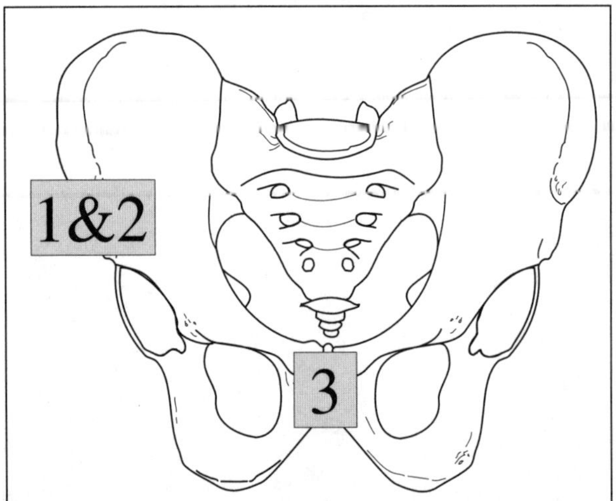

Anterior View: 1) ASIS (elevation); 2) ASIS (equidistant to midline); 3) Public Ramus

ANTERIOR VIEW

1. ASIS (elevation)
2. ASIS (equidistant to midline)
3. Pubic Ramus

   Biomechanics of Pelvis

Iliosacral Joint Movement vs. Sacroiliac Joint Movement

Iliosacral joint movement is described by movement of the ilium on the sacrum. Many reviews in the literature discuss this motion. Often, hypothetical axes of movement are defined. The author discusses the problems that present clinically. It is important to note that there is a distinct difference between different hypothetical models. Often, models that address iliosacral joint dysfunction are relevant only to joint mobility, i.e. the three planes of movement. Other models include articular balance, i.e., convex motion on concave surfaces, or concave movement on convex surfaces. This author refers to biomechanics as the three planar dimensions between the articular surfaces. Thus, axes are less relevant.

Sacroiliac joint movement is movement of the sacrum on the ilium. Multiple axes of movement are referred to in various hypothetical models by other researchers. Often, two transverse axes and two oblique axes are defined. For the relevance of this particular model, axes of motion are less important for defining sacroiliac movement.

Three criteria need to be presented in this model of the intra-articular space: (1) Quanta, or energy that moves in particle and/or wave form, fills and maintains this intra-articular space; (2) there is reciprocal movement between L5 and S1; (3) there is an auricular-shape articular surface for sacrum and ilium.

Occasionally, the prime movement occurs at the articular surface of ilium, which would then define the motion as iliosacral joint movement. Occasionally, the prime movement occurs at the articular surface of sacrum, which would then define the motion as sacroiliac joint movement. During dynamic movement, typically both iliosacral joint movement and sacroiliac joint movement are co-joined.

Whenever there is motion dysfunction of the iliosacral or sacroiliac joints, these two movement patterns are easily distinguishable. However, the pelvic joints, which include the pubic symphysis and iliosacral joints, are less distinguishable from joint motion at the femoral-acetabular (hip) joint.

CHAPTER 10

STRUCTURAL REHABILITATION WITH MUSCLE ENERGY AND "BEYOND" TECHNIQUE FOR TREATMENT OF BIOMECHANICAL DYSFUNCTION

3-Planar Movement

Muscle Energy Technique, developed by Fred Mitchell, D.O., is a positional technique that facilitates correction of biomechanical dysfunction by normalizing neuromusculoskeletal balance. Joint dysfunction is typically 3-*planar*; this technique utilizes 3-planar positions and gentle unidirectional isometric resistance to attain normal articular balance and joint mobility. The joint being treated is positioned on three planes in the *direction of least mobility* (greatest hypomobility). The movement direction for the sagittal plane position will be flexion or extension. The movement direction for the coronal plane will be right side bending or left side bending. The movement direction for the transverse plane will be right rotation or left rotation. These movements will be *stacked* on three planes. For example, the joint will first move into flexion up to the barrier; next, without returning to neutral position, the joint will move into right side bending; two planes have now been stacked. Next, without returning to neutral position, the joint will be positioned in right rotation. The movements have now been stacked on three planes.

What Is the Barrier?

The *barrier* is often considered to be an elusive concept. During treatment, each movement direction (three movement directions, on transverse, coronal, and sagittal planes) is up to the specific barrier. The barrier is *not* the end of existing range of motion in that direction on that plane. The barrier is the first, often subtle, soft tissue resistance.

Movement on each of the three planes, is passively performed. *Passive,* in the true meaning of the word, means there is no assistance for the movement and no resistance to the movement. The practitioner's hands are palpating, searching for the first indication of tissue tension. This tissue tension reflects resistance to passive movement.

There should be no tissue tension while the movement is performed passively. When there is tissue tension, this is the reflection of resistance by the muscle to the movement. The muscle fiber is now firing; it is in a state of contraction because the muscle spindle's threshold to stretch has been met. This very first indication of the most subtle tissue tension is the barrier.

This barrier, which is the position when soft tissue resistance to movement was first noticed via palpation, is engaged on all three planes. Yet the treatment (isometric resistance) will not occur at the barrier. The treatment will occur at the *interbarrier zone*. The interbarrier zone is a few degrees before the barrier. There is an interbarrier zone for each direction of movement on each of the three planes.

Treatment of a Joint with Biomechanical Dysfunction

Position of Treatment: The 3-Planar Interbarrier Zone

The joint is moved in one direction on one plane until the barrier is palpated (this barrier is the first sign of tissue tension). Immediately, the joint is returned in the opposite direction only 2 or 3 degrees. This is the interbarrier zone. Then the movement can be initiated on the second plane. As soon as the barrier is palpated (that first, most subtle tissue tension), the joint is returned in the opposite direction 2 or 3 degrees to the interbarrier zone for that plane. From that new position of an interbarrier zone on two planes, the joint will then be moved in the third and last direction on the third and last plane. As soon as the barrier is palpated, the joint will be returned 2 to 3 degrees in the opposite direction. Now the joint is stacked in the direction of greatest hypomobility on three planes at a 3-planar interbarrier zone.

Isometric Resistance

At this position, the 3-planar interbarrier zone, a gentle and unidirectional isometric resistance can be provided by the therapist. The resistance can be only 5 grams of force. The greater the resistance, the greater the co-contraction, the less effective the technique. The resistance may be maintained for 6 to 10 seconds. It is necessary to achieve a muscle contraction. The resistance should be unidirectional and uniplanar in order to prevent a co-contraction, which would inhibit the correction. After a complete contraction has been achieved, it is necessary to achieve a complete relaxation. It might take 10 seconds or more, with significant support of the body part, to attain the optimal relaxation for a chronic or severe patient.

After relaxation, this process is repeated a second time. The joint is moved further in the direction of hypomobility on a sagittal plane until the barrier is met. Then the joint returns 2 to 3 degrees to the interbarrier zone. At this point the joint is taken in the second direction on the coronal plane. As soon as the barrier is palpated, the joint is returned to the interbarrier zone. Then this process is repeated the third and last time. The joint is taken in the same direction of hypomobility on a transverse plane until the barrier is palpated and then returned to the interbarrier zone. The joint is now at the new interbarrier zone on three planes, but this time further into the range of motion than the last time. At this time, the therapist provides a gentle, unidirectional force of isometric resistance in the same direction as the previous repetition. After the 6 to 10 seconds, contraction is attained, followed by relaxation, the whole process is repeated a third and last time.

A common question is, "How does such a subtle isometric contraction achieve a correction?" The answer lies in the fields of quantum physics and neurophysiology. There is no doubt that a better result is attained with a lesser resistance. If indeed there is quanta (energy) within the joint spaces, and this space is maintained in a correct 3-planar orientation, then resistance is affecting energy.

Application for the Neurologic Patient

Because of spasticity present in the neurologic patient, the Muscle Energy and "Beyond" Techniques are performed with slight variations:

Modifications of Muscle Energy and "Beyond" Technique for the Neurologic Patient

1. It may be necessary to facilitate a muscle contraction by taping the muscle, using principles of proprioceptive neuromuscular facilitation.
2. It is necessary to maintain the contraction for 10 to 30 seconds, because the neural recruitment and the strength and endurance of the muscle contraction is not sufficient.
3. It is necessary to provide support to the body part to facilitate a complete relaxation between the contractions. This often takes longer for the neurologic patient: 10 to 30 seconds.
4. To attain optimal results, 5 to 6 repetitions of the contractions, each time in a new 3-planar interbarrier zone, may be necessary.

Principles of Muscle Energy and "Beyond" Technique

1. Palpate anatomical landmarks for an accurate structural diagnosis.
2. Engage the restrictive hypomobile barrier on each of the three planes:
   a. Sagittal (flexion/extension)
   b. Coronal (left/right sidebending)
   c. Transverse (left/right rotation)

3. Return to the interbarrier zone on each of the three planes.
4. Apply gentle isometric resistance (Tapping may be necessary to facilitate a muscle contraction for the neurologic patient.)
5. Patient muscle effort:
   a. Gentle amount of force
   b. Unidirectional/uniplanar resistance
   c. Duration of effort: 6 to 10 seconds (10 to 30 seconds for the neurologic patient)
6. Attain complete relaxation after the muscle contraction. (The neurologic patient may require facilitation for optimal relaxation.)
7. Reposition to the new restrictive barrier on each of the three planes. Treatment at the interbarrier zone.
8. Repeat steps 3 through 6, three times (three to five times for the neurologic patient).
9. Retest.

CHAPTER 11

THE PELVIC JOINTS: THE PUBIC SYMPHYSIS AND THE ILIOSACRAL JOINTS

The Pelvis

Often, biomechanical dysfunction of the pelvic joints is a primary problem. Certainly, whenever there is trauma, a fall or any other direct force, there can be a displacement of joint surfaces, with a resultant joint dysfunction. More often, biomechanical dysfunction is caused by soft tissue dysfunction of lower extremity muscles, pelvic floor diaphragm muscle and fascia, and pelvic bowl organs, vessels, and nerves. When there is a postural asymmetry relative to pelvic joint biomechanical dysfunction, there are specific indications:

1. The superior pubic ramus will be elevated or descended.
2. The ASIS will be medial or lateral.
3. The PSIS will be elevated or descended.
4. The ischial tuberosities will be elevated or descended.

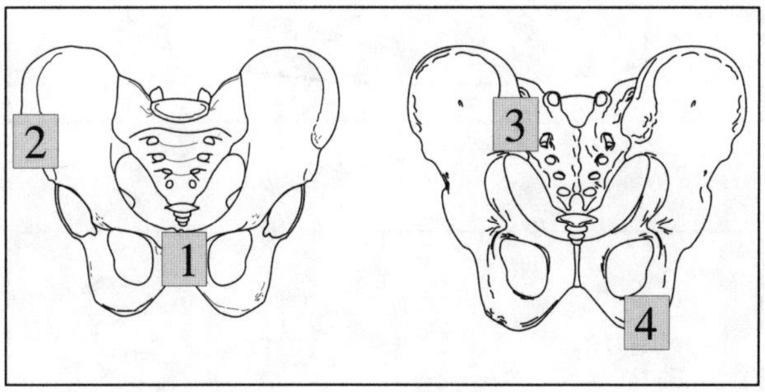

When there is a postural asymmetry of the pelvis secondary to soft tissue dysfunction (rather than primary joint dysfunction), there are reflections of this problem as follows:

1. The superior pubic ramus (near the pubic symphysis) will be anterior or posterior.
2. The ASIS will be elevated or descended.
3. The ischial tuberosities will be medial or lateral.

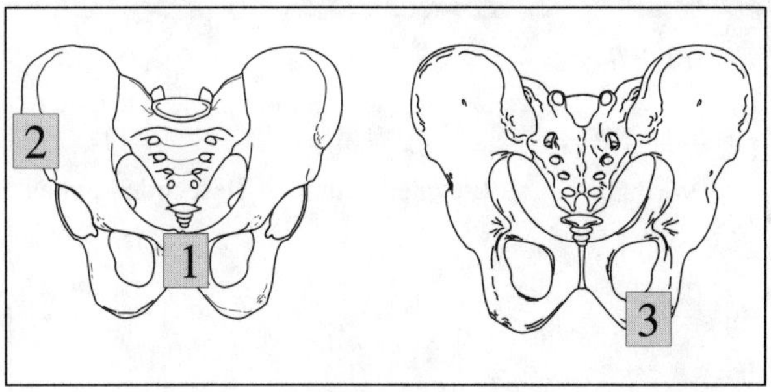

Although often all of the above (7) points are present, and confusion is the result, sometimes the above indications do stand alone. There is an abundance of information discussing soft tissue dysfunction of the pelvic floor and pelvic bowl. While it is not possible to elaborate on those soft tissue dysfunctions in this book, suffice it to state that they may contribute to pelvic joint biome-

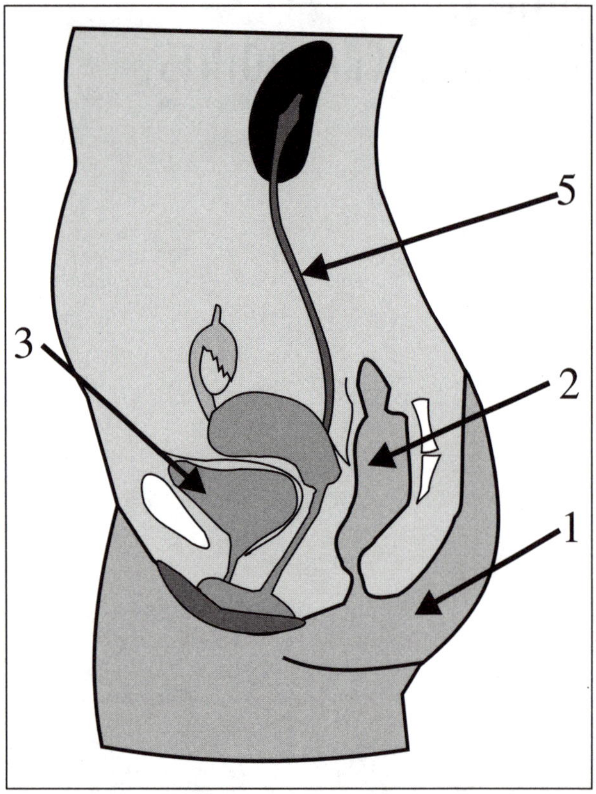

chanical dysfunction. The most typical of these soft tissue problems that cause pelvic joint biomechanical dysfunction appear to include:

1. Compression syndromes of the pelvic floor diaphragm
2. Sigmoid colon soft tissue dysfunction
3. Bladder soft tissue dysfunction
4. Cecum soft tissue dysfunction (not pictured)
5. Ureter soft tissue dysfunction

Note to the practitioner: Whenever there is a recurrence of joint dysfunction of the pelvic joints, assess the above for hypertonicity (muscle spasm) and hypomobility (upon mobility testing).

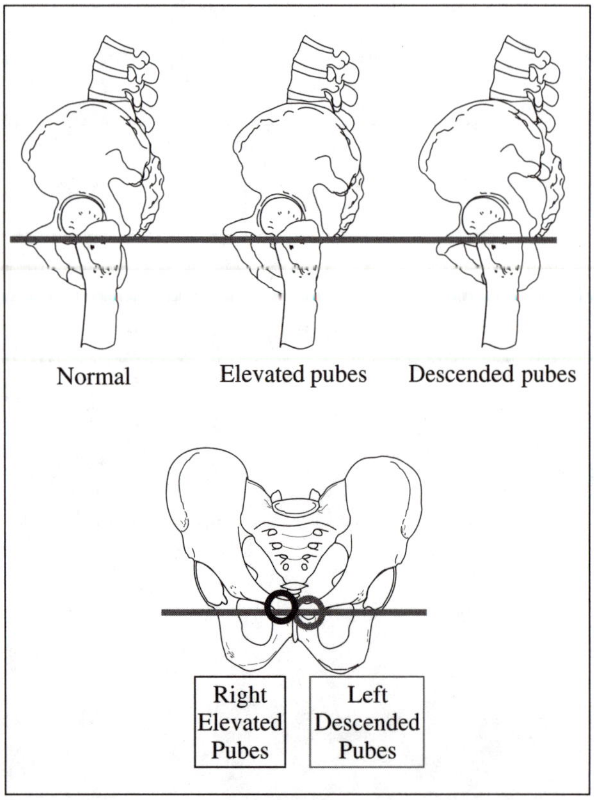

The Pubic Joints

All joints have three-planar movement, although the most significant plane of motion for the pubic symphysis is the sagittal plane. On a sagittal plane, there are inferior and superior glides of the moving pubis.

During hip flexion, the pubes glides superiorly. If stuck in that position, the diagnosis that describes the position of dysfunction is *an elevated pubes*. During hip extension, the pubes glides inferiorly. If the pubes gets stuck in an inferior position, the positional dysfunction that describes the problem is *a descended pubes*. The superior pubic ramus will be superior with an elevated pubes. The superior pubic ramus will be caudal with a descended pubes. With an elevated pubes, hip flexion may be within normal limits; full range of motion in hip extension will be inhibited. With a descended pubes, hip extension may be within normal limits, but hip flexion will be inhibited and limited.

The Iliosacral Joints

There are many biomechanical dysfunctions that are commonly diagnosed at the iliosacral joints. The most typical dysfunctions presented in this manual can be corrected with Muscle Energy and "Beyond" Techniques as described. An important biomechanical function of the iliosacral joints is their contribution to the stability and mobility of the trunk in standing, the stability and mobility of the pelvis in sitting, and the general level of biomechanical integrity of the spine. The pelvic joints, when functional, allow a normal balance at the lumbosacral junction where the spine meets its base of support. This base of support should be parallel to the ground and perpendicular to the spine. If there is pelvic obliquity (an elevation of one pelvic girdle observed by palpating the iliac crest), this base of support is not parallel to the floor, and the spine cannot be perpendicular both to the base of support and to the ground. The lumbar spine will be side bent and rotated, affecting the mobility of all the spinal intervertebral joints, as well as the articular balance of all of the vertebral bodies of the cervical, thoracic, and lumbar spine.

The biomechanical dysfunctions peculiar to the iliosacral joint include the flares (inflares and outflares), the upslips and downslips, and the posterior and anterior rotations (commonly known as anterior innominates and posterior innominates). The anatomical landmarks used for assessment and diagnosis of these biomechanical dysfunctions can be limited to the ASIS (anterior superior iliac spines), the PSIS (posterior superior iliac spines) and the ischial tuberosities. It is possible to attain a differential diagnosis of all the iliosacral joint dysfunctions with just the information from the assessment of the relative position of these anatomic landmarks and a mobility test to compare right and left mobility. More information might be confusing and is not usually necessary for diagnosis.

The ASIS will be positioned more medially whenever the ilium flares in. This represents an internal/medial rotation of the ilium. The anterior aspect of the iliosacral joint will be more closed; the posterior aspect of the joint will be gapped. The ilium, when stuck in this position of medial rotation, is diagnosed as an inflare. When the ASIS is more lateral to midline, this reflects a flaring out of the ilium. This is essentially a lateral/external rotation of the ilium. When the ilium is stuck in this externally rotated position, the dysfunction is described as an outflare.

Physiologic movements of the hip joints are affected by biomechanical dysfunctions of the iliosacral joints. With an inflare, there may be full adduction and internal rotation of the hip, but hip abduction and external rotation will be limited. With an outflare, the normal ranges of hip abduction and external rotation may be present, but hip adduction and internal rotation will be inhibited.

It is possible to differentially diagnose an inflare and an outflare of the iliosacral joint by assessing the medial/lateral placement of the ASIS bilaterally. When the ASIS is more medial, there is an inflare of that iliosacral joint. When the ASIS is more lateral, there is an outflare of that iliosacral joint.

Mobility testing, including the standing forward bending test (standing flexion test) and the squish test, will determine the side of the greater hypomobility in order to decide if the right or left side is more dysfunctional. For example, if the right ASIS is more medial than the left ASIS, and the standing forward bending test and the squish test are positive for joint hypomobility on the right side, then the patient has an inflare of the right iliosacral joint. In the other possible scenario, if the right ASIS is more medial than the left ASIS, but the standing forward bending test and the squish test are positive for joint hypomobility on the left side, then the patient has an outflare of the left iliosacral joint.

It is possible to have bilateral iliosacral joint dysfunctions, such as an inflare of the right iliosacral joint together with an outflare of the left iliosacral joint. In these cases, the mobility tests including the standing forward bending (flexion) test and the squish test would determine which joint, the right or left, is more hypomobile. The more hypomobile joint is treated first. Then the patient is reevaluated to reassess the medial/lateral placement of the ASIS.

The iliosacral joint has anterior and posterior rotations. The ASIS will be positioned more cephalad, while the PSIS on that side is more caudad, whenever the ilium is rotated posteriorly. Because of the oblique orientation of the joint, this is really a posteromedial rotation. When the ilium is rotated anterior, which is really an anterolateral rotation, the ipsilateral ASIS is more inferior, while the PSIS on that side is more superior. Anterior and posterior rotations of the pelvis are an essential biomechanical component of the gait cycle. During swing phase, there is a posterior rotation of the ilium on sacrum; during stance phase, between midstance and toe off, the ilium rotates in an anterior direction. With a posterior rotation (innominate), there may be full hip flexion, but the range of motion of hip extension is limited. With an anterior rotation (innominate), full hip extension may be attained, but the range of hip flexion will be inhibited.

Inflares, outflares, anterior rotations and posterior rotations are normal physiologic positions; however, the ilium is stuck in these positions. The upslips and downslips are pathologic biomechanical dysfunctions like subluxations/dislocations. With a fall on the buttocks, a force results that can move the ilium cephalad, where it becomes stuck. This is an upslip, or upslipped ilium. The ASIS and PSIS may both be superior on the ipsilateral side; the ischial tuberosity will be superior. With a fall off a horse, when the right foot gets stuck in the stirrup and the horse keeps moving forward, there is a caudad pull on the leg, which may pull the ilium caudad, with a resultant downslip. This downslipped ilium may present with an inferior ASIS and PSIS; there will be an inferior ischial tuberosity. With upslips and downslips, because the articular surface of the ilium is in a pathologic position (not stuck in a normal physiologic position), all physiologic ranges of hip motion will be limited. Often, there are bilateral upslips, typical in patients who ski, ice skate, or ride horses. Bilateral downslips are less common, but can be found with histories pertaining to sports injuries such as football tackles and ski injuries, and pedestrian car accidents. Patients occasionally present with an upslip together with a downslip.

CHAPTER 12

THE PELVIC JOINTS: TREATMENT OF BIOMECHANICAL DYSFUNCTION OF THE PUBES AND THE ILIOSACRAL JOINTS

The Pubic Symphysis

Elevated Pubes and Descended Pubes

Differential Diagnosis

- Palpate the pubic tubercles.
- Compare superior pubic ramis (cephalad/caudad).
- Positive squish test. Treat hypomobile side.

Note: Be precise. Assess elevation of bony landmarks: superior pubic ramus (do not be influenced by soft tissue tone).

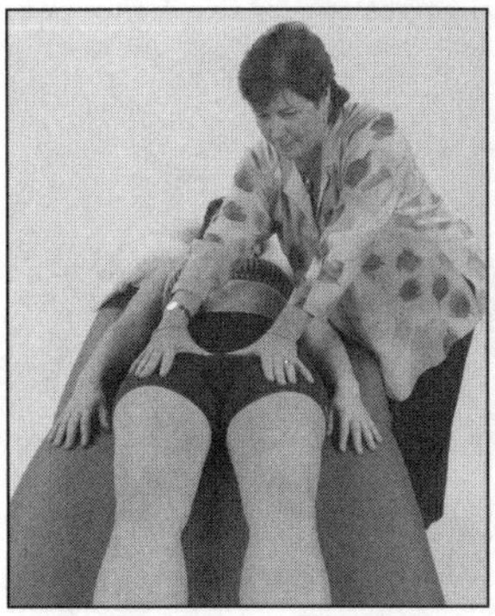

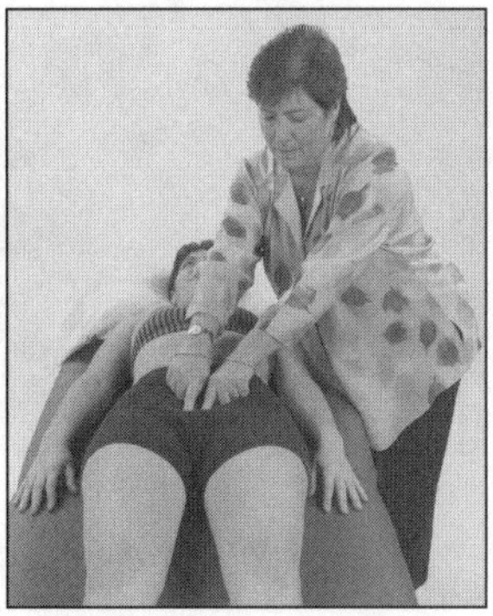

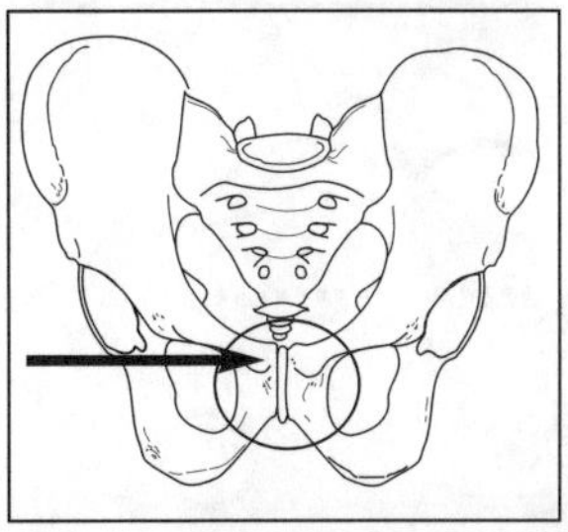

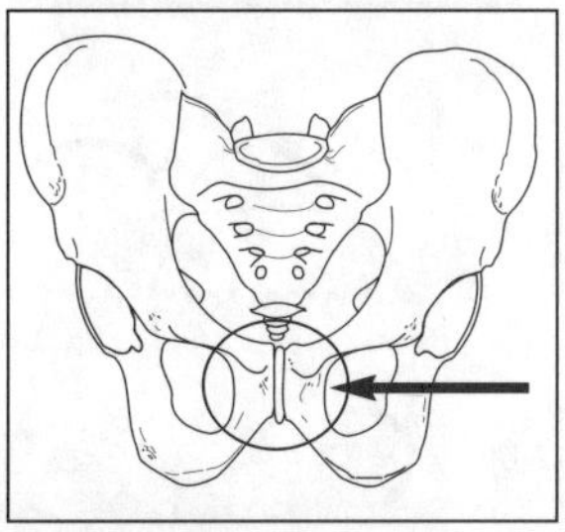

Muscle Energy and "Beyond" Technique for an Elevated Pubes

Diagnosis:

- Positive diagnostic anatomical landmark: Elevated ipsilateral superior pubic ramus.
- Positive squish test.

Treatment:

- Supine. Palpate ASIS for barrier.
- First barrier: Extend ipsilateral hip off the side of the table.
- Second barrier: Slight abduction.
- Third barrier: Slight external rotation.

Resist:

- Adduction

Note: Barrier can be palpated at pubic symphysis or at either ASIS. Typically, isometric resistance is applied to adduction to enhance technique with autogenic inhibition of hypertonic adductors.

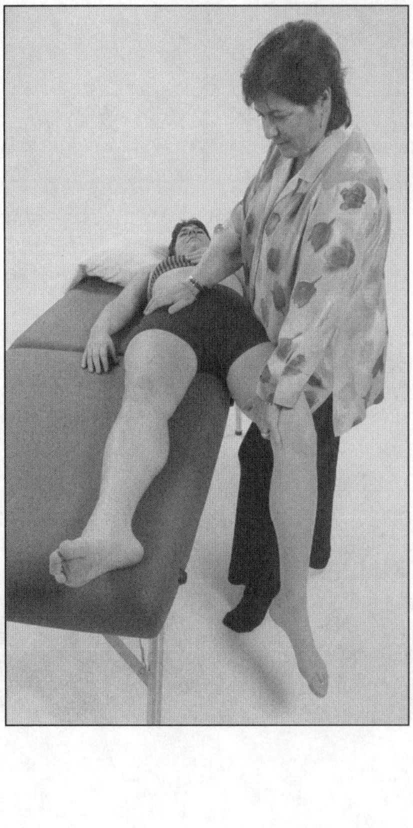

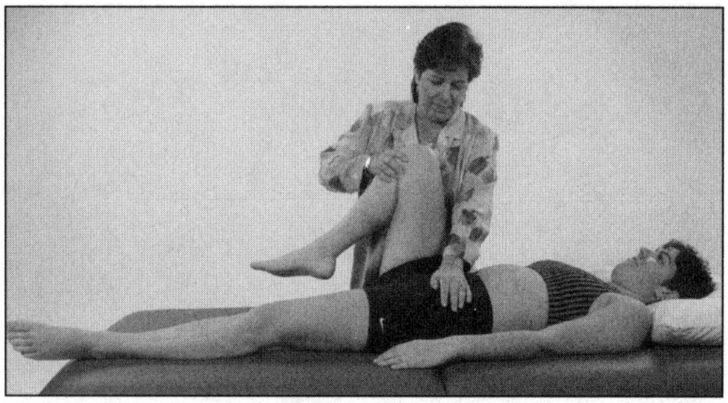

Muscle Energy and "Beyond" Technique for a Descended Pubes

Diagnosis:

- Positive diagnostic anatomical landmark: Caudal ipsilateral superior pubic ramus.
- Positive squish test.

Treatment:

- Supine. Palpate ASIS for barrier.
- First barrier: Flex hip on involved side.
- Second barrier: Slight adduction.
- Third barrier: Slight internal rotation.

Resist:

- Hip extension.

Note: Place thenar eminence on ischial tuberosity and mobilize superiorly during relaxation periods between isometric contractions if the descended pubes is not correcting.

Note: The only significant movements for pubic dysfunction are flexion and extension on a sagittal plane. If transverse plane (rotations) and/or coronal plane (abduction, adduction) movements are greater than a few degrees past midline, the barrier has been passed.

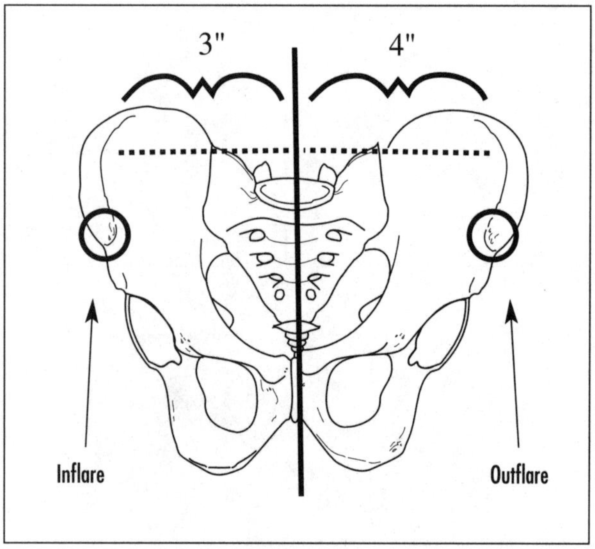

The Iliosacral Joints

Inflare and Outflare

Differential Diagnosis

- Compare ASIS (medial or lateral from mid line).
- Positive squish test. Treat the hypomobile side.

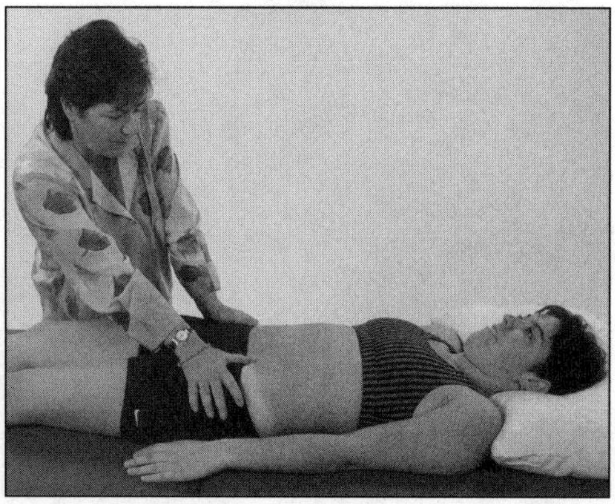

Muscle Energy and "Beyond" Technique for an Outflare

Diagnosis:

- Positive diagnostic anatomical landmark: Lateral ipsilateral ASIS.
- Positive squish test.

Treatment:

- Supine. Flex hip and knee. Foot on table. Palpate ASIS for barrier.

Barriers:

- Move knee medially (internally rotate and adduct) to barrier.

Resist:

- Abduction and external rotation: by patient pushing knee laterally.

Note: If the outflare is not correcting, place fingers under buttocks so that finger tips are medial to border of ilium. Pull/distract the ilium laterally during the relaxation periods between isometric contractions.

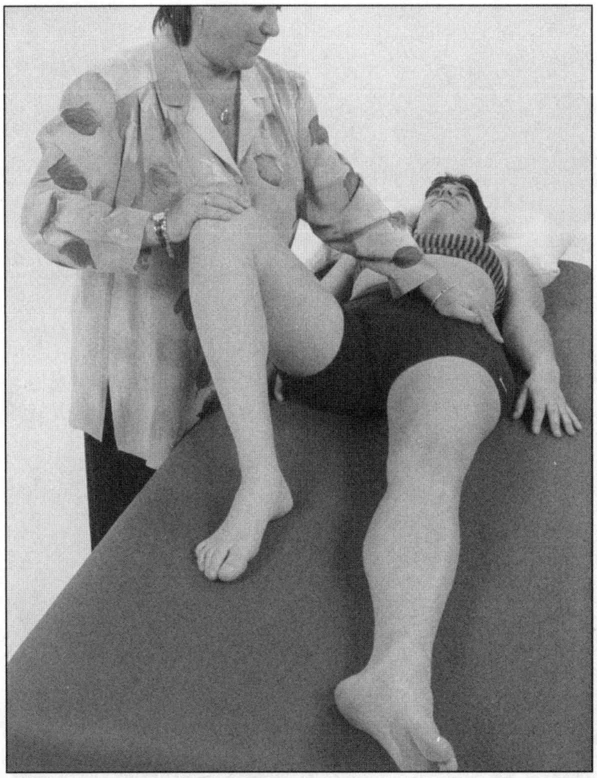

Muscle Energy and "Beyond" Technique or an Inflare

Diagnosis:

- Positive diagnostic anatomical landmark: Medial ipsilateral ASIS.
- Positive squish test.

Treatment:

- Supine. Flex hip and knee. Foot on table. Palpate ASIS for barrier.

Barriers:

- Move knee laterally (externally rotate and abduct) to barrier.

Resist:

- Adduction and internal rotation: by patient pushing knee medially

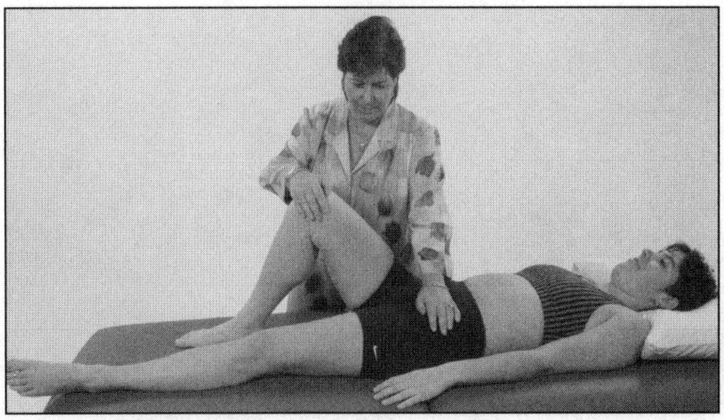

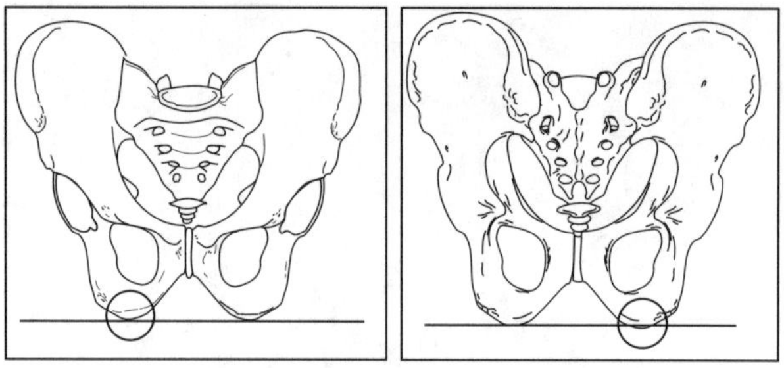

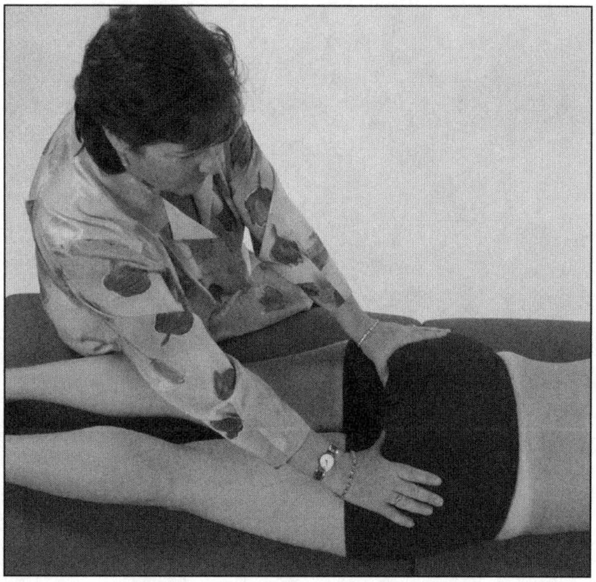

Upslip and Downslip

Differential Diagnosis

- Positive longitudinal distraction or compression test.
- Compare ASIS (cephalad/caudad).
- Compare PSIS (cephalad/caudad).
- Compare pubes (cephalad/caudad).
- Compare leg lengths.
- Compare ischial tuberosities.

Note: The elevation of the ischial tuberosities is the only conclusive diagnostic anatomic landmark.

Muscle Energy and "Beyond" Technique for an Upslip

Diagnosis:

- Positive diagnostic anatomical landmark: Elevated ipsilateral ischial tuberosity (possibly elevated ASIS, elevated PSIS, elevated pubes).
- Positive longitudinal distraction test.

Treatment:

Supine or prone. Therapist should sit; while standing it is difficult to maintain pure longitudinal leg traction. Find loose packed position of the hip joint (in external and internal rotation; in abduction and adduction). Apply caudad (longitudinal) traction on the ipsilateral leg. Distract to take up all the slack in caudal glide. Then perform a high velocity, low amplitude, caudal manipulation with approximately 5 lbs. force.

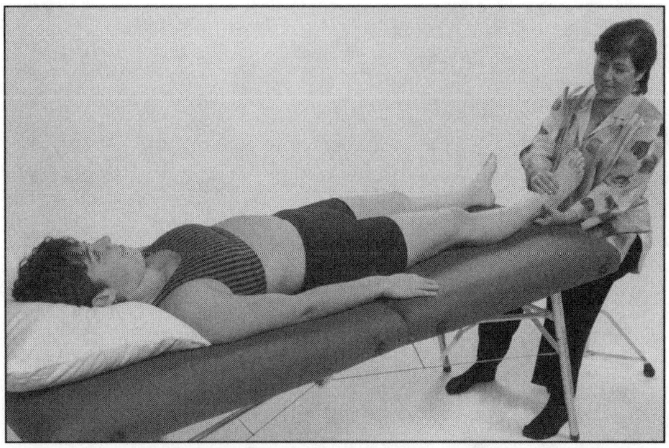

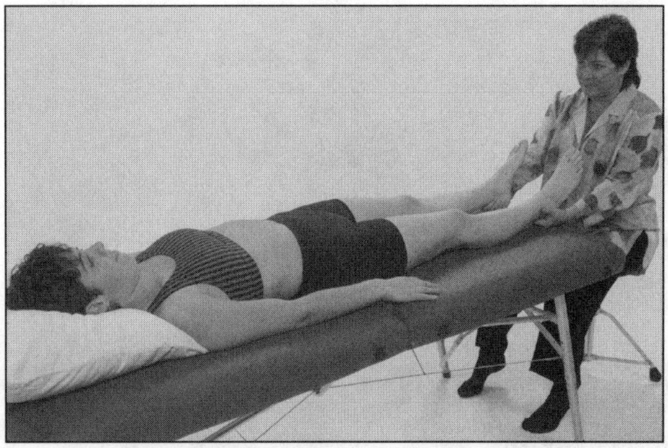

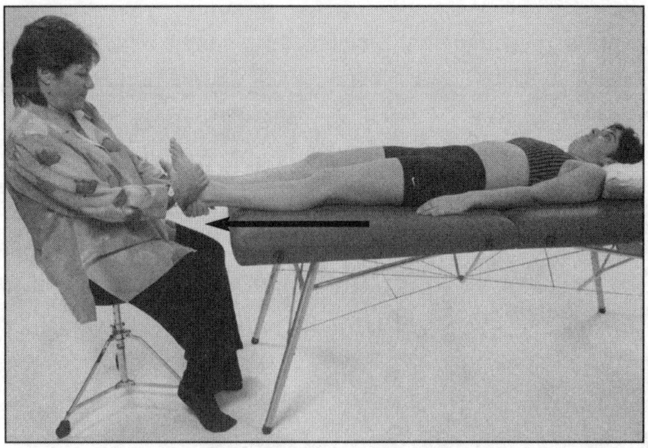

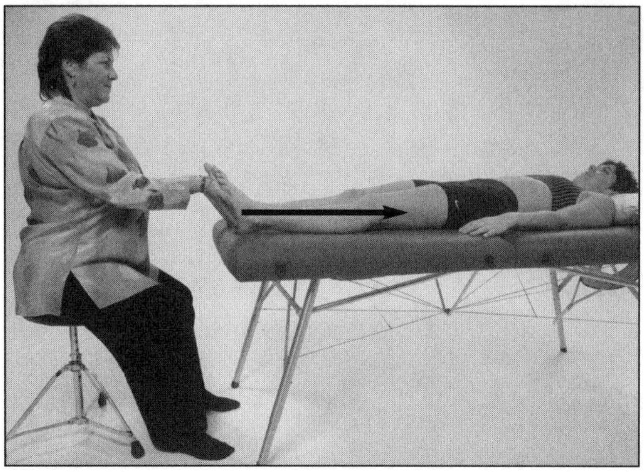

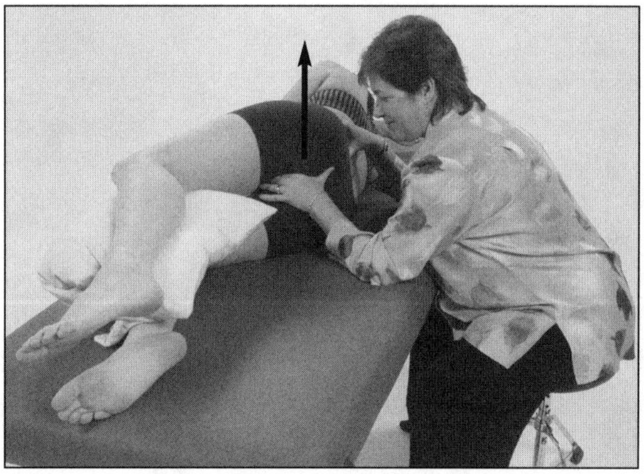

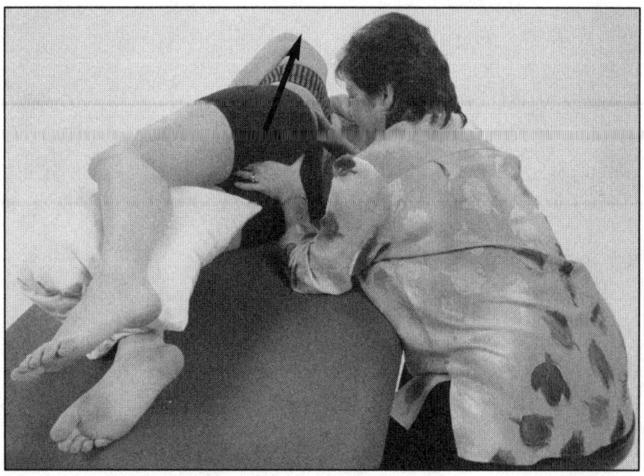

Muscle Energy and "Beyond" Technique for a Downslip

Diagnosis:

- Positive diagnostic anatomical landmark: Descended ipsilateral ischial tuberosity (possibly descended ASIS, descended PSIS, descended pubes).
- Positive longitudinal compression test.

Treatment:

Side lying with ipsilateral downslip side up. Therapist places caudad hand (thumb and web space) on the ipsilateral inferior ramus of pubes and the ischial tuberosity. (Don't pinch the bone. Grip the bone with the web space of the hand. Don't be in the groin space or in the hip joint space.) Place the other hand (web space) under the iliac crest of the ipsilateral side; the thumb can be medial to the PSIS.

Step 1: Distract the ilium laterally. Maintain the lateral distraction force until the tissue tension changes subside.

Step 2: Maintain lateral distraction and push ilium cephalad until the tissue tension changes subside.

Anterior Rotation and Posterior Rotation

Differential Diagnosis

Compare:

- ASIS caudad together with PSIS cephalad vs. ASIS cephalad together with PSIS caudad.
- Positive squish test. Treat the hypomobile side.

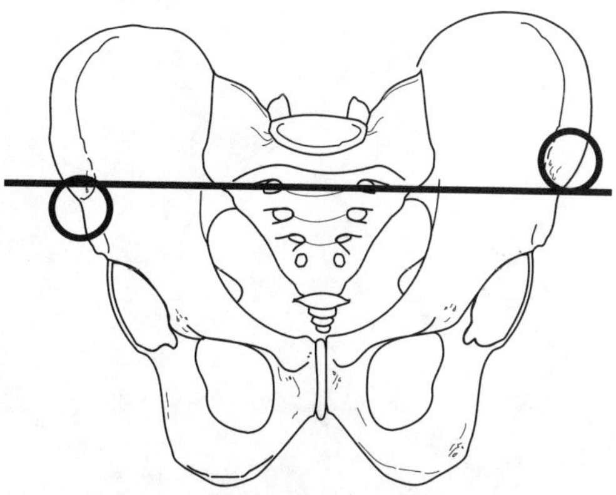

Muscle Energy and "Beyond" Technique for an Anterior Rotation

Diagnosis:

- Positive diagnostic anatomical landmarks: Caudad ipsilateral ASIS with cephalad ipsilateral PSIS.
- Positive squish test.

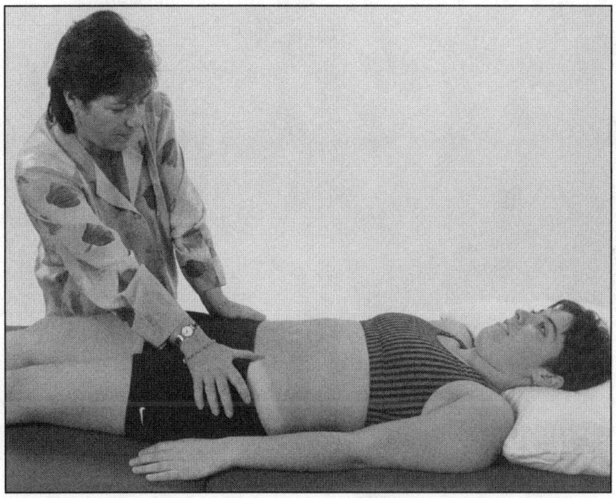

Treatment:

- Supine.
- First barrier: Flex hip and knee toward chest to barrier.
- Second barrier: Internal rotation.
- Third barrier: Adduction.

Resist:

- Hip extension.

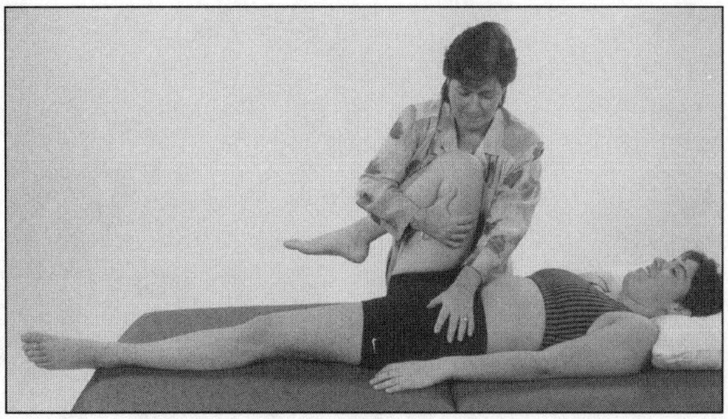

Note: A variation: The knee can be flexed; the therapist can control the leg from the knee; hip extension resistance can be performed at the knee.

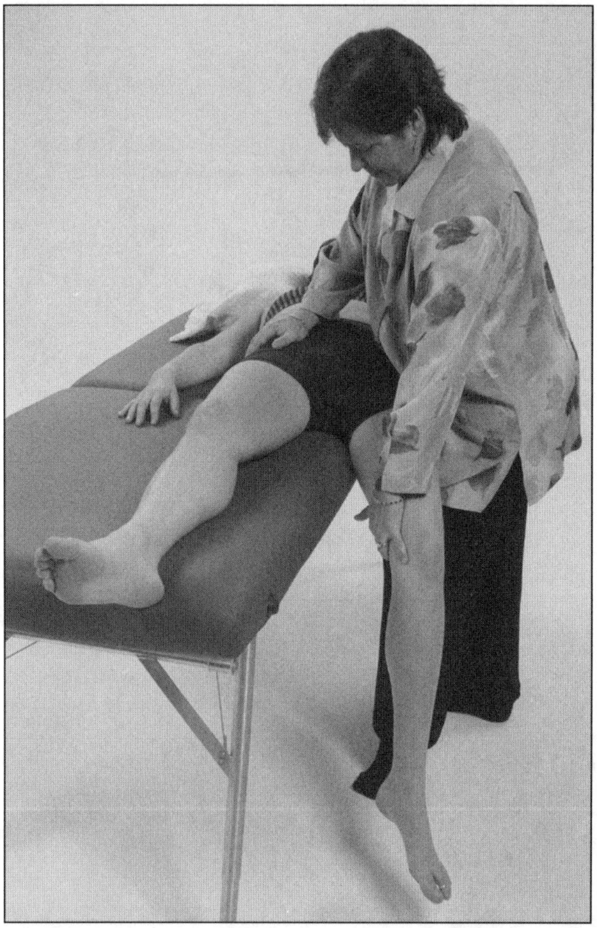

Muscle Energy and "Beyond" Technique for a Posterior Rotation

Diagnosis:

- Positive diagnostic anatomical landmarks: Cephalad ipsilateral ASIS with caudal ipsilateral PSIS.
- Positive squish test.

Treatment:

- Supine.
- First barrier: Extend ipsilateral hip off side of table to the barrier.
- Second barrier: External rotation.
- Third barrier: Abduction.

Resist:

- Hip flexion.

Note: The hip and buttocks must be enough off the edge to allow full hip extension without excessive abduction (beyond the barrier).

Sequence of Treatment for Biomechanical Dysfunction of the Pelvic Joints with Rationale

1. Treat the pubic symphysis dysfunctions first.
2. Treat the accessory sacroiliac joints. If these joints are not treated and remain hypomobile, corrections of iliosacral and sacroiliac joint dysfunctions are inhibited.
3. Treat the iliosacral joint dysfunctions.

   First treat the inflares and the outflares. Occasionally, the corrections are inhibited by an upslip or a downslip. If an inflare or an outflare is not correcting, look for an upslip or a downslip.

   Then treat the upslips and downslips. Occasionally these will not correct if there is an inflare or an outflare. In these cases, look for an outflare or an inflare.

   Note: Treatment for upslips and downslips is much more aggressive than treatment for flares. Therefore, attempt to treat the flares first, and there will be no need to manipulate/mobilize the upslips/downslips more than once with the more aggressive techniques.

   Then treat the anterior and the posterior rotations (innominates). Occasionally, a posterior innominate will not correct completely due to tension of the ipsilateral iliolumbar ligaments secondary to L5 biomechanical dysfunction. In these cases, perform fascial release to the ipsilateral iliolumbar ligaments, and repeat the correction of the posterior innominate after L5 has been biomechanically corrected. More commonly, the anterior and posterior innominate won't correct because of the hypomobile accessory sacroiliac joint, which is contributing to the L5 dysfunction via iliolumbar ligament tension.
4. When all pelvis (pubic symphysis and iliosacral joint) biomechanical dysfunction has been corrected, there should be symmetry of all anatomical landmarks: right/left superior pubic ramus; right/left ASIS; right/left PSIS; right/left ischial tuberosities. The longitudinal distraction and compression tests should reflect normal inferior and superior glide of right/left ilium. Glides of the PSIS on sacrum should reflect normal mobility.
5. When all anatomical landmarks are symmetrical (see #4 above), there should be no residual pelvic obliquity (elevation of one iliac crest). If there is any pelvic obliquity (different from a pelvic shear, when the pelvis is sheared to the right or left of midline, which is an L5/S1 postural phenomenon), assess right/left leg lengths.
6. To clinically assess right/left leg lengths, there should be symmetric anatomical pelvic landmarks.

   Compare measurements between greater trochanter and medial/lateral tibial plateaus, medial/lateral tibial plateaus and medial/lateral malleoli. If there is any discrepancy in right/left leg measurements, refer the patient for radiographic leg length measurements. The radiologist can specify the bony leg length discrepancy. If there is a true bony leg length discrepancy measured on radiographs, the patient must wear a shoe lift in order to maintain stable biomechanics of the pelvic joints. (Note: Historically, asymmetry of the pelvis has been treated with shoe lift therapy. Yet when a shoe lift is used to change asymmetry of the pelvis and there is no true bony leg length discrepancy, then biomechanical dysfunctions of the pelvic and sacral joints are locked in, causing greater problems in the future.)

CHAPTER 13

SPINAL MOTION

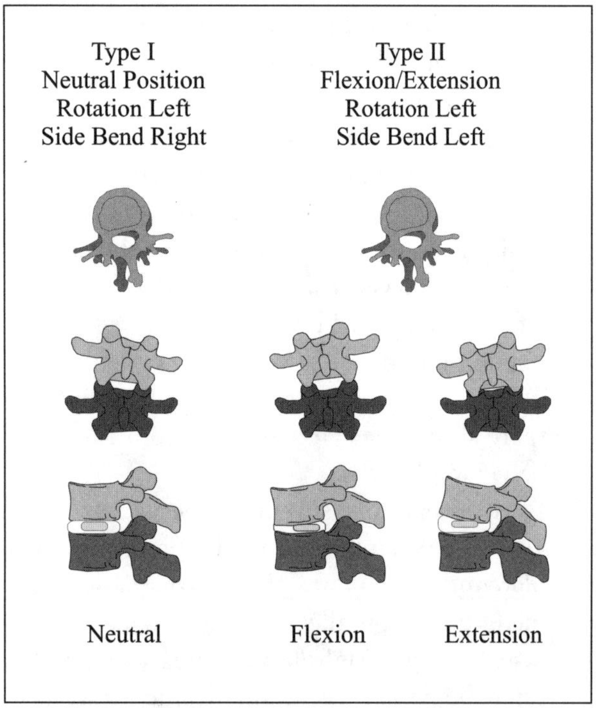

Fryette's Laws of Motion

- Law 1: When the AP spine is in neutral (i.e., the spine is not flexed or extended), the facets are not engaged, and the ligaments are not in tension. Side bending and rotation occur to the opposite side. This concerns multisegmental (two or more vertebrae) movement.
- Law 2: When the AP spine is flexed or extended, the facets are engaged, and the ligaments are in tension. Side bending and rotation occur to the same side. This concerns monosegmental motion of one vertebra.
- Law 3: When motion is introduced in one direction (on one plane), then motion in all other directions and in all other planes is reduced. Therefore, if motion in one plane is lost to dysfunction, another plane of motion must compensate (adapt) for it.

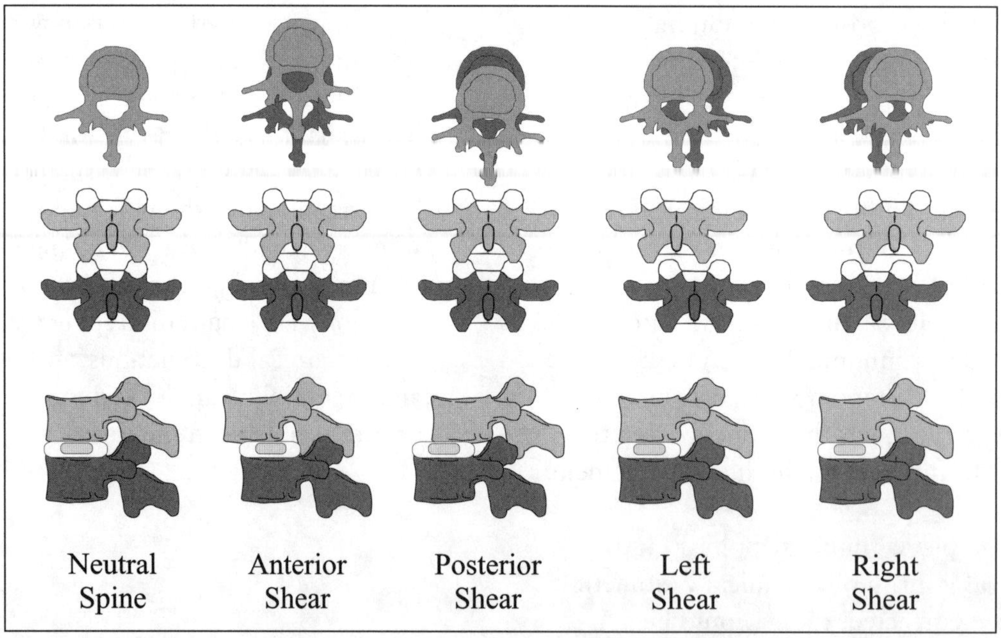

Type I Movement

Type I movement occurs at the thoracic and lumbar spine, not at the cervical spine. When the spine is upright, in static posture, there should not be any signs of asymmetry on a coronal (side bending) or on a transverse (rotation) plane. If there is prominence posteriorly of two or more transverse processes, this indicates that those vertebrae are rotated toward the side of the posterior prominence. Those vertebrae will also be side bent away from the side of the posterior prominence, because side bending occurs in the opposite direction of rotation when the spine is in sagittal plane neutral.

Type II Movement

Type II movement occurs at the cervical, thoracic and lumbar spine. When the spine is flexed or extended in static posture, there should not be any signs of asymmetry on a coronal (side bending) or on a transverse (rotation) plane. If there is prominence posteriorly of one transverse process, this indicates that that vertebra is rotated toward the side of the posterior prominence. That vertebra will also be side bent toward the side of the posterior prominence, because side bending occurs in the same direction as rotation during flexion and extension.

Type III Movement

Type III movement has been defined by (Weiselfish) Giammatteo as a *shear* or subluxation. These anterior and lateral shears occur at all joints of the body, including all spinal joints. These dysfunctions appear at the vertebral end plate, at the spinous process, and at the pedicle/transverse process. They respond well to Integrative Manual Therapy (treatment of anterior and lateral shears, or Type III dysfunction, as presented by DCR).

Biomechanics of the Facet Joints

1. During flexion, the facets are opened.
2. During extension, the facets are closed.
3. During right side bending, the left facet opens and the right facet closes.
4. During rotation, there is a glide of the superior facets.
5. Motion is described by the moving part on the fixed part.
6. Motion is described by the movement of the superior facet on the inferior facet.
7. Motion of rotation in the transverse plane is described by the direction of movement of the anterior surface in relation to the midline of the body, and labeled according to the movement of the superior vertebra on the inferior vertebra.

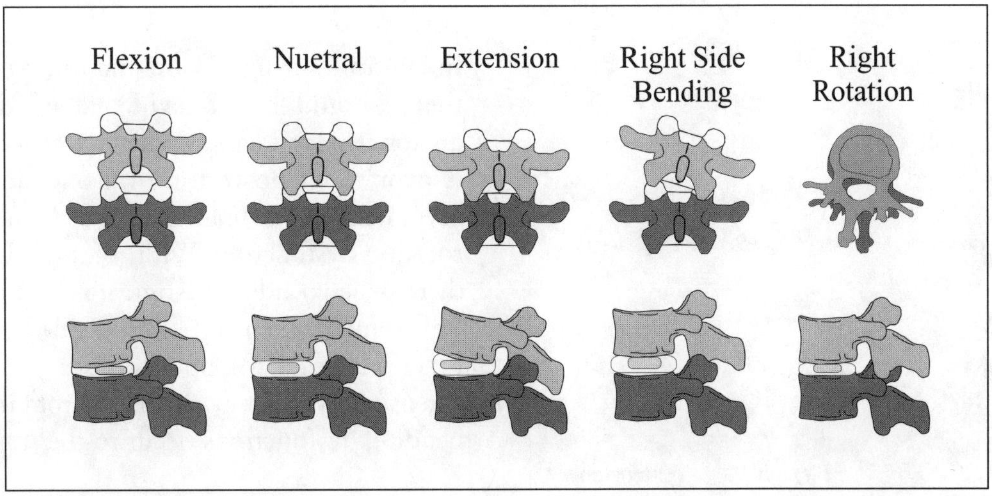

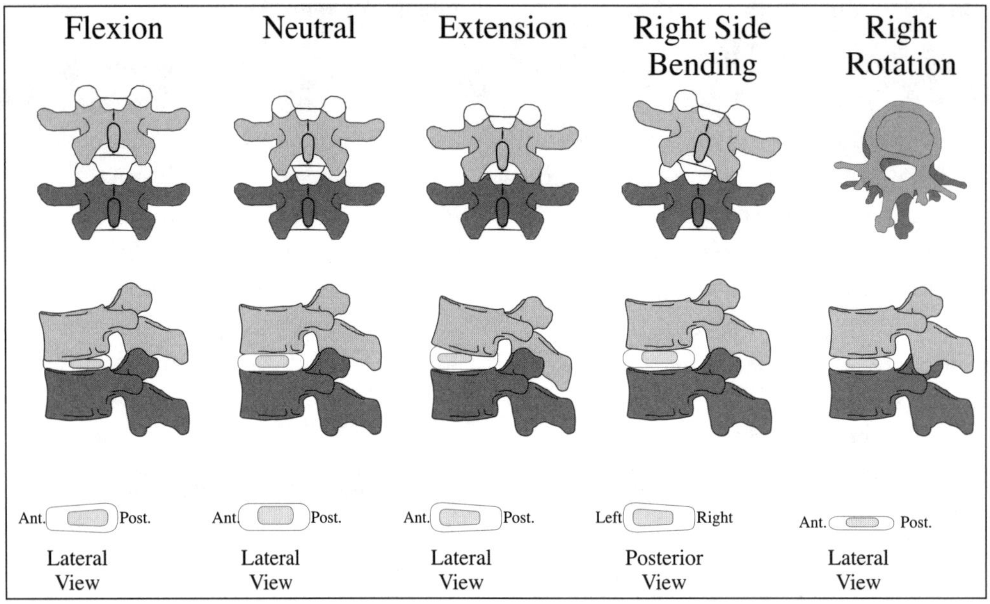

Integrative Facet Joint and Disc Theories

- When both facets are opened (flexion), the nuclear disc material moves posteriorly.
- When both facets are closed (extension), the nuclear disc material moves anteriorly.
- When the right facet is opened (left side bending), the nuclear disc material moves to the right.

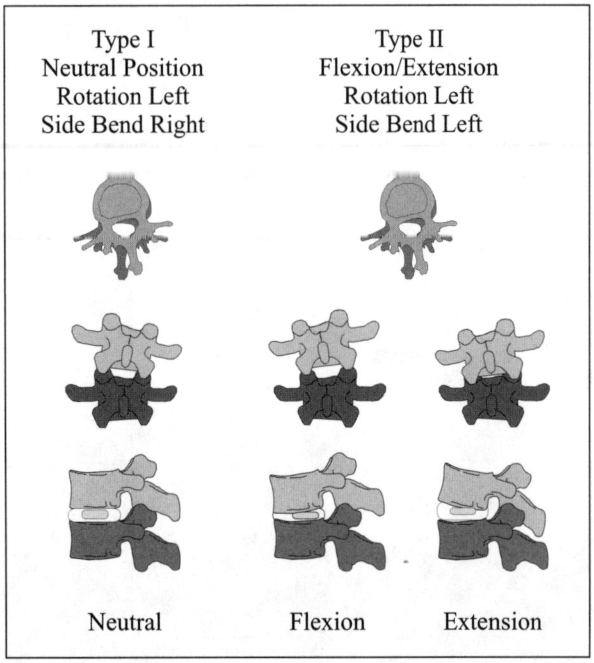

- When the vertebra rotates, there is an increase in the intradiscal pressure proportional to the degree of rotation.

Motion Types by Area

- Occipitoatlantal (O/A) joint: Type I, consider convex on concave theory
- Atlantoaxial (A/A) (C1/C2): Joint Rotation
- C2–C7: Type II
- C7–L5: Type I and Type II
- Lumbosacral junction (L5/S1): Type II, consider reciprocal movement

Application

In neutral, during flexion and during extension, there should be no evidence of asymmetric anatomical landmarks. When a transverse process is prominent posteriorly on one side, there is asymmetry of landmarks. This asymmetry reflects rotation dysfunction. With rotation dysfunction, there is side bending dysfunction. In Neutral Type I movement, the rotation and side bending dysfunctions occur to opposite sides. In flexion and extension Type II movement, the rotation and side bending dysfunctions occur to the same side.

CHAPTER 14

THE SACRUM

Biomechanics of the Sacrum

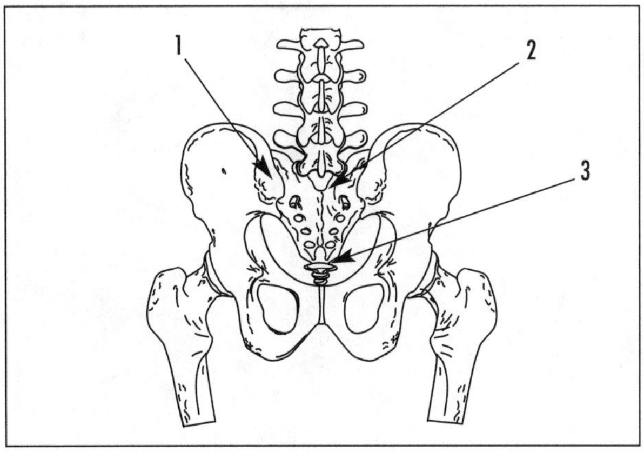

A great deal of clinical research has been performed relative to sacral function since the author's first book on Muscle Energy and "Beyond" Technique. Clinical results have been excellent, when compared with the good results of five years ago.

The sacral joints include:

1. Right and left sacroiliac joints
2. The lumbosacral junction
3. The sacrococcygeal joint

The osseous vertebral components of the sacrum, S1 through S5, become bone by age six. It has been stated by multiple researchers that fusion of the joints (S1 through S5) occurs by age 25 to 30. The author believes, together with other clinical experts in the field of manual therapy, that the bones of sacrum do not fully fuse. Similar to cranial sutures, when there is *no* rigor mortis, there is joint mobility between the sacral vertebrae. Many anatomy texts describe disclike material between the sacral vertebrae.

Consider the sacroiliac joints; they are often implicated in low back pain. The ligaments are short and strong, with the ability to bind and support. During rigor mortis, these ligaments are hard, unlike living tissue. Only during motion palpation is it possible to realize that these joints, both right and left sacroiliac joints, are able to move significantly (similar to the motion of a bridge when it expands or contracts with temperature changes, or when a tall building sways in the wind).

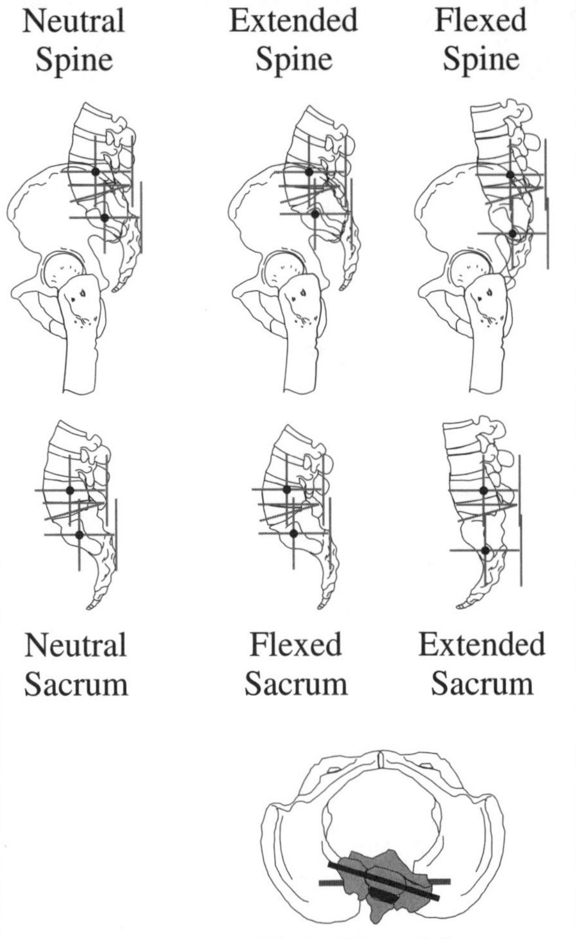

A Right Flexed Sacrum
& A Left Extended Sacrum

The Sacroiliac Joints

Consider movement on three planes: sagittal, coronal, and transverse planes. Combined 3-planar motion occurs at the sacroiliac joints. Yet, in order to assess and diagnose dysfunctional movement, movement may be evaluated in each plane. Dysfunction can be assessed as a uni-planar entity. When the uniplanar motion component is assessed and corrected, the total combined 3-planar movement will present as good movement patterns.

Sagittal plane movement of the sacroiliac joints occurs individually, as well as bilaterally. For example, the right sacroiliac joint can perform sagittal plane movement alone, without the accompanying sagittal plane movement of the left sacroiliac joint. Sagittal plane movement can be co-joined (right and left sides move simultaneously). Flexion and extension are sagittal plane movements. Both right and left sacroiliac joints can flex together. Both right and left sacroiliac joints can extend together. The right sacroiliac joint can flex while the left sacroiliac joint extends. The left sacroiliac joint can flex while the right sacroiliac joint extends.

When one of the sacroiliac joints flexes while the other one extends, the sacrum is in a scoliosis-like dysfunction. When the sacrum is "locked" in this position, consequences are significant for the total spinal column position.

Transverse plane motion does not occur in isolation at the sacroiliac joints. The vertical limbs of the sacral articular surface inhibit rotation around a vertical axis.

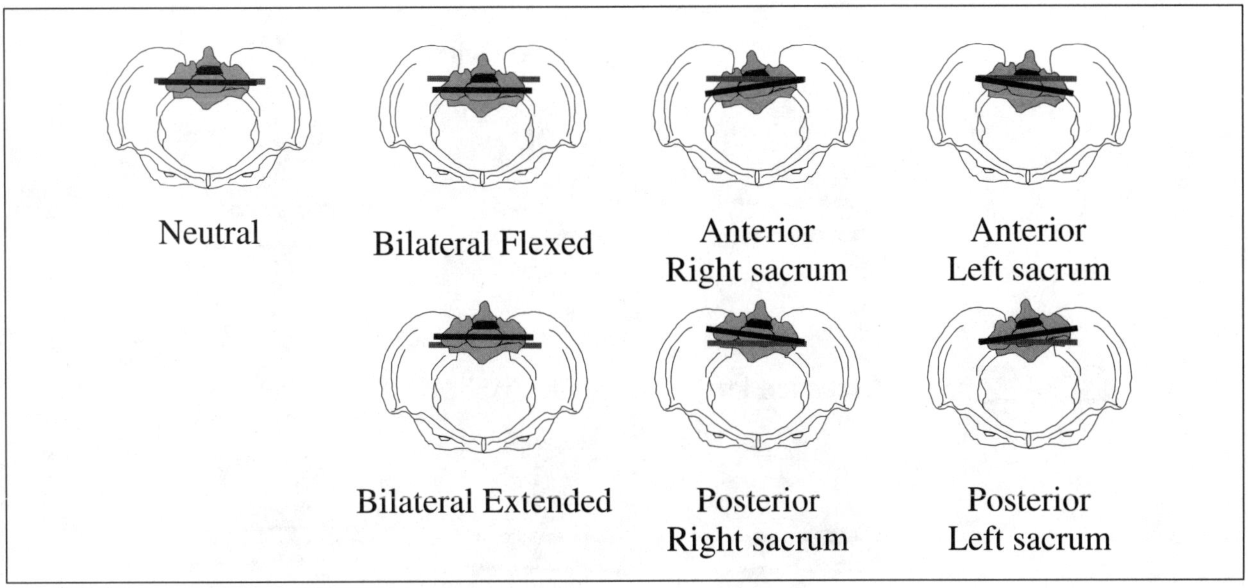

Coronal plane motion does not occur in isolation at the sacroiliac joints. The horizontal limbs of the sacral articular surface inhibit uniplanar side bending of the sacrum. Therefore, whenever the sacrum is stuck in transverse plane (in apparent rotation dysfunction), or stuck in coronal plane (in apparent side bending dysfunction), these are non-physiologic dysfunctions. A physiologic dysfunction is when a joint is stuck, yet it is stuck in a range of motion on a normal plane of function. A non-physiologic dysfunction is when a joint is stuck in a range of motion that is not a normal plane of function. Nonphysiologic joint dysfunctions cause more pain and disability than physiologic joint dysfunctions.

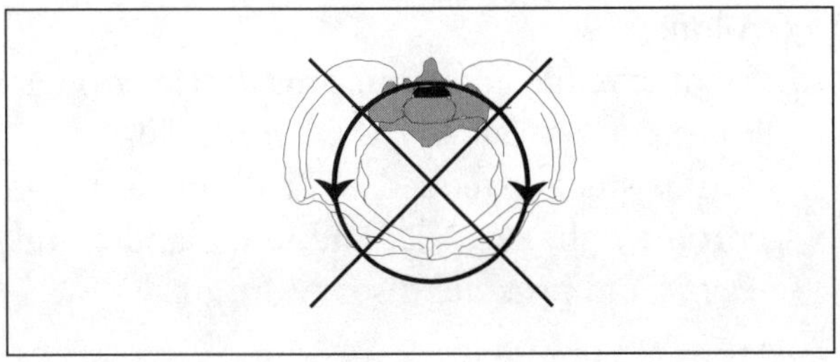

Sacroiliac Joint Dysfunction

Physiologic sacroiliac joint dysfunctions can include the following:

On a sagittal plane:

1. Bilateral flexed sacroiliac joints
2. Right flexed sacroiliac joint
3. Left flexed sacroiliac joint
4. Bilateral extended sacroiliac joints
5. Right extended sacroiliac joint
6. Left extended sacroiliac joint

Nonphysiologic sacroiliac joint dysfunctions can include the following:

On a transverse plane:

1. Left anterior rotation on a vertical axis
2. Right anterior rotation on a vertical axis
3. Left posterior rotation on a vertical axis
4. Right posterior rotation on a vertical axis

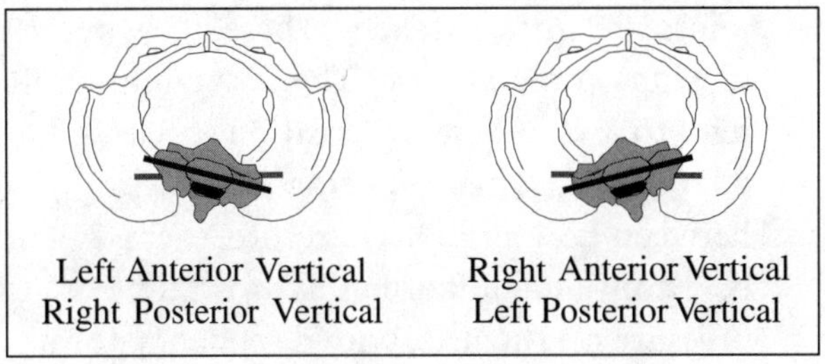

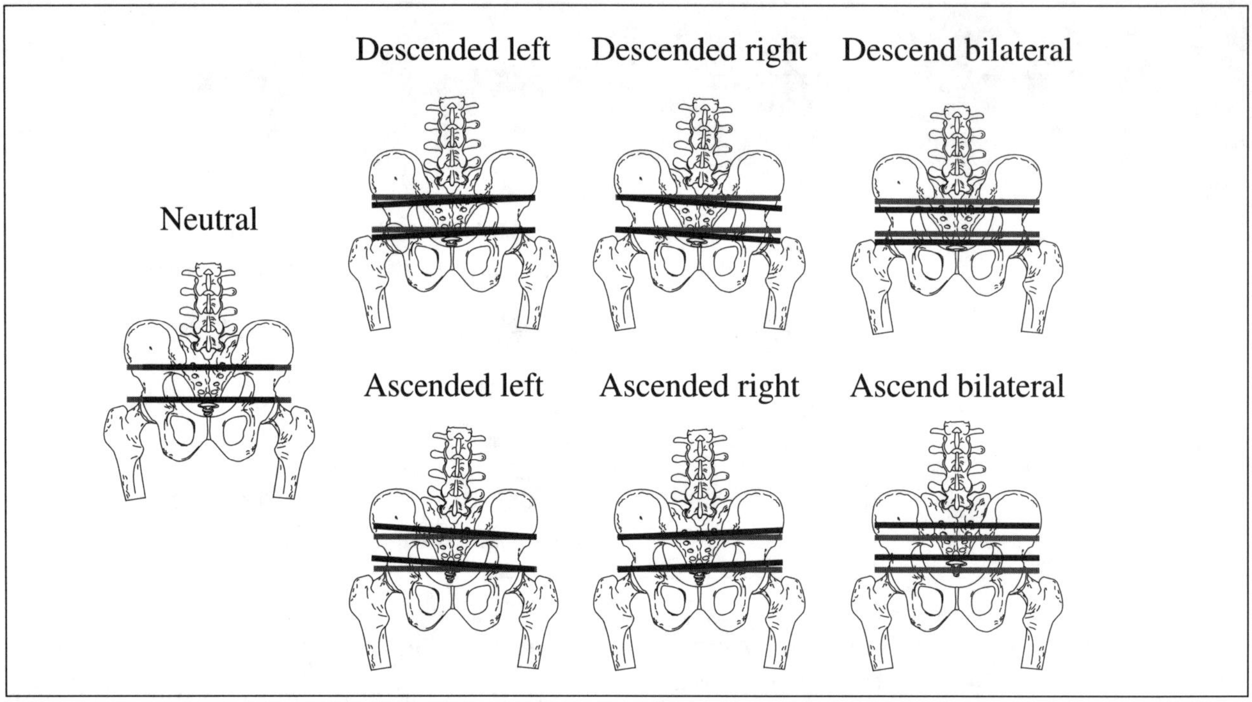

On a coronal plane:

1. Right descended
2. Left descended
3. Right ascended
4. Left ascended

All of the above sacroiliac joint dysfunctions can cause pain and disability. The sagittal plane dysfunctions are less compromising than the transverse plane and coronal plane dysfunctions. For example:

1. Bilateral flexed sacrum: Pelvic floor diaphragm tension is typical. Uterus and cervix cramping before and during menstrual cycle is common.
2. Bilateral extended sacrum: If the sacroiliac joints are stuck extended, there is always tension on the L5 disc. The L5 vertebra will tend to rest in a flexed position (because L5 adapts to S1; there is reciprocal motion between L5 and S1). Therefore, there will be a continual abnormal hydrostatic pressure on the disc, with a tendency for a posterior nucleus pulposus displacement. Prolonged posterior disc displacement of the L5 disc will contribute to hypomobility of the lumbosacral junction. Low back pain and persistent disability with a tendency for exacerbations of disc derangement and an ongoing cause of disc degeneration will result.
3. Right flexed sacrum: Right ovary discomfort is common. Also, cecum dysfunction is typical (the cecum is the most distal part of the ascending colon on the large intestines), as well as pain from ileocecal valve syndrome.
4. Left flexed sacrum: Common problems include anal-rectal sphincter syndrome and irritable bowel syndrome of the sigmoid colon.
5. Right extended sacrum: This dysfunction is less common. Mostly, the left sacroiliac joint is stuck extended. When there is a chronic right extended sacrum, bladder and uterus and prostate discomfort are common.

6. Left extended sacrum: Valve of Houston syndrome is common. The valve of Houston is the sphincter between the descending colon and the sigmoid colon, located slightly superior from the left ASIS.
7. Left anterior rotation on a vertical axis: This dysfunction is not common.
8. Right anterior rotation on a vertical axis: This dysfunction is common. Hip pain (right ipsilateral and/or bilateral hip pain) is typical. Bladder problems, including stress incontinence, and PMS are common. Women have this dysfunction much more than men.
9. Left posterior rotation on a vertical axis: This will cause severe low back pain and disability. Usually, there are L5 disc problems, with a unilateral or bilateral sciatic nerve problem. The sacral plexus is often impinged; hypotonia and atrophy of the ipsilateral gluteal and hamstrings muscle are common. Headaches are also common. Significant sigmoid colon and rectal dysfunction, as well as cervix, uterus, and prostate problems are typical.
10. Right posterior rotation on a vertical axis: This will cause intense low back pain and disability. Usually, there are L5 disc problems, with a unilateral or bilateral sciatic nerve problem. The sacral plexus is often impinged; hypotonia and atrophy of the ipsilateral gluteal and hamstrings muscle are common. Headaches are also common. Significant cecum problems, as well as cervix, uterus, and prostate problems are typical.
11. Bilateral posterior rotations: Unless this dysfunction occurs during delivery (when the head of the infant pushes the sacrum posterior), then this probably occurs as two individual unilateral dysfunctions (i.e., a right plus a left posterior vertical rotation). This causes the most severe pain and disability. Severe low back and buttocks pain, severe bilateral referred leg pain, severe discogenic dysfunction, plus severe headaches are common.
12. Right descended: Usually, this dysfunction occurs because of a direct or indirect cranial trauma. Occasionally, a tackle (e.g., football) can cause this dysfunction. This does not cause low back pain; yet the sacrum does impinge on the inferior sacral plexus (S3 and S4) and this results in right foot pain and paresthesia. Back stiffness is a common complaint.
13. Left descended: Usually, this dysfunction occurs because of a direct or indirect cranial trauma. Occasionally, a tackle (e.g. football) can cause this dysfunction. This does not cause low back pain; yet the sacrum does impinge on the inferior sacral plexus (S3 and S4) and this results in left foot pain and paresthesia. Back stiffness is a common complaint.
14. Bilateral descended: This dysfunction does not cause back pain. Rather, severe spinal stiffness will result. This is the most common cause of spinal cord fibrosis ((Weiselfish) Giammatteo, Brieg, De La Porte). Usually, complaints include bilateral foot pain and paresthesia. Children may be "toe walkers." Headaches are common. Pelvic floor and pelvic diaphragm soft tissue dysfunction is common, with bladder, uterus/prostate and sigmoid colon problems.
15. Right ascended: This dysfunction does not cause significant low back pain. Usually, there is pressure at the cranio-cervical junction, with some kind of odontoid process syndrome (i.e. alar ligament dysfunction).
16. Left ascended: This dysfunction does not cause significant low back pain. Usually, there is pressure at the craniocervical junction, with some kind of odontoid process syndrome (i.e., alar ligament dysfunction).

17. Bilateral ascended: This dysfunction will cause degenerative L5 disc problems. Usually, severe headaches occur, even migraine headaches, because of the pressure of the dens at the foramen magnum.

Because the dura mater inserts at S2, at the sacrococcygeal joint, and at the coccyx, biomechanical dysfunction of the sacroiliac joints can cause significant spinal syndromes. Also, because the sacral plexus can be torqued at the intervertebral joints, which in turn may cause tension up the spinal cord, neurologic deficits can be elicited because of sacroiliac joint problems. For example, rotation dysfunctions cause headaches, cervical syndrome, thoracic outlet and reflex sympathetic dystrophy, and more. Descended sacrums cause all types of neurologic disorders, including head pain and migraine headaches, visual deficits, speech and hearing impairment, learning disabilities, hyperactivity, tactile defensiveness, and more.

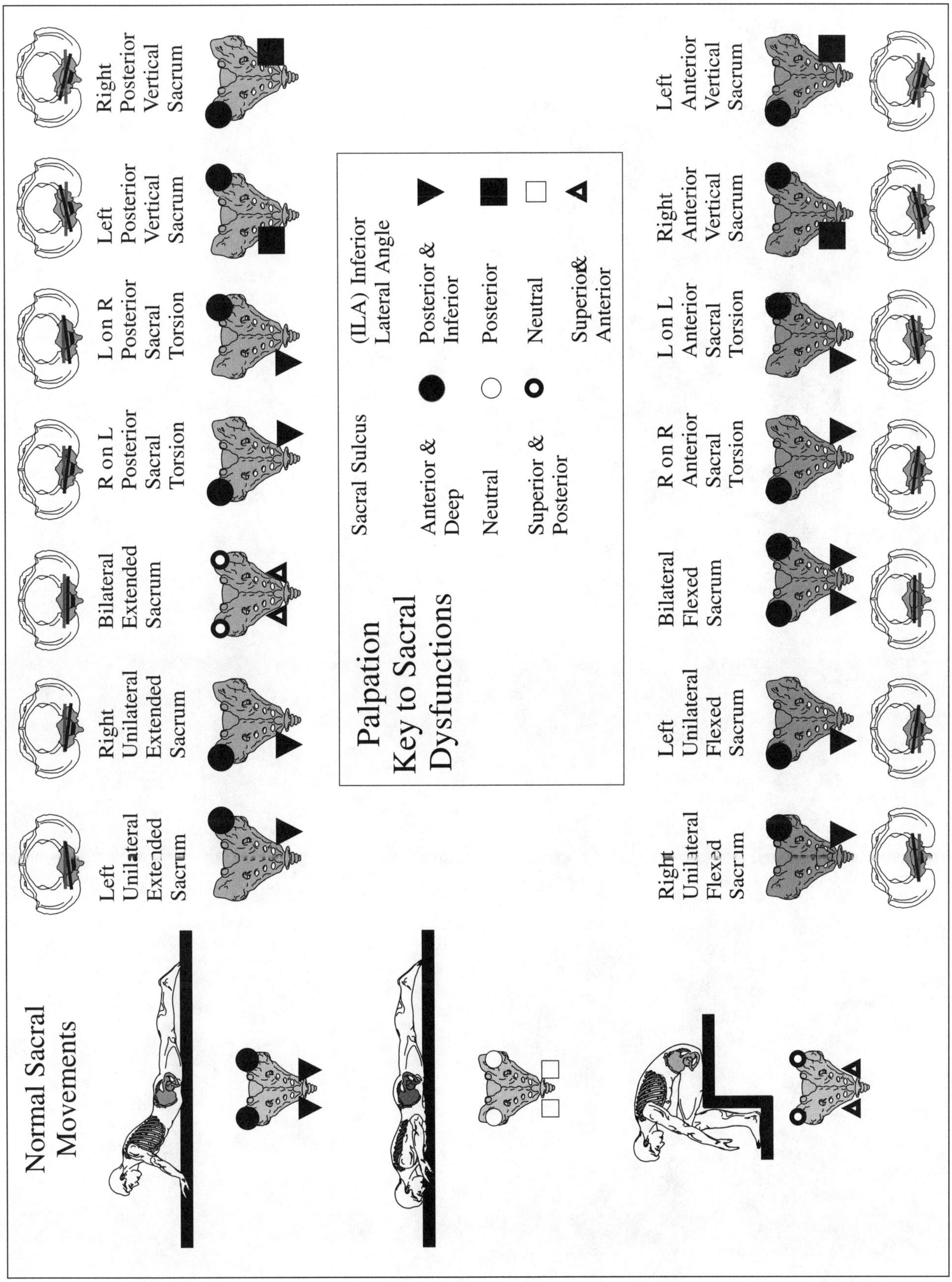

Normal Sacral Movements
Left Unilateral Extended Sacrum
Right Unilateral Extended Sacrum
Bilateral Extended Sacrum
R on L Posterior Sacral Torsion
L on R Posterior Sacral Torsion
Left Posterior Vertical Sacrum
Right Posterior Vertical Sacrum
Palpation Key to Sacral Dysfunctions
Sacral Sulcus
(ILA) Inferior Lateral Angle
Anterior & Deep
Posterior & Inferior
Neutral
Posterior
Superior & Posterior
Neutral
Superior& Anterior
Right Unilateral Flexed Sacrum
Left Unilateral Flexed Sacrum
Bilateral Flexed Sacrum
R on R Anterior Sacral Torsion
L on L Anterior Sacral Torsion
Right Anterior Vertical Sacrum
Left Anterior Vertical Sacrum

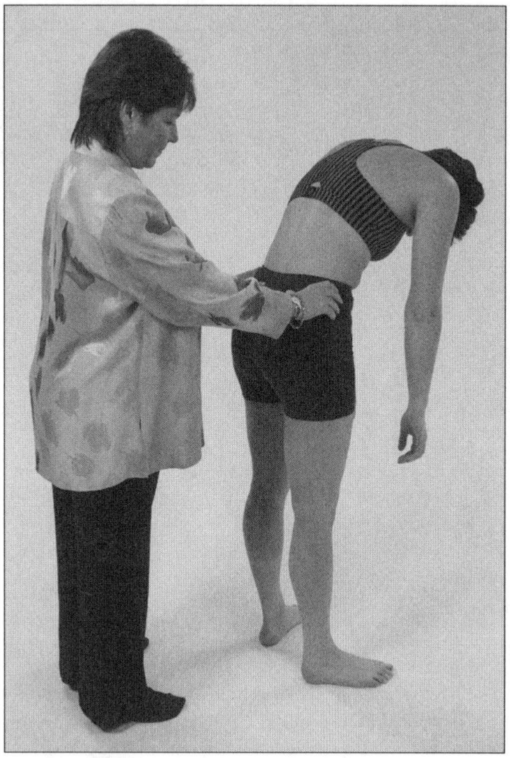

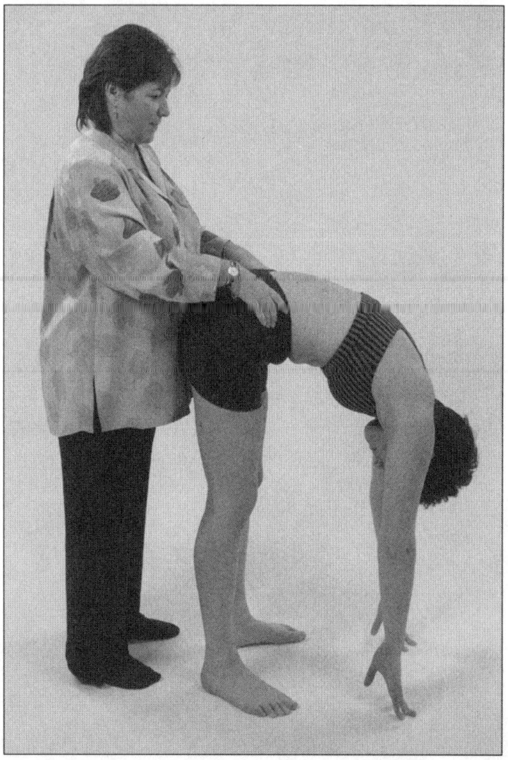

The L5/S1 Joint (The Lumbosacral Junction)

A great deal of explanation is offered in this book regarding causes and incidence of low back pain relative to this joint. The lumbosacral junction is the source of immense pain and disability. Eighty percent of all spinal surgery occurs at this joint. It appears that all traditional and nontraditional intervention has failed up to now. Yet, remarkable success is attained with this approach: Muscle Energy and "Beyond" Technique.

The lumbosacral junction is a complex, rather than a joint. There is exceptional range of motion. Flexion at this joint should be 50 degrees at least! Extension at this joint should be 40 degrees at least! Rotation as well as side bending are present at least 45 degrees in each direction! Why have we been mis-informed regarding the "normal" ranges of motion at L5/S1? Because the typical young adult and adult has low back pain. Limitations of motion precede symptoms. Therefore, almost the whole population has limitations of motion. (Statistics note that 100 percent of Americans will have neck and/or back pain sometime in life.) Only when this area is treated (with Muscle Energy and "Beyond" Technique) can we "see normal." It is not sufficient to eliminate pain; we must restore normal ranges of motion to eradicate low back pain syndromes.

The author and colleagues have discovered that all spinal motion is dependent upon the lumbosacral junction. At least 70 percent (or more) of all cervical, thoracic, and lumbar ranges of motion are restored when lumbosacral junction motion is restored! Difficulty in understanding lumbosacral junction movement is understandable. When the practitioner recognizes that L5 and S1 movements are reciprocal, then explanations make sense.

The lumbosacral junction does not have Type I motion. It only has Type II motion. This means there is not a Type I relationship between L5 and S1 (L5 has a Type I relationship with L4 as well as with the coccyx). Of significance: S1 cannot have a Type I relationship with L5 or with S2. S1 has isolated movement, which means S1 moves in a Type II movement pattern.

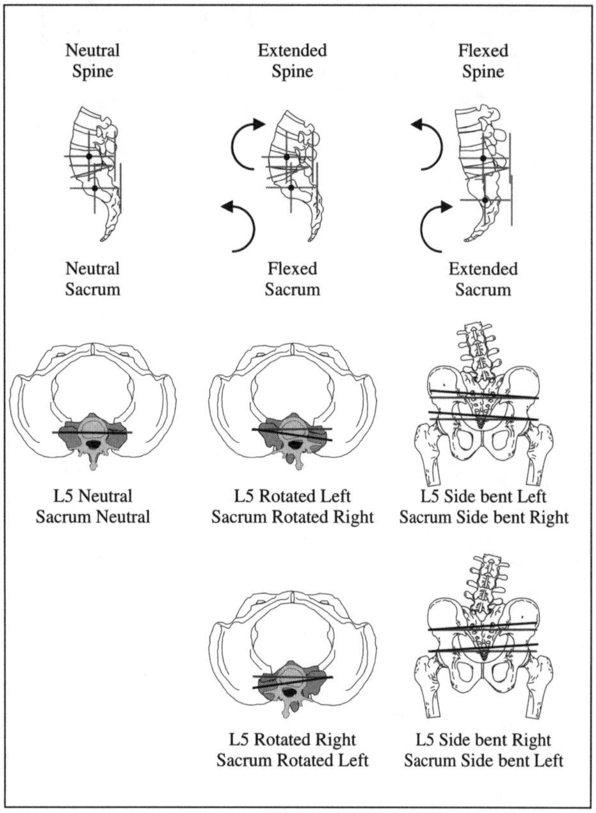

There is another relationship between L5 and S1, but it is not a Type I relationship. It is a relationship of reciprocal movement patterns. When L5 moves in a direction, S1 will always reflect a tendency to move in the opposite direction.

When the movement patterns are broken down for analysis, they appear as follows:

1. L5 will flex (an anterior motion) while S1 extends (there is a posterior glide of sacral base).
2. L5 will rotate right; S1 will rotate left.
3. L5 will side bend right; S1 will side bend left.

Yet as described already, there is no isolated uniplanar movement of S1 on a transverse or coronal plane. This is because S1 moves only in a Type II movement pattern, rather than a Type I movement pattern. This means that S1 rotation only occurs together with side bending, and S1 side bending only occurs with rotation. This also means that S1 side bending and rotation will only occur together with S1 flexion or extension. In other words, there is always some degree of flexion and extension in the relationship between L5 and S1. However, because there is no Type I movement at the lumbosacral junction, there is no rotation of S1 (and sacrum) on a vertical axis, and no side bending of S1 (and sacrum) on a coronal axis. Yet the three motions can occur simultaneously: S1 rotation cannot occur alone; S1 side bending cannot occur alone; S1 rotation plus side bending can occur together with S1 flexion or extension. This is called a 3-planar motion. This is also called a torsion. A torsion is a 3-planar motion. This torsion is a Type II (biomechanical) movement. A torsion has three component movements: sagittal plane, plus transverse plane, plus coronal plane. Therefore, a torsion will always have the following components:

1. Flexion or extension
2. Right rotation *or* left rotation
3. Right side bending *or* left side bending.

At the lumbosacral junction the following possibilities can occur at S1: sagittal plane (flexion or extension); or sagittal plane plus transverse plane plus coronal plane. S1 cannot be in pure rotation on a transverse plane unless there is also side bending plus flexion or extension. S1 cannot be in pure side bending on a coronal plane unless there is also rotation plus flexion or extension. S1 can be in any of the following motion patterns:

1. Flexion
2. Extension
3. Flexion plus right rotation and right side bending
4. Flexion plus left rotation and left side bending
5. Extension plus right rotation and right side bending
6. Extension plus left rotation and left side bending

When there is a torsion, or a Type II movement pattern (torsions and Type II movements patterns are one and the same), then rotation and side bending occur to the same side. Therefore, when S1 is flexed or extended, rotation and side bending will always be to the same side. The following motion restrictions will exist in opposite directions:

1. If S1 is stuck flexed, rotated, and side bent to the right: Sacrum will be limited towards extension, rotation and side bending to the left.
2. If S1 is stuck flexed, rotated, and side bent to the left: Sacrum will be limited towards extension, rotation and side bending to the right.
3. If S1 is stuck extended, rotated, and side bent to the right: Sacrum will be limited towards flexion, rotation and side bending to the left.
4. If S1 is stuck extended, rotated, and side bent to the left: Sacrum will be limited towards flexion, rotation and side bending to the right.

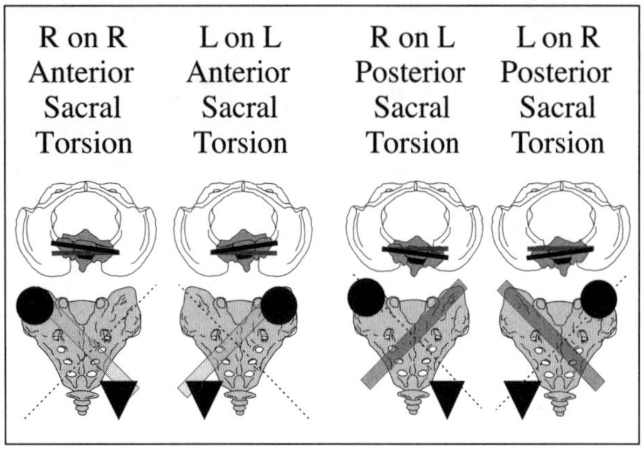

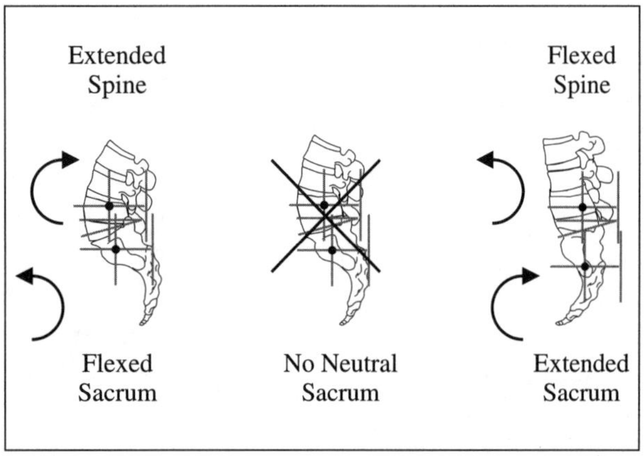

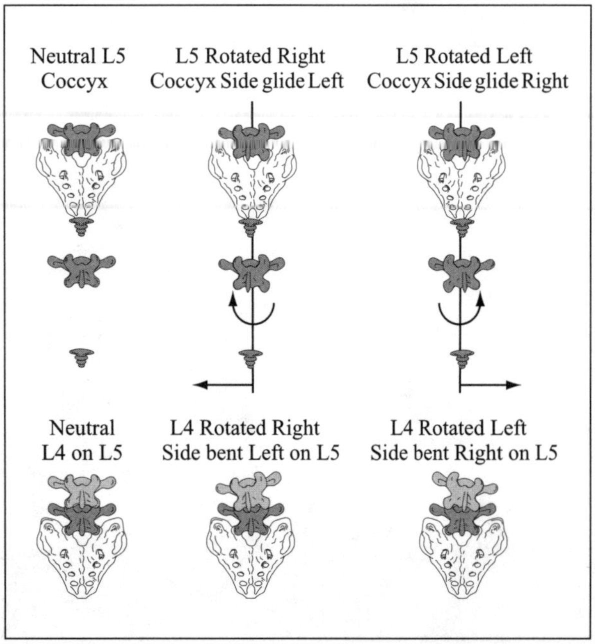

The following dysfunctions can occur at S1:

Anterior torsion:

1. Flexed, rotated, and side bent right
2. Flexed, rotated, and side bent left

Posterior torsion:

1. Extended, rotated, and side bent right
2. Extended, rotated, and side bent left

What about the lumbosacral junction? If S1 can be stuck in four possible torsion patterns (which are really four possible Type II dysfunctions), then what can happen to L5? The following possible biomechanical dysfunctions can occur at L5:

1. Neutral Type I dysfunction
2. Type II (out of neutral, i.e., flexed or extended) dysfunctions

The laws of biomechanics state the following:

1. In neutral (i.e., not flexed or extended), rotation and side bending occur to opposite sides. In neutral, rotation can occur around a vertical axis. In neutral, pure side bending can occur on a uniplanar coronal plane. (Note: S1 cannot rotate alone, S1 cannot side bend alone.)
2. In flexion or extension (i.e., out of neutral), rotation and side bending occur to the same side.

Although S1 cannot be in neutral, L5 can be in neutral. Type I (in neutral) dysfunctions are relationship dysfunctions. In other words, L5 requires a "partner" to be in Type I dysfunction. L5 cannot be in a Type I relationship with S1, because S1 cannot be in neutral! L5 can be in Type I relationships with L4 or coccyx (via tensions of the sacrospinal ligaments).

Therefore, what can happen to L5 biomechanically? The following are possible L5 Type I dysfunctions:

1. L4/L5 Type I dysfunctions:
   a. Neutral, rotated right, and side bent left
   b. Neutral, rotated left, and side bent right.

2. L5/Coccyx Type I dysfunctions:
   a. Neutral, rotated right, and side bent left
   b. Neutral, rotated left, and side bent right
3. L5 Type II dysfunctions:
   a. Flexed, rotated, and side bent right
   b. Flexed, rotated, and side bent left
   c. Extended, rotated, and side bent right
   d. Extended, rotated, and side bent left

What about the rest of the lumbosacral junction complex? Consider the following:

1. L5 disc
2. L5 spinous process
3. L5 nerve roots
4. L5 facet joints (Type II dysfunction)
5. L5 vertebral body (Type I relationship dysfunction)
6. L5 end plate (Type III shear dysfunction)
7. Spinal cord
8. Membranes: dura mater, arachnoid mater; pia mater of the cauda equina
9. Anterior longitudinal ligament
10. Posterior longitudinal ligament
11. Spinal artery and pial arterial plexus
12. Spinal vein and more

The Low Back and the Lumbosacral Junction

The low back can be divided biomechanically into four parts: the pelvis, which includes the right and left iliosacral joints and the pubic symphysis; the lumbosacral junction, which is the articulation between L5 and S1; the lumbar spine, which includes the vertebral bodies, discs, and the right and left facet joints between L1 and L5; and the right and left sacroiliac joints. Traditionally in manual therapy, the lumbosacral junction has not been included as a subcategory of the low back. It requires a unique assessment for differential diagnosis and should be treated as a unique entity with an individual identity.

There is a high incidence of low back pain in North America; often, severe cases undergo surgical intervention. Statistics representative of these surgical cases can provide needed learning. Approximately 85 percent of lumbar spinal surgery is performed at the lumbosacral junction. Of the remaining 15 percent of surgical cases, the majority of surgery occurs at L4/L5. Surgery may be designated for disc herniation and rupture, nerve root impingements, sequestered discs, spinal stenosis, spinal cord fibrosis, and more; but the location of surgery occurs most frequently at the lumbosacral junction.

To begin to understand the biomechanics and pathomechanics of the lumbosacral junction, focus is required on the sacrum, especially the S1 articular surface and L5. Joint mobility and articular balance between L5 and S1 will reflect tendencies towards disc pathologies, because the hydrostatic pressure affecting disc position correlates with articular balance of the fifth lumbar vertebra on the sacral base (flexed or extended, side bent, and rotated L5 on S1).

Mobility of the sacrum can be assessed, with recognition that full sacroiliac joint mobility is essential for normal movement of the sacral base; this movement of the sacral base is necessary for lumbosacral junction mobility. Lumbosacral biomechanical integrity requires full mobility and normal articular balance of L5 and the sacral base. When the sacrum is determined to be hypomobile, it is recognized that some of the limited mobility may be due to the biomechanical dysfunctions of the sacroiliac joints, and some of the limited sacral mobility may be due to lumbosacral junction dysfunctions. The sacrum contributes to the movement at the lumbosacral junction and to the movement at the sacroiliac joints.

Dysfunctions of the lumbosacral junction affect reciprocal movement. Reciprocal movement has commonly been assumed to mean the synchronous movement of the right arm together with the left leg, and the synchronous movement of the left arm together with the right leg during ambulation. This is a limited definition. The author's clinical research indicates that reciprocal movement is a neurologically driven, reflexive-like phenomenon, directed from supraspinal structures, which primarily affects proximal rather than distal structures. This means that reciprocal movement directly and primarily affects the proximal trunk; the appendages (the arms and legs) are secondary to the movement of the trunk in regard to reciprocal movement. The author's clinical research has provided evidence that reciprocal movement occurs at the lumbosacral junction, is palpable, and is not primarily a function of the extremities. There is evidence documented by other researchers that reciprocal movement occurs at the lumbosacral junction on a sagittal plane: During gross lumbar spine flexion, when L5 moves anteriorly into flexion, the sacral base glides posteriorly; when the lumbar spine moves into gross extension and L5 moves posteriorly, the sacral base glides anteriorly. Anterior and posterior glides can be palpated as accessory joint movements. Anterior glide can be palpated at the sacral base during lumbosacral extension when the lumbar spine is extending. This is easily palpated during a prone press-up with the pelvis and hips in neutral. Posterior glide can be palpated at the sacral base during lumbosacral flexion, when the lumbar spine is flexing. This is easily palpated during sitting flexion, being aware that lumbar flexion is segmental movement down the spinal kinetic chain and does not include hip flexion.

It is unlikely that reciprocal movement would occur only on the sagittal plane and not affect the transverse plane as well. Although rotation on a vertical axis does not appear to be present as normal movement at the sacroiliac joints, it is evident at the lumbosacral junction. Clinical evidence by the author indicates that reciprocal movement occurs on the transverse plane at the lumbosacral junction: When L5 rotates to the left, the sacral base rotates to the right. Movement on a coronal plane at the lumbosacral junction appears less clear, possibly because the sacroiliac joints have definite coronal plane movement, which may compromise lumbosacral side bending. An indication that coronal plane movement at the lumbosacral junction is reciprocal: In standing, when the right knee is flexed, the right hip is dropped, and there is an apparent right side bending of the sacrum, which correlates with left side bending of L5. It does appear clinically that side bending is reciprocal at the lumbosacral junction; i.e., when L5 side bends to the right, the sacral base side bends to the left. More investigation is required.

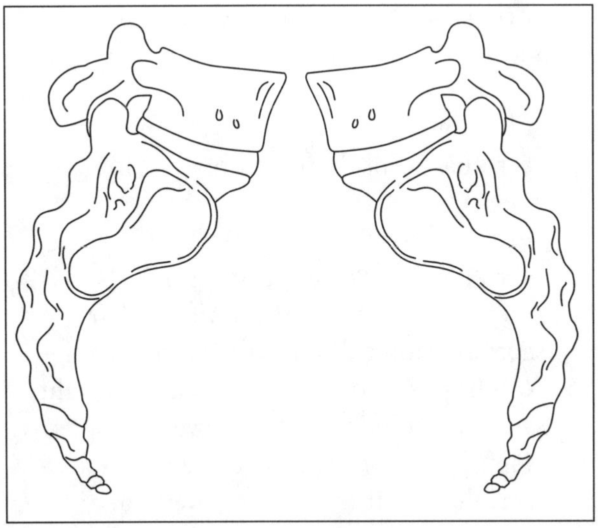

The articular surfaces of the Sacroiliac Joint and their relation to the movements of the sacrum.

The Auricular-Shaped Articular Surfaces

During assessment for diagnosis and treatment of the sacrum, it is important to be aware of the reciprocal movement at the lumbosacral junction. When treating biomechanical dysfunction of the sacrum, it is also important to remember the auricular-shaped surfaces of the articular surfaces of the sacrum, which articulate with the auricular-shaped surfaces of the ilia.

Kapandjie and others have documented the possible variations of the sacral angle of these auricular-shaped articular surfaces. Each joint surface includes a vertical limb, a sacral angle, and a horizontal limb. The sacral angle is at the level of the second sacral segment, the center of gravity, through which a transverse axis hypothetically is the source of biomechanical flexion and extension of the sacrum. This sacral angle has been found on cadavers to vary greatly in degree, less than and greater than 90 degrees.

Biomechanical Dysfunctions of Sacrum

Axes of motion defined by the author are included below. The most common biomechanical dysfunctions of the sacrum include:

1. A bilateral flexed sacrum (a bilateral sacroiliac joint dysfunction; middle transverse axis)
2. A bilateral extended sacrum (a bilateral sacroiliac joint dysfunction; middle transverse axis)
3. A right unilateral flexed sacrum (a right sacroiliac joint dysfunction; superior transverse axis)
4. A left unilateral flexed sacrum (a left sacroiliac joint dysfunction; superior transverse axis)

5. A right unilateral extended sacrum (a right sacroiliac joint dysfunction; superior transverse axis)
6. A left unilateral extended sacrum (a left sacroiliac joint dysfunction; superior transverse axis)
7. A right anterior vertical rotation (a right sacroiliac joint dysfunction; vertical axis)
8. A left anterior vertical rotation (a left sacroiliac joint dysfunction; vertical axis)
9. A right posterior vertical rotation (a right sacroiliac joint dysfunction; vertical axis)
10. A left posterior vertical rotation (a left sacroiliac joint dysfunction; vertical axis)
11. An ascended sacrum (typically bilateral, occasionally right or left, sacroiliac joint dysfunction)
12. A descended sacrum (typically bilateral, occasionally right or left, sacroiliac joint dysfunction)
13. A right anterior sacral torsion on a right oblique axis (a lumbosacral junction dysfunction; right oblique axis)
14. A left anterior sacral torsion on a left oblique axis (a lumbosacral junction dysfunction; left oblique axis)
15. A right posterior sacral torsion on a left oblique axis (a lumbosacral junction dysfunction; left oblique axis)
16. A left posterior sacral torsion on a right oblique axis (a lumbosacral junction dysfunction; right oblique axis)
17. An accessory sacroiliac joint dysfunction (typically bilateral, occasionally right or left, a dysfunction of the sacroiliac joints between the posterior superolateral aspect of the sacrum and the anteromedial aspect of the iliac crest at the PSIS)

Definition of Sacral Motion and Sacroiliac Joint Dysfunction in a 3-Planar Concept

The author has attempted to describe and define sacroiliac joint dysfunction as a 3-planar concept. It is important to remember that such a definition essentially is a fabrication for the purpose of diagnostics and therapy. The problem of sacroiliac motion, in this model, is a problem of the 3-planar dimension of the intra-articular space.

A different manner of presenting biomechanical dysfunction of the sacral joints would be according to the plane of movement/plane of movement dysfunction. For example, sagittal plane motions and movement dysfunctions of the sacral joints include:

1. Bilateral flexed sacrum
2. Right unilateral flexed sacrum
3. Left unilateral flexed sacrum
4. Bilateral extended sacrum
5. Right unilateral extended sacrum
6. Left unilateral extended sacrum

Transverse plane movement dysfunctions of the sacral joints include:

1. Anterior rotation of sacrum to the left on a vertical axis
2. Anterior rotation of sacrum to the right on a vertical axis
3. Posterior rotation of sacrum to the left on a vertical axis
4. Posterior rotation of sacrum to the right on a vertical axis

Note: Because there is no Type I motion of sacrum (S1), there is no "normal" (physiologic) rotation motion of sacrum on a transverse plane. Thus, all positions of sacrum on a transverse plane are pathologic.

Coronal plane movement dysfunctions of the sacral joints include:

1. Right descended sacrum
2. Left descended sacrum
3. Right ascended sacrum
4. Left ascended sacrum

Note: Because there is no Type I motion of sacrum (S1), there is no normal (physiologic) side bending motion of sacrum on a coronal plane. Thus, all positions of sacrum on a coronal plane are pathologic.

Lumbosacral Junction Movements Differentiated from Sacroiliac Joint Movements

The above defined sacroiliac joint movements do not define motion and movement dysfunction at the lumbosacral junction. S1 movement is similar to any other vertebra. The only difference between S1 and other vertebrae that move on three planes is that S1 moves in reciprocal movement patterns with L5. Motion dysfunction of S1 is a Type II dysfunction, which can be any of the following torsion patterns:

1. Anterior sacral torsion to the right
2. Anterior sacral torsion to the left
3. Posterior sacral torsion to the right
4. Posterior sacral torsion to the left

The Sacroiliac Joint and the Lumbosacral Junction

Sacroiliac joint dysfunctions primarily affect the sacroiliac joint and have a secondary influence on the lumbosacral junction, contributing to joint hypomobility and limited ranges of motion between L5 and S1, a compromised L5/S1 disc space, increased intradiscal pressure, and hypomobility of L5 facet joints.

The primary sacroiliac joint dysfunctions are treated with an awareness of the significance of the auricular-shaped articular surfaces of sacrum, and with an understanding that sacroiliac joint dysfunction affects the reciprocal movement at the lumbosacral junction.

The primary lumbosacral junction dysfunctions may compromise sacroiliac joint movement, and are treated with an awareness of the significance of the reciprocal movement between L5 and S1, while recognizing the auricular-shaped articular surfaces of sacrum.

Accessory Sacroiliac Joints

The accessory sacroiliac joint is a simple joint, possibly a syndesmosis. For successful diagnosis and treatment of accessory joint hypomobility, the manual pracatitioner does not require an awareness of the auricular-shaped articular surfaces of the sacrum, nor an understanding of the reciprocal movement of the lumbosacral junction. But joint mobility must be attained at the accessory sacroiliac joints between the PSIS and sacrum in order to attain full excursion of the sacrum so that dysfunctions of the lumbosacral junction and sacroiliac joints can be assessed, diagnosed, and treated.

The Sacral Complex

There is interaction within this complex system, which includes the lumbosacral junction, bilateral sacroiliac joints, bilateral accessory sacroiliac joints (when present), and L5 (as it is affected by Type I movement in anteroposterior (sagittal plane) neutral and Type II movement in flexion and in extension (sagittal plane). Assessment discovers that any biomechanical dysfunction within this complex can affect movement of the sacrum throughout the total range of motion into flexion and extension, and on coronal and transverse planes.

Mobility and Palpation of Anterior and Posterior Glides of the Sacral Base

Assessment requires observation and palpation skills with a focus on the symmetry or asymmetry of anatomic landmarks, and also on the mobility of the sacrum: the anterior glide of the sacral base during lumbosacral extension, and the posterior glide of the sacral base during lumbosacral flexion.

It is more difficult to assess and diagnose dysfunctions in outer ranges of motion. There are mild, moderate, and severe biomechanical dysfunctions that occur within this complex. A mild biomechanical dysfunction will cause mild joint hypomobility and mild limitations in ranges of motion, with mild protective muscle spasm of the surrounding musculature. A mild dysfunction is discovered during end ranges of lumbosacral flexion and extension, presenting asymmetric anatomic landmarks and limited posterior and anterior glides of the sacral base. A moderate biomechanical dysfunction will cause moderate joint hypomobility, with moderate limitations in ranges of motion and moderate protective muscle spasm of surrounding musculature. A moderate dysfunction is discovered during mid ranges of lumbosacral flexion and extension, presenting with asymmetric anatomic landmarks and limited posterior and anterior glides of the sacral base. A severe biomechanical dysfunction will cause severe joint hypomobility, with severe limitations in ranges of motion and severe protective muscle spasm of surrounding musculature. A severe dysfunction is discovered at rest and during inner ranges of lumbosacral motion, presenting with asymmetric anatomic landmarks and decreased anterior glide during extension and limited posterior glide during flexion.

Joint mobility of the sacrum can be assessed at the lumbosacral junction, at the sacroiliac joints, and at the accessory sacroiliac joints. Joint mobility at the lumbosacral junction is best tested as anterior glide and posterior glide of the sacral base. Orthopedic physical therapists often use a spring test of L5 to discern posterior sacral glide of the sacrum, but this is not specific enough to the sacral base movement; there could be good spring test of L5 anteriorly, when the sacral base is stuck posteriorly. A spring test on the sacral base can discern anterior glide, but must be performed throughout the full potential range of lumbosacral extension.

Joint mobility of the sacroiliac joint can be discerned in numerous ways. Orthopedic physical therapy has contributed multiple joint mobility tests for the sacroiliac joint. The squish test included in this manual is the posteromedial compression of the ilium on sacrum, and can be performed from multiple angles (with a more superiorly directed force or more inferiorly directed force) in order to discern the mobility of the total joint.

Joint mobility of the accessory sacroiliac joint can be discerned with traditional joint mobility testing. In prone, the posterior superior iliac spine is gripped, while the superolateral aspect of the sacrum is relatively fixed, and glides of the PSIS are attempted in all directions: superior, inferior, medial, lateral, and oblique. Joint hypomobility can be easily palpated at the accessory sacroiliac joints.

Incidence of Sacral Dysfunction

The sacrum is a most crucial area to be mobilized and liberated. Biomechanical dysfunction of the sacrum will cause tension restrictions of the dura mater, which inserts at S2, and spinal cord tension via the sacral plexus nerve roots, which exit at the intervertebral neural foramina of the sacrum. In motor vehicle accidents, the trauma occurs at the lumbosacral pelvic region, and a common denominator for all whiplash and traumatic brain-injured patients is sacral dysfunction. The significance of sacral mobility for the recovery of the neurologic patient cannot be overemphasized.

Biomechanical Movement of the Sacrum

Different researchers describe sacral motion in different manners. Nelson describes movement of the sacrum within the innominates (ilia): (1) Flexion of the sacrum is described as an anterior sacral tilt; a downward (caudal) movement of the upper limb together with a backward (posterior) movement of the lower limb; (2) extension of the sacrum is described as a posterior sacral tilt; forward (anterior) movement of the lower limb with upward (cephalad) movement of the upper limb.

The Lumbosacral Junction and Reciprocal Movement

Often, reciprocal movement is defined as the movement of the arms and legs in an asynchronous movement pattern. In fact, Jacqueline Perry, in her book on gait analysis, describes the reciprocal movements of the arms and legs in this manner.

The author defined reciprocal motion at the lumbosacral junction in the early 1980's. Because of her work with reflexogenic movement patterns, it was easily discerned that arms and legs cannot be the "seat" of a neuroreflexogenic pattern.

In the late 1980s, Frank Lowen discovered the reciprocal movement pattern of neural tissue, which is the driving force behind the neuroreflexogenic movement pattern of the arms and legs, as well as at the lumbosacral junction. Since Lowen's discovery that neural tissue has a right and left reciprocal movement pattern (a biologic rhythm, called a "motility," with a circadian rhythm pattern of 3 to 4 cycles per minute), he has further defined this entity with the author during many years of clinical research with neurologic clients.

At the lumbosacral junction there is reciprocal movement. During spinal movement (when the spine flexes or extends down the spinal kinetic chain): during flexion of L5, the sacral base glides posteriorly into extension; during extension of L5, the sacral base glides anteriorly into flexion. In spinal flexion and extension: During right rotation of L5, the sacral base rotates left; during right side bending of L5, the sacral base side bends left.

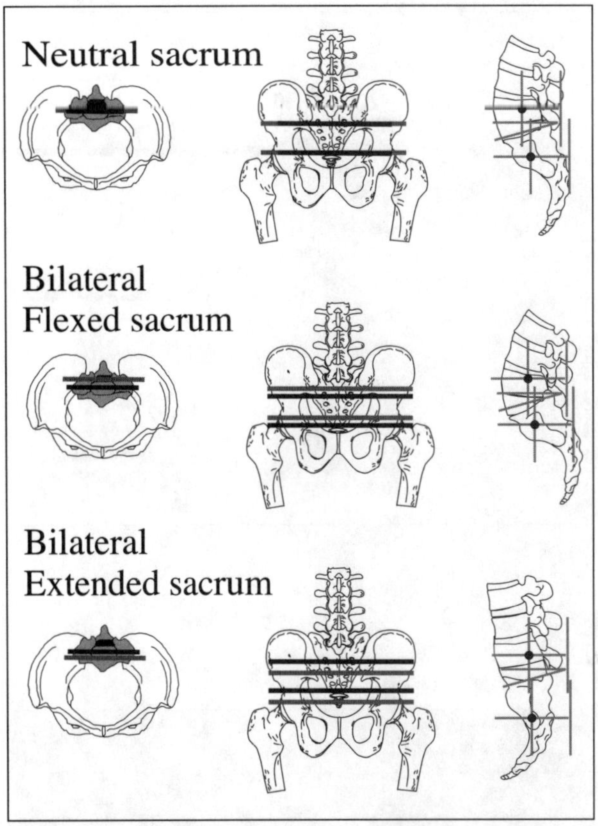

Three Anatomic Landmarks for Differential Diagnosis of Biomechanical Dysfunctions of the Sacrum (Sacroiliac Joint and Lumbosacral Junction)

1. Sacral sulcus
2. Inferior lateral angle
3. Lateral border of sacrum

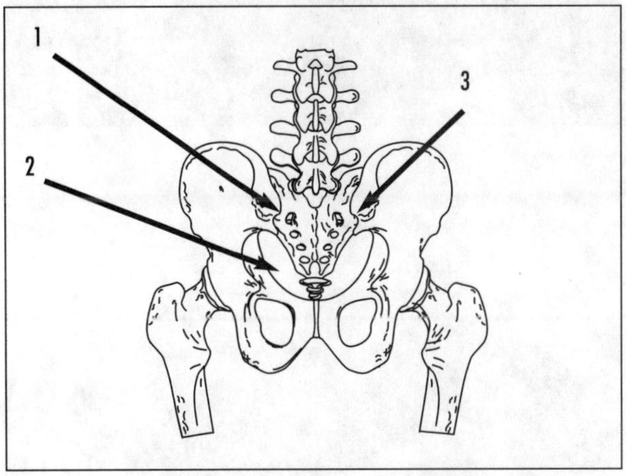

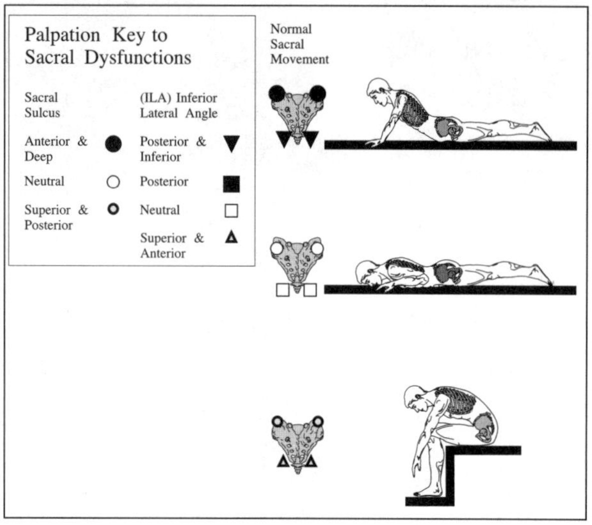

Four Observations of Anatomic Landmarks for Differential Diagnosis of Biomechanical Dysfunction of the Sacrum

1. Which sacral sulcus is deeper?
2. Which inferior lateral angle is more posterior *and* more inferior?
3. Which inferior lateral angle is more posterior but *not* inferior?
4. Which lateral border of the sacrum is more prominent posteriorly?

Palpation of Anatomic Landmarks

To palpate depth of the sacral sulcus to assess dysfunction during lumbosacral extension:

1. Patient is prone.
2. Therapist places pads of thumbs on both PSIS.
3. Shift thumbs medially; drop into sacral sulcus. Don't press hard.
4. Compare the difference in depth of right and left sulci.

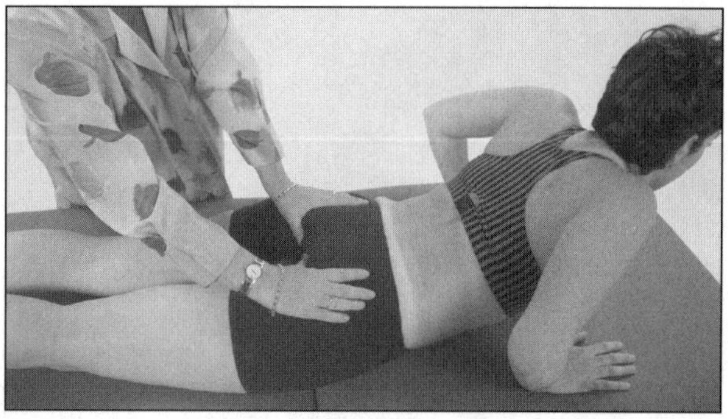

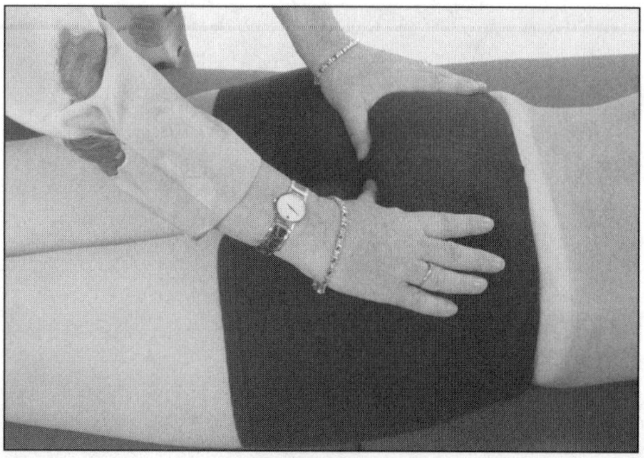

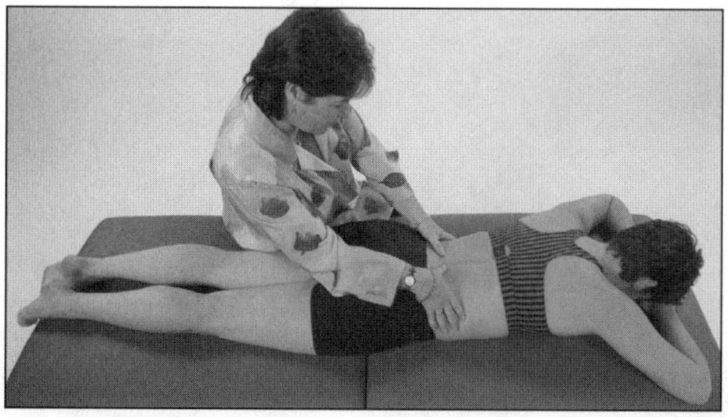

To palpate depth of the sacral sulcus to assess dysfunction during lumbosacral flexion:

1. Patient is sitting.
2. Therapist places pads of thumbs on both PSIS.
3. Shift thumbs medially; drop into sacral sulcus. Don't press hard.
4. Compare the difference in depth of right and left sulci.

Note: This hand hold requires good palpation skills. Beginners should use thumbs together to palpate depth of sacral sulci.

Note: Assessment in prone extension will assess sacral base mobility during lumbosacral extension; assessment in sitting flexion will assess sacral base mobility during lumbosacral flexion.

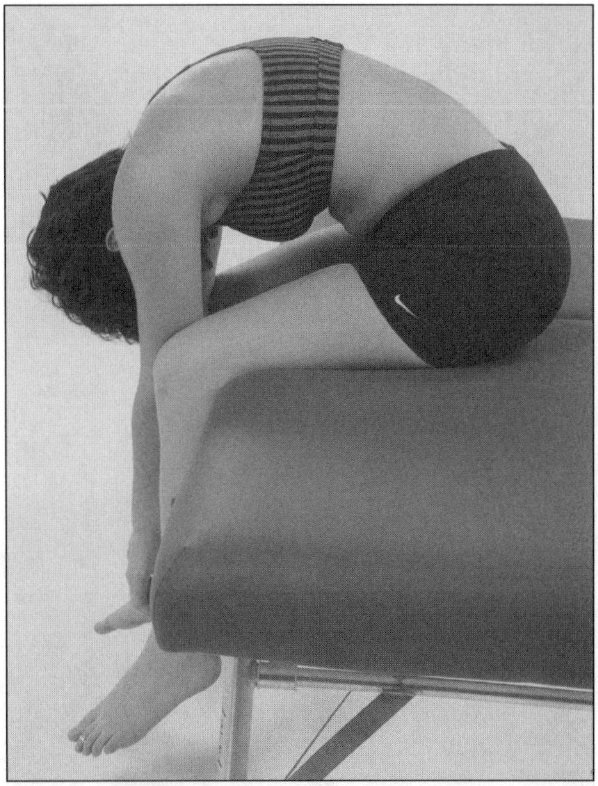

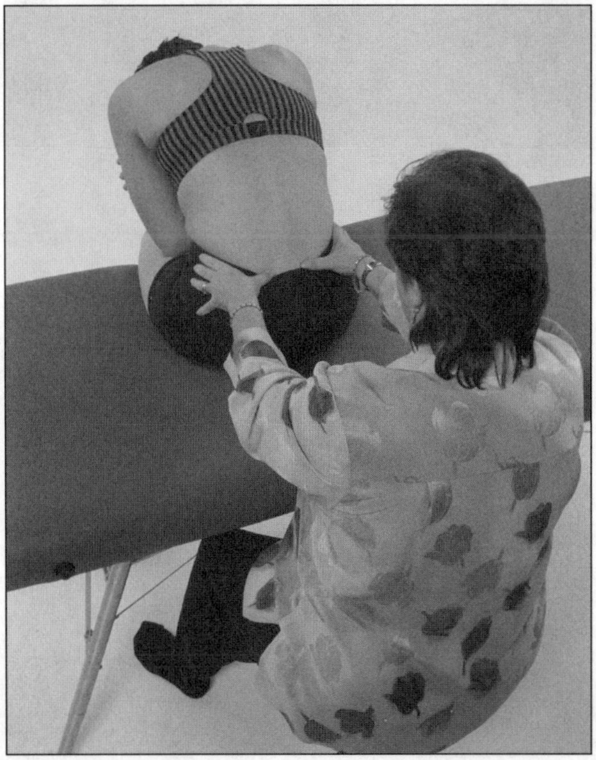

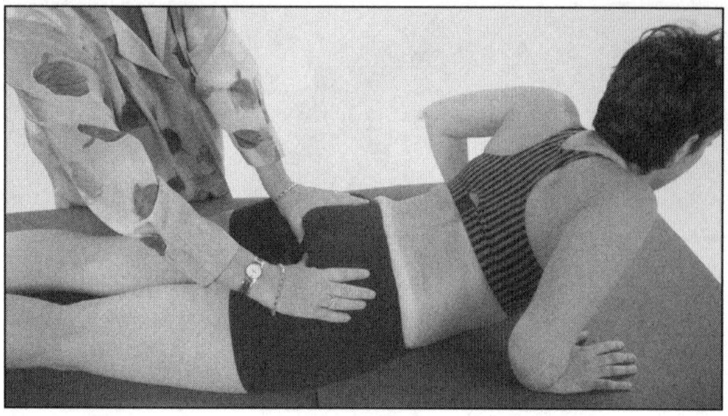

To palpate inferior aspect of inferior lateral angles to assess dysfunction during lumbosacral extension:

1. Patient is prone.
2. Therapist finds the top of the gluteal crease.
3. Move 2 to 2½ inches inferiorly.
4. Place hands on either sides of buttocks, fingers parallel to midline, thumbs perpendicular to midline.
5. Thumbs are 2 to 2½ inches caudal to top of gluteal crease.
6. Separate tips of thumb to 1 inch only.
7. Move thumbs 1 inch anteriorly, toward the floor.
8. Rotate pads of thumbs to face cephalad.
9. The bony ridge of the inferior aspect of the ILA should be just superior to the pad of each thumb.

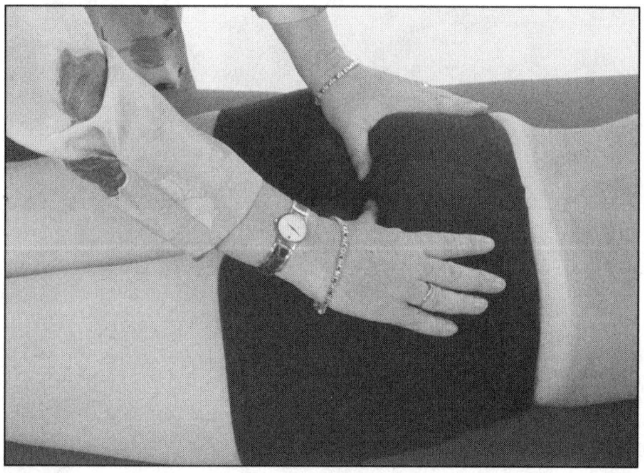

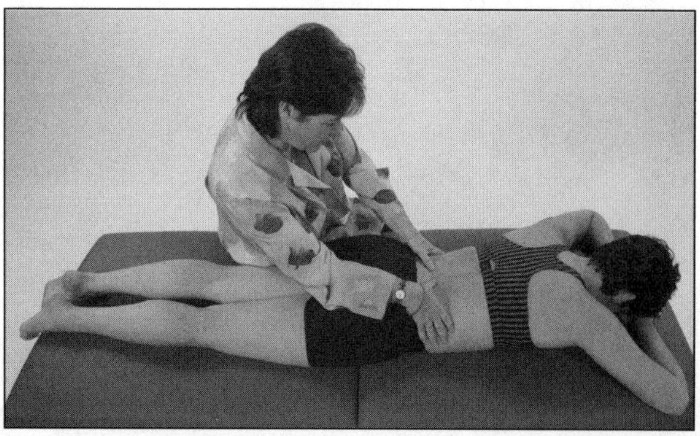

To palpate inferior aspect of inferior lateral angles to assess dysfunction during lumbosacral flexion:

1. Patient is sitting.
2. Therapist places two fingers (both third digits, or index and third digit of same hand) under buttocks with fingertips $1\frac{1}{2}$ inches apart.
3. Lift pads of distal phalanges superior to palpate inferior lateral angles.

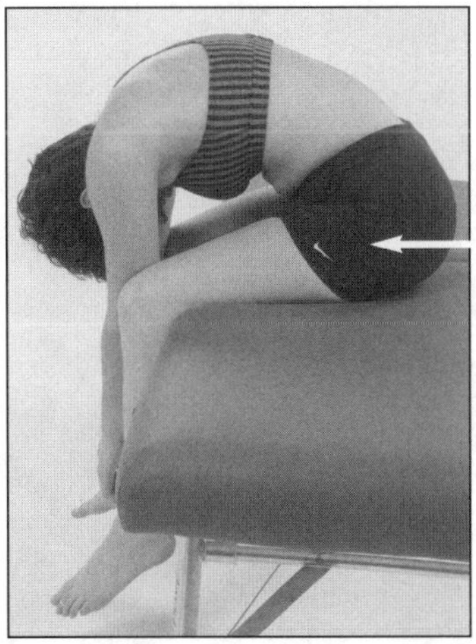

To palpate posterior prominence of lateral border of sacrum:

1. Patient is prone.
2. Therapist places pad of distal phalanx of third finger into sacral sulcus.
3. Place pad of distal phalanx of index finger at middle of lateral border of sacroiliac joint line.
4. Place pad of distal phalanx of thumb on inferolateral aspect of sacrum.
5. Compare right and left side for posterior prominence of lateral border of sacrum.

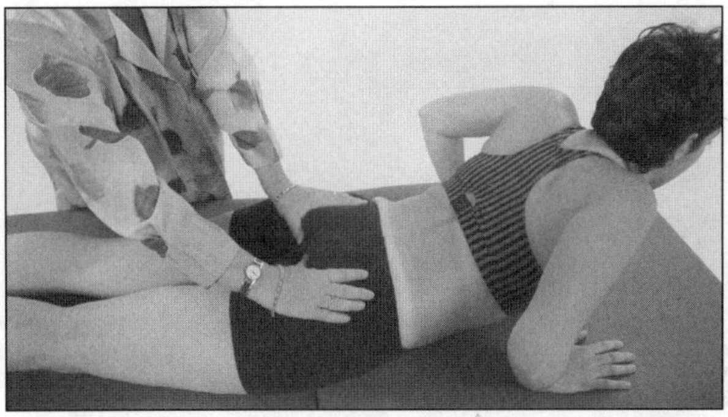

Note: Palpation during lumbosacral extension will uncover posterior vertical rotations. To discover anterior vertical rotations, palpate for posterior prominence of lateral border of sacrum in sitting flexion.

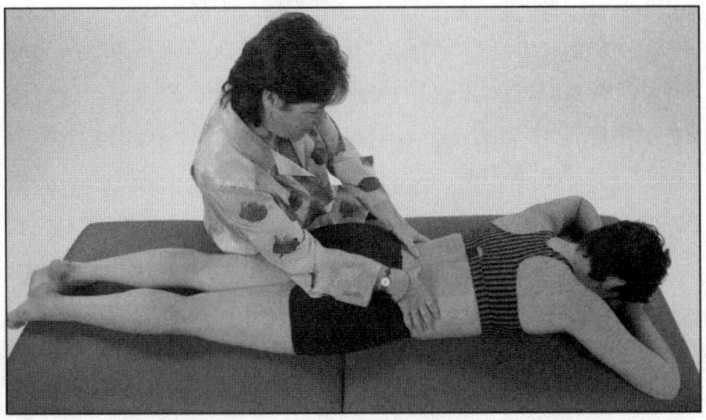

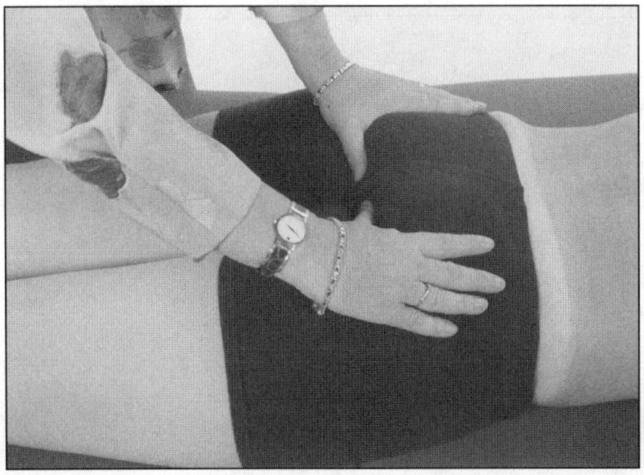

Dynamic Testing for Differential Diagnosis of Sacroiliac and Lumbosacral Dysfunction

All three assessments including:

1. Depth of sacral sulci
2. Inferior/posterior prominence of inferior lateral angles
3. Posterior prominence of lateral border of the sacrum: Must be assessed throughout full, existing ranges of motion, into lumbosacral extension and lumbosacral flexion.

Which Direction Is Assessed and Treated First, Lumbosacral Extension or Lumbosacral Flexion?

At the lumbosacral junction there is an excursion in anterior and posterior sacral glide during extension and flexion, which is necessary for movement. For basic function, the most important activities of daily living include the ability to stand up straight and the ability to lie down (to sleep). Standing up straight and lying prone or supine require inner ranges of motion of lumbosacral extension for comfort. Lumbosacral extension means extension of L5 with flexion of S1. Lumbosacral extension is assessed and diagnosed in prone extension. Thus, for the client to attain comfortable upright standing and comfortable lying down in the recumbent position, the posterior and extended sacral dysfunctions are commonly assessed and treated before the anterior and flexed problems. On the other hand, some people sit and bend a lot during the day; they require treatment of anterior and flexed sacral dysfunctions in order to sit comfortably.

In summary, when there is low back pain, for functional activities a person must be able to stand upright. The person must also be able to lie down and rest. Standing upright and lying down to rest require some lumbosacral extension. Thus, lumbosacral extension is commonly treated before lumbosacral flexion. Lumbosacral flexion is required for sitting.

Differential Diagnosis Affecting Extension

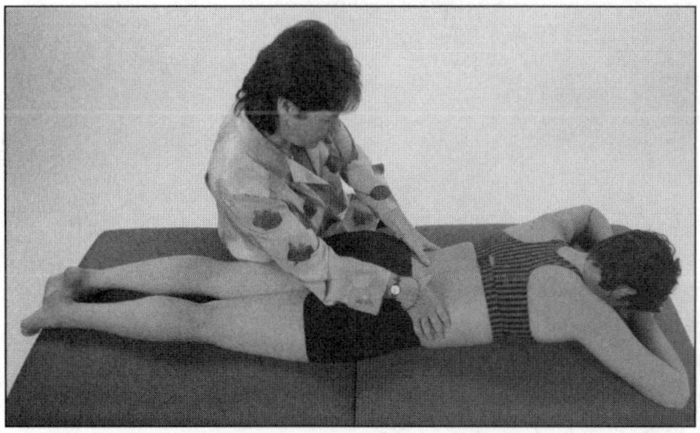

The patient is in prone. The patient's hands are at shoulder level. The therapist directs: "Push down on your hands. Lift your upper body up. Keep the hips and buttocks on the bed. Relax the buttocks. Don't lift up with your head. Lift with your arms." (This will facilitate relaxation of the paravertebral muscles for easier palpation.) The therapist palpates during extension until an asymmetry is perceived in the anatomic landmarks, with both hands, in the following manners:

For easier palpation:

1. Phase one:
   Place thumbs in the sacral sulci.
   Perceive which sacral sulcus is deeper, right or left side.
   Don't press; "sink" into the sulci gently.

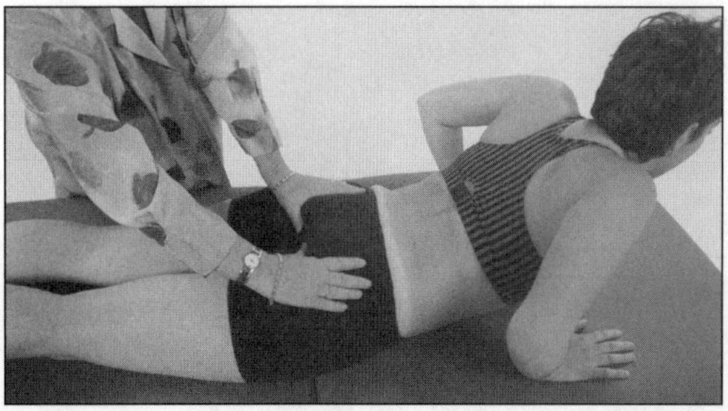

2. Phase two:
   Place pads of thumbs on the inferior bony border of the ILAs (inferior lateral angles of the sacrum).
   Perceive which ILA is more inferior, right or left side.
   Don't press; take up the slack of the soft tissue and rest against the bone.
3. Phase three:
   Place pads of thumbs on the posterior aspects of the ILAs.
   Perceive which ILA is more posterior, right or left side.
4. Phase four:
   Place tips of third fingers, index fingers, and thumbs along the length of the sacrum, just medial to the joint lines.
   Perceive which sacral margin is more posterior, right or left side.

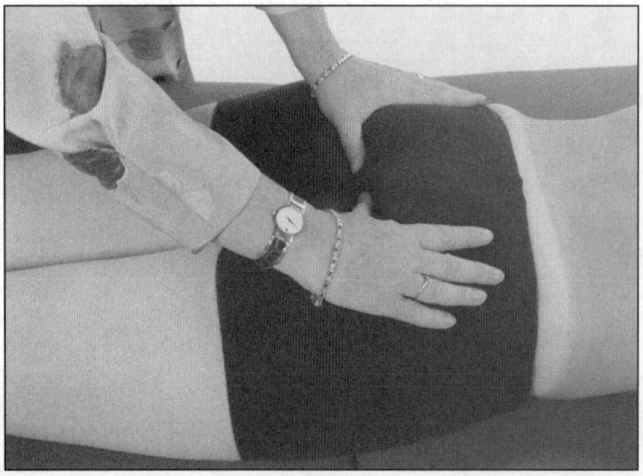

For more advanced palpation:

1. Place the tips of the third fingers in the sacral sulci.
2. Place the tips of the index fingers into the middle of the sacral margins just medial to the sacroiliac joint lines.
3. Place the pads of the thumbs inferior to the ILAs.
4. Perceive which sacral sulcus is deeper, which ILA is more inferior, which sacral margin is more posterior, the right or left side.

Treatment as a Dynamic Process: Into Extension

1. Step one:
   The patient is in prone.
   The therapist palpates the anatomic landmarks while the patient performs prone extension.
   The therapist then interprets the findings of the anatomic landmarks. The therapist makes a differential diagnosis.
   The therapist then corrects the dysfunction with treatment.
   After treatment, there will be a change in the presentation of the anatomic landmark asymmetry and an increase in the range of lumbosacral extension.
2. Step two:
   Repeat step one.
   The therapist palpates the anatomic landmarks as the patient pushes further into lumbosacral extension.
   The therapist interprets the new findings and makes a differential diagnosis.
   The therapist then corrects the dysfunction with treatment.
   The patient will be diagnosed while prone, pressing further into extension of the lumbosacral junction.
   Continue to repeat step one until full lumbosacral junction extension has been attained.

Clinical research indicates that there can be approximately 60 degrees of lumbosacral junction extension.

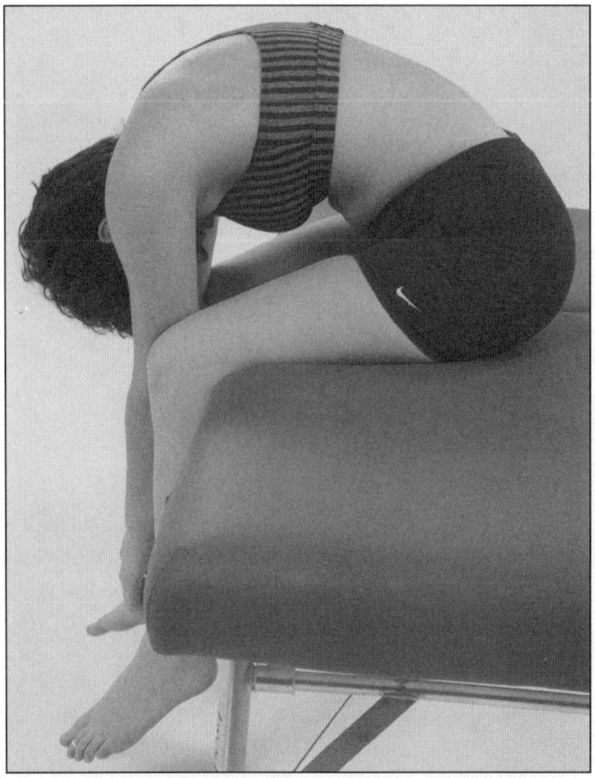

Differential Diagnosis Affecting Flexion

The patient is sitting. The therapist directs: "Flex your spine. Tuck your chin in and keep your chin tucked in at all times. Bend down the spine, one segment at a time, slowly." The therapist palpates until he or she perceives, with both hands, an asymmetry in the anatomic landmarks during flexion. The therapist palpates during flexion until an asymmetry is perceived in the anatomic landmarks, with both hands in the following manner:

1. Phase one:
   a. Place both thumbs in the sacral sulci.
   b. Perceive which sacral sulcus is deeper, right or left side.
2. Phase two:
   a. Place third fingers underneath patient, so that the distal pads of the third fingers are level with the ILAs (about 1 to 2 inches anterior).
   b. Move the finger pads superiorly.
   c. Take up the slack of the soft tissue to palpate the bony inferior borders of the ILAs.
   d. Perceive which ILA is more inferior, right or left side.
3. Phase three:
   a. Place the pads of the thumbs on the posterior aspects of the ILAs.
   b. Perceive which ILA is more posterior, right or left side.
4. Phase four:
   a. Place tips of third fingers, index fingers, and thumbs along the lengths of the sacral margins, just medial to the sacroiliac joint lines.
   b. Perceive which sacral margin is more posterior, the right or left side.

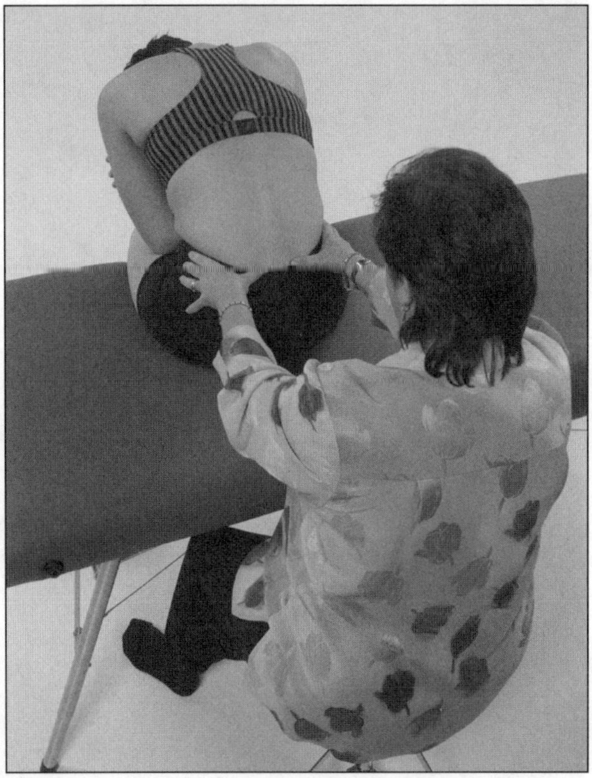

Treatment as a Dynamic Process: Into Flexion

1. Step one:
   a. The patient is sitting.
   b. The therapist palpates the anatomic landmarks while the patient flexes down the spinal kinetic chain.
   c. The therapist then interprets the findings of the anatomic landmarks.
   d. The therapist makes a differential diagnosis.
   e. The therapist then corrects the dysfunction with treatment.
   f. After treatment, there will be a change in the presentation of the anatomic landmark asymmetry and an increase in the range of lumbosacral flexion.
2. Step two:
   a. Repeat step one.
   b. The therapist palpates the anatomical landmarks as the patient flexes down the spinal kinetic chain.
   c. The therapist interprets the new findings and makes a differential diagnosis.
   d. The therapist then corrects the dysfunction with treatment.
   e. The patient will be diagnosed while flexing the lumbosacral junction.
   f. Continue to repeat step one until full lumbosacral junction flexion has been attained.

If there is asymmetry of the sacral anatomic landmarks, there is limitation in lumbosacral movement.

Note: Anatomic landmarks are assessed in neutral and throughout ranges of lumbosacral flexion and extension.

CHAPTER 15

TREATMENT OF BIOMECHANICAL DYSFUNCTION OF THE SACROILIAC JOINTS WITH MUSCLE ENERGY AND "BEYOND" TECHNIQUE

Rationale of Treatment for Flexed and Extended Sacrum

Consider the vertical and horizontal limb of the auricular-shaped articular surface of the sacrum. There is a vertical limb plus a horizontal limb of the articular surface of the sacrum. The intersection of the vertical and horizontal limbs is at the sacral angle. The sacral angle is at S2. Interestingly, S2 is also the center of gravity of the body, and where the dura mater inserts into the sacrum.

There is significance for all sacroiliac joint biomechanics regarding the vertical limb and the horizontal limb of the sacrum's joint surface. Although (according to cadaver dissections) the sacral angle varies for each body, there is a common phenomenon to be considered. The vertical limb facilitates movement; the horizontal limb places constraints upon movement.

During corrections of biomechanics with Muscle Energy and "Beyond" Technique, consider the sacral angle to be 90 degrees. In other words, consider the vertical limb to be upright (which it is *not),* and consider the horizontal limb to be perpendicular to the vertical limb (which it is *not*). The results of the techniques will be optimal.

Apply the Law of Inertial Mass

When treating flexed and extended dysfunctions of the sacrum the amount of pressure is important. In physics there is a concept: *the law of inertial mass.* Imagine pushing a boat in water: A slowly increasing force upon the boat would make it move, rather than a high-velocity, high-force push. When "pushing" upon the sacrum for corrections of biomechanical dysfunction, imagine a boat stuck in mud! Begin with a low force, slowly increasing in force. When the sacrum (boat stuck in mud) begins to move, follow it, increasing the amplitude, as well as the force. These techniques should be painless!

Flexed Sacrum

Position of Dysfunction

The sacrum is stuck flexed with the sacral base in anterior glide. In order to attain this position, the vertical limb moves caudal (down); then the horizontal limb moves posterior (backward).

Rationale of Treatment

In order to correct a flexed sacrum, the vertical and horizontal limbs must be repositioned. If the sacrum is stuck inferior on the vertical limb and posterior on the horizontal limb, then the sacrum must be mechanically repositioned by pushing on the sacral apex. Pushing anterior first will reposition the horizontal limb (moving it from its backward position to a forward position); then pushing cephalad will reposition the vertical limb (moving it from an inferior position to a superior position).

Extended Sacrum

Position of Dysfunction

The sacrum is stuck extended with the sacral base in posterior glide. In order to attain this position, the horizontal limb moves anterior; then the vertical limb moves superior.

Rationale of Treatment

In order to correct an extended sacrum, the vertical and horizontal limbs must be repositioned. If the sacrum is stuck anterior on the horizontal limb and superior on the vertical limb, then the

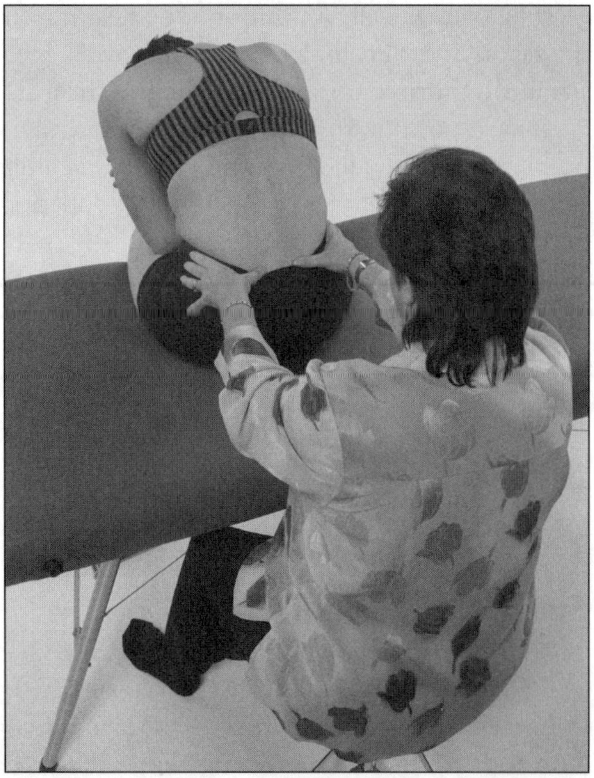

sacrum must be mechanically repositioned by pushing on the sacral base. Pushing caudal first will reposition the vertical limb (moving it from its superior position downward to an inferior position); then scooping posteriorly (with an anteroinferior directed force) will reposition the horizontal limb (moving it from a forward position backward to a more posterior position).

Bilateral Flexed Sacrum

It is important to remember, in regard to flexed and extended dysfunctions, that there is an auricular-shaped articular surface of the sacrum, and there is reciprocal movement at the lumbosacral junction.

Differential Diagnosis

1. Diagnose in sitting flexion
2. Limitations of lumbosacral flexion
3. Sacral sulcus deep bilateral
4. No asymmetry of anatomical landmarks
5. Good anterior glide of sacral base during lumbosacral extension (during extension of the lumbar spine)
6. Poor posterior glide of sacral base during lumbosacral flexion (during flexion of the lumbar spine)

Note: Palpation for diagnosis can be in two phases.

Phase One: Palpate depth of sulci with thumbs during flexion.

Phase Two: Palpate inferior aspects of ILAs with fingers during flexion.

Bilateral Flexed Sacrum

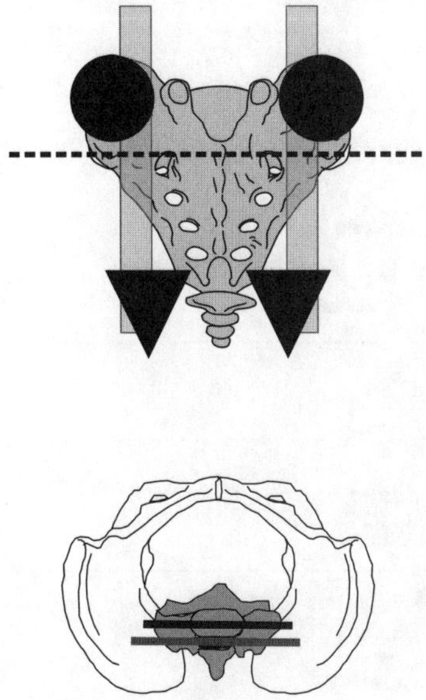

Treatment

Patient is prone.

1. Abduct 15 degrees and internally rotate *both legs* to gap the joints to facilitate correction.
2. Therapist stands at low back facing head.
3. Thenar eminence between ILAs.
4. Push (hand between the ILAs).

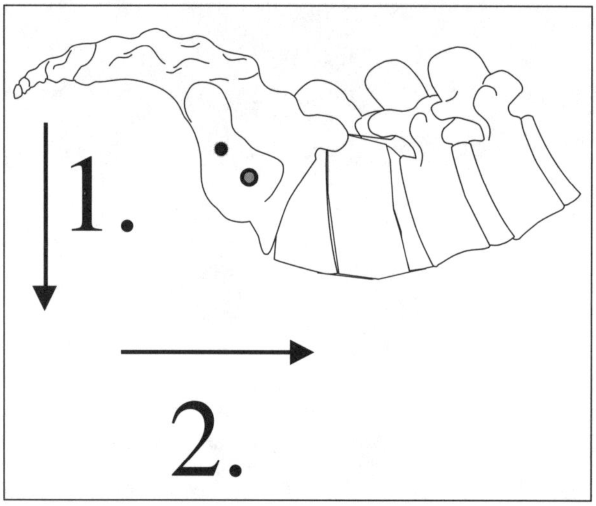

Phase One: Anterior pressure for 45 seconds to reposition the horizontal limb.
Phase Two: Cephalad pressure for 45 seconds to reposition the vertical limb.

A bilateral flexed sacrum is difficult to diagnose. The therapist is typically frustrated because of lack of asymmetry of landmarks. Focus on limited lumbosacral flexion and loss of posterior glide of sacral base during flexion. Often, it is difficult to flex into a low car. Sitting long distances in a car or sitting for long durations can contribute to significant pain in the buttocks region.

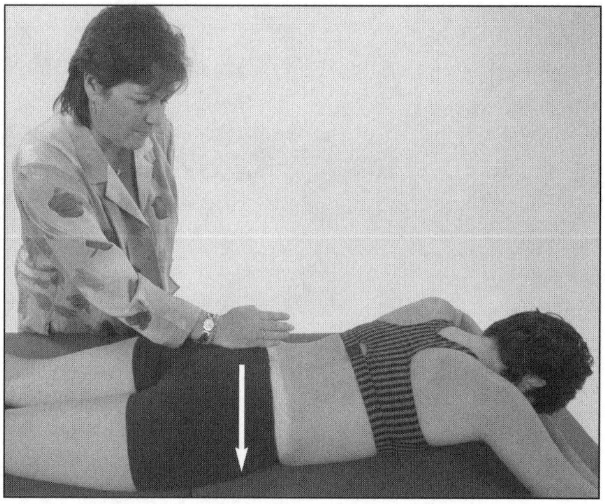

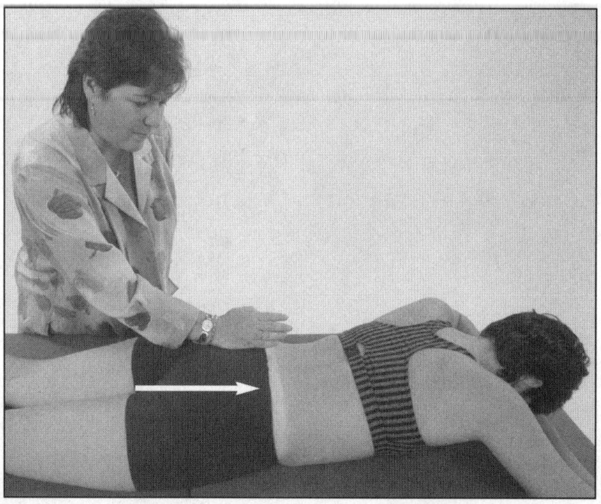

Bilateral Extended Sacrum

It is important to remember in regard to flexed and extended dysfunctions, that there is an auricular-shaped articular surface of the sacrum, and there is reciprocal movement at the lumbosacral junction. A bilaterally extended sacrum is a severe biomechanical dysfunction, which causes increased intradiscal pressure with bilateral L5 radiculopathy. Often, these patients go into spinal surgery. Some of the pain and disability will be eliminated with surgery, simply because the disc compression is alleviated. Yet this problem is quickly eradicated with Muscle Energy and "Beyond" Technique.

1. Diagnose in prone extension.
2. Limitation of lumbosacral extension.
3. Sacral sulci shallow bilaterally, difficult to palpate. Sulci are partially obliterated.
4. No asymmetry of anatomical landmarks.
5. Good posterior glide of sacrum during lumbosacral flexion (during flexion of the lumbar spine).
6. Poor anterior glide of sacrum during lumbosacral extension (during extension of the lumbar spine).

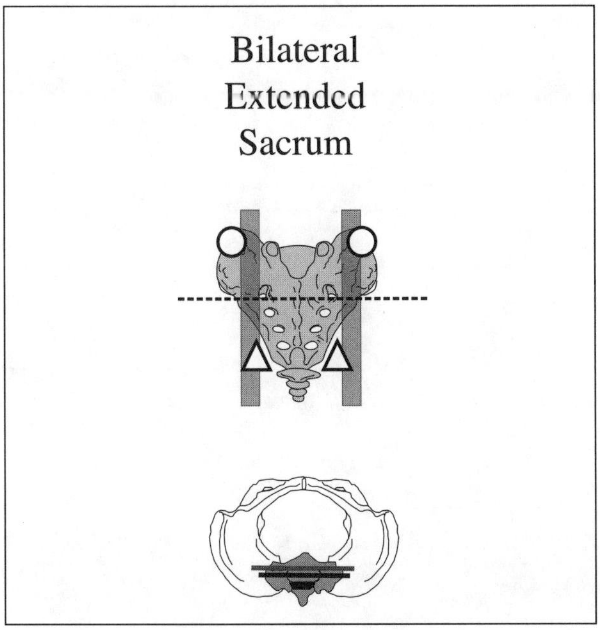

Note: All posterior and extended sacral dysfunctions, including at the sacroiliac joints and at the lumbosacral junction, are assessed in prone extension.

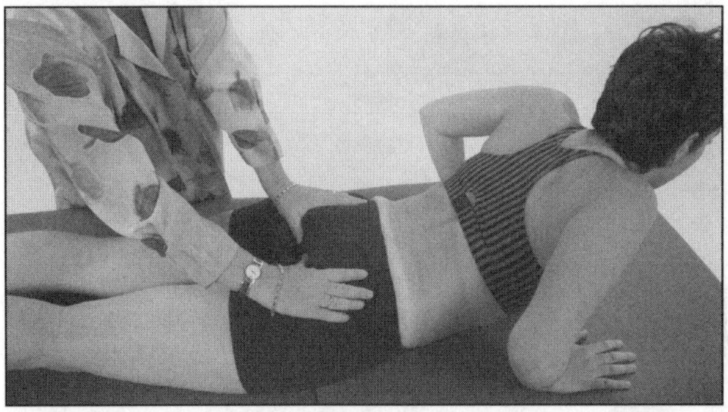

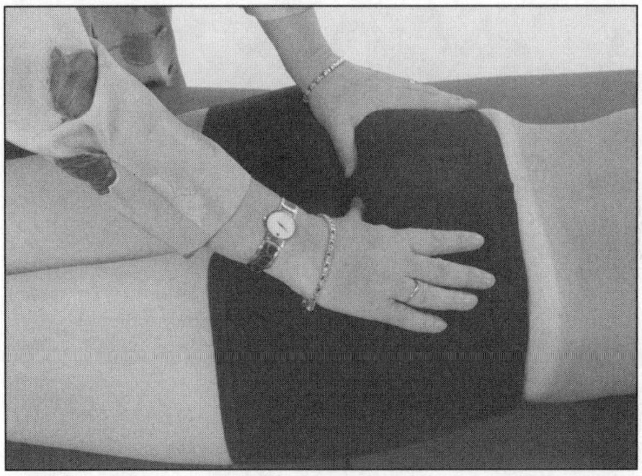

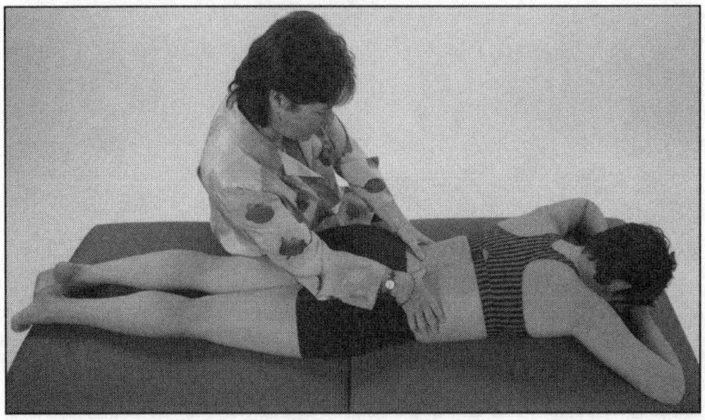

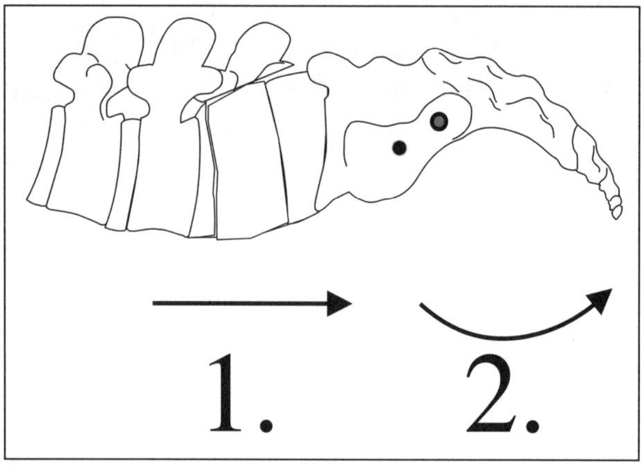

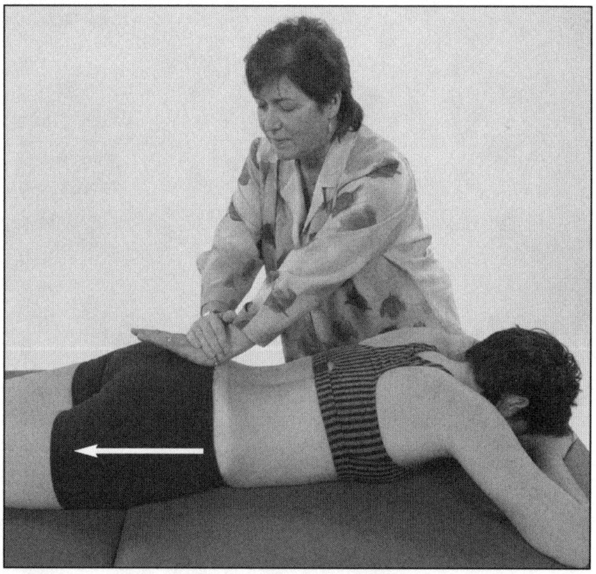

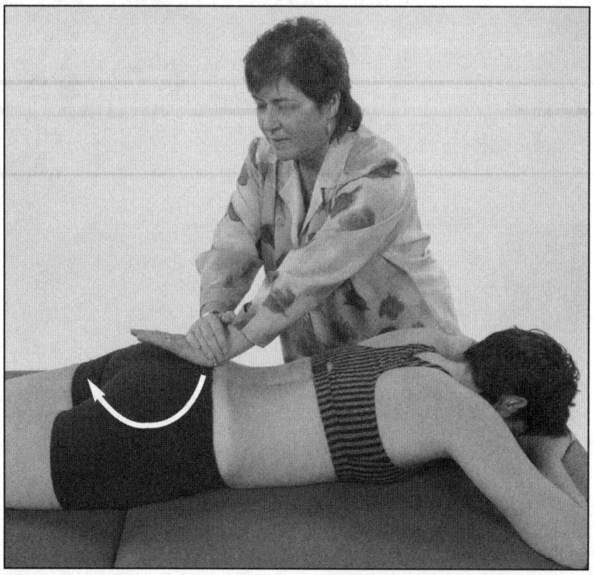

Treatment

Patient is prone.

1. Abduct 15 degrees and internally rotate both legs to gap the joints.
2. Therapist stands at head facing low back.
3. Thenar eminence on mid sacral base.
4. Pressure on sacral base.
   Phase One: Caudad pressure for 45 seconds to reposition the vertical limb.
   Phase Two: Scoop posteriorly with an anteroinferior-directed force for 45 seconds to reposition the horizontal limb.

Remember to apply the law of inertial mass (imagine pushing a boat stuck in mud) when pushing on the sacrum. A bilateral extended sacrum is difficult to diagnose, and the therapist often experiences frustration at inability to palpate depths of dysfunctionally shallow sacral sulci. Focus on limitation of lumbosacral extension and loss of anterior glide of sacral base during extension.

Unilateral Flexed Sacrum

As in the above explained dysfunctions, it is important to remember, in regard to flexed and extended dysfunctions, that there is an auricular-shaped articular surface of the sacrum, and there is reciprocal movement at the lumbosacral junction.

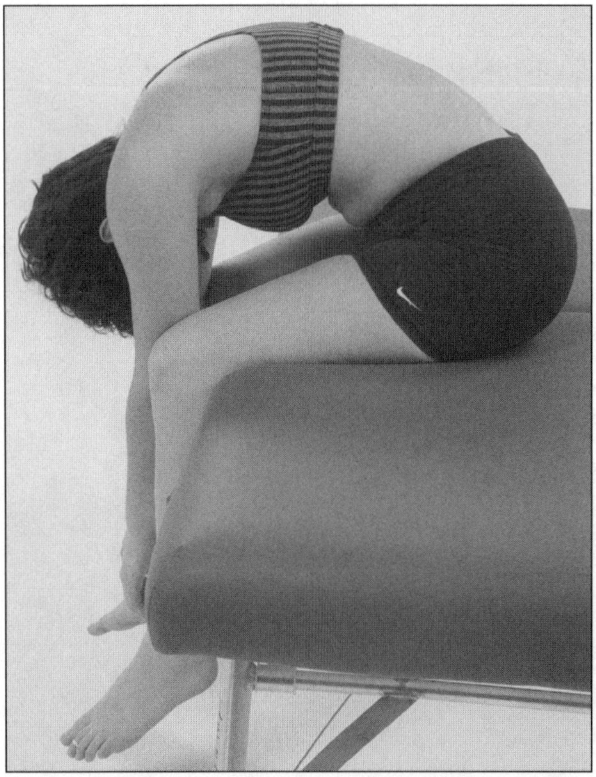

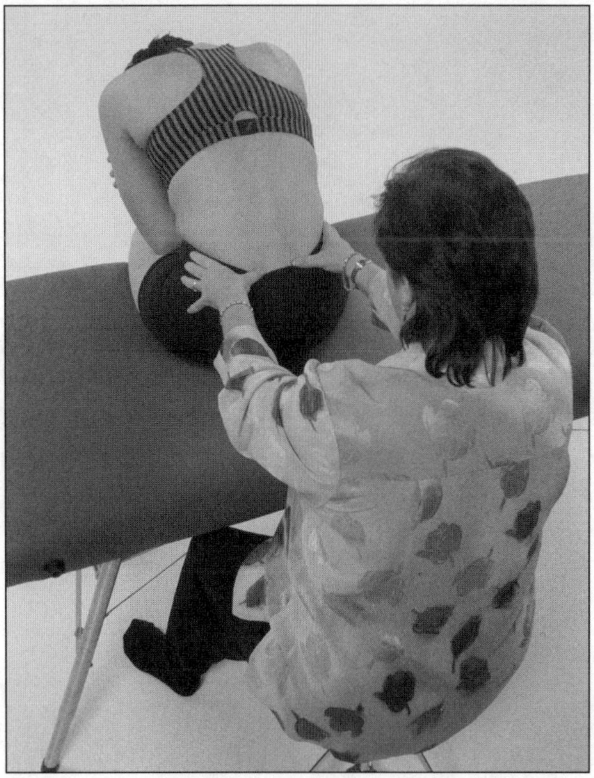

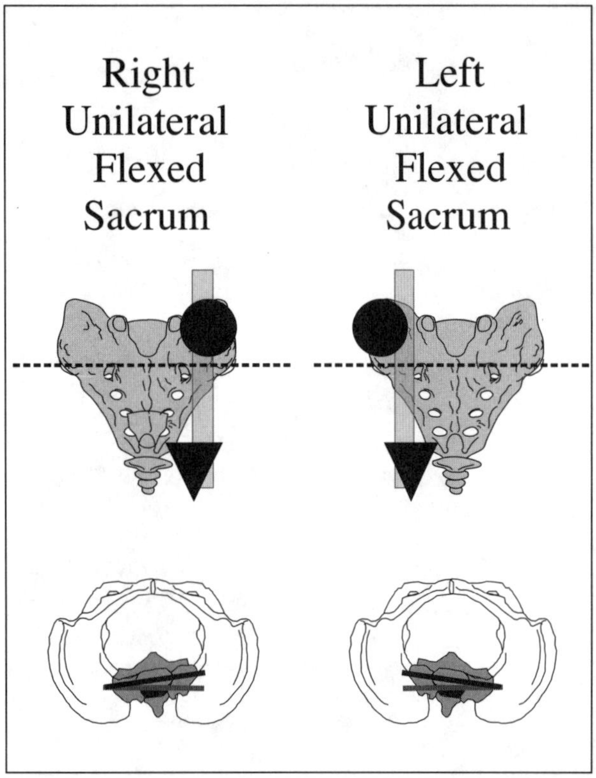

Differential Diagnosis:
Right Unilateral Flexed Sacrum

1. Diagnose in sitting flexion.
2. Limitation of lumbosacral flexion.
3. Asymmetry of anatomical landmarks.
4. Anatomic landmarks (deep sulcus and posterior/inferior ILA) are asymmetrical on same (right) side.
5. Mobility tests: Sitting flexion test and squish test are positive on right.
6. Sacral sulcus deep on right.
7. ILA posterior and caudal on right.
8. Good anterior glide of sacral base during lumbosacral extension (during extension of lumbar spine).
9. Poor posterior glide of sacral base on right side (during lumbosacral flexion during flexion of lumbar spine).

Treatment

Patient is prone.

1. Abduct 15 degrees and internally rotate ipsilateral right leg to gap joint.
2. Therapist stands at low back and faces head.
3. Pressure with hypothenar eminence and fifth finger on ipsilateral right ILA.
   Phase One: Anterior pressure 45 seconds to reposition right horizontal limb of the articular surface of the sacrum.
   Phase Two: Cephalad pressure 45 seconds to reposition right vertical limb of the articular surface of the sacrum.

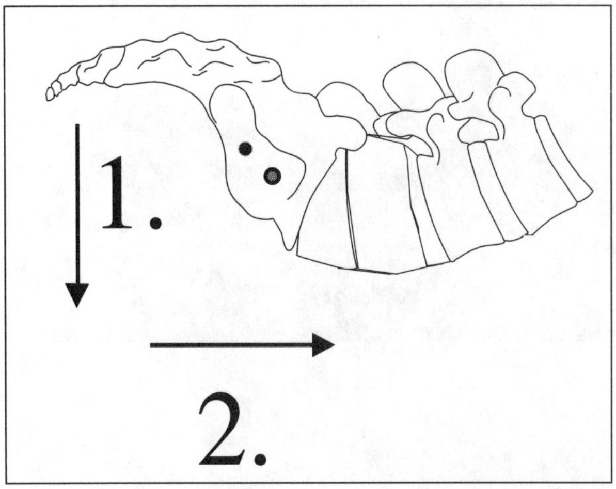

Remember to apply the law of inertial mass (imagine pushing a boat stuck in mud) when pushing on the sacrum. Approximately 85 percent of unilateral flexed sacrums are right unilateral flexed sacrums. Only about 15 percent of this category of dysfunctions are left unilateral flexed sacrums. There is more pain on sitting. Occasionally, impingement upon the lower sacral plexus can contribute to foot pain and paresthesia.

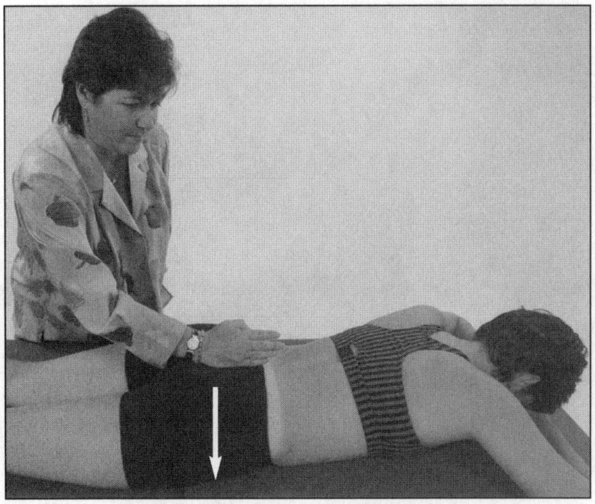

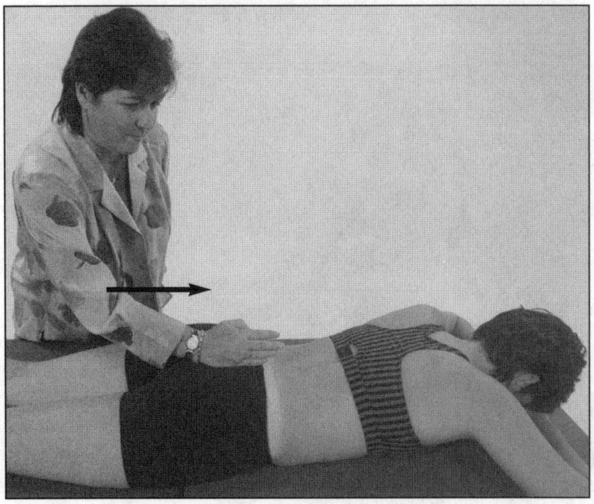

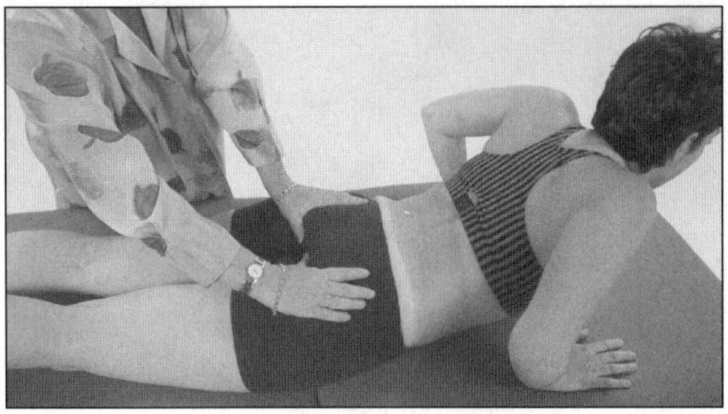

Unilateral Extended Sacrum

As in the above explained dysfunctions, it is important to remember in regard to flexed and extended dysfunctions, that there is an auricular-shaped articular surface of the sacrum, and there is reciprocal movement at the lumbosacral junction.

Differential Diagnosis:
Left Unilateral Extended Sacrum

1. Diagnose in prone extension.
2. Limitation of lumbosacral extension.
3. Asymmetry of anatomical landmarks.
4. Anatomical landmarks (deep sulcus and posterior/inferior ILA) are both found on right side of the body (The problem is on the contralateral left side of the body. This is often confusing for the practitioner!).
5. Mobility tests: Sitting flexion test and squish test are positive on the left.

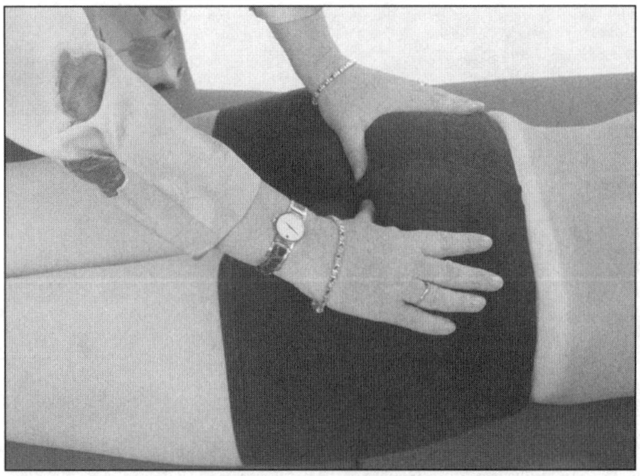

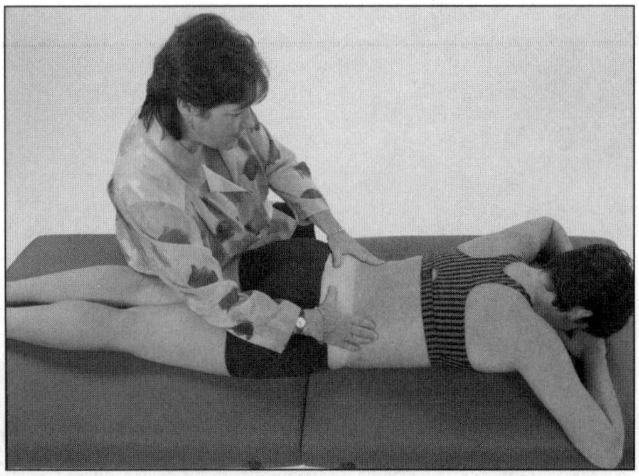

6. Sacral sulcus deep on right (The problem is on the left side).
7. ILA posterior and caudad on right (The problem is on the left side).
8. Good posterior glide of sacral base during lumbosacral flexion (during flexion of lumbar spine).
9. Poor anterior glide of sacral base on left side during lumbosacral extension (during extension of lumbar spine).

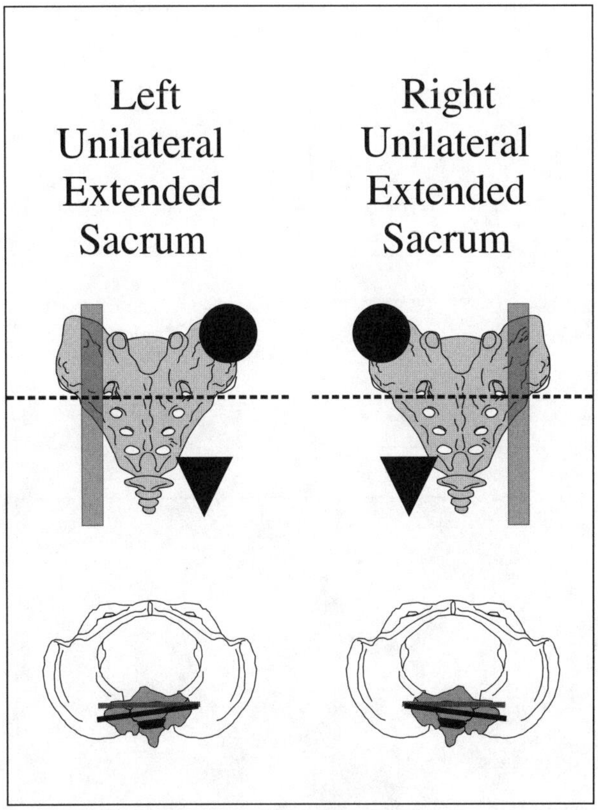

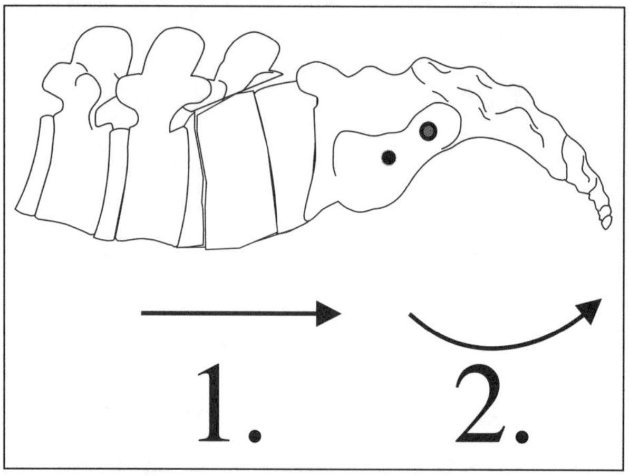

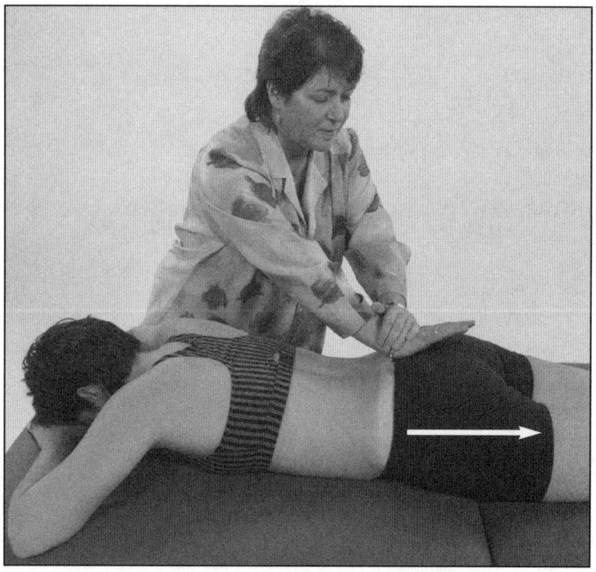

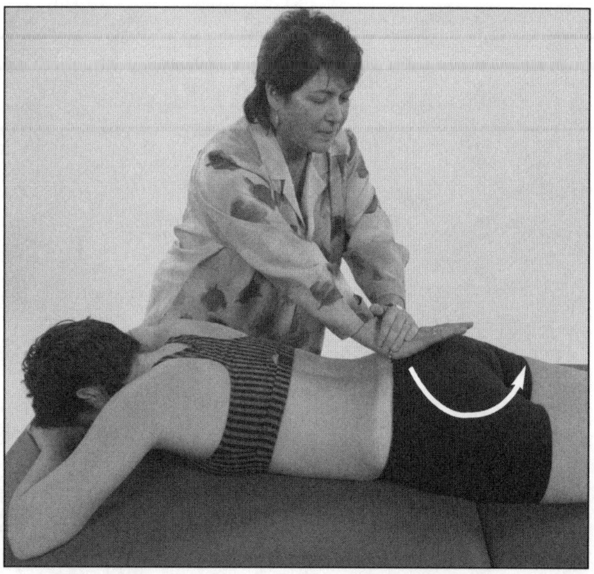

Treatment

Patient is prone.

1. Abduct 15 degrees and internally rotate left ipsilateral leg to gap joint.
2. Therapist stands at head facing low back.
3. Pressure from thenar eminence on left sacral base.
   Phase One: Caudad pressure 45 seconds to reposition left vertical limb.
   Phase Two: Scoop posteriorly with an anteroinferior directed force 45 seconds to reposition left horizontal limb.

Remember to apply the law of inertial mass (imagine pushing a boat stuck in mud) when pushing on the sacrum. The occurrence of the left unilateral extended sacrum is approximately 85 percent. Right unilateral extended sacrums are only about 15 percent of total incidence. The unilateral flexed sacrum and the unilateral extended sacrum may appear similar at first because of the anatomical landmarks. For example: When there is a right unilateral flexed sacrum, or a left unilateral extended sacrum, the anatomic landmarks are similar: For both, the right sacral sulcus is deeper than the left, and the right ILA is more posterior and more inferior on the right (as compared to the left). How to discern between a right unilateral flexed sacrum, when compared to the possibility of a left unilateral extended sacrum? The answer is simple: A flexed sacrum can only be discovered in lumbar flexion; an extended sacrum can only be discovered in lumbar extension. Consider reciprocal movement. Therefore, to recognize a right unilateral flexed sacrum from a left unilateral extended sacrum, consider the reciprocal motion of L5 with S1. If the flexed sacrum is stuck anterior at the sacral base, it cannot glide posterior. But the sacral base must glide posterior in order to attain lumbosacral flexion. Thus, in lumbar flexion, because there is a limitation of lumbosacral flexion, there will be evidence of the inability to perform posterior glide of the sacral base. Why? Because the sacral base is stuck in anterior glide when there is a flexed sacrum. Find the flexed sacrum in sitting flexion, rather than in prone extension! Therefore, to recognize a left unilateral extended sacrum from a right unilateral flexed sacrum, consider the reciprocal motion of L5 with S1. If the extended sacrum is stuck posterior at the sacral base, it cannot glide anterior. But the sacral base must glide anterior in order to attain lumbosacral extension. Thus, in lumbar extension, because there is a limitation of lumbosacral extension, there will be evidence of the inability to perform anterior glide of the sacral base. Why? Because the sacral base is stuck in posterior glide when there is an extended sacrum. Find the extended sacrum in prone extension, rather than in sitting flexion!

Vertical Rotations of the Sacrum

Rationale

The sacrum does not perform rotations around a vertical axis. The sacral segments are united (later "fused") so that the sacral segments (S1 through S4 or S5) all move together as a unit. Their movement is as a single segment, which means that the sacrum could be assessed as L6 (a sixth lumbar vertebra) except for two issues: (1) There is reciprocal movement between L5 and S1. (2) There is an auricular-shaped articular surface on the sacrum that dictates movement. Therefore, the sacrum is unique and needs to be assessed as an entity with unique movement patterns.

Rotation of the sacrum cannot be performed without side bending. This is true of all vertebral segments. In Type I movement, when two or more vertebral segments in AP neutral rotate, they side bend together in the opposite direction of the rotation. In Type II movement, in flexion or extension, each vertebra, when it rotates, will side bend to the same side. Because side bending of the sacrum occurs on the vertical limb plus the horizontal limb of the auricular-shaped articular surface, there is a torsion, which is a combined movement of rotation with side bending together. (This distinguishes sacral movement from other vertebrae moving in rotation and side bending.)

If the sacrum is stuck rotated on a vertical axis, this is a nonphysiologic position, similar to an upslip or downslip of the ilium, i.e., a subluxation. This can be treated with a manipulation that corrects this pathologic position of the sacrum. (The ilium is manipulated to correct the relationship of ilium and sacrum, although this is a sacroiliac joint dysfunction.)

The barrier to movement is observed at the lumbosacral junction and at the sacroiliac joints, in all physiologic ranges of motion, because this is essentially a dislocation of the sacrum. What does this mean? It is a nonphysiologic position, the vertical rotation of the sacrum (otherwise known as a rotation of the sacrum on a vertical axis). Therefore, the sacrum is not only stuck. It is stuck in a position, on an axis, where it should never be.

Treatment involves positioning the sacrum into its barrier and then manipulating the ilium. Treatment for a left posterior vertical rotation: The patient is supine. The sacrum is rotated right. Then a posteromedial manipulation of the left ilium is performed. This will correct the relationship of the articular surfaces of the left ilium and the left dysfunctional sacroiliac joint.

Treatment for a right posterior vertical rotation: The patient is supine. The sacrum is rotated left. A posteromedial manipulation of the right ilium is performed. This will correct the relationship of the articular surfaces of the right ilium and the right dysfunctional sacroiliac joint.

Anatomic landmarks are positive for a left posterior vertical sacral rotation as follows: In lumbosacral extension, there is a deep right sulcus and a posterior prominence of the left sacral margin. Anatomic landmarks are positive for a right posterior vertical sacral rotation as follows: In lumbosacral extension there is a deep left sulcus and a posterior prominence of the right sacral margin.

Posterior vertical sacral rotations are the most disabling of all sacroiliac dysfunctions. Often, a compression of the sacral plexus will result in hypotonia: Atrophy of the ipsilateral gluteus maximus and occasionally decreased tone of the ipsilateral hamstrings are typical.

Treatment for an anterior vertical rotation to the left: The patient is prone. The sacrum can be rotated right. An anterolateral manipulation of the right ilium is performed to correct the relationship of the articular surfaces of the right ilium and the right dysfunctional sacroiliac joint. Treatment for an anterior vertical rotation to the right: The patient is prone. The sacrum can be rotated left. An anterolateral manipulation of the left Ilium can be performed to correct the relationship of the articular surfaces of the left ilium and the left dysfunctional sacroiliac joint.

Anatomic landmarks are positive for an anterior vertical sacral rotation to the left as follows: In lumbosacral flexion, there is a deep right sulcus and a posterior prominence of the left sacral margin. Anatomic landmarks are positive for an anterior vertical sacral rotation to the right as follows: In lumbosacral flexion there is a deep left sulcus and a posterior prominence of the right sacral margin.

Correction of anterior vertical sacral rotation may result in decreased urinary incontinence, decreased menstrual discomfort, and more, possibly secondary to a normalization of pressure from the sacrum on the tissue within the pelvic bowl. Hip pain is often alleviated with these corrections.

It is important to note that nomenclature (i.e., terminology) is often problematic with an anterior vertical sacral rotation. When the sacrum is rotated to the left and it is an anterior rotation (rather than a posterior rotation), the problem is a right sacroiliac joint problem. When the sacrum is rotated to the right and it is an anterior rotation (rather than a posterior rotation), the problem is a left sacroiliac joint problem.

Rotations of the sacrum on a vertical axis are pathologic dysfunctions, rather than physiologic dysfunctions. Because S1 is never in neutral (S1 is always flexed or extended), S1 and the sacrum do not rotate around a vertical axis. Trauma is required to cause such a dysfunction. This dysfunction can be found at any time, in inner ranges, mid ranges, and outer ranges of motion. This dysfunction can be discerned in neutral. The anterior vertical sacral rotation can be found in flexion. The anterior vertical sacral rotation cannot be found in extension. (As compared to the posterior vertical sacral rotation dysfunction, which can be found in neutral or in extension, but cannot be found in flexion.)

Posterior Vertical Rotation of Sacrum

Horizontal plane/vertical axis

Note: This is a severe biomechanical dysfunction that causes L5 discopathy and radiculopathy, and can cause lower motor neuron dysfunction with atrophy of the ipsilateral buttock musculature.

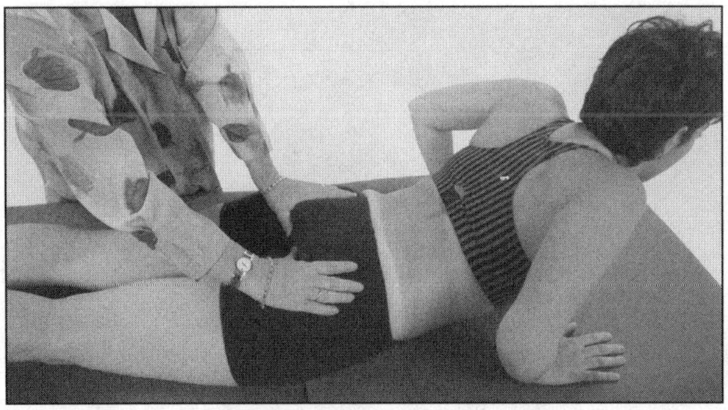

Differential Diagnosis:
Right Posterior Vertical Rotation of Sacrum

This is a dysfunction of the right sacroiliac joint, and the sacrum is rotated to the right.

1. Diagnose in prone extension.
2. Limitation of lumbosacral extension.
3. Mobility tests: Positive right ipsilateral sitting flexion test and squish test.
4. Asymmetry of anatomical landmarks.
5. Opposite (left) sacral sulcus is deeper.
6. Ipsilateral (right) ILA is posterior but not inferior.
7. Ipsilateral (right) lateral sacral border is posterior (opposite left lateral sacral border appears anterior).
8. Ipsilateral sacrotuberous ligament tension.
9. Follow dynamic movement of sacrum: increased asymmetry during extension.

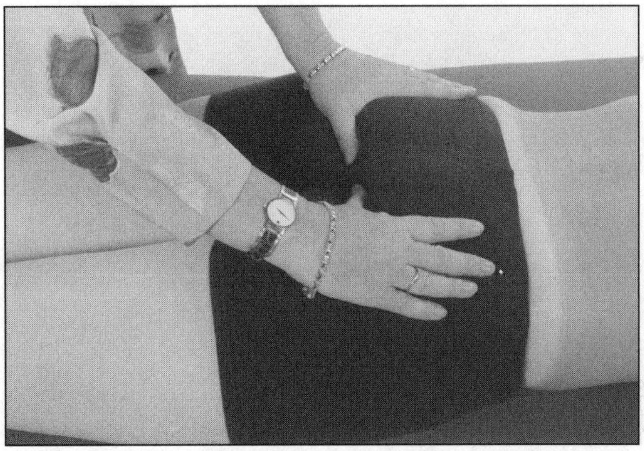

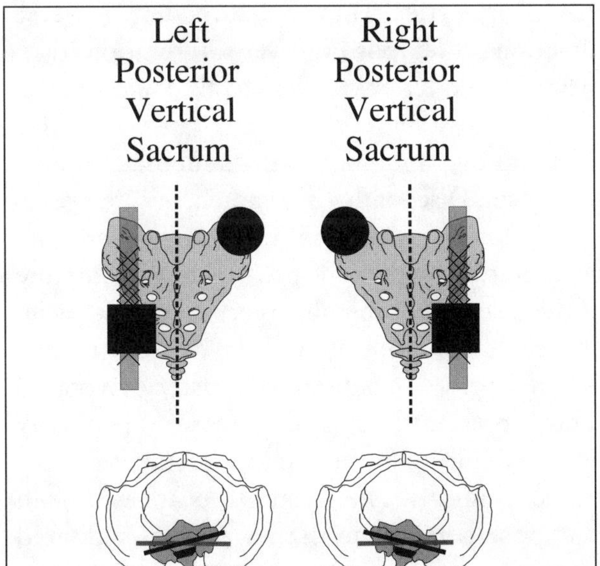

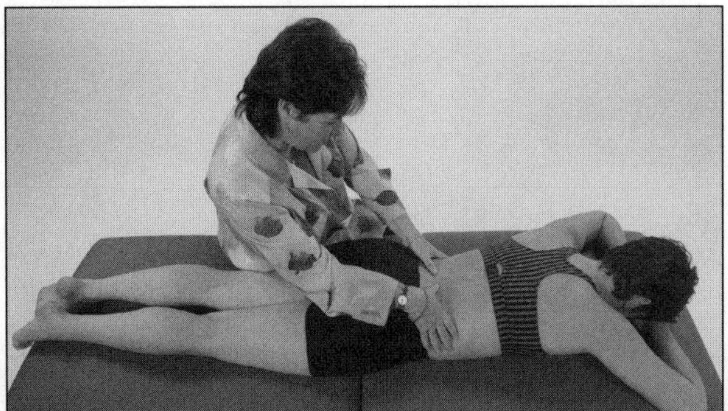

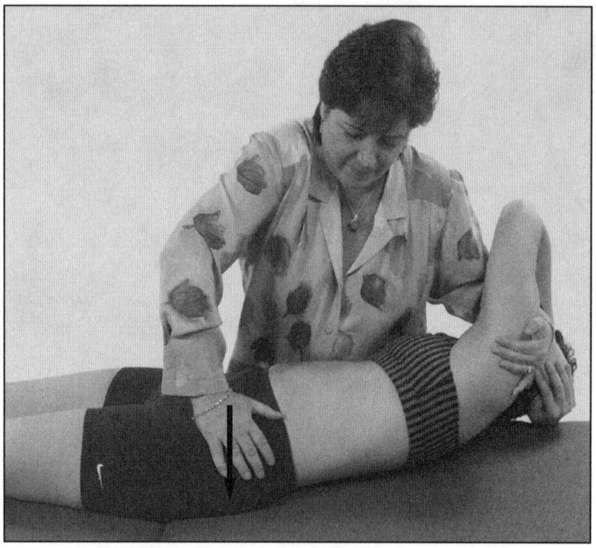

Treatment

1. Patient is supine.
2. Therapist stands on opposite (left) side of dysfunctional right sacroiliac joint.
3. Patient's hands clasped are behind head.
4. Therapist's upper right hand is placed through the patient's right arm space. Therapist utilizes right arm (cephalad arm) to rotate patient's spine to the left toward therapist, away from dysfunctional right side (anterior rotation without side bending).
5. Rotate patient to barrier of sacroiliac joint (palpate at joint).
6. Therapist places left hand (caudal hand) over right ASIS of dysfunctional side.
7. Therapist pushes posteromedial with patient's exhalation. Take up slack, perform high velocity/low amplitude/low force thrust of ilium posteromedial, to reposition ilium on the sacrum (although the sacrum is maintained in left anterior rotation during the technique).

Note: This picture is for a posterior vertical rotation to the right.

Eighty-five percent of posterior vertical rotations are left sacroiliac joint dysfunctions, i.e., posterior vertical rotation to the left. Reverse direction for left posterior vertical rotation treatment. The posterior vertical sacral rotation is possibly the worst biomechanical dysfunction. Severe pain and hypomobility of the lumbosacral junction occur. Discopathy is common. Severe hydrostatic pressure abnormalities cause severe disc degeneration, with predisposition toward rupture of the disc. Often spinal surgery gives brief relief of symptoms; more often the symptoms return. When a person has bilateral posterior vertical sacral rotations, this dysfunction probably occurred during more than one traumatic experience. Clinically, the sacrum appears as though it is posterior to the ilia. A ridge is palpated between ilia and sacrum; sacrum borders are posterior to ilia borders.

Anterior Vertical Rotation of Sacrum

Horizontal plane/vertical axis

Note: This dysfunction often causes hip joint pain and pelvic bowel signs, including bladder incontinence and dysmenorrhea.

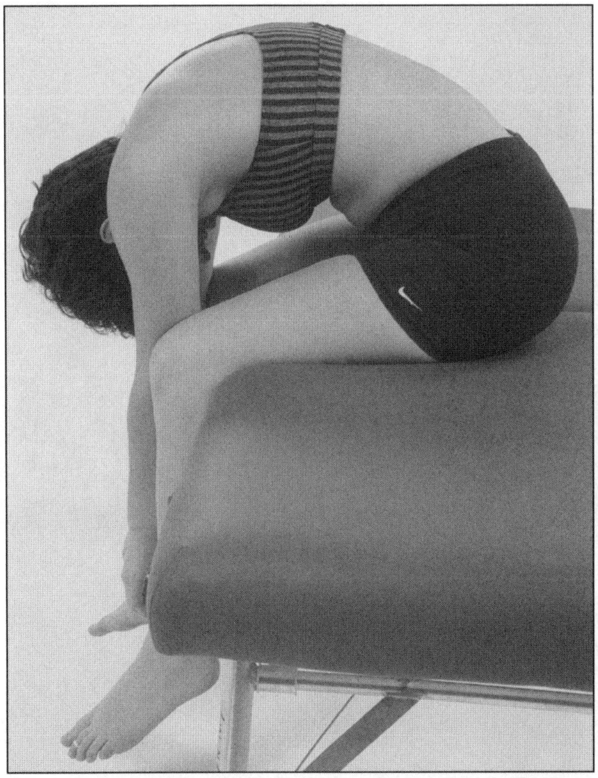

Differential Diagnosis:
Right Anterior Vertical Rotation of Sacrum

This means the right sacroiliac joint is dysfunctional, but the sacrum is rotated to the left.

1. Diagnose in sitting flexion.
2. Limitation of lumbosacral flexion.
3. Mobility tests: Ipsilateral sitting flexion test and squish test are positive.
4. Asymmetry of anatomical landmarks.
5. Ipsilateral (right) sacral sulcus is deeper.
6. Opposite (left) ILA is posterior but not inferior.
7. Ipsilateral (right) lateral sacral border is anterior (opposite left lateral sacral border appears posterior).
8. Ipsilateral sacrotuberous ligament tension.
9. Follow dynamic movement of sacrum: increased asymmetry during flexion.

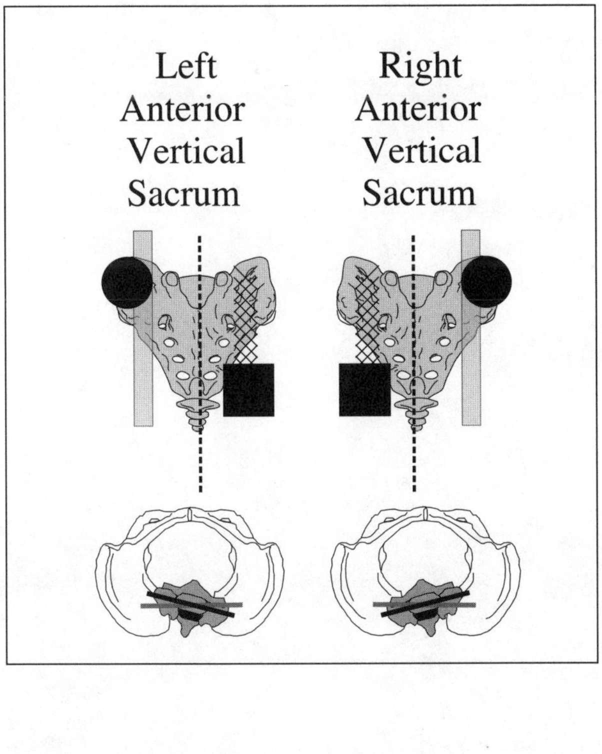

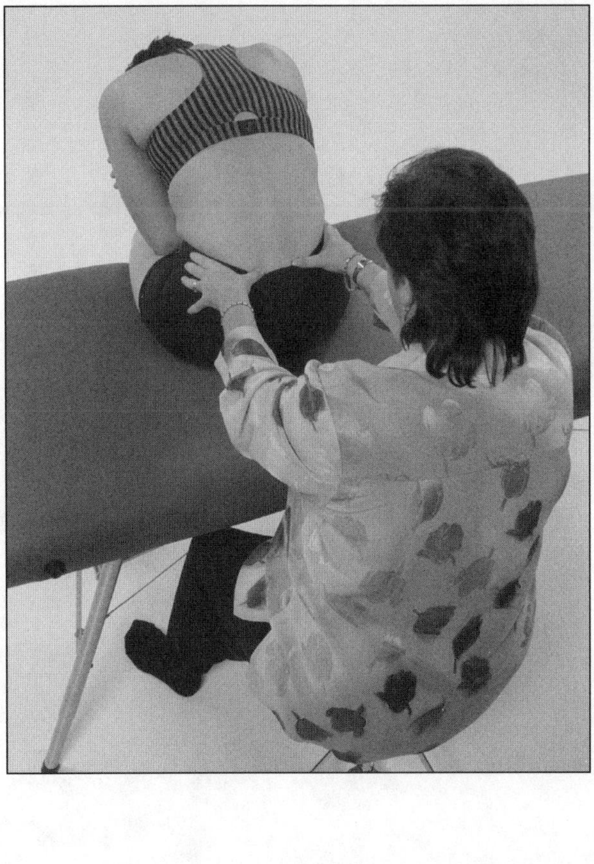

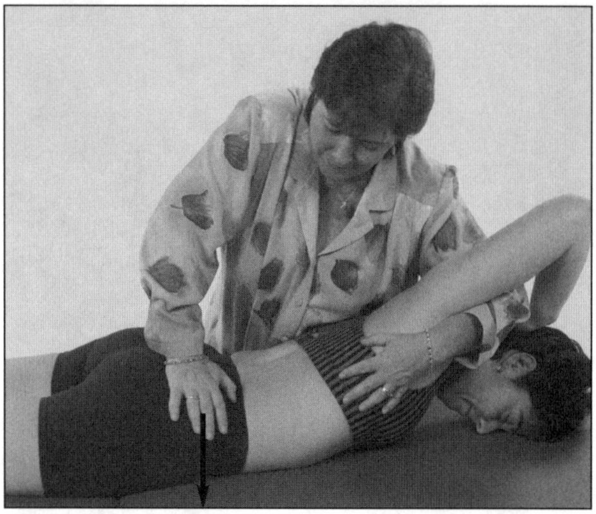

Treatment

1. Patient is prone.
2. Therapist stands on opposite (left) side of dysfunctional right sacroiliac joint.
3. Abduct 15 degrees and externally rotate ipsilateral (right) leg.
4. Patient's hands are clasped behind head; therapist's left hand (cephalad hand) moves between patient's right elbow space.
5. Therapist uses left arm (cephalad arm) to rotate patient's spine posterior to the right toward therapist (posterior rotation without side bending).
6. Therapist pushes anterolateral onto ipsilateral (right) posterior ilium with exhalation. Take up slack. (Therapist places caudal right hand over ipsilateral right PSIS and posterior iliac margin of dysfunctional right side.)
7. Therapist performs high velocity/low amplitude/low force (5 lbs) thrust of right ilium in an anterolateral direction while the sacrum is maintained in right posterior rotation.

Often, it is necessary to perform myofascial release, direct or indirect, on the ipsilateral iliolumbar ligament, sacrotuberous ligament, and sacrospinous ligament in order to attain a complete correction.

The Descended Sacrum

The descended sacrum is a clinical phenomenon that occurs secondary to head-on injuries, such as in flexion-acceleration and flexion-deceleration injuries in motor vehicle accidents. Assessment of this dysfunction requires advanced palpatory skills and diagnostic training.

Objective Findings

1. Positive Myofascial Mapping of the sacrotuberous ligaments and the lower sacral plexus nerve roots
2. Positive Myofascial Mapping of the medial iliolumbar ligaments
3. Positive Myofascial Mapping of the sacroiliac joints
4. Positive local listening (Barral) from the sacrum to the frontal and/or parietal regions of the cranium.
5. Normal inferior glide of the sacrum at the sacroiliac joint.
6. Hypomobile/limited superior glide of the sacrum at the sacroiliac joint

A bilateral descended sacrum is much more common than a unilateral descended sacrum. The sacrococcygeal joint and coccyx are compromised. There is compression of the inferior sacral plexus nerve roots contributing to pain and paresthesia of the feet. Patients presenting with a descended sacrum often do not complain of low back pain. There is a high incidence of back stiffness, headaches, and other cranial symptoms. Many neurosurgeons, like Alf Brieg, have contributed to the literature defining spinal cord fibrosis. This author has documented evidence that indicates that the descended sacrum predisposes toward spinal cord fibrosis. Treatment is a high velocity/low amplitude/low force mobilization/manipulation in a superior direction. Each joint (right and left) is treated separately.

The most significant problems that result from a descended sacrum are not back stiffness, general decreased spinal flexibility, pain and paresthesia of feet and ankles, and headaches. Rather, the descended sacrum causes brain dysfunction. Because the dura mater inserts at S2, the dural sheaths of the nerve roots as well as the

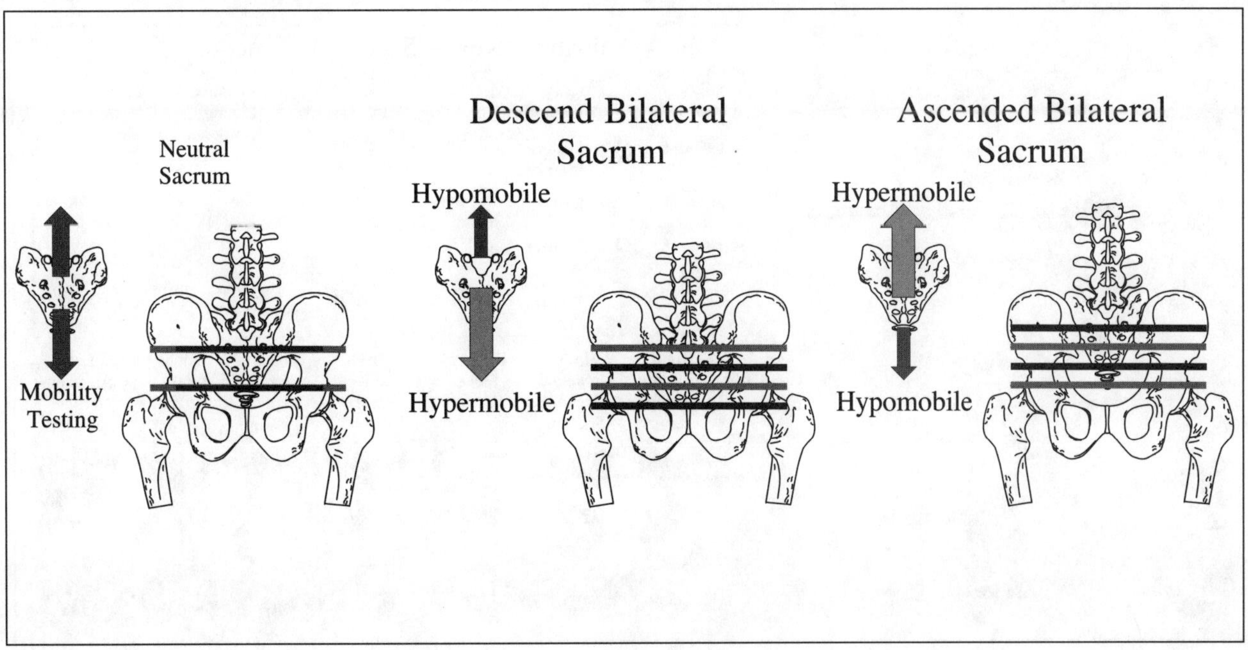

envelope of the brain are compromised. Also, because the sacral plexus is compromised in this dysfunction, with a direct line of tension to the brain stem and brain, multiple secondary and other functional impairments will occur secondary to this dysfunction.

Always mobilize the hypomobile accessory sacroiliac joint dysfunction before correcting a descended sacrum; otherwise treatment reactions may be intense.

Note: Because a descended sacrum results from cranial trauma, the therapist should have training in Cranial Therapy before attempting to treat this dysfunction.

The descended sacrum for the sacroiliac joint is similar to the downslip for the iliosacral joint. It is in the category of subluxations. This dysfunction can inhibit the correction of other sacroiliac joint and lumbosacral junction dysfunctions.

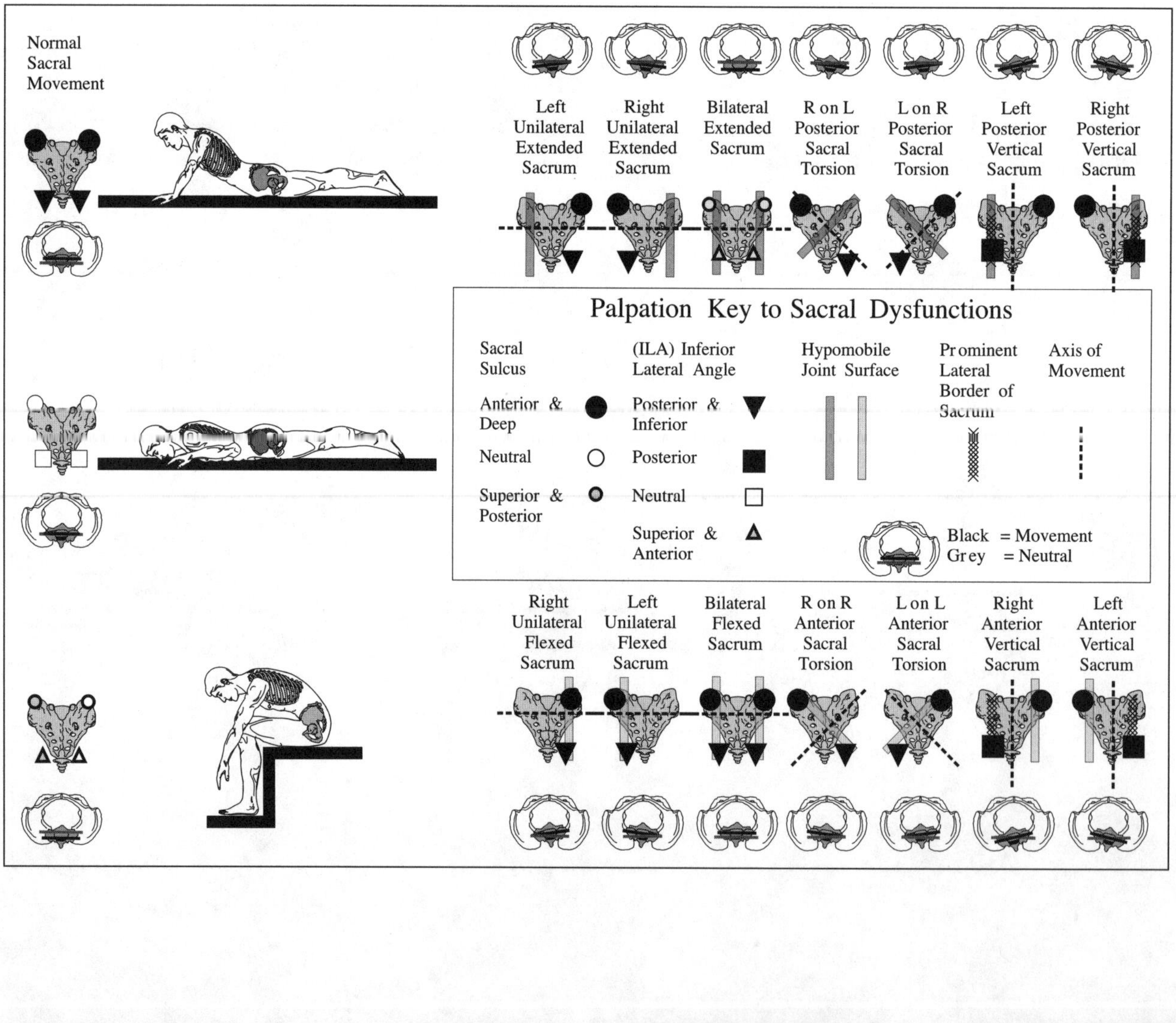

CHAPTER 16

THE LUMBOSACRAL JUNCTION: SACRAL TORSIONS

The Hypothetical Oblique Axes of Motion

An axis is hypothetical. A left oblique axis is a vector from superior left through inferior right. A right oblique axis is a vector from superior right through inferior left.

Image: (1) when the sacrum moves in an anterior direction, sacral base glides anterior in relation to L5; the vector for the axis of motion is anterior to sacrum; (2) when the sacrum moves in a posterior direction, sacral base glides posterior in relation to L5; the vector for the axis of motion is posterior to sacrum.

Sacral torsions are primary rotational dysfunctions of the lumbosacral junction. The sacroiliac joint will be compromised.

Name the Torsion

When the right sacral sulcus is deeper than the left sulcus, the sacrum is rotated left. If the sacrum is moving on an oblique axis, the movement is a torsion, rather than a rotation.

1. Deep right sacral sulcus: left sacral torsion
2. Deep left sacral sulcus: right sacral torsion

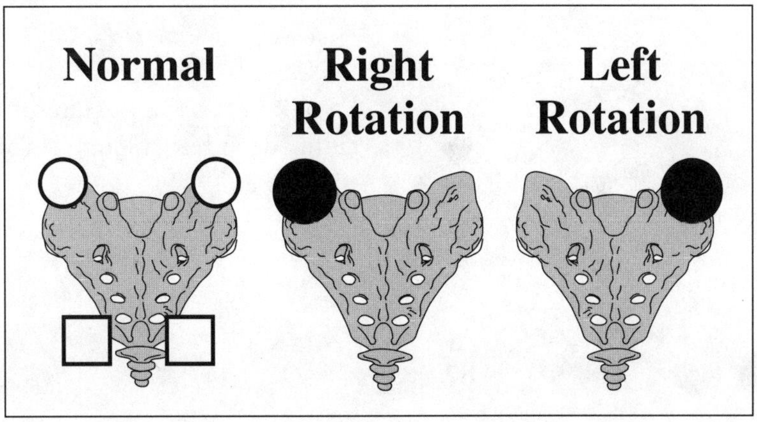

Name the Axis

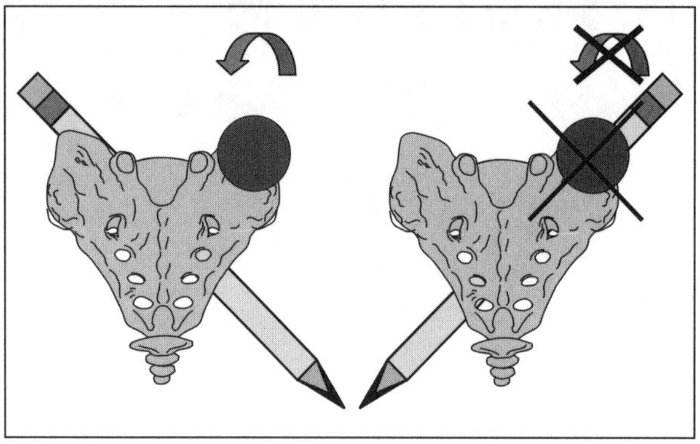

Step 1: When the sacrum torques in an anterior direction to the left (deep right sulcus), it torques around a left oblique axis. (Exercise: Take a pencil/stick and a piece of paper cut like the shape of a sacrum. Use the pencil/stick as an axis. Place the axis anterior to the sacrum. Rotate the sacrum anterior to the left on a left oblique axis. This is possible. Then attempt to rotate the sacrum anterior to the left on a right oblique axis. This is not possible.)

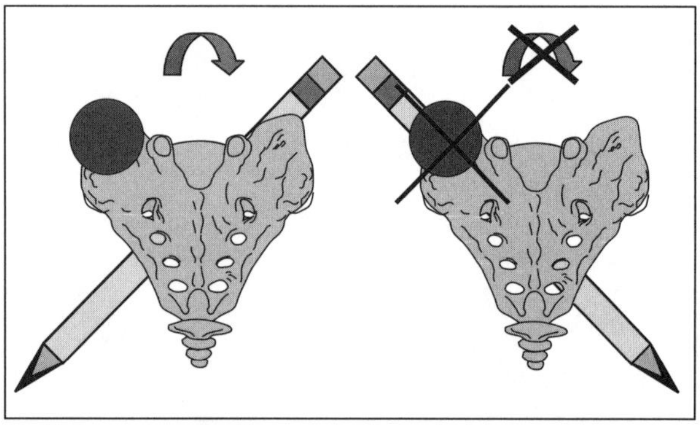

Step 2: When the sacrum torques anterior to the right (deep left sulcus), it torques around a right oblique axis.

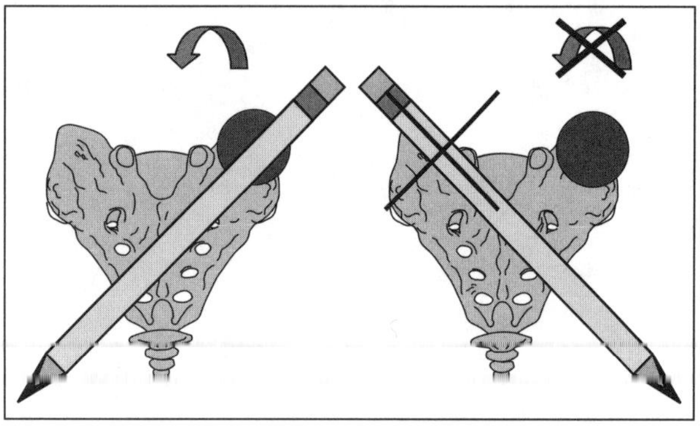

Step 3: When the sacrum torques posterior to the left (deep right sulcus), it torques around a right oblique axis. (Exercise: Take a pencil/stick and a piece of paper cut like the shape of a sacrum. Use the pencil/stick as an axis. Place the axis posterior to the sacrum. Rotate the sacrum posterior to the left on a right oblique axis. This is possible. Then attempt to rotate the sacrum posterior to the left on a left oblique axis. This is not possible.)

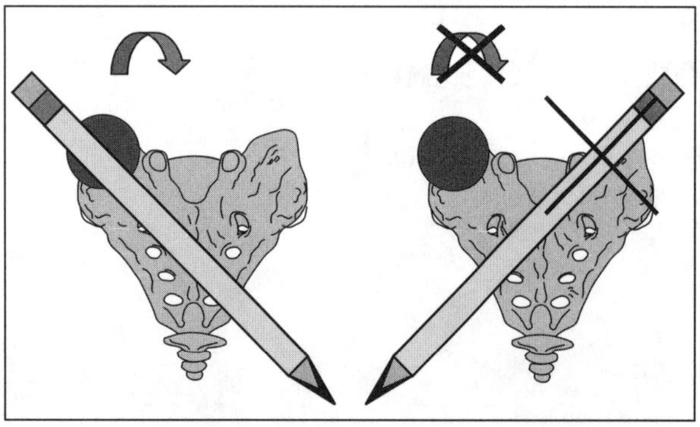

Step 4: When the sacrum torques posterior to the right (deep left sulcus), it torques around a left oblique axis.

Anterior Torsions Diagnosed in Flexion

Left Sacral Torsion on a Left Oblique Axis (Anterior Torsion)

The right sacral sulcus is deeper because the sacrum is anteriorly torqued (rotated) toward the left. The sacrum (S1) will present Type II movement. Rationale: Anterior glide of the sacral base is S1, out of AP neutral, in flexion; therefore, rotation and side bending are to the same side. If the sacrum is rotated (torqued) left, it is side bent left. To side bend left, the sacrum will move caudal on the left vertical limb, and then posterior on the left horizontal limb (similar to a left unilateral flexion of sacrum). Therefore, the left ILA will be inferior and posterior.

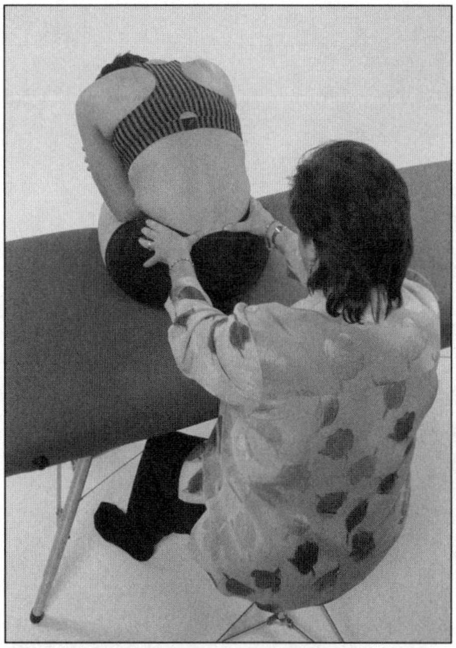

Left sacral torsion on a left oblique axis (LOL)
Deep right sulcus with inferior and posterior left ILA

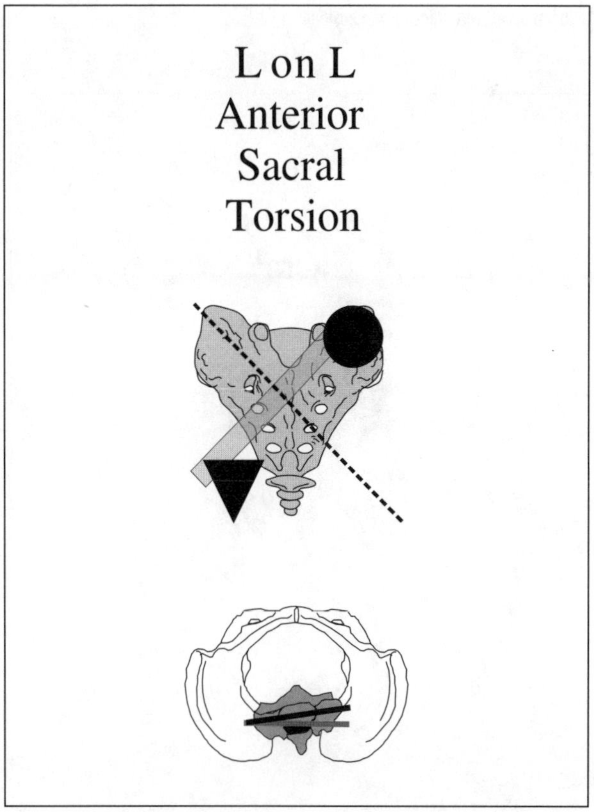

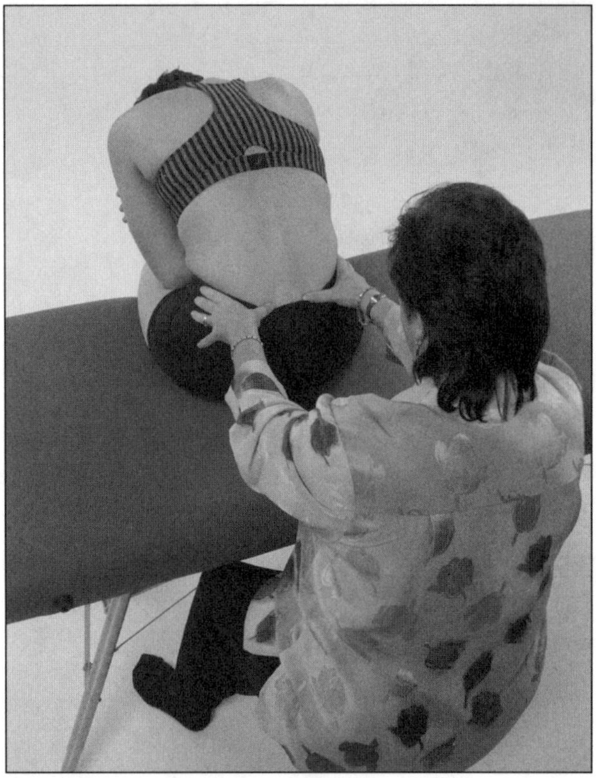

Right sacral torsion on a right oblique axis (ROR)
Deep left sulcus with inferior and posterior right ILA

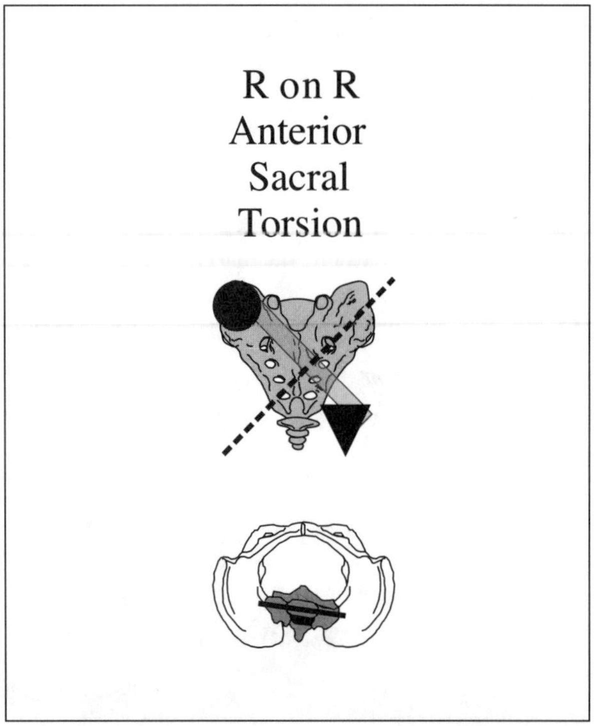

Right Sacral Torsion on a Right Oblique Axis (Anterior Torsion)

The left sacral sulcus is deeper because the sacrum is torqued (rotated) anterior toward the right. The sacrum (S1) will present Type II movement. Rationale: Anterior glide of sacral base is S1, out of AP neutral, in flexion, so that rotation and side bending are to the same side. If sacrum is rotated (torqued) right, it is sidebent right. To sidebend right, the sacrum will move caudal on the right vertical limb, and posterior on the right horizontal limb. Therefore, the right ILA will be inferior and posterior.

Posterior Torsions Diagnosed in Extension

Left Sacral Torsion on a Right Oblique Axis (Posterior Torsion)

The right sacral sulcus is deeper because the sacrum is torqued (rotated) posterior toward the left. The sacrum (S1) will present Type II movement (posterior glide of sacral base is S1 out of AP neutral, in extension) so that rotation and side bending are to the same side. If the sacrum is rotated left, it is also side bent left. To side bend left, the sacrum will move caudal on the left vertical limb and posterior on the left horizontal limb. Therefore, the left ILA will be inferior and posterior. The L5 vertebra will be rotated and side bent to the right, with a compromise of the right lumbosacral junction space. The L5 right transverse process will be prominent. This is a severe dysfunction in rotation for the left L5 disc space.

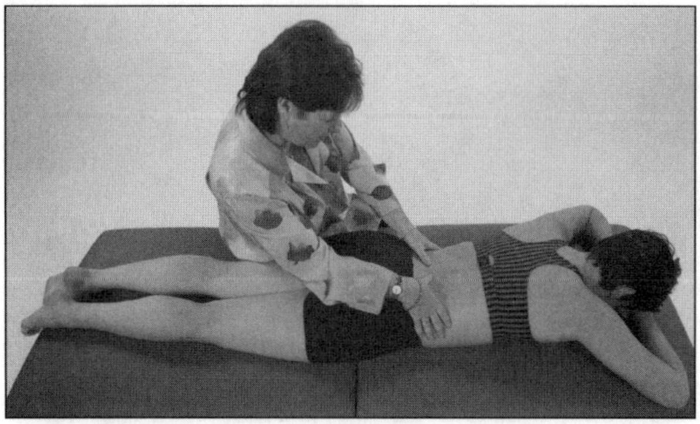

Left sacral torsion on a right oblique axis (LOR)
Deep right sulcus with inferior and posterior left ILA

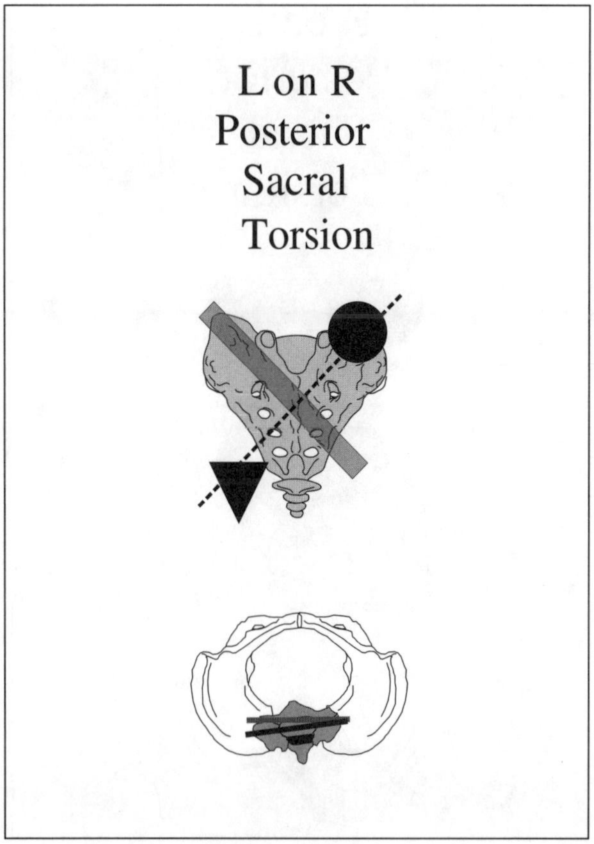

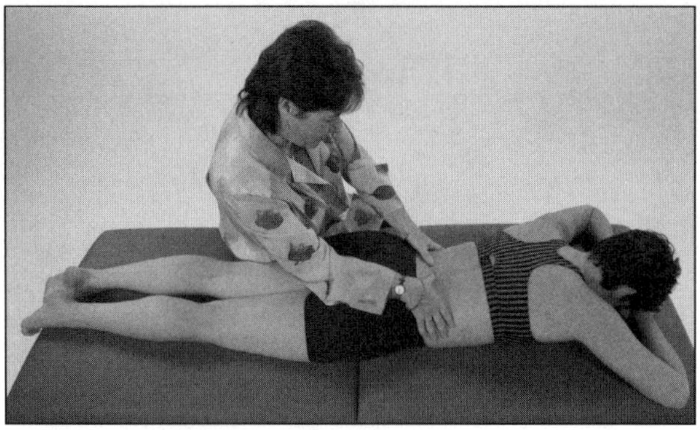

Right sacral torsion on a left oblique axis (ROL)
Deep left sulcus with inferior and posterior right ILA

Right Sacral Torsion on a Left Oblique Axis (Posterior Torsion)

The left sacral sulcus is deeper because the sacrum is torqued (rotated) posterior toward the right. The sacrum (S1) will present Type II movement (posterior glide of sacral base is S1 out of AP neutral, into extension), so that rotation and side bending are to the same side. If the sacrum is rotated right, it is side bent right. To side bend right, the sacrum will move caudal on the right vertical limb and posterior on the right horizontal limb. Therefore, the right ILA will be inferior and posterior. The L5 vertebra will be rotated and side bent to the left, with a compromise of the left lumbosacral junction space. The L5 left transverse process will be prominent. This is a severe dysfunction in rotation for the right L5 disc space.

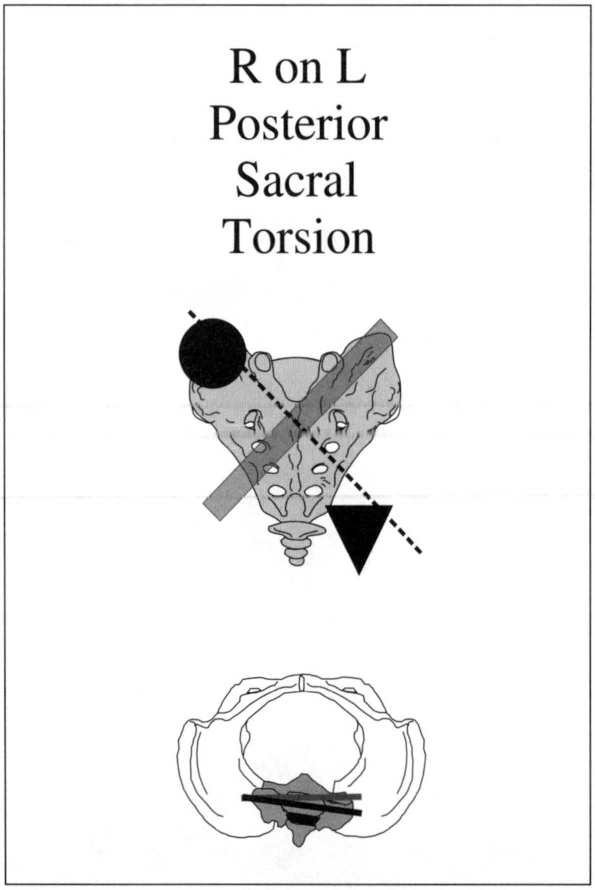

Incidence of Torsions

Approximately 85 percent of anterior torsions are left sacral torsion on a left oblique axis (LOL); 85 percent of posterior torsions are left sacral torsions on a right oblique axis (LOR). This incidence correlates with the incidence of right-handed dominant people in the population, which is 85 percent.

Differential Diagnosis of Torsions

Although the process of "naming" a torsion may appear complicated, the following step-by-step procedure will help:

1. Look for the anatomic landmarks. Ask question number one: Which sacral sulcus is deeper (more anterior)? Then ask question number two: Which inferior lateral angle (ILA) is more inferior as well as more posterior?
2. If these landmarks are present (i.e., there is a sacral sulcus on one side that is deeper than the other side, *and* there is an ILA that is *both* more inferior *plus* more posterior than the other side), *then* ask the next question: Are the anatomic landmarks on

opposite sides of the body? In other words, the deeper sacral sulcus is on one side of the body, while the more inferior and posterior ILA is on the other side of the body. If this is the situation, there is a sacral torsion present.

3. The next step is how to name the torsion. There are two questions in naming a sacral torsion
   Which direction is the torsion?
   On which oblique axis is the torsion?
4. Before naming the torsion, it is necessary to know:
   Is the sacral torsion an anterior torsion?
   Is the sacral torsion a posterior torsion?
   • If the anatomic landmarks described above are found in sitting flexion, the sacral torsion is an anterior sacral torsion.
   • If the anatomical landmarks described above are found in prone lumbar extension, the sacral torsion is a posterior sacral torsion.
5. Now the torsion can be "named."
   Part one: In which direction is the sacral torsion? Right or left?
   • If the right sacral sulcus is deeper (more anterior) than the left sacral sulcus, the sacral torsion is named left sacral torsion. Why? Because the sacrum is rotated toward the left side.
   • If the left sacral sulcus is deeper (more anterior) than the right sacral sulcus, the sacral torsion is named right sacral torsion. Why? Because the sacrum is rotated toward the right side.
   Part two: On which oblique axis is the sacrum torqued? Right or left?
   • Why is the axis an oblique axis? When there is uniplanar rotation, the axis is a vertical axis. When there is a 3-planar torsion, the axis is an oblique axis. Because the sacrum does not rotate around a vertical axis, all motion with rotation is 3-planar. This means that if the sacrum is in rotation, it must flex or extend as well as side bend. In other words, motions of the sacrum include movement on a sagittal plane, plus motion on a coronal plane, plus the rotation movement on a transverse plane. Torsions are 3-planar movements. Torsions occur around an oblique axis.
   • According to the directions above, determine: Is the sacral torsion an anterior torsion, or a posterior torsion? If the sacral torsion is found in sitting flexion, it is an anterior sacral torsion. If the sacral torsion is found in prone lumbar extension, it is a posterior torsion. If the sacral torsion is an anterior torsion, name the oblique axis *the same* as the name of the sacral torsion. If the sacral torsion is a posterior torsion, name the oblique axis *opposite* to the name of the sacral torsion. This means:
   · If there is an anterior sacral torsion, there are two possibilities in naming the oblique axis: right sacral torsion on a right oblique axis (ROR); left sacral torsion on a left oblique axis (LOL).
   · If there is a posterior sacral torsion, there are two possibilities for naming the oblique axis: right sacral torsion on a left oblique axis (ROL); left sacral torsion on a right oblique axis (LOR).

Approximately 85 percent of anterior sacral torsions are left sacral torsions on a left oblique axis (LOL). Anatomic landmarks: The right sacral sulcus is deeper (more anterior) than the left sacral sulcus. The left ILA is more inferior and more posterior than the right ILA. This sacral torsion is found in sitting flexion.

Approximately 15 percent of anterior sacral torsions are right sacral torsions on a right oblique axis (ROR). Anatomic landmarks: The left sacral sulcus is deeper (more anterior) than the right sacral sulcus. The right ILA is more inferior and more posterior than the left ILA. This sacral torsion is found in sitting flexion.

Approximately 85 percent of posterior sacral torsions are left sacral torsions on a right oblique axis (LOR). Anatomic landmarks: The right sacral sulcus is deeper (more anterior) than the left sacral sulcus. The left ILA is more inferior and more posterior than the right ILA. This sacral torsion is found in prone lumbar extension.

Approximately 15 percent of posterior sacral torsions are right sacral torsions on a left oblique axis (ROL). Anatomic landmarks: The left sacral sulcus is deeper (more anterior) than the right sacral sulcus. The right ILA is more inferior and more posterior than the left ILA. This sacral torsion is found in prone lumbar extension.

The "quick steps" in the following table may help discover how to name the sacral torsion.

Step 1: Anatomic landmarks are on opposite sides of the body. The deep sulcus is on the opposite side of the inferior/posterior ILA.

Step 2: Name the torsion (right/left) with a deep sulcus. If the sulcus is deep on the right side, there is a left sacral torsion. If the sulcus is deep on the left side, there is right sacral torsion.

Step 3:	Find landmarks in flexion. Anterior torsion		Find landmarks in extension. Posterior torsion	
Step 4:	Anterior torsion Name axis same as name of torsion (right/left).		Posterior torsion Name axis opposite name of torsion (right/left).	
	Right sacral torsion on right oblique axis	Left sacral torsion on left oblique axis	Right sacral torsion on left oblique axis	Left sacral torsion on right oblique axis

Example: Naming a Left Sacral Torsion on a Left Oblique Axis

Step 1: Anatomic landmarks are on opposite sides of the body.
The deep sacral sulcus is on the right side. The inferior/posterior ILA is on the left side.

Step 2: The right sacral sulcus is deeper than the left sacral sulcus.
Name the torsion: left sacral torsion.

Step 3: These anatomic landmarks were found in sitting lumbar flexion.
Name the torsion: anterior sacral torsion.

Step 4: Because this is an anterior sacral torsion, the oblique axis is named the same as the sacral torsion. This is a left sacral torsion.
Name the axis: left oblique axis.

Step 5. Name the torsion: left sacral torsion on a left oblique axis.

Example: Naming a Left Sacral Torsion on a Right Oblique Axis

Step 1: Anatomic landmarks are on opposite sides of the body:
The deep sacral sulcus is on the right side. The inferior/posterior ILA is on the left side.

Step 2: The right sacral sulcus is deeper than the left sacral sulcus.
Name the torsion: left sacral torsion.

Step 3: These anatomic landmarks were found in prone lumbar extension.
Name the torsion: posterior sacral torsion.

Step 4: Because this is a posterior sacral torsion, the oblique axis is named the opposite of the sacral torsion. This is a left sacral torsion.
Name the axis: right oblique axis.

Step 5. Name the torsion: left sacral torsion on a right oblique axis.

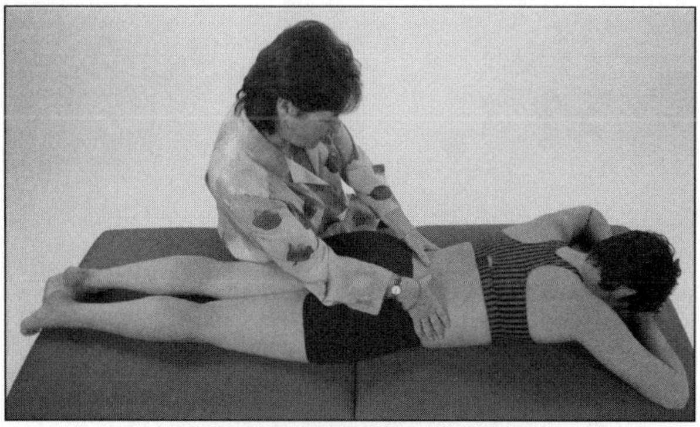

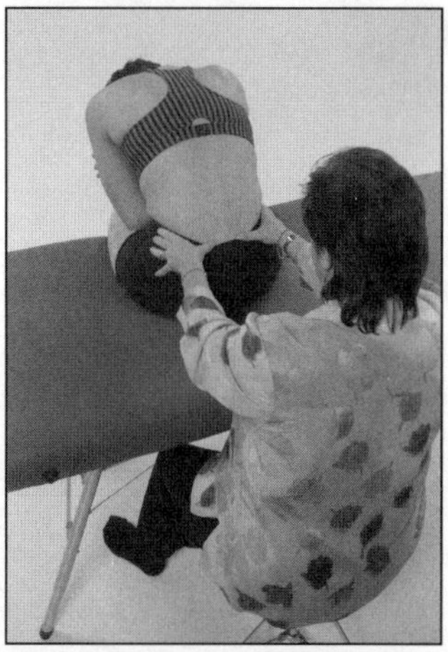

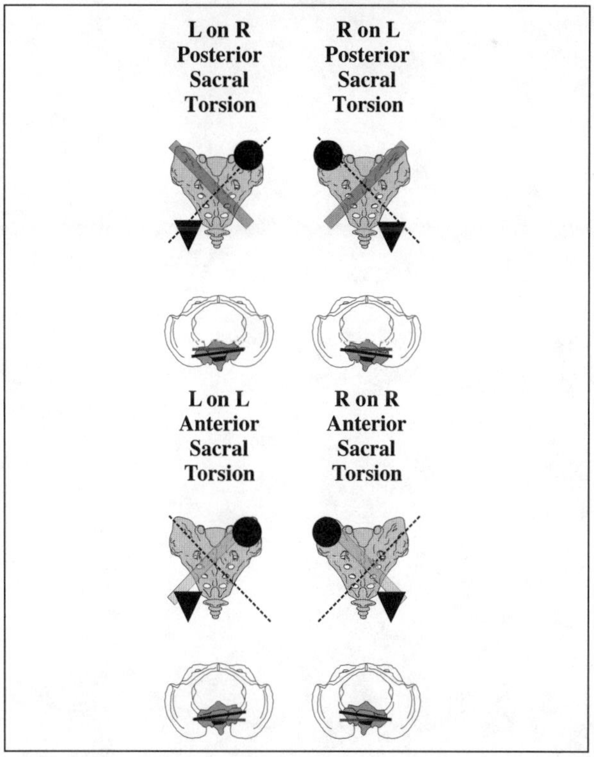

Why Name the Torsion?

The significance of naming the sacral torsion using the instructions above relates to treatment. There are three basic instructions for treatment of a sacral torsion:

1. Have the client lie on the side of the involved axis. This means if the sacral torsion is on a right oblique axis, the client lies on the right side for therapy to correct this problem. If the sacral torsion is on a left oblique axis, the client lies on the left side for therapy to correct this problem.
2. If the sacral torsion is found in sitting lumber flexion, it is an anterior torsion. Because Muscle Energy and "Beyond" Technique is a direct technique, moving into the barrier, the sacrum requires extension for correction of this problem.
3. If the sacral torsion is found in prone lumbar extension, it is a posterior torsion. Because Muscle Energy and "Beyond" Technique is a direct technique, moving into the barrier, the sacrum requires flexion for correction of this problem.

In chapters that describe therapy for correction of these problems, note that there is reciprocal motion between L5 and S1. When L5 flexes, the sacrum extends. When L5 extends, the sacrum flexes. There are two possible scenarios:

1. When a sacral torsion is found in sitting lumbar flexion, L5 is flexed. S1 is supposed to extend. When S1 does not extend, it is because S1 is stuck flexed. Thus, the sacral torsion is an anterior sacral torsion, because S1 is stuck flexed.
2. When a sacral torsion is found in prone lumbar extension, L5 is extended. S1 is supposed to flex. When S1 does not flex, it is because S1 is stuck extended. Thus, the sacral torsion is a posterior sacral torsion, because S1 is stuck extended.

CHAPTER 17

MUSCLE ENERGY AND "BEYOND" TECHNIQUE FOR THE LUMBOSACRAL JUNCTION

It is important to remember, in regard to sacral dysfunctions, that there is an auricular-shaped articular surface of the sacrum, and there is reciprocal movement at the lumbosacral junction.

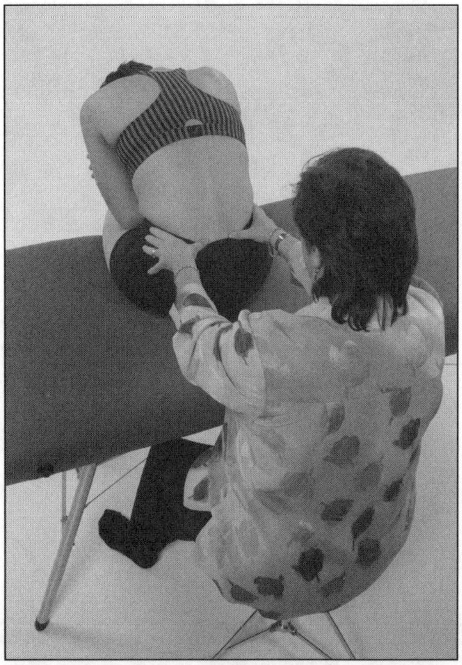

Anterior Sacral Torsion

Left Sacral Torsion on a Left Oblique Axis (LOL)

1. Diagnose in sitting flexion
2. Limitation of Lumbosacral flexion
3. Asymmetry of anatomical landmarks
4. Landmarks are asymmetrical on opposite side (right/left)
5. Sacral sulcus deep on right (sacral base rotated/torsioned left)
6. ILA posterior and inferior on left

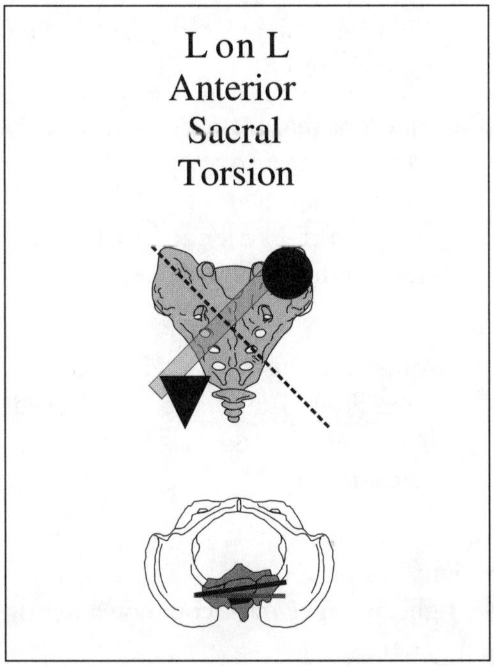

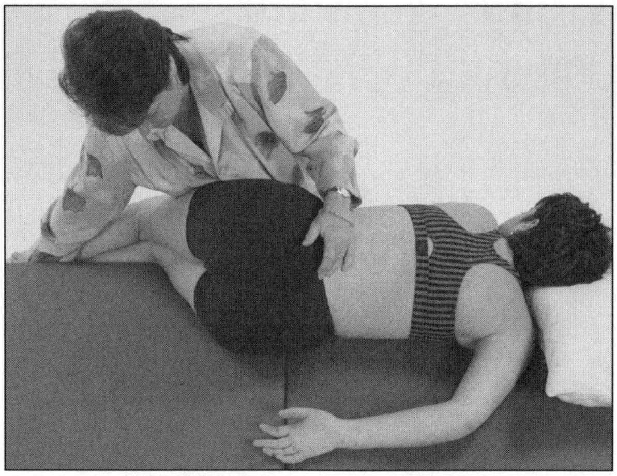

Rationale

1. Sacral base stuck anterior (flexed) on the right (Type II)
2. Sacral base rotated anteriorly to the left around a left oblique axis (deep right sulcus)
3. Sacrum side bent left (inferior and posterior left ILA)

Treatment

Have patient lie on side of involved axis (left side down) with lower arm behind back. Find in flexion, treat in flexion (i.e., the flexed fetal-like position). Knees are just off edge of bed. Have top leg positioned comfortably on the medial aspect of the bottom leg.

First barrier (sagittal plane):

Flex knees and hips to chest and palpate for barrier at L5/S1 interspace (flexion of L5).

Second barrier (transverse plane):

Rotate shoulders to left (chest gets close to table) with right upper arm off the bed and reaching to the floor. Rotate down the kinetic chain until palpation of barrier at L5/S1 interspace (rotation of L5).

Third barrier (coronal plane):

Drop both feet together to the floor to side bend up the kinetic chain until palpation of barrier at L5/S1 interspace (side bending of L5).

Resist:

Isometrics: Push both feet to ceiling, or to the floor, 6 to 10 seconds. Repeat 3 to 5 times (5 to 6 times for the neurologic patient).

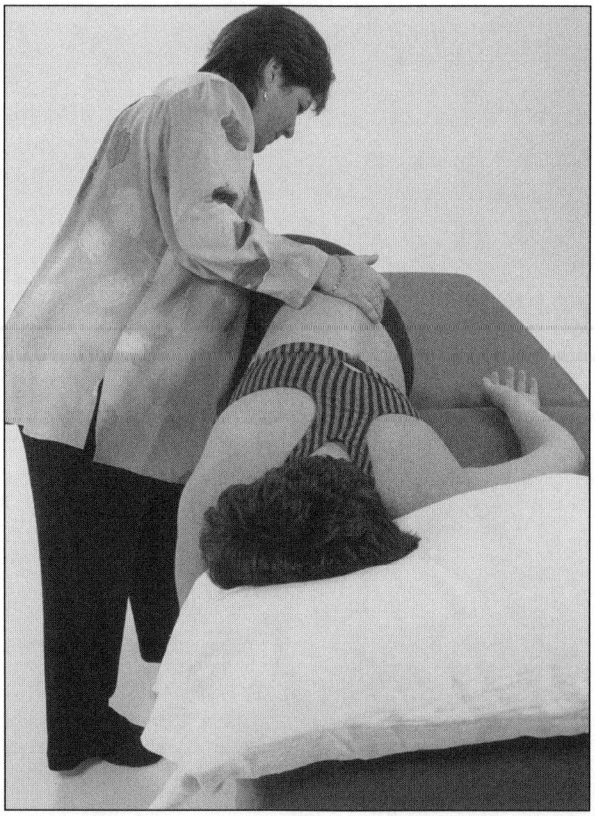

An ROR is treated in the same manner, except the patient lies on the side of the involved axis, i.e., the right side. The most common anterior torsions are LOL.

Note: Patient is lying on the left side, therefore axis involved is the left oblique axis. Patient is in the flexed position, therefore treatment is for an anterior torsion. This picture shows treatment of LOL (85 percent of anterior torsions are LOL, 15 percent are ROR).

Application: Rationale for Treatment of Anterior Sacral Torsions

Remember the reciprocal movement at L5/S1. When L5 flexes by flexing the hips and knees, the sacral base glides posterior. (When L5 extends, the sacral base glides anterior.) This treatment technique positions L5 into the barriers, in order to facilitate a correction of S1, by using the tendency of reciprocal movement; i.e., if L5 is positioned in one direction, the sacral base will tend to move in the opposite direction because of reciprocal movement. Because the sacral base is anterior in an anterior torsion, L5 is flexed during the technique in order to facilitate posterior glide of the sacral base. Because in LOL, the sacral base is stuck rotated left, L5 is stuck rotated right. Therefore, L5 is rotated to the left with the technique, in order to facilitate right rotation of the sacral base. Rotation and side bending occur to the same side. Therefore, L5 is stuck side bent right and must be side bent left with the technique in order to facilitate right side bending of the sacrum.

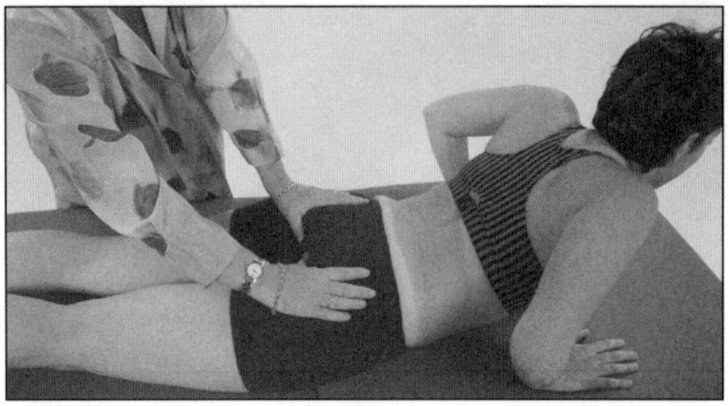

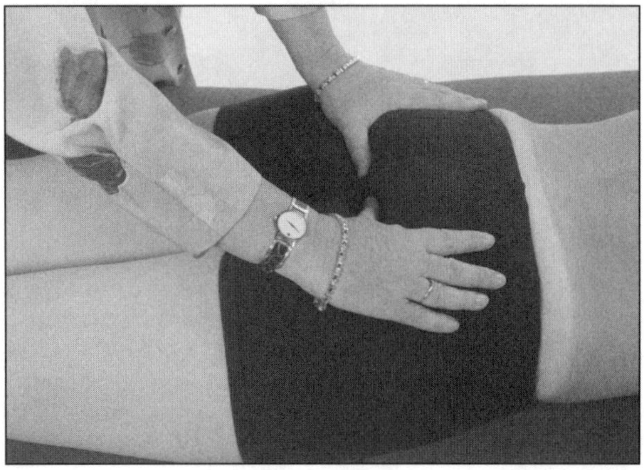

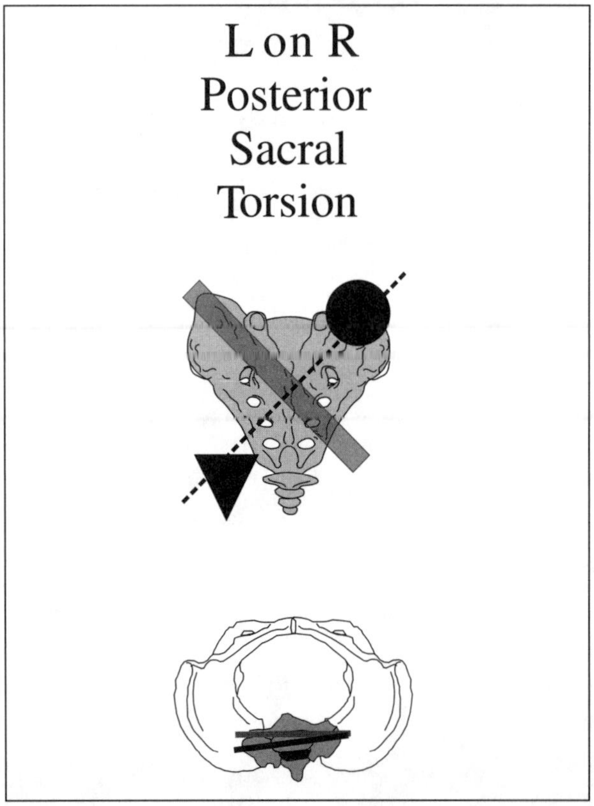

Posterior Sacral Torsion

Diagnosis: Left Sacral Torsion on a Right Oblique Axis (LOR)

1. Diagnose in prone extension
2. Limitation of lumbosacral extension
3. Asymmetry of anatomical landmarks
4. Landmarks are asymmetrical on opposite sides (right/left)
5. Sacral sulcus deep on right (sacral base rotated/torsioned left)
6. ILA posterior and inferior on left

Rationale

1. Sacral base posterior (extended) on left (Type II)
2. Sacral base rotated posterior to the left around a right oblique axis (deep right sulcus)
3. Sacrum side bent left (inferior and posterior left ILA)

Treatment

Have patient lie on side of involved axis (right side down). Find in extension, treat in extension.

First barrier (sagittal plane):
Extend lower leg until palpation of barrier at L5/S1 interspace (extension of L5).

Second barrier (transverse plane):
Rotate trunk to the left (backward rotation) down the kinetic chain to L5 until palpation of barrier at L5/S1 interspace (rotation of L5).

Third barrier (coronal plane):
Upper leg has hip flexed 45 degrees. Knee is on bed with distal leg off bed. Knee is straight. Lift leg toward the ceiling, only an inch to attain side bending barrier. Palpate L5/S1 interspace (side bending of L5).

Resist: Isometric resistance 6 to 10 seconds. Resist abduction plus external rotation. Direct patient to push foot to ceiling and turn toes to ceiling (piriformis). Repeat 3 to 5 times (5 to 6 times for the neurologic patient).

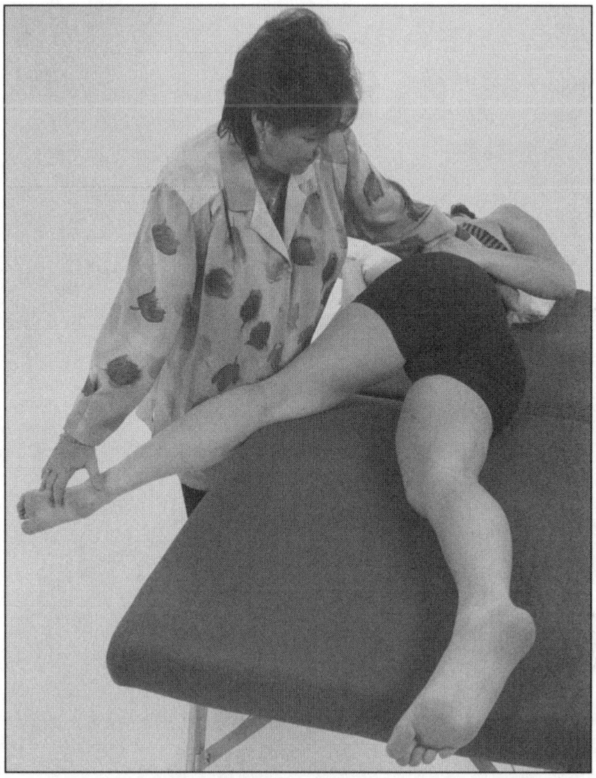

Note: Patient is lying on the right side; therefore, the involved axis is the right axis. The patient is lying in the extended position; therefore, treatment is for a posterior torsion. This picture presents treatment for an LOR.

An ROL is treated in the same manner, except the patient lies on the side of the involved axis: on the left side. Eighty-five percent of posterior torsions are LOR.

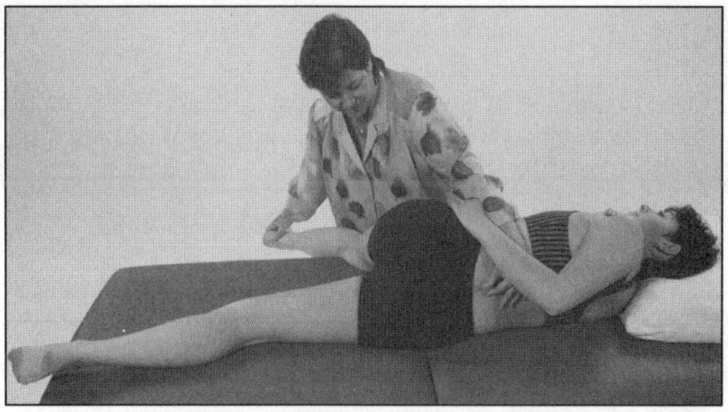

Application: Rationale for Treatment of Posterior Sacral Torsions

Remember the reciprocal movement at L5/S1. (When L5 flexes, the sacral base glides posteriorly. When L5 extends, the sacral base glides anteriorly.) This treatment technique positions L5 into the barriers, in order to facilitate a correction of S1, by using the tendency of reciprocal movement; i.e., if L5 is positioned in one direction, the sacral base will tend to move in the opposite direction because of reciprocal movement. Because the sacral base is posterior in a posterior torsion, L5 is extended during the technique (by extending the lower leg) to facilitate anterior glide of the sacral base. Because in an LOR, the sacral base is stuck rotated left, L5 is stuck rotated right. L5 is rotated to the left with the technique, in order to facilitate right rotation of the sacral base. Rotation and side bending occur to the same side. Therefore, L5 is stuck side bent right, and must be side bent left with the technique in order to facilitate right side bending of the sacrum.

CHAPTER 18

THE REAL SCENARIO: HOW TO TREAT THE LUMBOSACRAL JUNCTION

In treating the lumbosacral junction, its full ranges of motion must be recognized. The author believes that normal ranges of motion are not recognized by health care practitioners. Most practitioners assume that the motion of L5/S1 is far less than it really is. Why? Because almost all clients, and almost all practitioners have restrictions!

How did such a situation occur? Low back pain is the most common orthopedic disorder. Statistics are amazing: 80 percent of all Americans will suffer from back pain sometime in their lives. When neck pain is considered, 100 percent of Americans will suffer from spinal pain. Signs always precede symptoms. A symptom is low back pain. A sign is the limitation of motion of the low back.

It is understood that the major problem of the low back is the lumbosacral junction. Eighty-five percent of spinal surgery for spinal pain occurs at the lumbosacral junction. Yet the typical health care practitioner does not know what normal movement is at this region!

If most Americans have limitations of motion at the lumbosacral junction, who is the person analyzed for normal motion? How can there be a recognition of normal motion at the L5/S1 junction? Only when the typical person is treated does the typical person have normal motion. At that time, the typical person becomes normal in that regard. The author discovered "normal" after thousands of clients were treated with Muscle Energy and "Beyond" Technique for spinal pain and restrictions.

Norms of lumbosacral junction motion, as defined by the author, are:

1. At least 50 degrees of lumbosacral flexion
2. At least 40 degrees of lumbosacral extension
3. At least 45 degrees of L5 rotation
4. At least 45 degrees of L5 side bending

In order to attain long-lasting elimination of low back pain, the practitioner should restore normal movement at L5/S1. When mobility is not restored at the lumbosacral junction, there may be a temporary decrease of low back pain, but there will probably be a recurrence.

There are other unique "normals" discovered by this author:

1. When normal pelvic joint biomechanics are restored, usually the person can flex sufficiently to touch the floor! In other words, lumbosacral flexion is not vital to functional physiologic spinal flexion. Rather, the pelvis is vital to functional physiologic spinal flexion.
2. When normal lumbosacral (L5 and S1) biomechanics are restored with Muscle Energy and "Beyond" Technique, at least 70 percent of all spinal motion is restored at the lumbar spine, the thoracic spine, and the cervical spine! In other words, cervical, thoracic, and lumbar biomechanics are not vital for functional spinal motion: extension, rotation, and side bending. Rather, L5/S1 extension (when L5 extends and S1 flexes) is vital for functional physiologic spinal extension, rotation, and side bending.

Usually, the patient/client comes to the office with varying intensity of low back pain. Start with the pelvic joint biomechanics, as described in this book. Begin with the pubes. Correct elevated and descended pubes. Then correct the accessory sacroiliac joints. Perhaps 20 percent of the total population will have accessory joints. Of those, 70 percent will be bilateral; 30 percent will be unilateral (right or left). If the accessory sacroiliac joints are difficult to mobilize, search for evidence of a pelvic compression reflex. When the PSIS's are mobile, look for flares (inflare and /or outflare). Then assess for upslips or downslips. Remember to use the longitudinal compression test and the longitudinal distraction test for upslips and downslips. Then correct rotation dysfunctions of ilium (anterior rotation and/or posterior rotation).

Once the pelvic joints are mobile, anatomic landmarks will be equal. When the anatomic landmarks are equal, yet a pelvic obliquity remains, assess for true anatomic leg length discrepancy. True anatomic leg length discrepancies are not common. A small percentage of clients have histories of severe fractures of the pelvis, femur, or tibia. Malunion could account for an anatomic difference in bone growth. This will occur when the fracture happened in childhood. Usually, the leg length discrepancy will be less than 1 cm in those cases. If the person is a neurologic client from birth, for example, cerebral palsy with hemiplegia, then there may be a true anatomic leg length discrepancy. Because the affected hemiplegic leg did not have equal and normal pressures in standing and ambulation, muscle pulls were not equal and normal. Therefore, bone growth was unequal. Thus in the population of neurologic clients who had the neurologic insult in childhood before final growth, it is common to discern some anatomic leg length discrepancy. But the majority of all orthopedic and non-neurologic clients from birth will not have true anatomic leg length discrepancies. There will be many functional leg length discrepancies when pelvic joint biomechanics are not corrected.

There are many discussions about shoe lift therapy. This author believes that shoe lifts are not required when there are functional leg length discrepancies. The best approach is to correct pelvic biomechanics. Otherwise, the person is locked into dysfunction with the shoe lift, and other secondary problems result.

Does the client have normal ranges of motion for a body with good pelvic joint biomechanics? Can the person touch the floor? Do not allow the buttocks to move in a posterior direction during spinal flexion! When assessing spinal flexion, only movement anterior to the center of gravity is assessed. Although it is normal to have spinal motion occur with hip motion, true spinal flexion restrictions will not be uncovered and found in that manner.

If there is evidence of good pelvic joint biomechanics, then progress to the sacrum! Correction of biomechanical dysfunction of the sacrum means: sacroiliac joints plus S1. Because S1 always moves relative to L5, correction of biomechanical dysfunction of sacrum means: sacroiliac joint, S1 (torsions), and L5.

Occasionally, in approximately 30 percent of all clients, there will be minimal to moderate restrictions of the spine when the sacrococcygeal joint is not corrected. In less than 10 percent of clients, there will be more significant restrictions of spinal motion if sacrococcygeal joint dysfunction is not corrected. This means 90 percent of all clients need the sacrum (sacroiliac joints plus L5) corrected before the sacrococcygeal joint is corrected.

The practitioner has a choice:

1. First assess lumbosacral extension in prone extension.
2. First assess lumbosacral flexion in sitting flexion.

Usually, clients are first assessed in lumbosacral extension, looking for limitations of L5/S1 extension. This means L5 extends while the sacrum flexes. Clients are next assessed in lumbosacral flexion, looking for limitations of L5/S1 flexion. This means L5 flexes while the sacrum extends. Why are clients first assessed in lumbosacral extension, rather than in lumbosacral flexion? Because 10 degrees of lumbosacral extension is required for supine, prone, standing, and ambulation. This way, the client can at least rest! Also, disc dysfunction of L5 occurs in lumbosacral extension. When motion of lumbosacral extension is increased, the neurologic signs of L5 discopathy can be eliminated. This is an important factor! Correct sacrum biomechanics sufficiently to restore some lumbosacral extension, and there will be spontaneous reduction of the L5 disc!

In summary, assess and diagnose in prone extension, in order to correct biomechanical dysfunction affecting lumbosacral extension first. Then L5 disc herniations can spontaneously reduce.

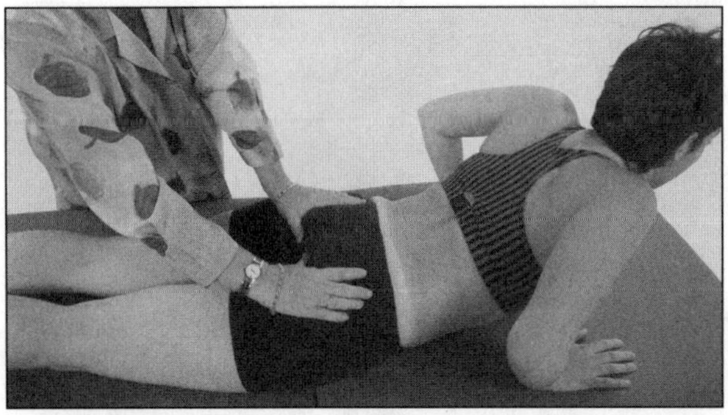

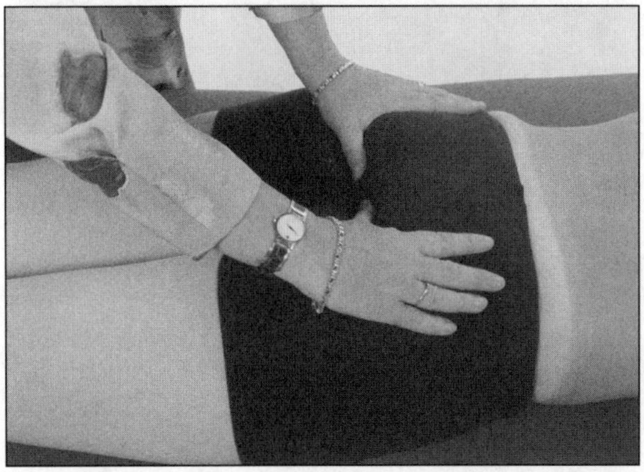

The Scenario: Treating the Lumbosacral Junction

The client is in prone. The client attempts to press up, leaving hips on the table. The practitioner "looks" at the anatomic landmarks. The practitioner finds the following: right deep sacral sulcus; right inferior posterior ILA.

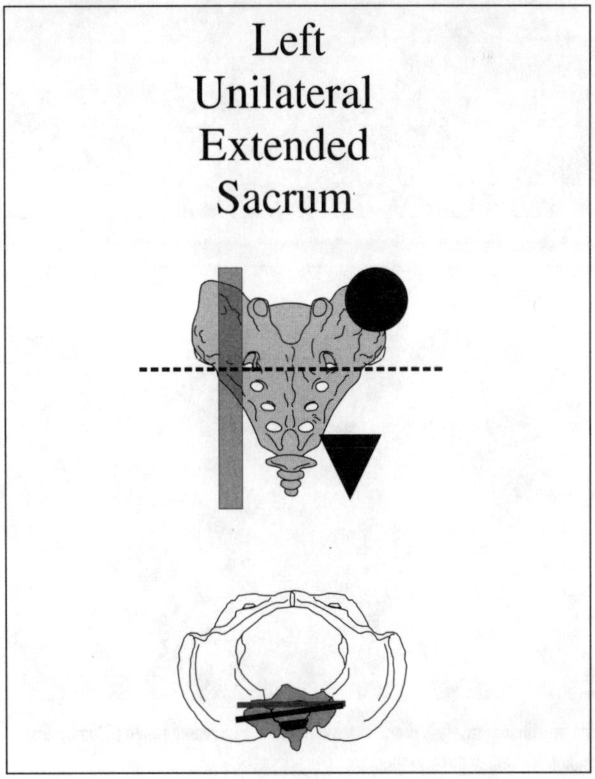

The practitioner corrects a left unilateral extended sacrum.

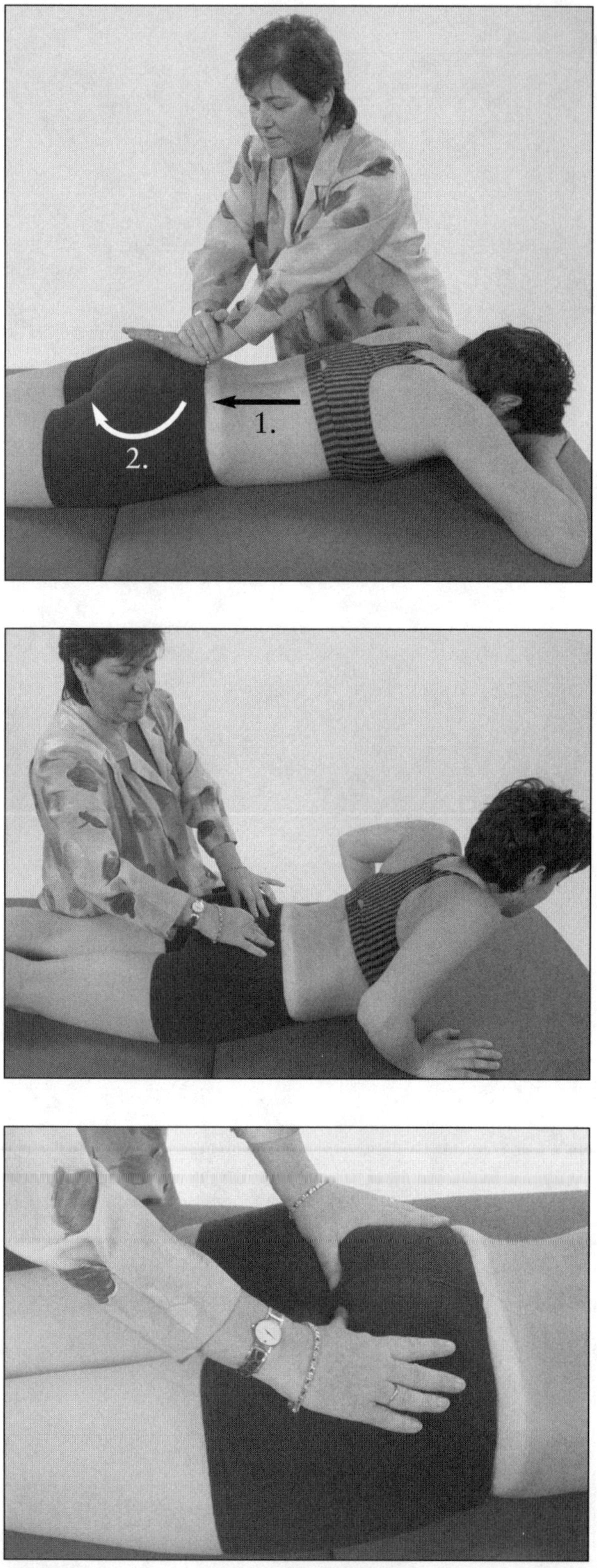

The practitioner assesses prone extension: There is increased range of lumbosacral motion and the anatomic landmarks are different.

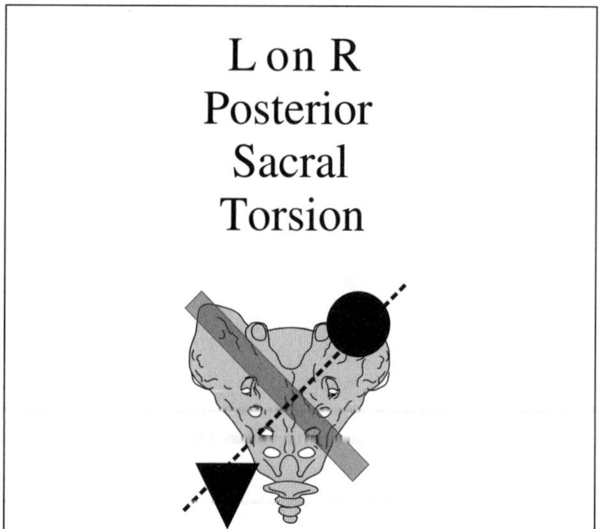

The practitioner finds the following: right deep sacral sulcus; left inferior posterior ILA.

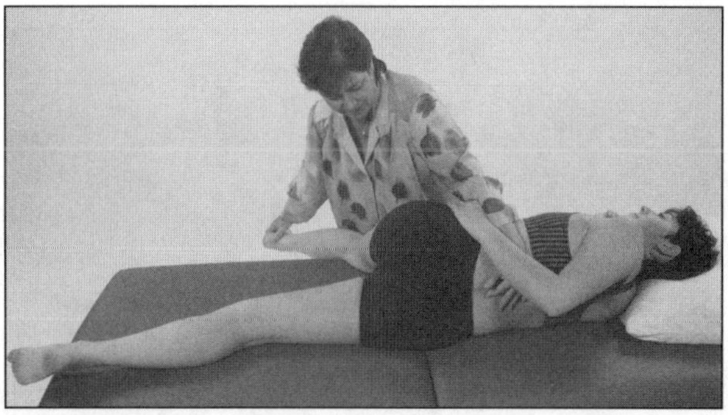

The practitioner corrects a posterior sacral torsion: left sacral torsion on a right oblique axis (LOR).

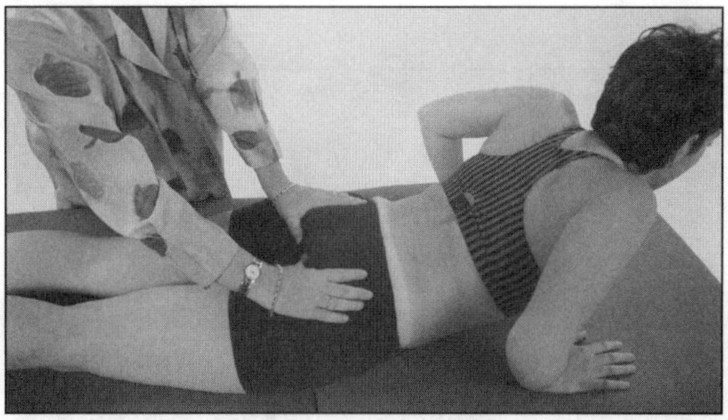

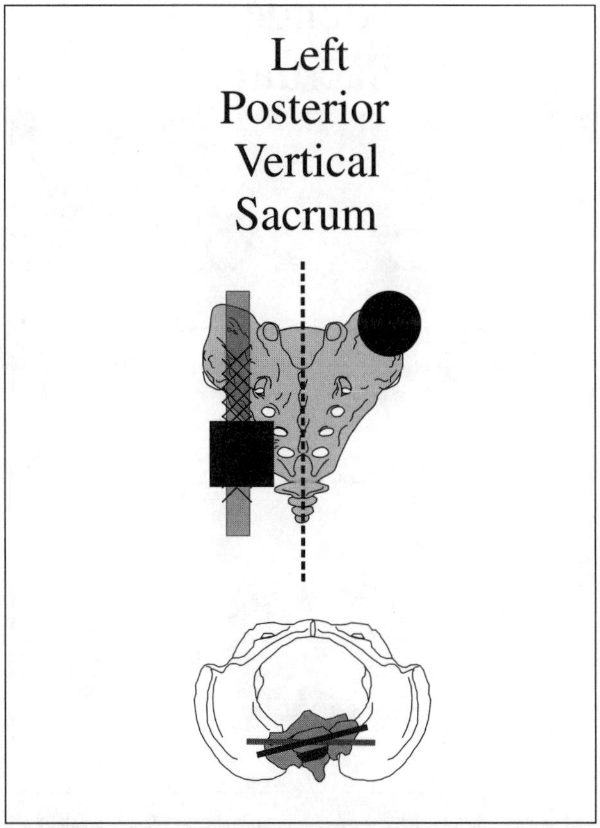

The practitioner finds the following: right deep sacral sulcus; no inferior posterior ILA; posterior prominence of left lateral sacral border

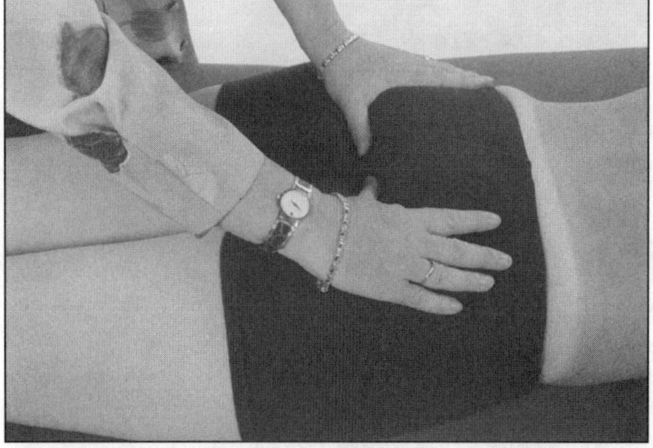

The practitioner assesses prone extension: There is increased range of lumbosacral motion and the anatomic landmarks are different.

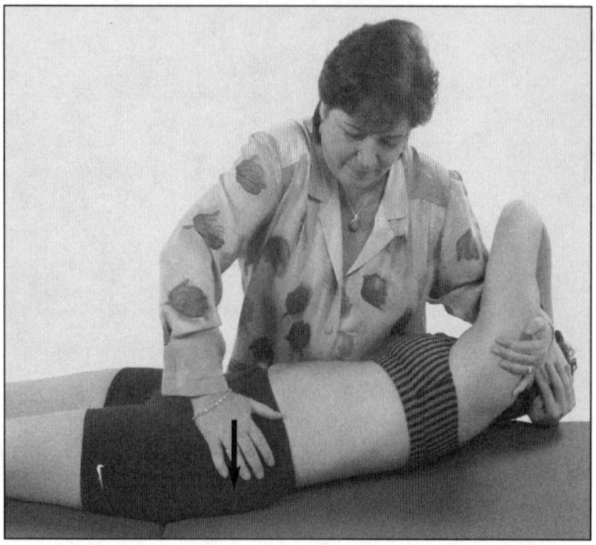

The practitioner corrects a posterior vertical rotation of the sacrum to the left.

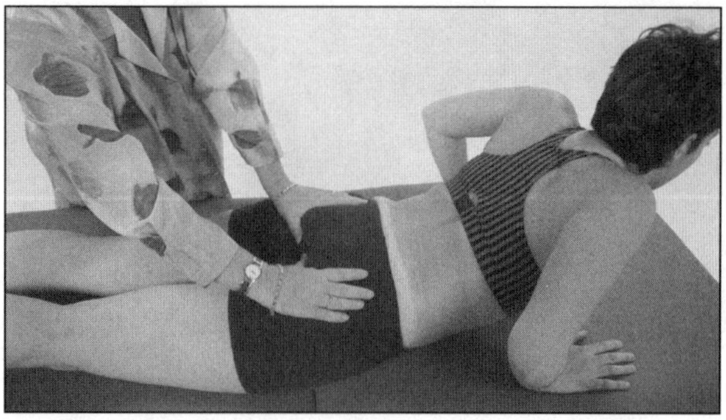

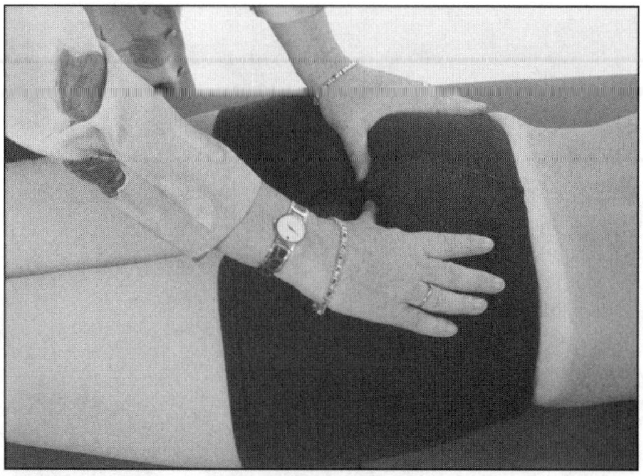

The practitioner assesses prone extension: there is increased range of lumbosacral motion and the anatomic landmarks are different.

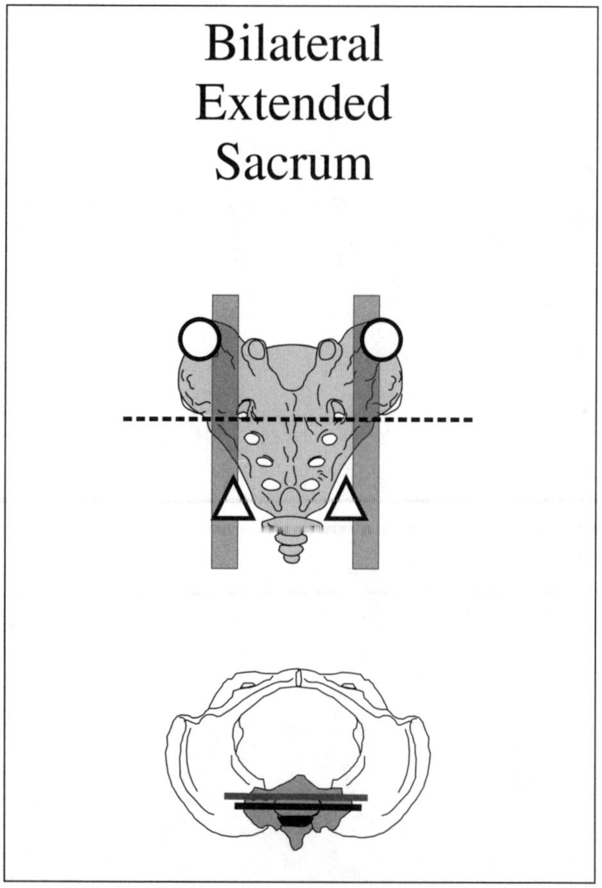

The practitioner finds the following: no deep sacral sulcii; no inferior posterior ILA; but there is no anterior glide of sacral base during lumbosacral extension.

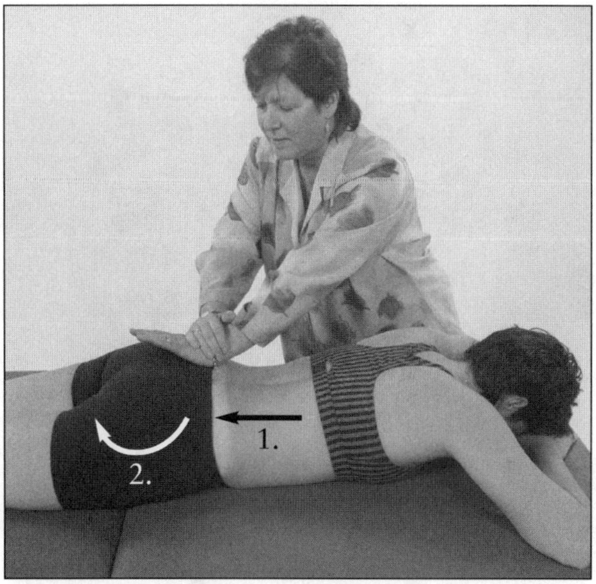

The practitioner corrects a bilateral extended sacrum.

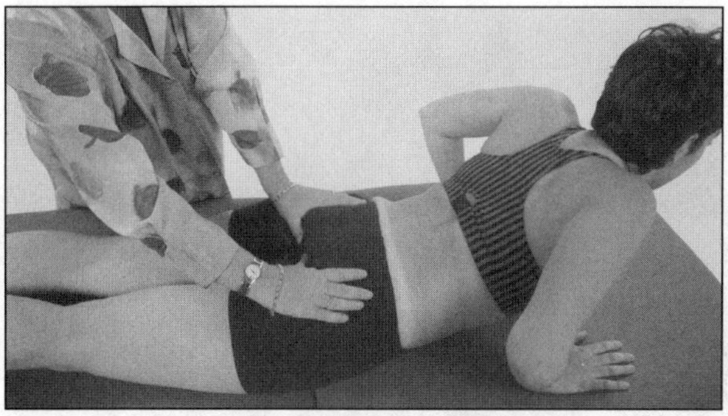

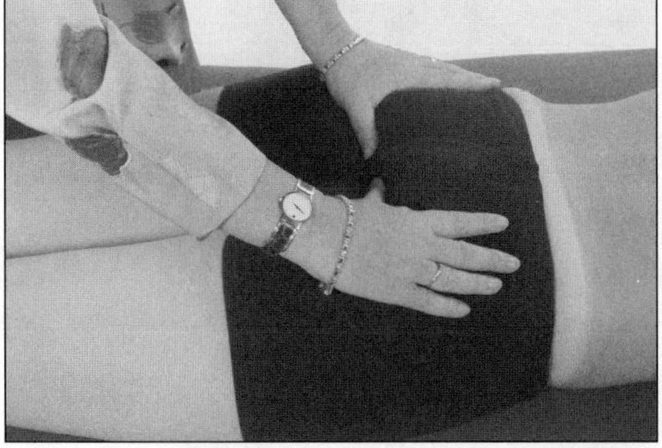

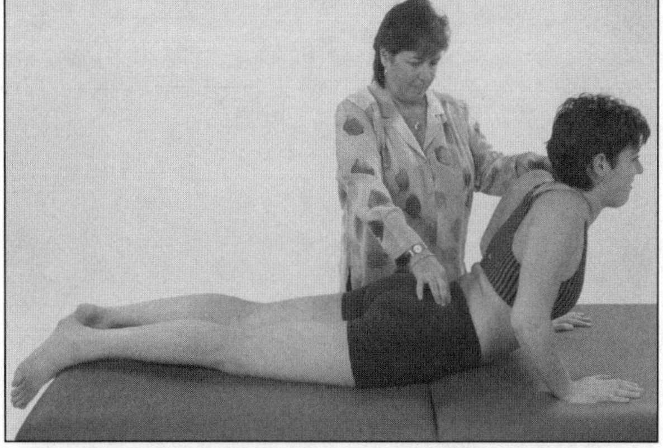

The practitioner assesses prone extension: There is increased range of lumbosacral motion and the anatomic landmarks are equal. There appears to be 40 degrees of potential lumbosacral junction extension.

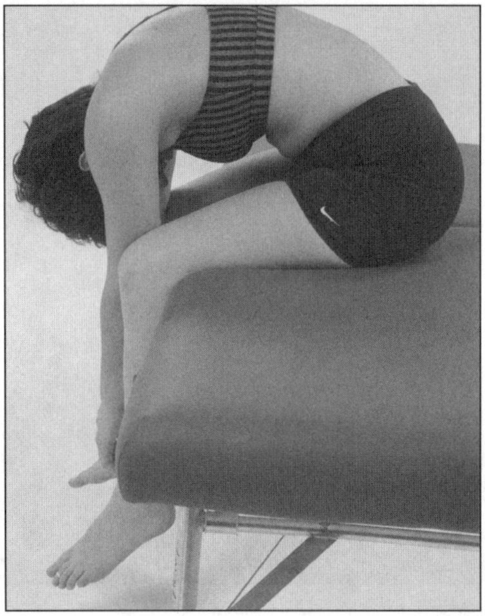

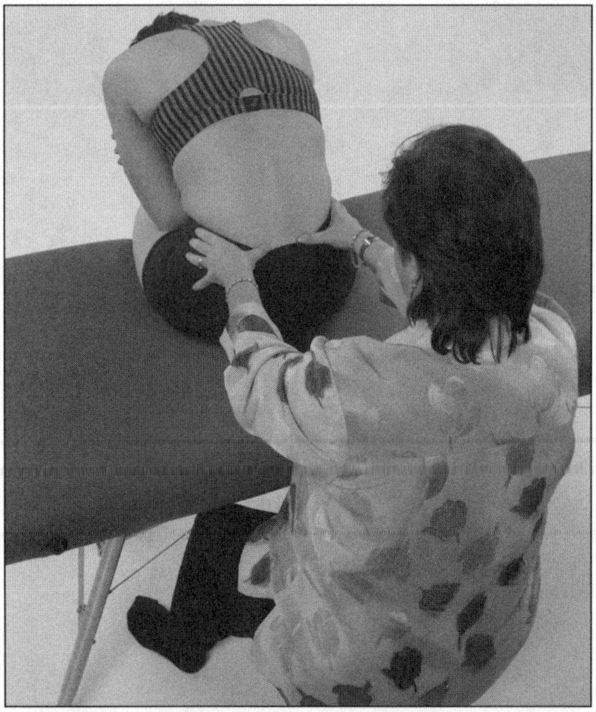

The practitioner determines that continued therapy will be focused at increasing lumbosacral flexion. For example, assess and diagnose in sitting flexion, in order to correct biomechanical dysfunction affecting lumbosacral flexion.

The client sits on the edge of the treatment table. The client tucks in the chin and slowly flexes down the kinetic chain. The practitioner, sitting on a stool behind the client, palpates the anatomic landmarks, searching for the biomechanical dysfunction that is causing limitations of lumbosacral flexion.

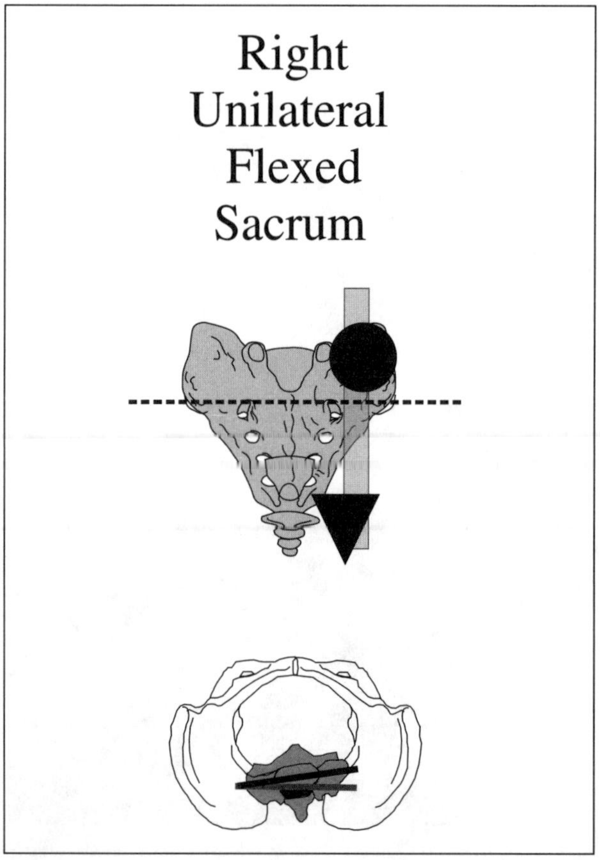

The practitioner finds the following landmarks: right deep sulcus; right inferior posterior ILA.

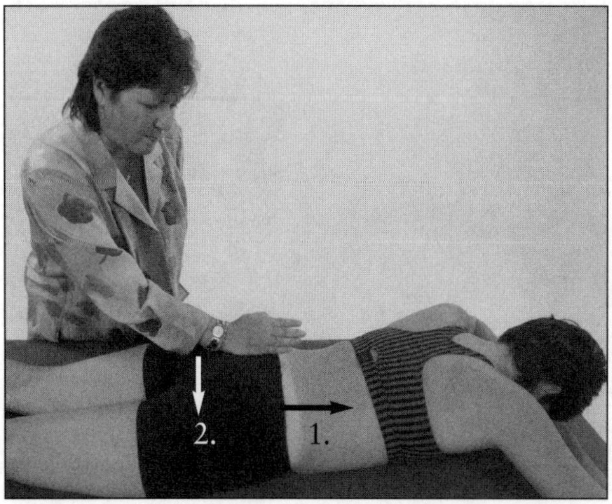

The practitioner treats a right unilateral flexed sacrum. The client can flex further down the kinetic chain, with greater lumbosacral flexion, and the landmarks are different.

The practitioner finds the following landmarks: right deep sulcus; left inferior posterior ILA.

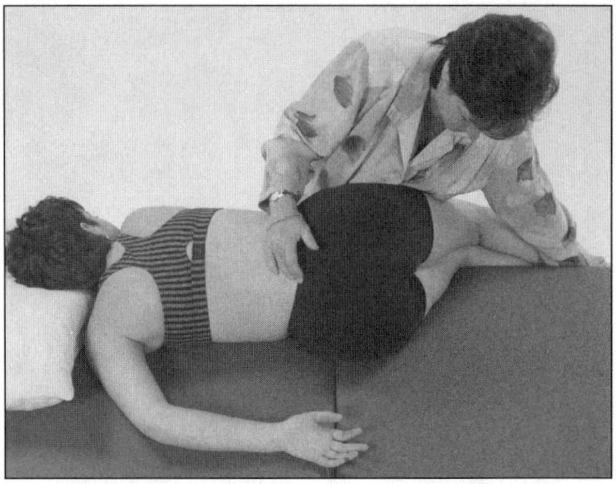

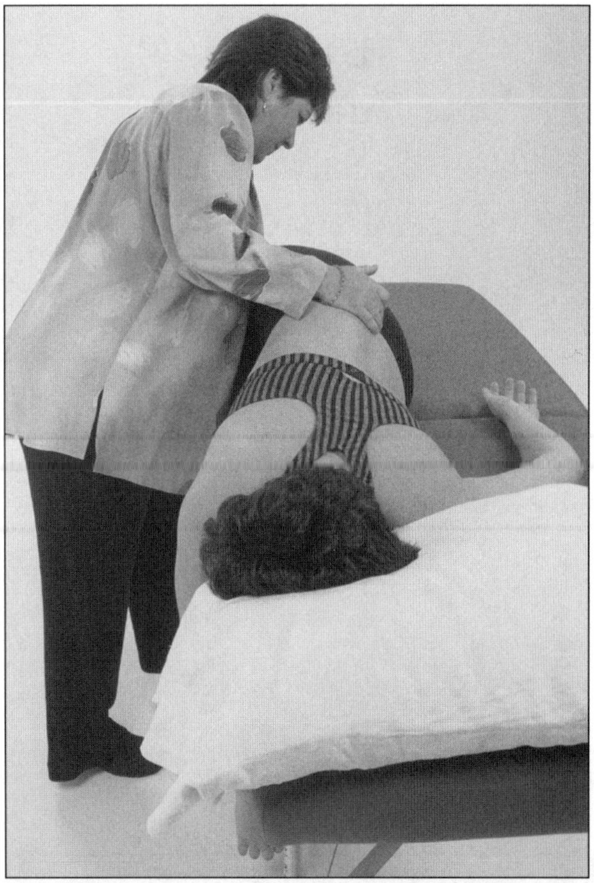

The practitioner treats a left sacral torsion on a left oblique axis (LOL).

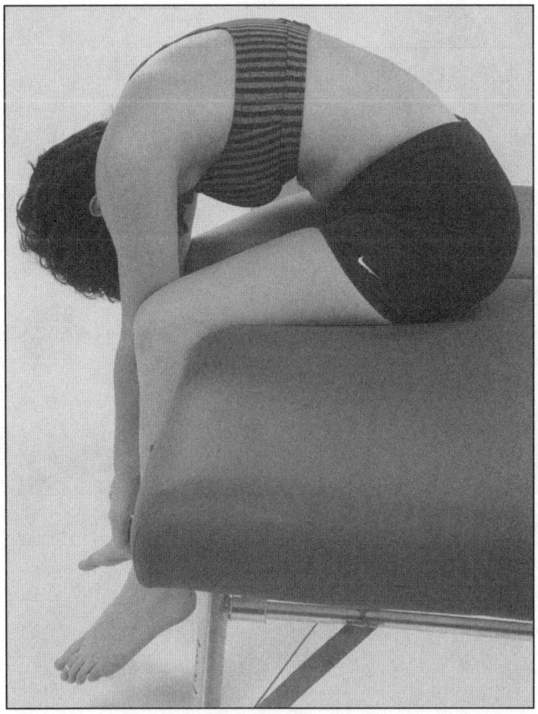

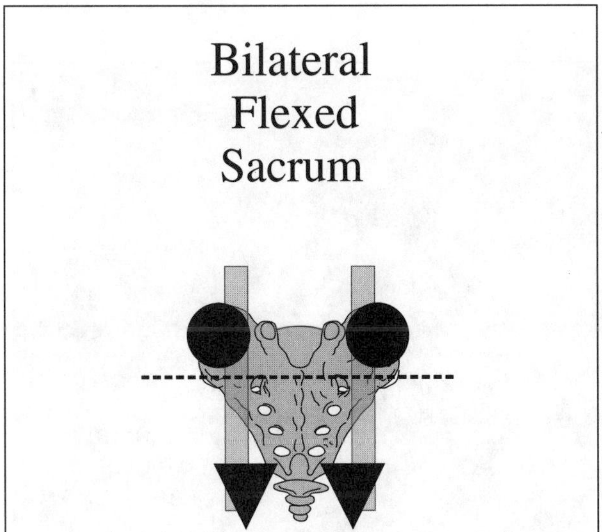

The practitioner finds the following landmarks: no deep sulcus; no inferior posterior ILA; but no posterior glide of sacral base during lumbosacral flexion.

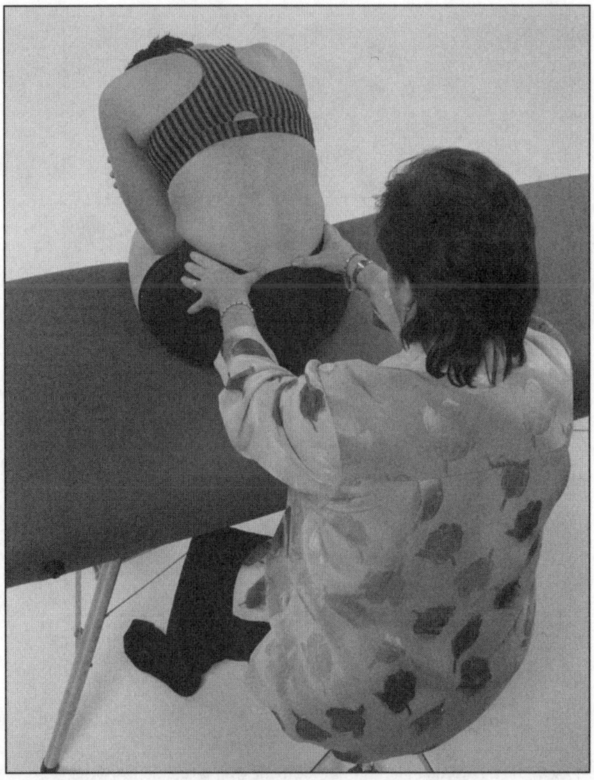

The client can flex further down the kinetic chain, with greater lumbosacral flexion, and the landmarks are different.

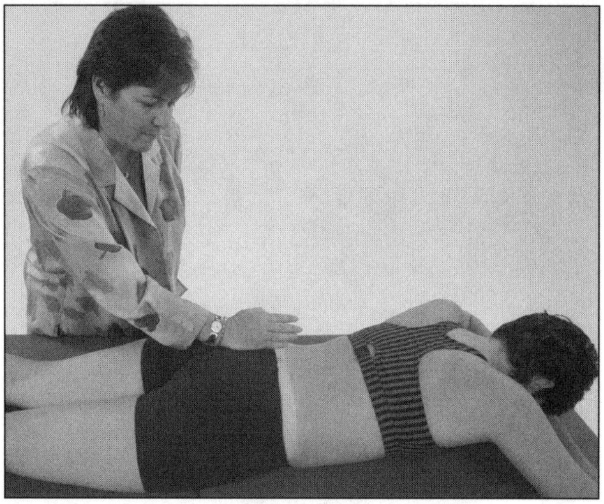

The practitioner treats a bilateral flexed sacrum.

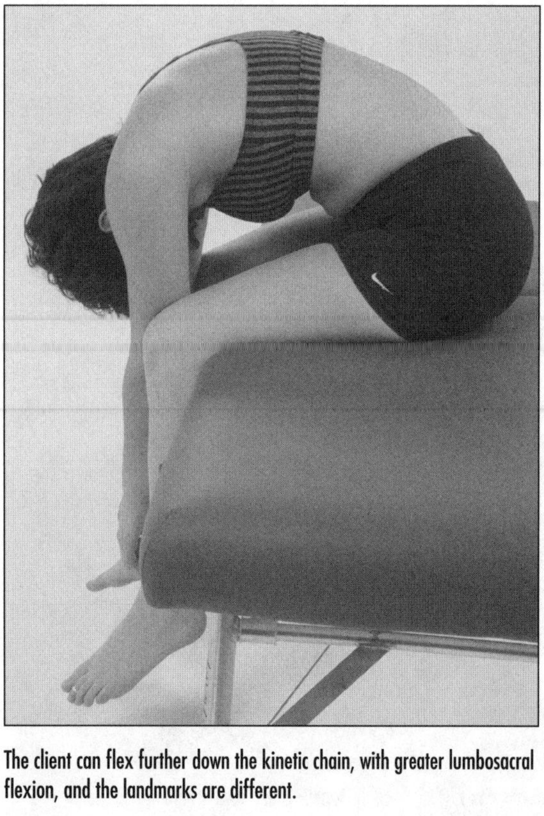

The client can flex further down the kinetic chain, with greater lumbosacral flexion, and the landmarks are different.

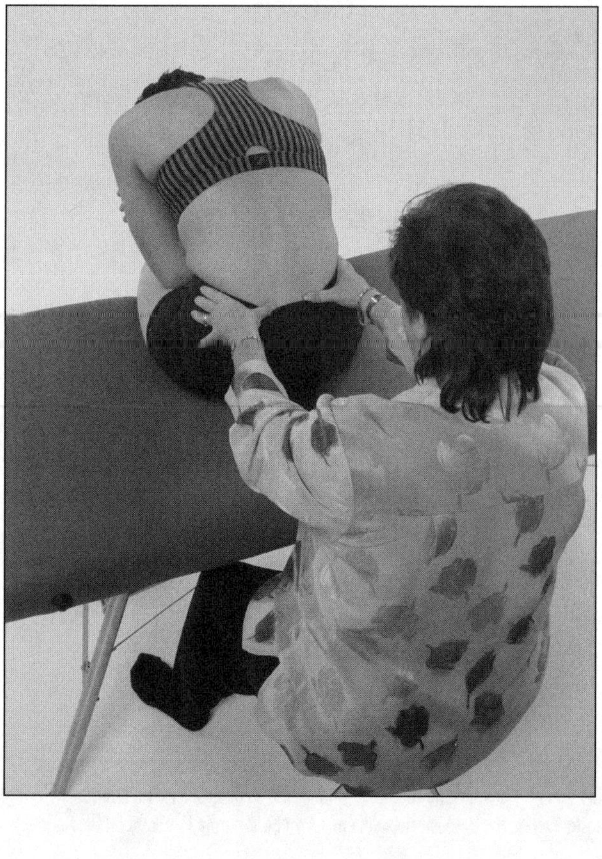

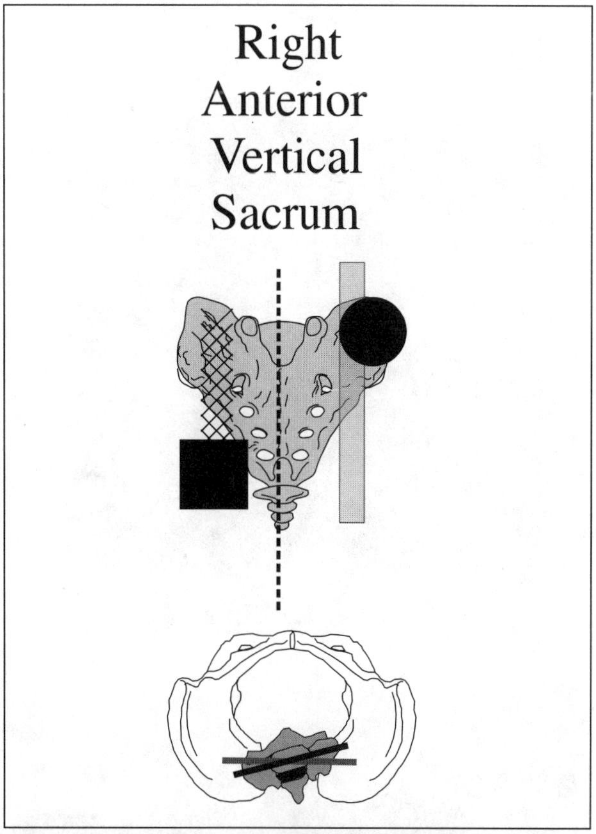

The practitioner finds the following landmarks: right deep sulcus; no inferior posterior ILA; left posterior prominence of the left lateral border of sacrum (which is really right anterior depression of the right lateral border of the sacrum).

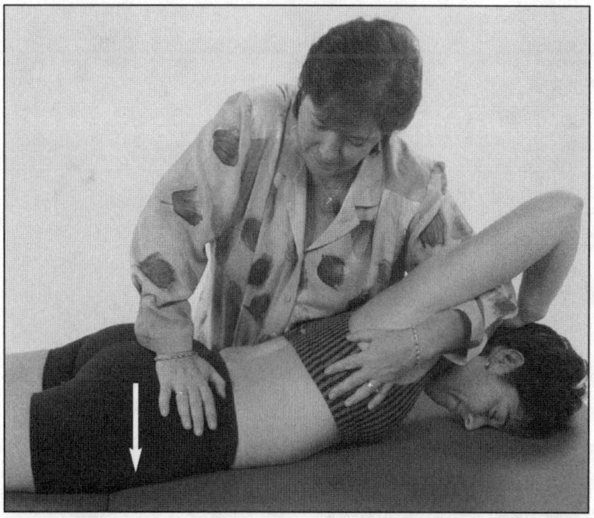

The practitioner treats an anterior vertical rotation of the sacrum to the left (which is a right sacroiliac joint dysfunction).

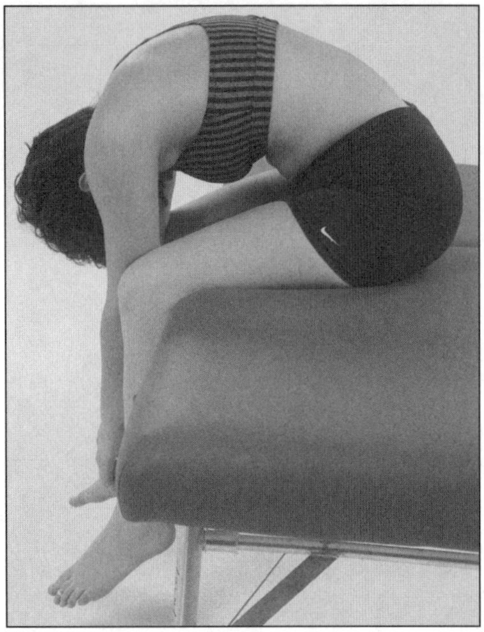

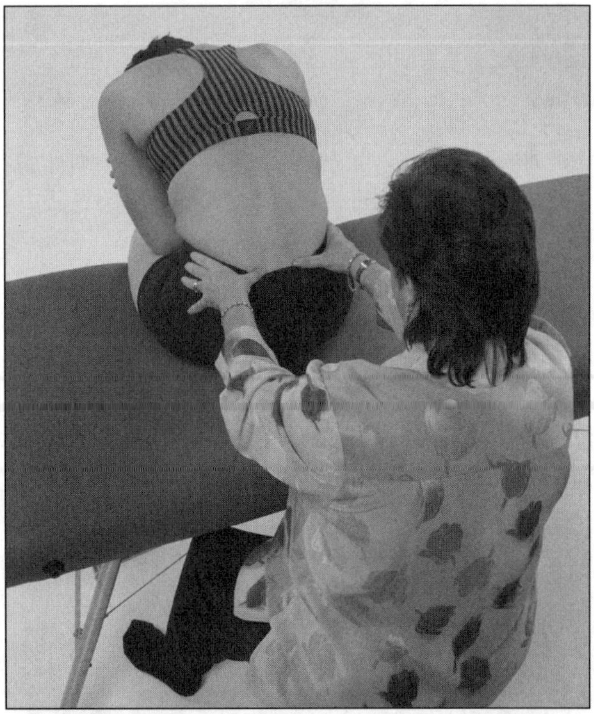

The client can flex further down the kinetic chain, with greater lumbosacral flexion, and the landmarks are different.
The practitioner determines that there is 50 degrees of lumbosacral flexion.
Therapy for correction of biomechanical dysfunction of the sacrum is discontinued.
The practitioner begins to assess Type I movement of the lumbar spine.

CHAPTER 19

BIOMECHANICAL DYSFUNCTION OF THE SACROCOCCYGEAL JOINT

About the Coccyx and the Pelvic Floor

The coccyx usually does not have a primary biomechanical dysfunction. More often than not (depending upon the client population, 50 percent to 75 percent), the original joint dysfunction occurs at the sacrum. The coccyx is dependent upon the sacrum for movement and posture.

Whenever there is flexion of the sacrum, the coccyx apex moves posterior.

Whenever there is extension of the sacrum, the coccyx apex moves anterior.

Whenever there is a rotation dysfunction of the sacrum (anterior or posterior rotation on a vertical axis), the coccyx apex deviates to the side of the rotation dysfunction. Therefore, if the sacrum is stuck in a posterior rotation to the right on a vertical axis, then the coccyx apex will deviate to the right.

Whenever there is a descended sacrum, there is a direct stretch/tension placed on the pelvic floor diaphragm by the coccyx apex.

All of the combined sacral dysfunctions are cumulative, and there is a resultant force placed upon the coccyx. Yet sacrococcygeal dysfunction is not difficult to correct. The sequence of treatment is important:

1. Biomechanical dysfunction of both sacroiliac joints and the lumbosacral junction should be addressed first, before biomechanical dysfunction of the sacrococcygeal joint is addressed.
2. Often, soft tissue dysfunction of the pelvic floor diaphragm should be addressed before biomechanical dysfunction of the sacrococcygeal joint is addressed.

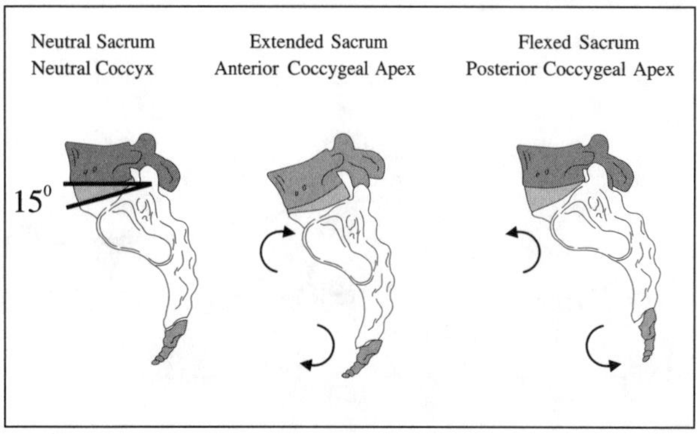

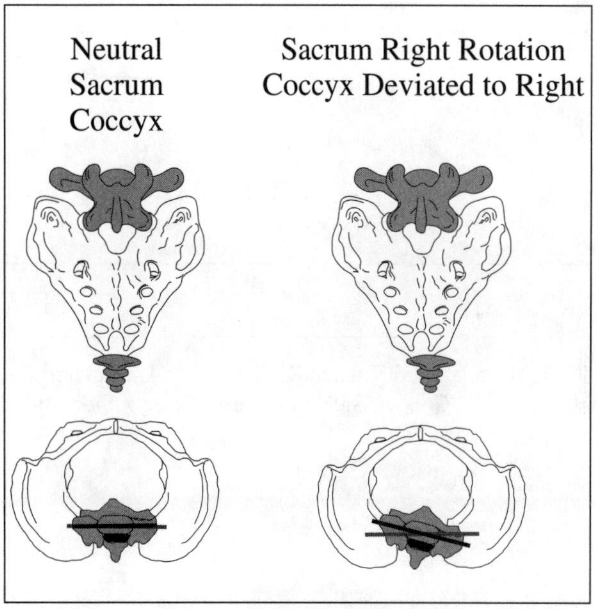

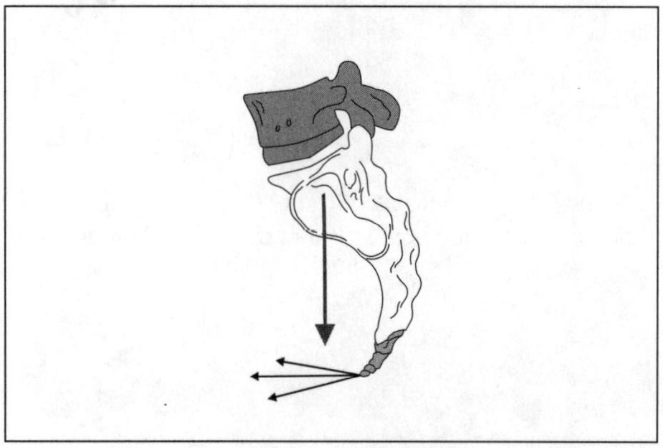

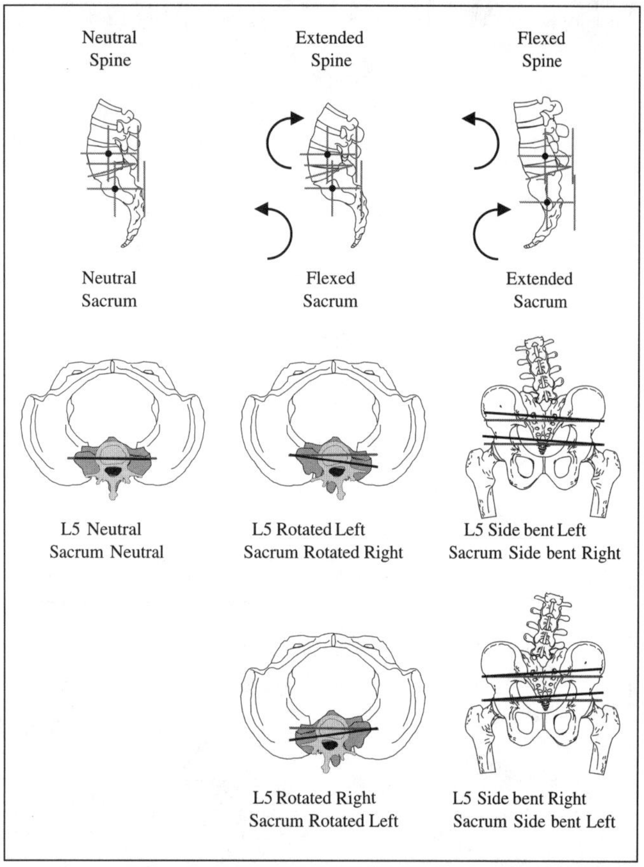

Dynamics of the Sacrococcygeal Joint: Type I Movement of L5 and Coccyx

The sacrococcygeal joint performs as a unit, moving together with the L5 vertebra, in Type I movement. This occurs through tensions of the sacrospinous ligament.

This means that L5 and the coccyx rotate and side bend together, as a functional unit. They rotate and side bend to opposite sides. L5 and the coccyx will both rotate to the left, as they both side bend to the right.

L5 has Type I movement with the coccyx. L5 has Type II movement with the sacrum at the sacral base (S1).

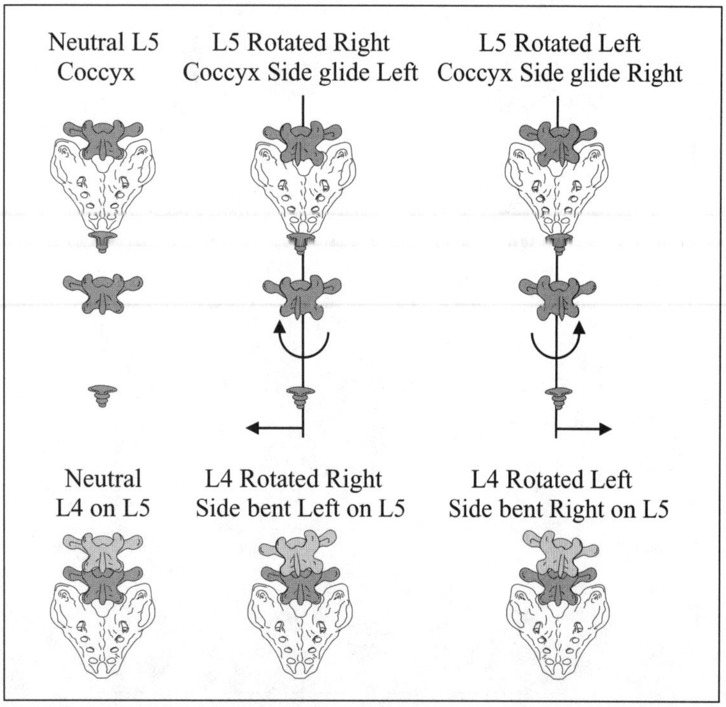

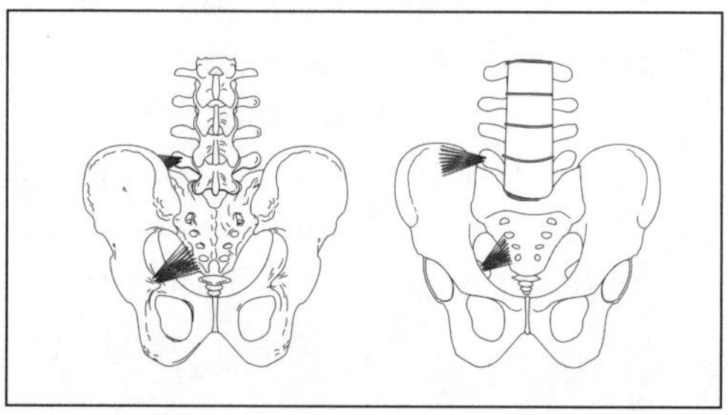

There is also co-joined movement of L5 with the coccyx during flexion and extension. When L5 flexes, the coccyx flexes; during L5 extension, there is coccygeal extension. Therefore, whenever the coccyx is maintained in a flexed position, tensions of the coccygeal ligaments are able to influence L5 into flexion.

There is a unique movement pattern at the sacrococcygeal joint. The coccyx does not have physiologic motions in the sagittal, coronal, and transverse planes. Rather, the motions are accessory joint movements.

Movements of the coccyx are unique, in the following manner:

1. There is anterior glide of the base (proximal aspect) of the coccyx instead of physiologic flexion.
2. There is posterior glide of the base (proximal aspect) of the coccyx instead of physiologic extension.
3. There is right side gliding of the base (proximal aspect) of the coccyx instead of physiologic left side bending.
4. There is left side gliding of the base (proximal aspect) of the coccyx instead of physiologic right side bending.
5. Rather than rotate right, the coccyx deviates with a tilt to the right. This occurs because of tension of the filum terminale.
6. Rather than rotate left, the coccyx deviates with a tilt to the left. This occurs because of tension of the filum terminale.

When there is physiologic motion at L5, there is accessory movement occurring at the sacrococcygeal joint because of the accessory movements of the coccyx base. The coccyx can be treated with Type I Muscle Energy and "Beyond" Technique in order to restore biomechanical function of the sacrococcygeal joint.

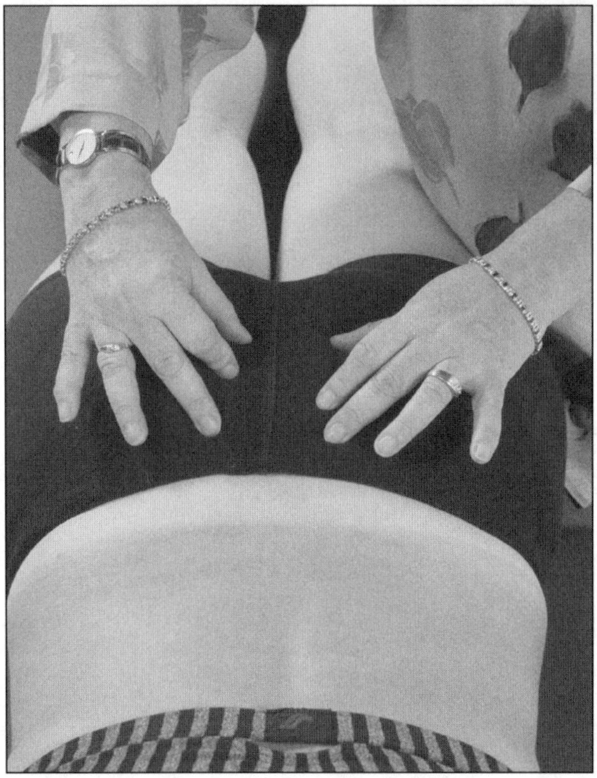

Treatment of Type I Biomechanical Dysfunction of L5 with Coccyx

Assessment

In the prone position, palpate L5 transverse processes and palpate the sacrococcygeal joint. There will be a tendency toward rotation of the sacrococcygeal joint line, which will mimic the direction of rotation of L5. When the left transverse process of L5 is prominent posteriorly, the left sacrococcygeal joint line will be prominent posteriorly.

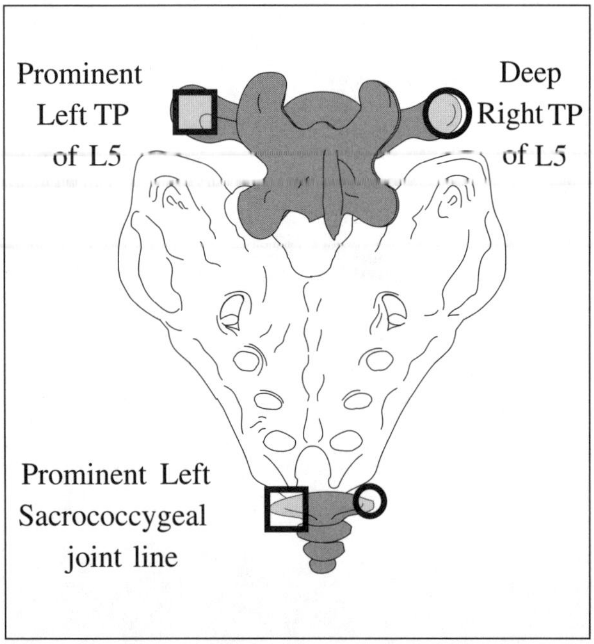

Treatment

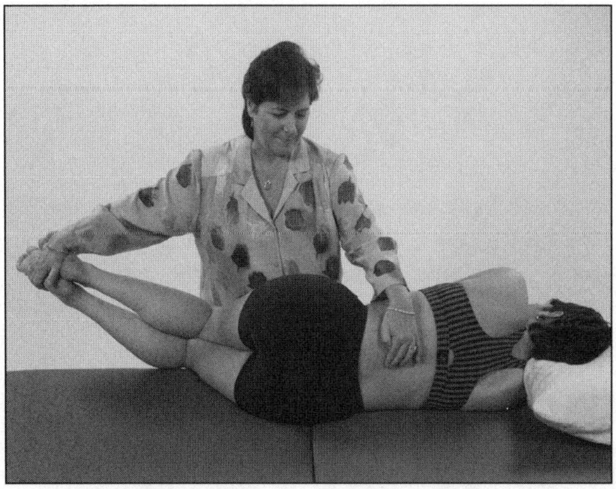

The patient lies on the side opposite the posterior prominence (similar to treatment of other Type I dysfunctions). Palpate at the sacrococcygeal joint line. The hips are flexed to 35 degrees, with 90 degrees of knee flexion. Lift the feet toward the ceiling in order to side bend, up the kinetic chain, toward the side of the posterior prominence of L5 transverse process and the sacrococcygeal joint. Perform isometric resistance for 6 seconds. Repeat this process for 3 to 5 repetitions.

Principles for Biomechanical Correction of the Sacrum: Assess the Sacrum Prior to Treatment of the Coccyx

Assess landmarks of the sacrum. When there is evidence of biomechanical dysfunction of the sacrum, even when the pain at the coccyx is significant, assess and treat the sacrum first. Occasionally, in less than 20 percent of clients, you will be required to correct the sacrum up to a point, then correct the coccyx, which will be an obstacle to further correction of the sacrum. Usually it will be more efficient to correct the sacrum first. Otherwise, the coccygeal joint problems will be resistant to optimal correction.

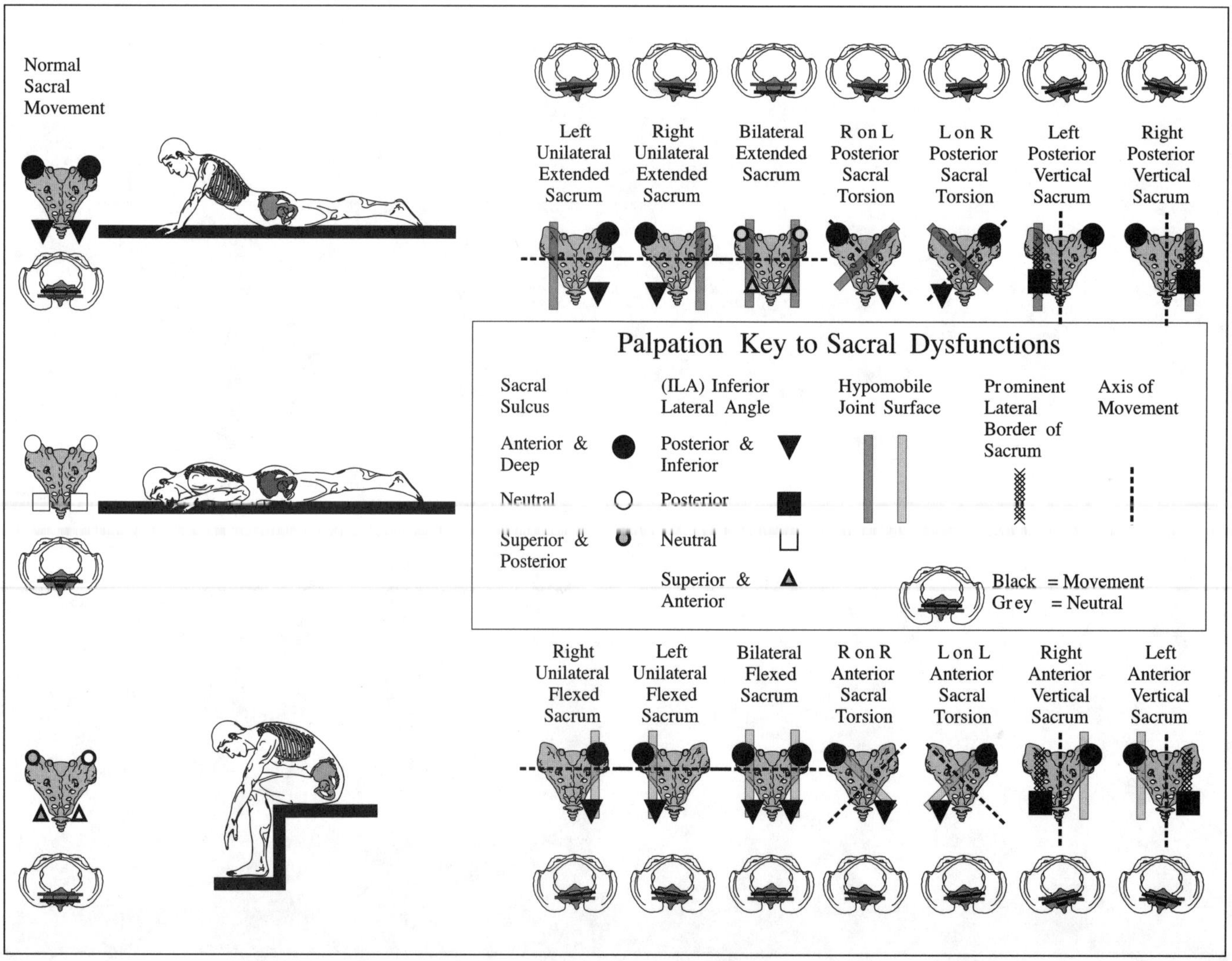

CHAPTER 20

THE L5 VERTEBRA

There is a unique vulnerability in L5. The stability of the lumbosacral junction is created by the sacrum, but compromised because of excessive mobility of L5 (to compensate for this stability).

The lumbosacral junction actually presents up to 30 percent of total low back movements and directly contributes to 70 percent of total spinal motion.

This means that when the lumbosacral junction is fused, up to 30 percent of low back movements on three planes may be sacrificed, but up to 70 percent of total spinal movements on three planes of the cervical, thoracic, and lumbar regions may be compromised [(Weiselfish) Giammatteo, clinical research]. This is due to unique reciprocal movement at the L5/S1 junction [(Weiselfish) Giammatteo, *Manual Therapy with Muscle Energy Technique for the Pelvis, Sacrum, Cervical, Thoracic and Lumbar Spine,* ANA Publishing, 1994].

There is reciprocal motion of L5 and S1, which occurs on three planes. This reciprocal movement requires rotation and side bending mobility of L5 that is significantly beyond the movement capacity of all other vertebrae.

Although the sacrum has side bending capability, it is significantly decreased because of the auricular shape of the articular surfaces. Therefore, the side bending that occurs at the lumbosacral junction is mostly produced by the coronal plane motion of L5 ((Weiselfish) Giammatteo, clinical research).

There is a co-joined functional coronal plane motion (side bending) of L5/S1, which is necessary to adapt to the cephalad forces at heel strike that are transcribed up the leg [(Weiselfish) Giammatteo, clinical research].

As for rotation, the sacrum has less than 5 degrees (similar to the rotation available at occiput for occipitoatlantal movement).

When there is a rotation displacement at the lumbosacral junction, it is directed at the L5 vertebra, which is the reason for the exceptional mobility at L5. There is also a requirement for more intradiscal motility and mobility. For L5 requirements, the disc has a looser organization of fascia and a more easily displaced nucleus that can be affected by larger variations in the hydrostatic pressure, which occurs secondary to L5 motions.

Questions for Research

This area of L5 mobility requires greater investigation. It is a source of disability and cost to health care that is unequaled. With knowledge of the co-joined interplay between a stable sacrum and a mobile L5, there are questions that require solutions, such as:

1. Why is the sacrum the center of gravity?
2. When is the sacrum really affected by lower extremity movements?
3. Why is L5 motion required to compensate for sacral stability?
4. When does the L5 disc turn into a shock absorber?
5. When does the L5 disc act as a reciprocal motion practical transducer?
6. What does L5 require to energetically compensate for rotation compromise of the sacral base?
7. When does the L5 vertebral body respond directly to forces from the spine?

8. Why are there transducers of forces at the lumbosacral junction for reciprocal movement to be dispersed throughout the spinal column?

These questions, and many more, will require the collaborative efforts of therapists, physicians, biochemists, biophysicists, architects, and others involved in research of forces with transducers and levers.

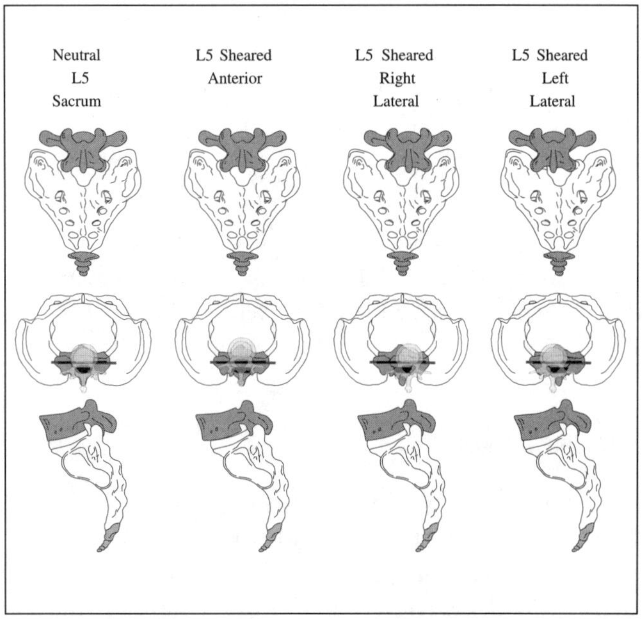

The Anterior and Lateral Shear of L5

There is a phenomenon that occurs at L5 called an anterior shear.

L5 can flex and extend on a sagittal plane; it can side bend right and left on a coronal plane; it can rotate right and left on a transverse plane. These movements are all documented and used in clinical, practical application by orthopedists, osteopaths, chiropractors, physical therapists and others. Other translatory movements, which Caillet diagrammatically portrayed as a shear, are an anterior (and/or lateral) displacement of the vertebral body which occurs on a frontal plane without any known physiologic axis. This has been termed a *subluxation* by chiropractors, and has implications for vertebral column pain and disability. This dysfunction requires more subtle intervention. The articular balance of L5 must be restored in order to allow normal physiologic three-planar movement.

This dysfunction is the typical "spondylo" of the lumbosacral junction. It has been named a *Type III shear* by the author. It is corrected by a technique called *Type III technique*. This technique is a modified Myofascial Release technique.

The Descended Vertebra Phenomenon

Cranial Trauma Causes the Descended Vertebra

An exceptional phenomenon, the descended vertebra, has been noted by (Weiselfish) Giammatteo and her colleagues. Traditional biomechanical analysis of 3-planar motion does not isolate this dysfunction. The descended vertebra is a dislocation of the vertebra's sagittal plane position.

All cranial traumas that result in a caudal-directed energetic force can cause this positionally longitudinal traction-like dysfunction. Trauma can result in a force that can displace the vertebra. The force caused by the trauma is equal to mass times acceleration. The application of force consists of the total mass (including actual mass, friction, and resistance) and the acceleration velocity. The effect of the force is dependent on the size and density of the object (the vertebra) being traumatized. The force load is directed in a longitudinal direction with a cranial injury. When the trauma is directed in a caudal direction, the displaced anatomic structure is affected according to the direction of the force.

The author notes the possible caudal displacement of the vertebra, which results in a cephalad limitation during mobility testing that is not a physiologic dysfunction. The requirements of the vertebral body for cephalad and caudal glides should be further investigated.

A similar dysfunction, defined by the author and colleagues as a *descended vertebra,* has been

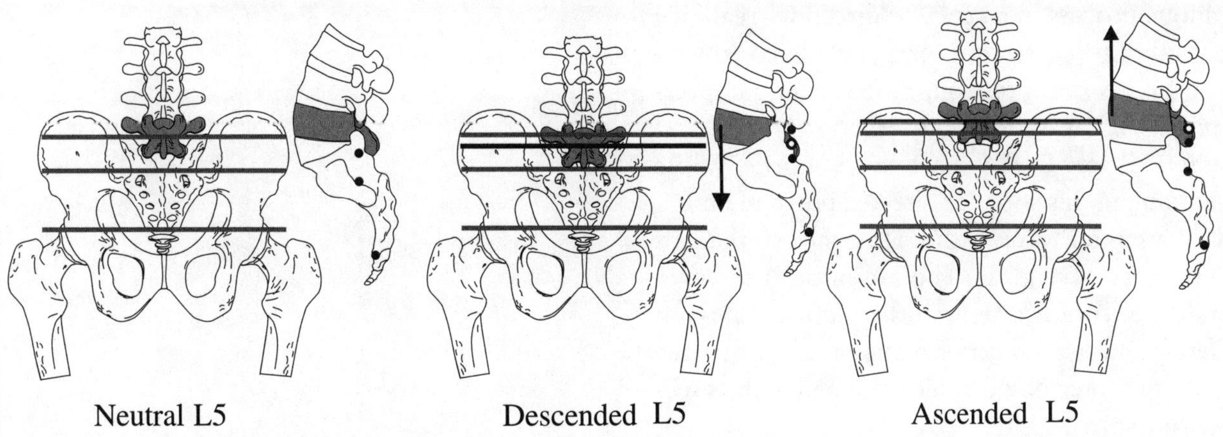

researched clinically at the sacrum (the descended sacrum). Assessment and diagnostic criteria for intervention were described by (Weiselfish) Giammatteo in *Manual Therapy with Muscle Energy Technique for the Pelvis, Sacrum, Cervical, Thoracic and Lumbar Spine* (ANA Publishing, 1994). Treatment of the descended sacrum is effective in restoration of normal axes of motion for the sacrum (required for sacroiliac joint and lumbosacral junction movements).

Assessment of the L5 Descended Vertebra

The caudal displacement of L5, and occasionally other vertebral bodies, occurs secondary to trauma directed onto the cranium in an inferior direction. This entity can be assessed with a simple mobility test: cephalad mobility of L5, with a focus on the vertebral body and the articular masses of the facet joints. There is an impaction of L5 on the sacrum, which further contributes to the significance of this L5/S1 lesion. There is a traction on L4 from the L5 vertebral body. This may affect the vertebral mobility of L4, which may affect the L4 intradiscal pressure and does affect the mobility of the L4 facet joints that articulate with L5.

The cephalad mobility test is simple to perform. Align the L5 spinous process as much as possible with sacral tubercles. Then push the L5 spinous process in a superior direction, realizing that inhibitory pressures may include L5/S1 vertebral body compression; L5/S1 facet compression; L4/L5 vertebral body traction forces; and right and left L4/L5 facet dysfunction. While maintaining this superior directed pressure on L5, place a lateral force to the right, then to the left, in a translatory glide. The inhibitory pressures could be from the right and/or left sides, from inferior and from superior pathologies. Then place a rotatory force on L5 while maintaining the superior pressure on L5.

With these mobility tests, there is pretest palpatory information regarding L5 mobility restrictions. When treatment has successfully restored movement capacity, gross physiologic changes may affect L5 rotation. (This is different from treatment of the descended sacrum which mostly affects changes in gross physiologic flexion; (Weiselfish) Giammatteo, clinical research.) With an increase in L5 rotation after treatment of an L5 descended vertebra, there will be a normalization of the intradiscal pressures of the L5 disc.

Treatment of the L5 Descended Vertebra

The treatment is a simple mobilization when performed by manual therapists with good palpatory skills. The L5 descended vertebra should be corrected before the descended sacrum is corrected, in order to eliminate traction of the disc fibers of L5.

1. The patient is prone. A thumb can be positioned under (inferior to) the L5 spinous process.
2. The thenar eminence of the second hand can be placed on top of the thumb under L5.
3. A superiorly directed force (from the thenar eminence onto the thumb) can push L5 in a superior direction until all mobility and excursion is compromised, i.e, "take up the slack."

4. A superiorly directed grade 5 mobilization (with very low amplitude, force less than 2 lbs., and high velocity) may be utilized.
5. There will be an immediate restoration of L5 motion on the axes of all three planes. This means that L5 movement can now be restored on three planes.

Precautions and Contraindications

1. Only the therapist skilled in assessment, diagnosis, and treatment of the descended sacrum should attempt this manipulation. The descended L5 vertebra treatment always precedes treatment of the descended sacrum.
2. The therapist should have an understanding of the craniosacral system. It is possible that a treatment reaction will occur, which is difficult to alleviate without intervention of craniosacral therapy. This L5 descended vertebra dysfunction is secondary to cranial trauma and is biomechanically insignificant when compared to the significance of the trauma to the craniosacral system.
3. It is necessary to be familiar with the accessory sacroiliac joints, described in (Weiselfish) Giammatteo's first book, *Manual Therapy with Muscle Energy Technique for the Pelvis, Sacrum, Cervical, Thoracic, and Lumbar Spine.* These joints are best described by Joan Walker, Ph.D., P.T., in "The Sacroiliac Joint, A Review of the Literature" (*J.A.P.T.A.*, December 1992). The joints are also mentioned by Gregory Grieves and in Diane Lee's book, *The Pelvic Girdle.* Whenever there is hypomobility of the PSIS, which indicates hypomobile accessory sacroiliac joints, and a descended sacrum, it is unwise to treat a descended L5 vertebra, because there is already existing traction dysfunction of the iliolumbar ligaments, which may be exacerbated.
4. The last precaution is: Maintain subtlety. This author rarely recommends manipulation (high velocity direct force, even when there is low amplitude with very low force) to the spine because of possible negative effects on the disc, which may already be ruptured, sequestered, or severely herniated. Caution is advised and manipulation should be practiced only by very experienced practitioners.

The Descended Vertebra of Other Vertebral Segments

The descended L5 vertebra often occurs together with the descended sacrum ((Weiselfish) Giammatteo) as a complex that results from cranial trauma, usually directed to the frontals and/or parietals. Usually cranial therapy is not sufficient intervention to correct this dislocation phenomenon. The same cranial trauma occasionally causes this descended vertebra phenomenon at other spinal vertebrae, most often found at C7 and T1, T12 and L1 ((Weiselfish) Giammatteo, clinical research). The prevalence of the descended L5 vertebra, which is second in incidence to the descended sacrum ((Weiselfish) Giammatteo, clinical research), is much more common than anywhere else in the spinal column. When this phenomenon occurs at other vertebrae, the same pretest and treatment can be performed on those vertebrae. There is often an appearance of a bilateral extended vertebra at these segments, which may need treatment once the descended vertebra is corrected.

CHAPTER 21

THE LUMBOSACRAL JUNCTION AND GAIT

Certain traits pertinent to ambulation are recognized with gait analysis. These traits have been recognized with force plates evidence research, which has been gathered during the past several decades. Stance phase and swing phase are well documented. Heel strike, mid-stance, toe off, and the full swing phase cycle are described in articles and texts about foot biomechanics, knee and hip mechanics, soft tissue performance, and many other significant gross and fine motor aspects of the gait cycle.

Clinical Research

Regarding gait and the lumbosacral junction, there is not a great deal of documentation. Research has been limited with regard to functional biomechanics of L5/S1 during ambulation. The interesting phenomenon of reciprocal motion at the lumbosacral junction has not been associated with gait in the past. There was no awareness that motion of L5/S1 is essentially co-joined with the lower extremities' movements in a unique manner. Valuable information about gait and the lumbosacral junction currently under investigation by researchers ((Weiselfish) Giammatteo, Kain, and others) can be interpreted for many purposes: the mechanics of gait for prosthetics, lower extremity dysfunctions in hemiplegia and more. Investigating the source of lumbosacral dysfunction, which is an entity often found on the operating tables in North America, has mostly been limited to investigation of primary disc problems. The information in this book on the lumbosacral junction could possibly change the present status of health care in North America.

There is a transducer of forces at the lumbosacral junction ((Weiselfish) Giammatteo), which may be in the disc material and may involve more anatomic tissues and structures.

The biomechanics of L5 are simple, but when co-joined with the sacrum, a complex motion occurs during ambulation, which must be viewed with respect to levers, force transducers, and anatomic displacement of structures. With tissue analysis, EMG, and other electrodiagnostic forms of investigation, it may be possible to further understand the complexity of this region.

Reciprocal Movement

There is reciprocal movement of L5 and S1 during ambulation. When L5 rotates to the right, there is a torsion of the sacrum to the left. When L5 rotates to the left, there is a torsion of the sacrum to the right. A torsion is a 3-planar movement; these motions of the sacrum include sagittal (flexion), coronal (side bending), and transverse (rotation) planes of movement.

Whenever the sacrum torques to the left, there is left rotation and left side bending. The transverse plane motion is left rotation. The coronal plane motion is left side bending. This is a difficult movement to analyze, because the sacrum has an auricular-shaped surface that functions during side bending movements for two purposes. One purpose is the side bending of the sacrum-simply to side bend. The author believes that the second purpose is to provide a resilient compilation of forces transcribed up the leg, which act as the lever for sacral movement. The side bending component of the 3-planar torsion will lock in the axis of motion. When the foot hits the ground, forces from earth up through the leg are present. The sacrum is anchored at the horizontal limb, locked in because of the side bending component.

There is a unique action of the muscles on the sacrum at this time, especially the piriformis. Together, the action of side bending with the pull of the piriformis facilitates the transduction of energies that occurs at the sacrum during ambulation, in stance phase.

Side bending occurs at the sacrum as a 2-phase motion: There is a vertical motion of the sacral articular surface inferior on the vertical limb of the joint surface; there is a co-joined motion on the horizontal limb of the joint surface for completion of side bending.

During ambulation, the sacrum is in a torsion placement of flexion and rotation plus side bending, which occur together on an oblique axis. What is the L5 response, or rather, how does L5 work in conjunction with this sacral torsion? Because there is reciprocal movement of L5 with S1, L5 will rotate to the opposite side of sacrum. Because both L5 and S1 have Type II movement during flexion, L5 side bends and rotates to the same side. Therefore, if the sacrum is rotated and side bent left, then L5 is rotated and side bent right.

This demands incredible movement capacity at the lumbosacral junction. Because of the reciprocal motion, there is excursion of L5 and S1 within a potential amplitude, on each plane, that requires symmetry, synchrony, and force.

When L5 is flexed, the sacral base extends (i.e., there is posterior glide of the sacral base). When L5 is extended, the sacral base flexes (i.e., there is anterior glide of the sacral base).

There is a need to understand: Is the sacrum flexed or extended during gait? Is the sacrum in neutral AP (sagittal plane) position during gait? Is L5 flexed or extended during gait? Is L5 in neutral AP (sagittal plane) during gait? Are L5 and S1 always reciprocal, or is the reciprocal movement evident only in flexion and/or extension, and/or neutral? These questions are significant and have been raised by (Weiselfish) Giammatteo and colleagues. These questions have not been raised by clinicians working toward gait and ambulation rehabilitation in most clinical settings.

Type II Movement of L5 and of S1

There is only Type II motion at the lumbosacral junction ((Weiselfish) Giammatteo, clinical research). When L5 is flexed, S1 is extended. When L5 is extended, S1 is flexed. We stand upright in Type II position at the lumbosacral junction. Therefore, during prone, supine, standing, and ambulation, L5 is slightly extended with S1 mildly flexed. As we move through the gait cycle, the sacral base moves into an anterior torsion that requires the extension of L5. At heel strike, there is an anterior glide of sacral base with a concurrent rotation and side bending to the side opposite the heel strike side (i.e., the sacral torsion is toward the swing phase side). At mid stance, there is maximal anterior torsion (sagittal plus coronal plus transverse plane motions) of the sacrum, with maximal extension of L5 during its Type II 3-planar motion.

The L5 Disc

This indicates that the L5 disc is not displaced posteriorly during the gait cycle, when the forces are transcribed up the leg. There should be an anterior placed nucleus pulposus, thereby unaffected by the compressive forces (from the ground up the leg), which might cause trauma to a posteriorly displaced disc (which would occur if L5 were flexed, rather than extended, during heel strike through mid stance). Then, as we move from mid stance through toe off, there is a glide of the sacral base back to almost neutral, but not quite to a neutral position, because there is no neutral position for the sacral base. Whenever we begin the swing phase, there is the increase once again of anterior glide of the sacral base, with rotation and side bending of the sacrum toward the leg now in swing phase.

The L5 disc should not be affected and traumatized during standing, ambulation and running, if the biomechanics are correct. Since L5 is extended, the disc should be anterior. Since L5 is side bent toward the leg in stance phase, the nuclear material in the disc will deviate away from

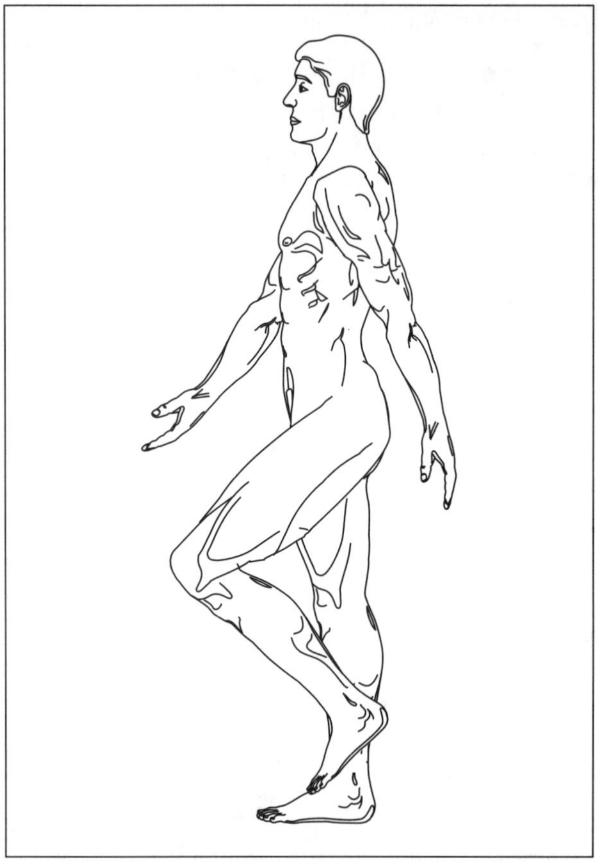

the compressive ground/leg forces. Although the intradiscal pressures will increase with rotation, the forces are on the vertebral body and facets contralateral to the side of disc deviation.

Stance Phase and Swing Phase

There is always co-joined motion of L5 with S1 during ambulation. In swing phase, L5 will extend, rotate, and side bend toward the leg in stance phase.

The sacrum always rotates and side bends toward the leg in swing phase.

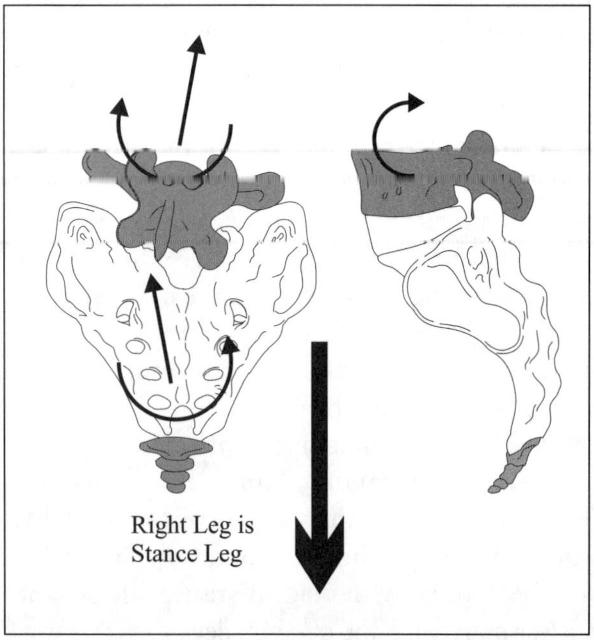

L5 always rotates and side bends toward the leg in stance phase.

Stance phase constitutes an extension of the forces from the ground up and can be viewed as a sacral function; swing phase is the result of forces from the trunk affecting the leg and can be viewed as a function of L5.

The gait cycle requires both the forces of the sacrum and L5 to work synchronously, and is evidently a neurologically induced rather than a mechanically oriented force process. These are reflexive-like aspects of ambulation, providing further evidence that walking is a neurologically oriented task, rather than a biomechanical functional process. This means that biomechanics are not, in fact, the most significant aspect of gait, but rather an integral portion of this total action. With forces transduced from the cephalad structures and tissues in coordination with forces transduced from the limb, there must be an energetic mechanism that responds to the neurologic program of walking.

Questions

- When does L5 flex, and when does sacral base extend? Not during prone, supine, standing or ambulation, unless there is dysfunction of these structures, which has been commonly found in low back pain patients.
- When does L5 flex? During sitting, with hip flexion. In fact, L5 flexion is co-joined with hip flexion. When we slouch, when we bend, when we perform somersaults, L5 will flex together with extension of S1. Otherwise, during supine, prone, and upright positions, L5 is mildly extended in the healthy person, with a mild anterior glide of the sacral base.
- What is significant for ambulation?
- An exceptional mobility for the sacrum?
- An exceptional mobility for L5?
- An exceptional synchronized motion of both bony surfaces which can be displaced in full amplitude excursion on all three planes?
- An energetic transducer within the L5/S1 complex, possibly the disc, which can formulate motion from energetic forces from above and from below?
- A disc that can allow shock absorption of motion but can extend in anterior, lateral, and posterior borders for the production of motion required at the lumbosacral junction during ambulation?

Many of the above questions, which are probably only a portion of the concern of this region, have not been addressed in relation to gait.

CHAPTER 22

BIOMECHANICAL DYSFUNCTION OF THE LUMBAR SPINE

About Spinal Motion

This author and colleagues have spent more than two decades studying various models of movement and motion dysfunction of the spinal column. Until developing this current model, no other hypothesis fit the clinical phenomena presented in clinical research.

Until there was acceptance that spinal motion from C3 through L5 was not the determining factor for spinal movement, hypothetical models were not appropriate.

The pelvis, sacrum, coccyx (to a lesser degree), and L5 vertebra are the determining factors for spinal movement. Without pelvic joint mobility and biomechanical integrity, sacral biomechanics cannot be optimally addressed. Therefore, the pelvic joints are not the determining factor for spinal movement. The sacrum is the determining factor for spinal movement. Yet it is usually not possible to attain optimal results for the lumbosacral junction when there is biomechanical dysfunction of the pubic symphysis, the iliosacral joints, and the accessory sacroiliac joints.

The coccyx is not the determining factor for spinal movement. Yet, if there is a tethered spinal cord, when the filum terminale is stuck to the sacrum and coccyx, the spine will refuse to move (a protective mode). Therefore, soft tissue dysfunction of the coccyx relative to the filum terminale must be addressed prior to attaining optimal results with sacrum biomechanics.

L5 is not the determining factor for spinal motion. Yet, because S1 moves reciprocally with L5, often the sacrum cannot be optimally corrected (for biomechanical dysfunction) without release of L5. Certainly when there is a descended L5 vertebra, the posterior vertical rotations of the sacrum and the descended sacrum, cannot be corrected.

What is the determining factor for spinal movement? The sacroiliac joints and S1 are the determinants for spinal motion. (It can be more simply stated that the sacroiliac joints and the lumbosacral junction are the determinants for spinal movement.)

When pelvic biomechanics are corrected, every person should be able to touch the floor! This indicates that there is no remaining primary joint dysfunction of the pubic symphysis, the right and left iliosacral joints, and the right and left accessory sacroiliac joints. Perhaps the person will not have wonderful intervertebral joint segmentation with extension. But the person will be able to touch the floor. (Note: Remember the pelvic compression reflex!)

When biomechanical integrity has been restored for the right and left sacroiliac joint and S1 (or the lumbosacral junction), at least 70 percent of all spinal movement should be restored! Therefore, cervical flexion, cervical extension, cervical right rotation, cervical left rotation, cervical right side bending, cervical left side bending, thoracic flexion, thoracic extension, thoracic right rotation, thoracic left rotation, thoracic right side bending, thoracic left side bending, lumbar flexion, lumbar extension, lumbar right rotation, lumbar left rotation, lumbar right side bending, lumbar left side bending-all these motions should present at least 70 percent within normal limits.

No matter what other problem exists (remember the pelvic compression reflex; if there is a pelvic compression reflex, there cannot be biomechanical integrity of sacral joints!), there will still be 70 percent normal spinal motions-even if there is a bulging or herniated disc; even if there are nerve root impingements; even if there are bone bruises (osteopenia, osteomalasia, osteoporosis); even if there are multiple problems affecting tissues and structure of the spine. If all biomechanical dysfunction of sacral joints can be corrected, there will be 70 percent of normal ranges of motion throughout the spinal column.

What, therefore, can be considered the function of the rest of the spinal column? There cannot be a primary function of movement between C3 through L5, if spinal motion can be restored (whether or not there is biomechanical dysfunction present between C3 through L5).

Consider the possibilities. C3 through L5 contribute to the following functions:

1. Some movement of the spine, in their respective spinal regions
2. Protection of the spinal artery, spinal vein, and other spinal circulatory vessels
3. Protection of the spinal cord
4. Protection of the spinal canal
5. A lever
6. An anchor (for ribs, muscles and other soft tissues)
7. A backbone
8. A center
9. An amazing avenue for energy transcription

Type I Dysfunction of the Lumbar Spine

This dysfunction is multisegmental, involving two or more vertebrae. The dysfunction is considered adaptive, compensatory, maintained due to hypertonic long side bending muscles. It is found in AP neutral. Two or more vertebral bodies, as a group are stuck in side bending and rotation. Side bending and rotation occur to the opposite sides. (There can be a single segment Type II within a Type I dysfunction.) Type I dysfunction of T9 to L5 is treated in the side lying position; T1 to T9 is treated in sitting. What is this Type I dysfunction? The spine should be symmetric in resting, anatomic neutral. If there is evidence of side bending and rotation in two or more vertebrae during resting, anatomic neutral, there is Type I dysfunction of these vertebral segments. The evidence of side bending and rotation is the asymmetry, i.e., the unilateral posterior prominence of the paravertebral muscles due to rotation of those vertebral segments.

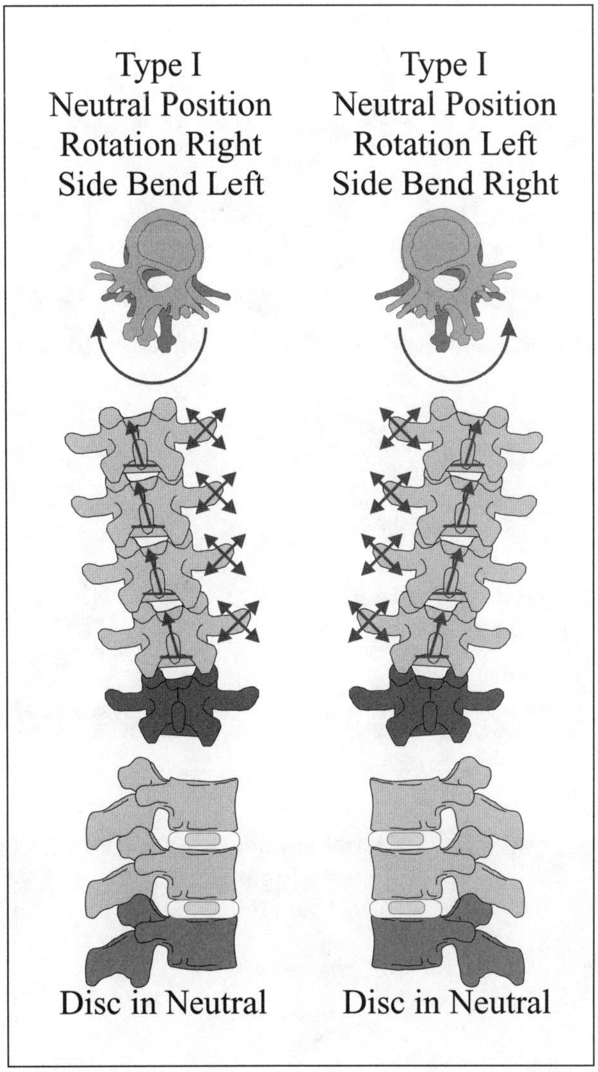

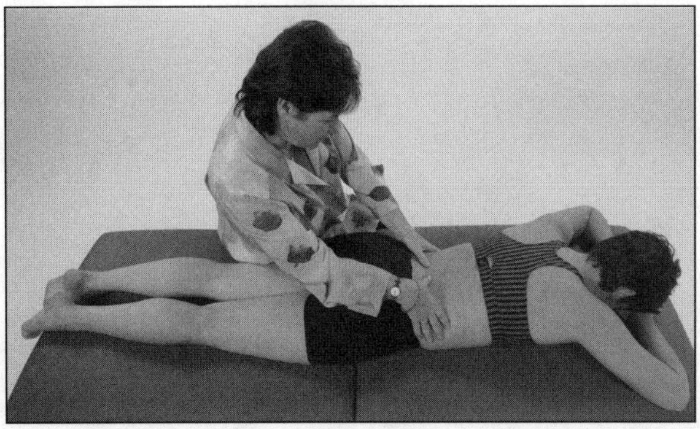

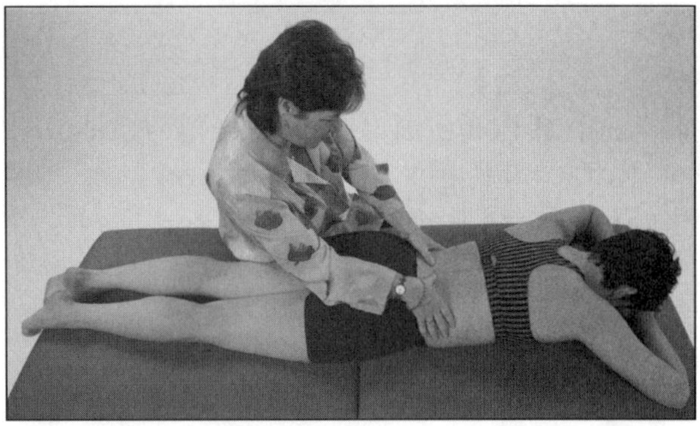

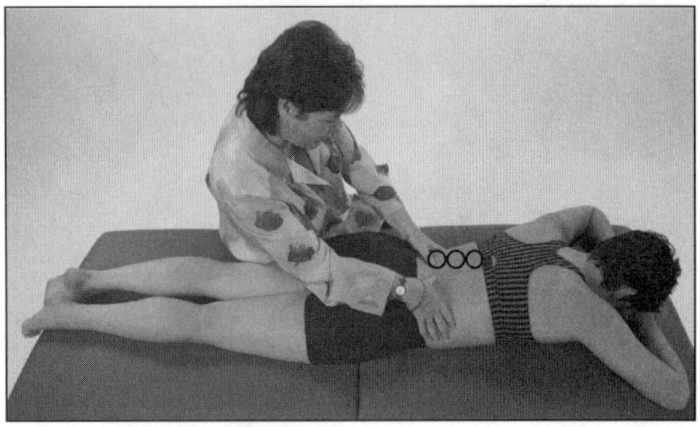

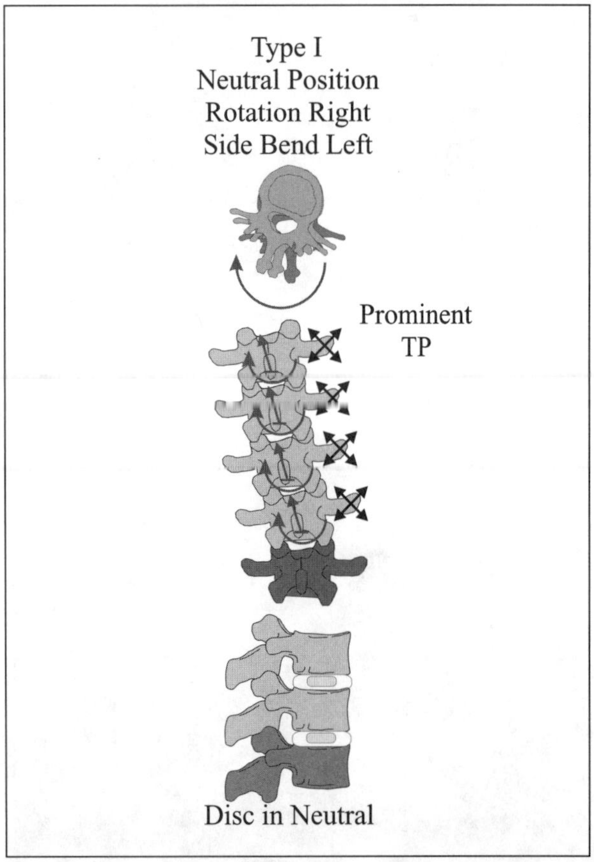

Positional Diagnosis

1. Palpate transverse processes in neutral.
2. Discover two or more posteriorly prominent transverse processes about one finger breadth lateral to midline. (Press lightly over the hypertonic paravertebral muscles. It is not necessary to palpate the bony prominence.)
3. Posterior prominence indicates rotation dysfunction of those segments.
4. The rotation of these vertebrae is to the side of the posterior prominence.
5. Side bending is to the opposite side of rotation (opposite the side of the posterior prominence), e.g., if two or more transverse processes were prominent posteriorly on the right: Neutral, side bent left, rotated right = NSLRR = NSL

Movement Barrier

The movement barrier is opposite the positional diagnosis. In neutral, if those vertebrae of the spine are found side bent to the left and rotated to the right, then they cannot side bend to the right and rotate to the left. The movement barrier is: Neutral, side bending right, rotation left = NSRRL = NSR

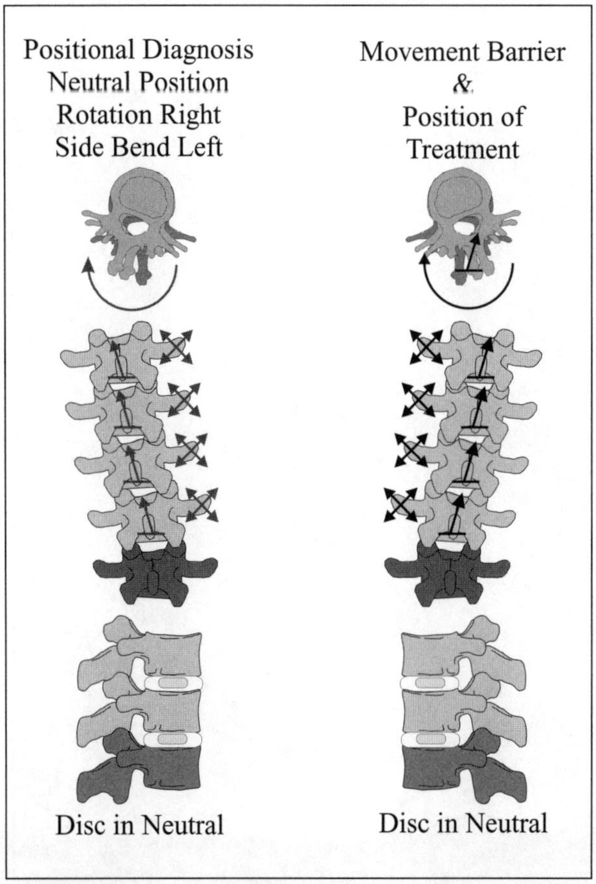

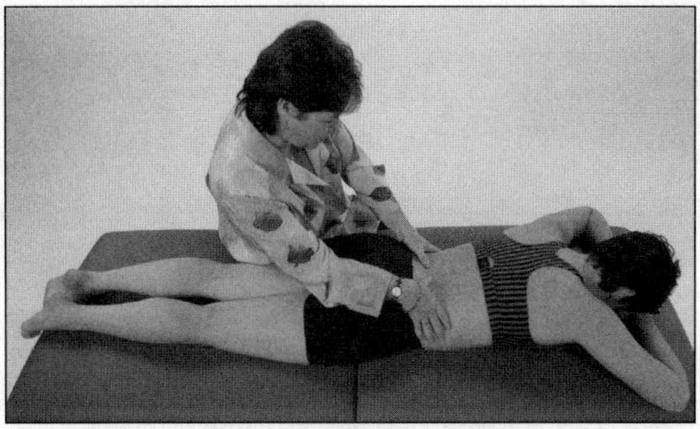

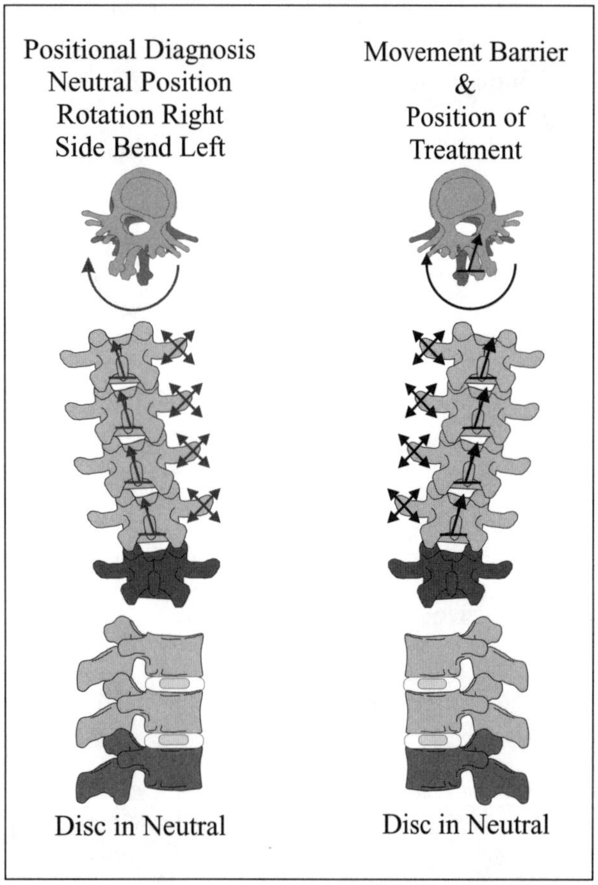

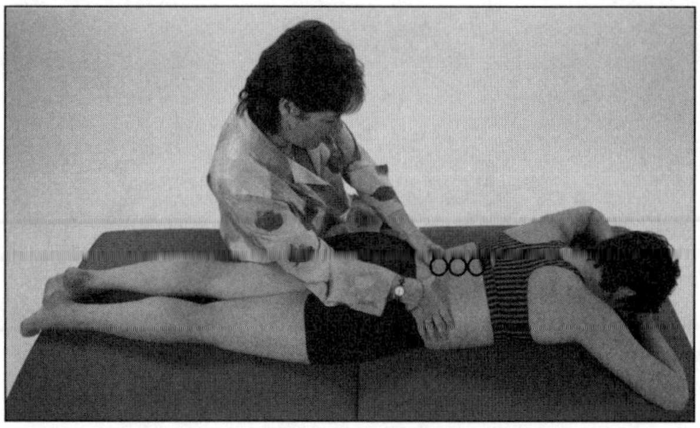

Position of Treatment

The position of treatment is the same as for the movement barrier, but opposite to the positional diagnosis. If the movement barrier is NSRRL, then the position of treatment is NSRRL; i.e., to treat the Type I dysfunction of the spine that is NSL, the spine would be positioned in neutral, in right side bending (and left rotation). Only side bending is considered important in treatment of Type I dysfunctions, therefore the spine is positioned in right side bending (i.e., the rotation component of the movement barrier is eliminated from the treatment position). The position of treatment is: Neutral, side bending right = NSR

Note: The three dots in this picture are over the three posteriorly prominent right transverse processes on L3, L4, and L5. These three vertebrae are rotated to the right; the transverse processes are therefore prominent on the right side. Therefore, these three vertebrae (L3, L4, and L5) are side bent to the left. The apex of the curve is L3; L3 is the vertebra most rotated and most side bent. The positional diagnosis is NSL.

Treatment

In Type I dysfunctions, there is no interbarrier zone. Therefore, the spine is side bent to the barrier palpated at the apex of the curve (which is the vertebra most rotated and side bent). Muscle Energy and "Beyond" Technique is a direct technique; treatment is into the movement barrier. If the positional diagnosis is NSL, the movement barrier is NSR, the position of treatment is NSR.

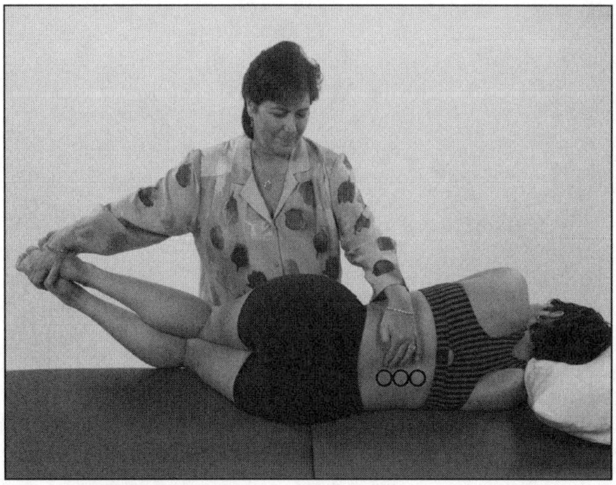

Isometric Resistance

Side bending is usually the resisted movement. Resisting side bending toward the concavity will provide an autogenic inhibition to the hypertonic side benders; resisting side bending toward the convexity will provide a reciprocal inhibition to the hypertonic side benders. Resistance in either direction is appropriate.

Treatment of Type I (Neutral) Lumbar Dysfunction

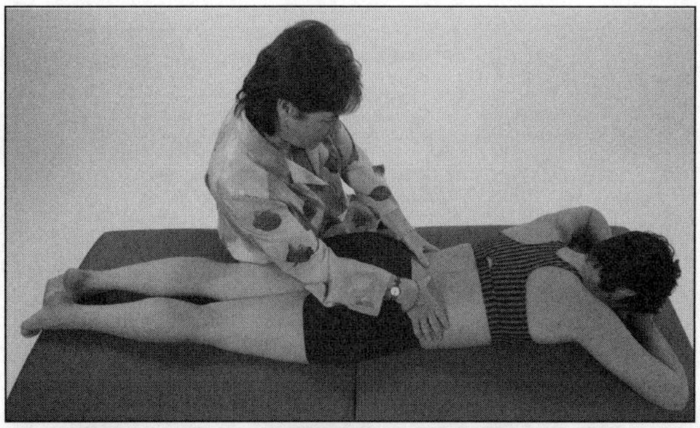

Diagnosis

Patient is prone in neutral position. Two or more adjacent transverse processes are prominent posteriorly on the same side.

Application

There is limitation in physiologic side bending to the side of the prominent transverse processes and limitations in physiologic rotation away from the side of the prominent transverse process. There should be an increase in physiologic side bending and rotation after the technique.

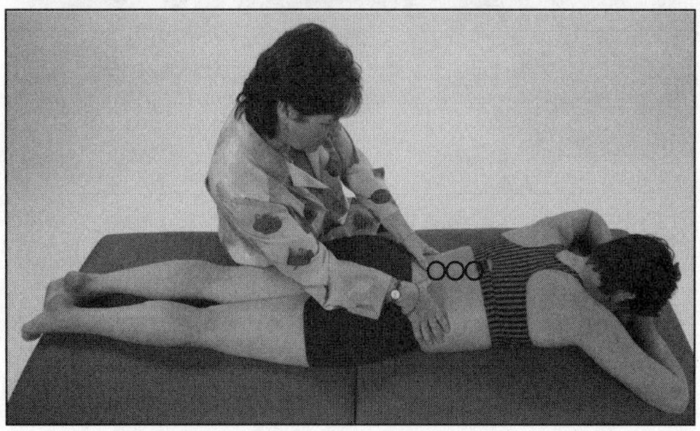

Rationale

This treatment is directed toward relaxation and elongation of the quadratus lumborum and erector spinae (side benders) muscles on the side of the concavity.

Treatment

Patient lies on the side of the concavity, the spine in AP neutral position (i.e., the spine is not flexed or extended), with the posteriorly prominent transverse processes up (convexity up). Flex hips about 30 degrees. Flex knees to 90 degrees. Palpate the middle and most restricted segment at the apex of the curve.

Barrier

Lift the patient's feet toward the ceiling, side bending up the kinetic chain; palpate barrier at segment of apex of curve. (Rotational component is ignored in this treatment. There is no interbarrrier zone.)

Isometric Resistance

Isometric contraction of 6 to 10 seconds. (Push feet straight down toward the floor, or push both feet up to the ceiling.) Complete relaxation. Repeat 3 to 5 times.

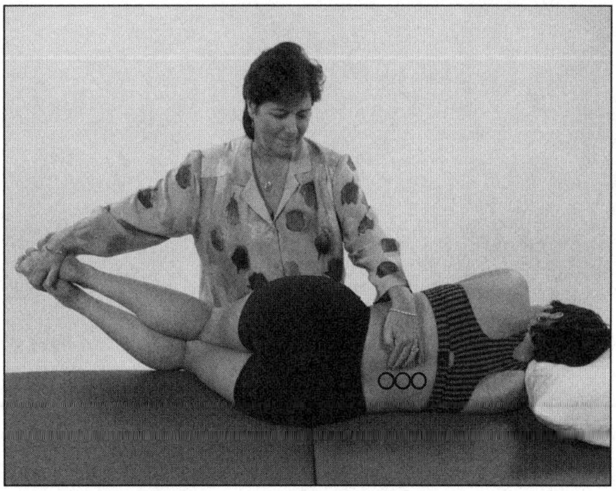

Note: This picture presents treatment of a Type I dysfunction of L2, 3, 4. The patient is lying on the side of the concavity; therefore L2, 3, 4 are stuck side bent to the right. Therefore L2, 3, 4 are stuck rotated to the left. The positional diagnosis is NSR. The movement barrier and the position of treatment is NSL. This patient would present limitations in motions of left side bending and right rotation of the lower lumbar spine.

It is important to note that often a latissimus dorsi in protective muscle spasm presents an appearance of a Type I lumbar dysfunction.

Type II Dysfunction of the Lumbar Spine

Type II dysfunctions of the lumbar spine are common. Flexed Type II's are more common than extended Type II's. Unless an adult has already had this therapy, there will almost always be a flexed Type II (FRSR or FRSL) at L5. Flexed Type II's of L4 are also common. L4 will be flexed, and rotation and side bending will present opposite the presentation of L5. For example, *if L5 is FRSR, then L4 will be FRSL.*

Other lumbar vertebrae may be flexed or extended. Often biomechanical dysfunction of L2, L3, and L4 is secondary to visceral dysfunction within the abdominal cavity.

L1 often presents as a crossover. Therefore, if L1 is stuck as a Type II flexed dysfunction, then T12 is Type II extended dysfunction. Or, if L1 is stuck as a Type II extended dysfunction, then T12 is a Type II flexed dysfunction.

Type II Flexed Lesions

Example

The left facet will not close. It is stuck open. The left facet is stuck flexed. During flexion, the facets will move evenly, and there will be no evidence of dysfunction. During extension, the right transverse process will become more posteriorly prominent, because the left facet will not close. The segment is stuck flexed and therefore cannot extend. This is a flexed lesion that will cause limitation of physiologic extension of the spine. The segment is side bent to the right because the left facet is stuck open. It cannot side bend to the left, and it will cause limitation of physiologic left side bending whenever the spine is extended. The segment is rotated to the right (rotation and side bending occur to the same side when flexed). Therefore, it cannot rotate to the left, and it will cause limitation of physiologic left rotation whenever the spine is extended. What sign indicates a Type II flexed lesion? There is one sole transverse process that is posteriorly prominent when the spine is extended. This is a unisegmental (one

Left Facet Stuck Open
Type II Dysfunction
Rotation & Side Bend Right

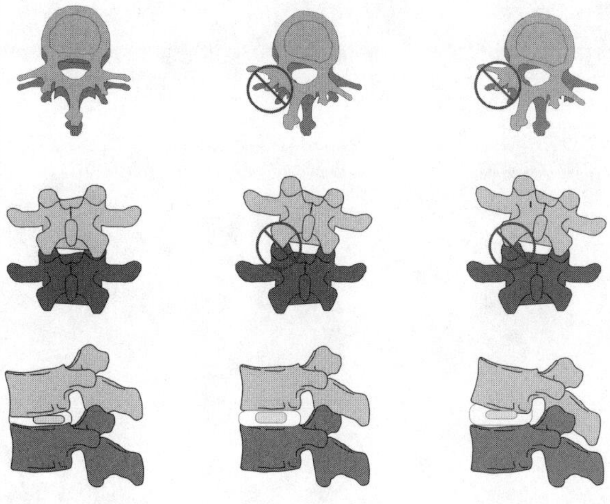

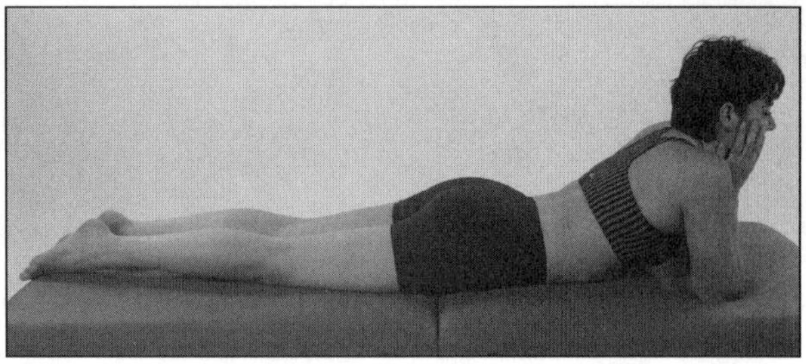

vertebra only) dysfunction. (Two or more consecutive and identical rotation dysfunctions found in spinal extension would not be a Type II dysfunction.)

Positional Diagnosis

1. Palpate transverse processes in extension.
2. Discover one individual posteriorly prominent transverse process. (Press lightly over the paravertebral muscles. Palpate over the hypertonic muscle. It is not necessary to palpate the bony prominence.)
3. Posterior prominence indicates rotation dysfunction of that vertebra.
4. The rotation of this vertebra is to the side of the posterior prominence.
5. Side bending is to the same side of the rotation (to the side of the posterior prominence); e.g., if the left facet is stuck open (flexed), the posterior prominent transverse process is on the right
   Flexed, rotated right, side bent right = FRRSR = FRSR

Note: In this position of sustained spinal extension, transverse processes can be palpated, looking for posterior prominence of one individual transverse process, for a Type II flexed lesion. It is easier to palpate if the chin is resting in the hands, rather than as presented in this picture, because in this manner the paravertebral muscles are not relaxed.

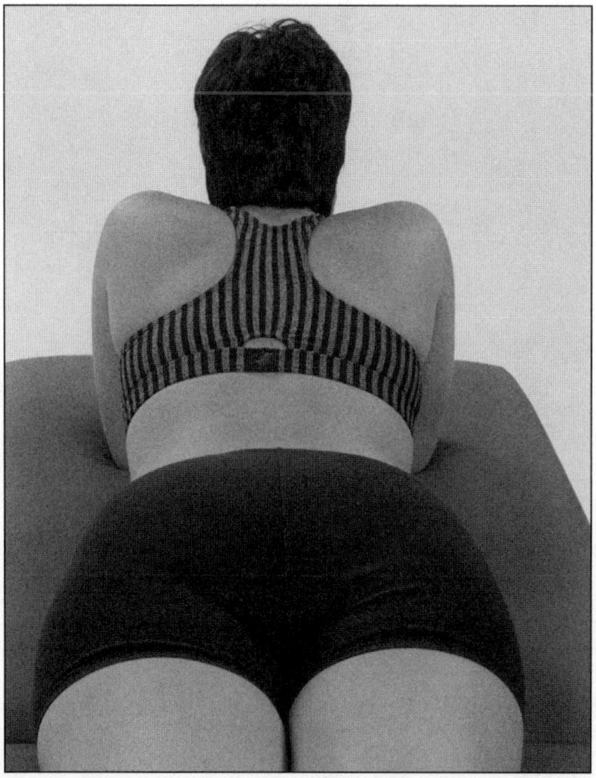

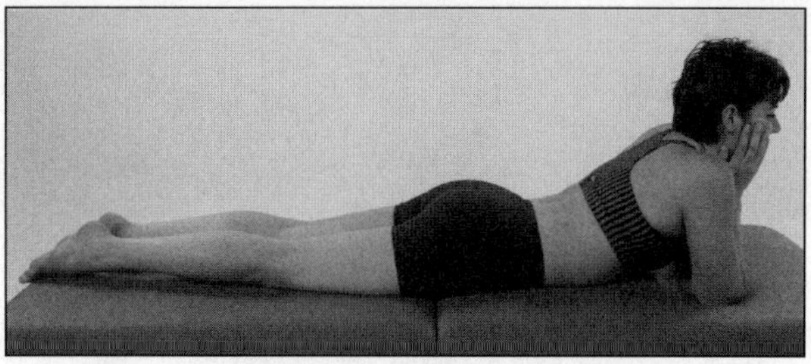

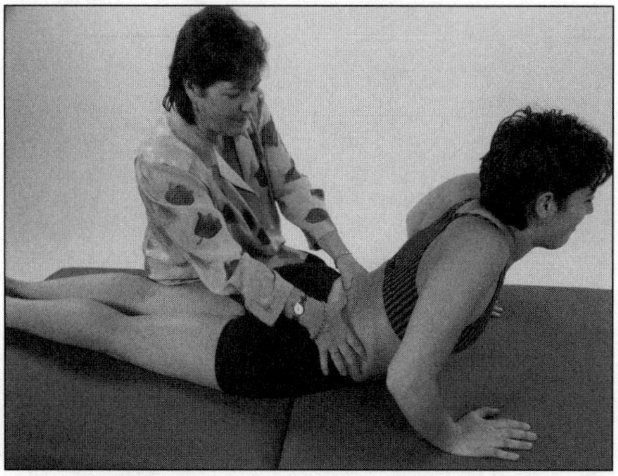

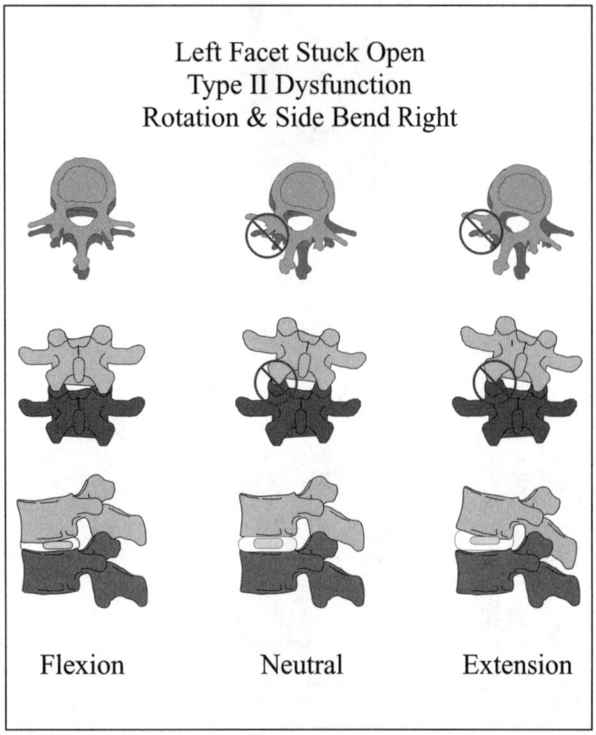

Movement Barrier

The movement barrier is opposite the positional diagnosis. That vertebra is stuck flexed, rotated, and side bent right, because the left facet is stuck open. That segment cannot extend, rotate left, and side bend left. The movement barrier is: extension, rotation left, side bending left = ERLSL.

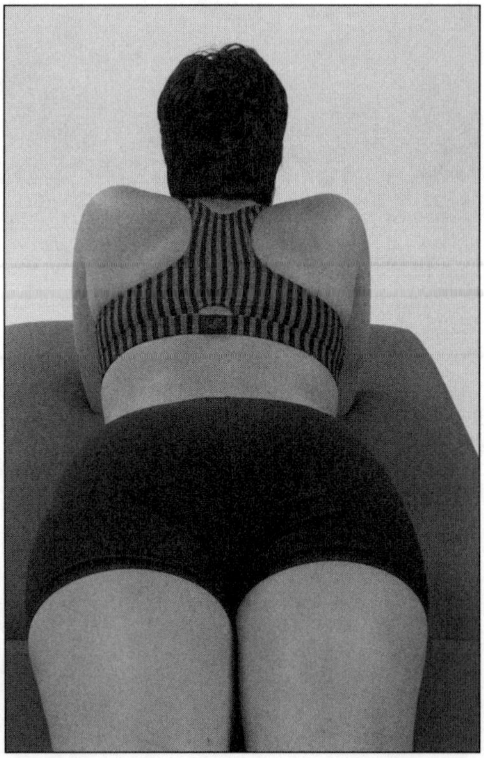

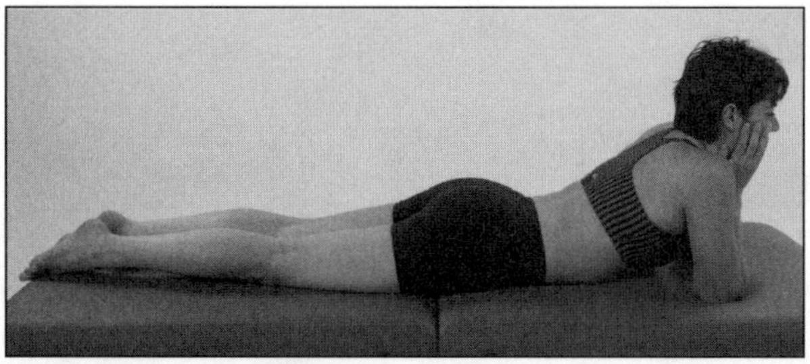

Extension, rotation and side bending left = ERSL

Position of Treatment

Position of treatment is the same as the movement barrier, but opposite the positional diagnosis. If the movement barrier is ERSL, then that vertebra must be positioned in the movement barrier positions: extension, left rotation, and left side bending. The position of treatment is: extension, rotation left, side bending left.

Observe: The lower leg extends; extension up the spinal kinetic chain attains the sagittal plane (extension) barrier. Rotation down the spinal kinetic chain attains the transverse plane (rotation) barrier. Flex the hip and knee; the knee rests on the bed. Raise the foot toward the ceiling to side bend up the spinal kinetic chain to attain the coronal plane (side bending) barrier. Apply isometric resistance: Push the foot toward the floor.

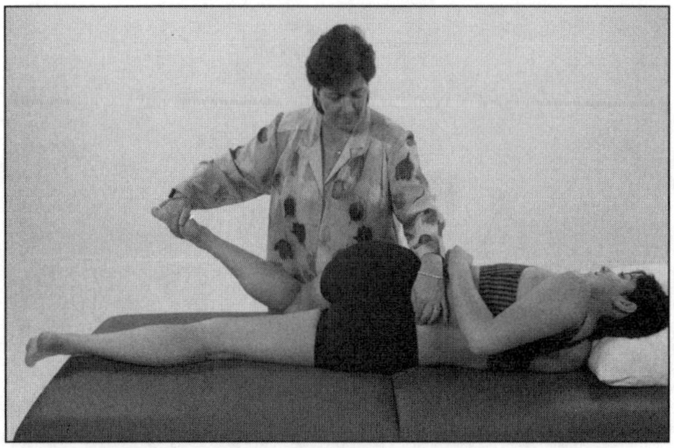

Extension, rotation and side bending left = ERSL

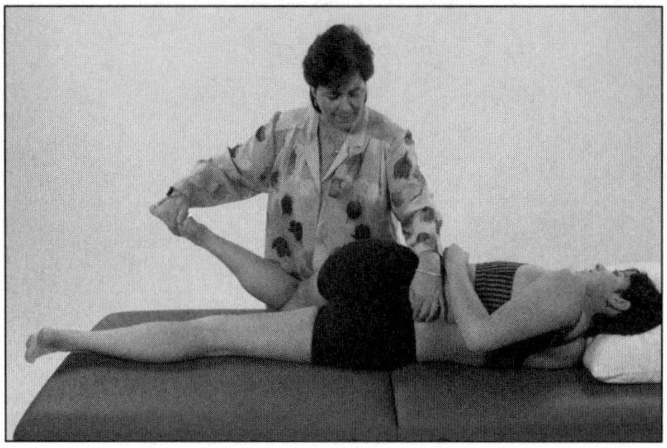

Treatment

In Type II dysfunctions the hypomobile vertebra is positioned in a 3-planar interbarrier zone: sagittal plane (extension); transverse plane (rotation to the movement barrier); coronal plane (side bending to the movement barrier). The barrier is palpated at the segment. The patient lies on the side of the prominent transverse process.

Find flexed Type II dysfunctions in physiologic extension.

The sagittal plane extension barrier is attained by extending the lower leg, extending up the kinetic chain to the barrier. The transverse plane rotation barrier is attained by rotating the trunk posteriorly, rotating down the kinetic chain to the barrier. The coronal plane side bending barrier is attained by lifting the top foot toward the ceiling, while the hip and knee are flexed, side bending up the kinetic chain to the barrier.

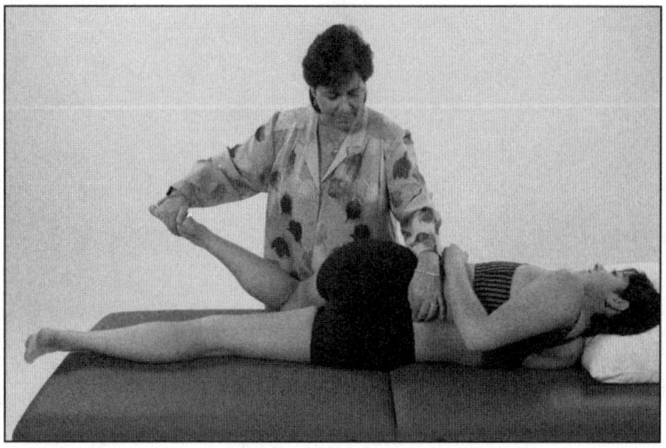

Isometric Resistance

The patient is side lying with the lower leg extended. The top leg is flexed at the hip and knee. The foot is lifted toward the ceiling to side bend up the kinetic chain to the side bending barrier. The resistance is to the upper foot; the patient pushes the upper foot to the floor.

Treatment of Type II Lumbar Dysfunction

Consider the significance of a Type II flexed dysfunction of L5. If L5 is flexed, then there is posterior displacement of the nucleus pulposus.

If L5 is side bent right, then the nucleus pulposus is displaced to the opposite left side.

If L5 is side bent right, then L5 is rotated right. (In Type II biomechanics, rotation and side bending occur towards the same side.)

If L5 is rotated, there is an increase in intra discal pressures.

Thus, at any time, with some trauma, there can be a further herniation of an already bulging disc. If there is a Type II flexed dysfunction of L5, there is also a bulging disc: posterior and to the side opposite the direction of side bending and rotation. In order to treat a bulging and/or herniated disc, correct biomechanical dysfunction of the segments above and below the disc (focus vertebral segment superior!). Then the abnormal hydrostatic pressure on the disc (Yes! Even the L5 disc) will normalize, and there should be a spontaneous reduction of the disc. Also, if the practitioner utilizes extension-exercises (i.e., Robin McKenzie extension press-ups) for exercise-induced reduction of discs, then correction of biomechanics (elimination of flexed, rotated and side bent vertebral dysfunction to normal Type II movement) will allow improved extension for movement therapy.

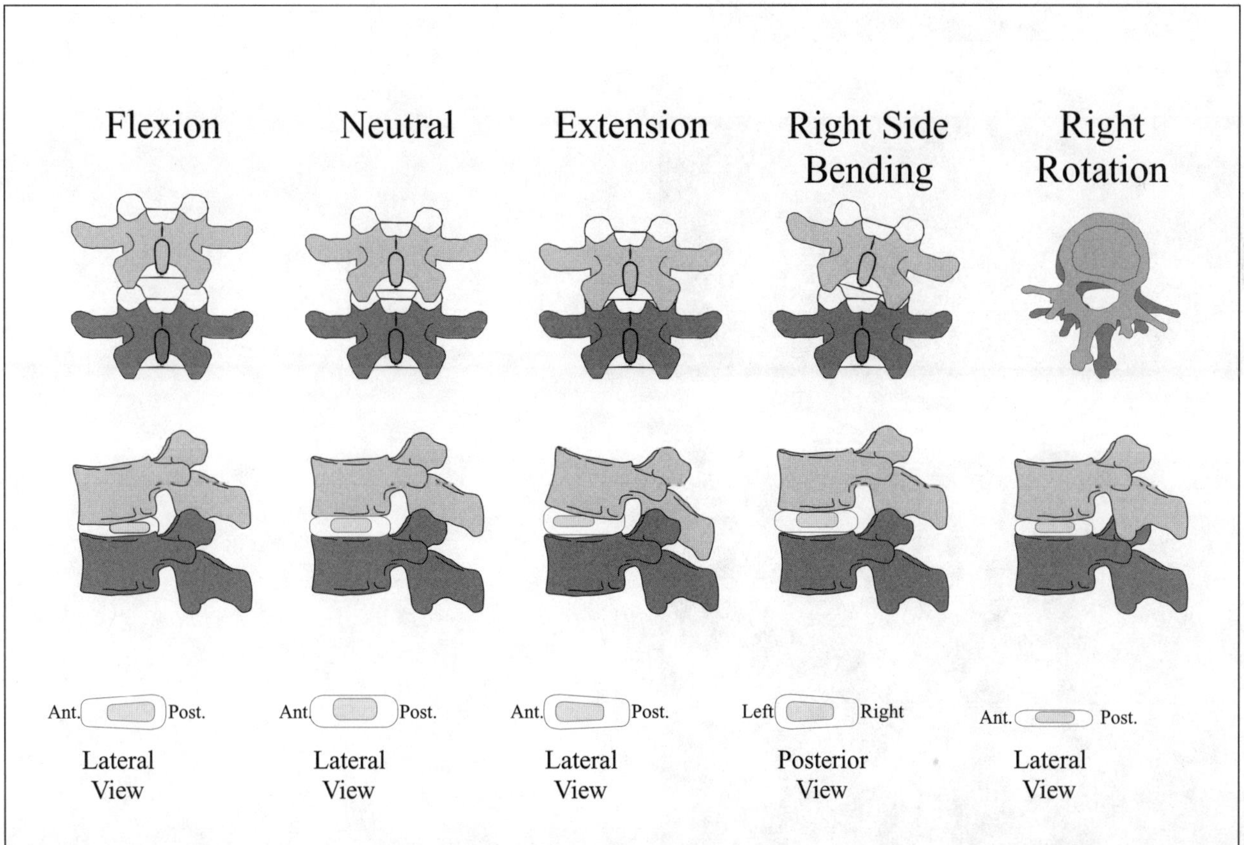

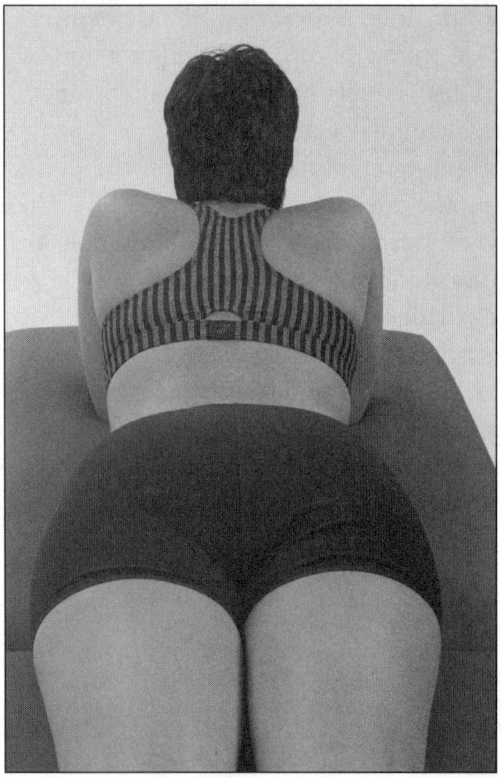

Flexed Lesions

Diagnosis

1. Diagnose in prone extension.
2. Limitations in lumbar extension.
3. Patient lies in prone puppy extended position with chin resting in hands.
4. Palpate transverse processes bilaterally from L5 to T9 (above T9 is performed in sitting extension).
5. Find one sole posteriorly prominent transverse process.
6. Ask patient to press up into full extension and palpate segment during dynamic movement.

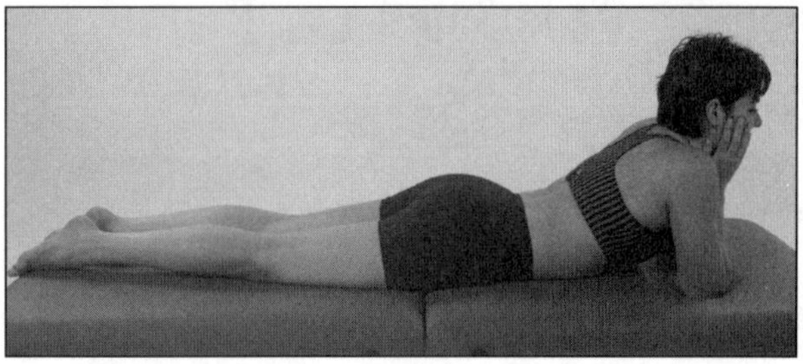

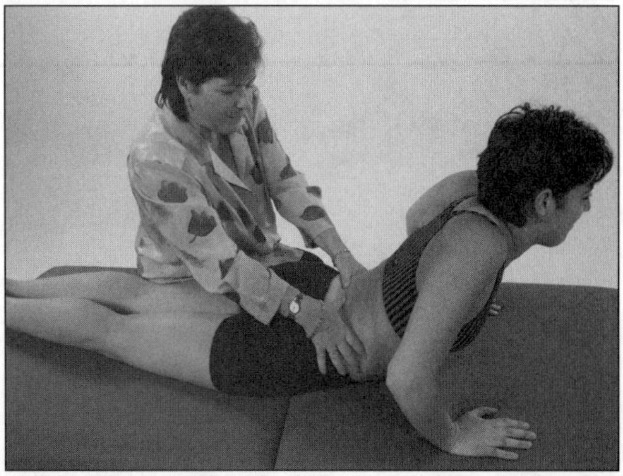

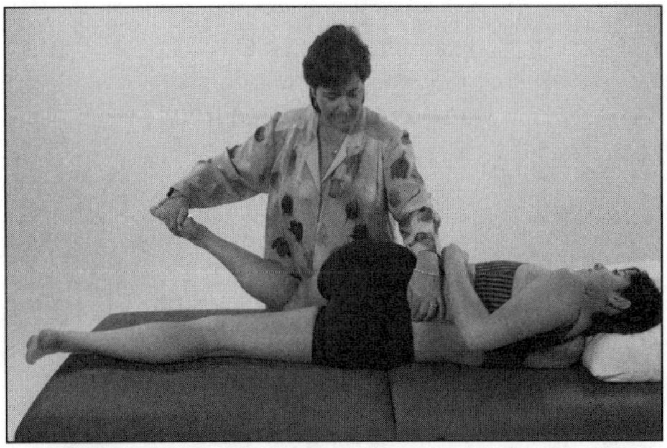

Rationale

Flexed lumbar lesions are treated in an extended position similar to that for treatment of posterior torsions of the sacrum. The patient lies on the side of the posterior transverse process.

Treatment

Patient is side lying with prominent transverse process down to floor (lie on side of prominent transverse process). (Position patient similar to the posterior sacral torsion treatment position.) Palpate barrier at segment (hypomobile vertebra).

First Barrier:

Extend lower leg. Extend up the kinetic chain until palpation of barrier. Return to interbarrier zone.

Second Barrier:

Rotate trunk and shoulder/arm backwards (posteriorly). Rotate down the kinetic chain until palpation of barrier.

Third Barrier:

Hip of top leg is flexed 45 degrees; knee is flexed to 90 degrees. With knee resting on table, lift foot up to ceiling, side bending up the kinetic chain until palpation of barrier. Return to interbarrier zone.

Resist:

Isometric contraction: Push foot to floor for 6 to 10 seconds. Repeat 3 to 5 times.

Note: This patient is lying on her right side, which is the side of the prominent transverse process. She is being treated for an FRSR.

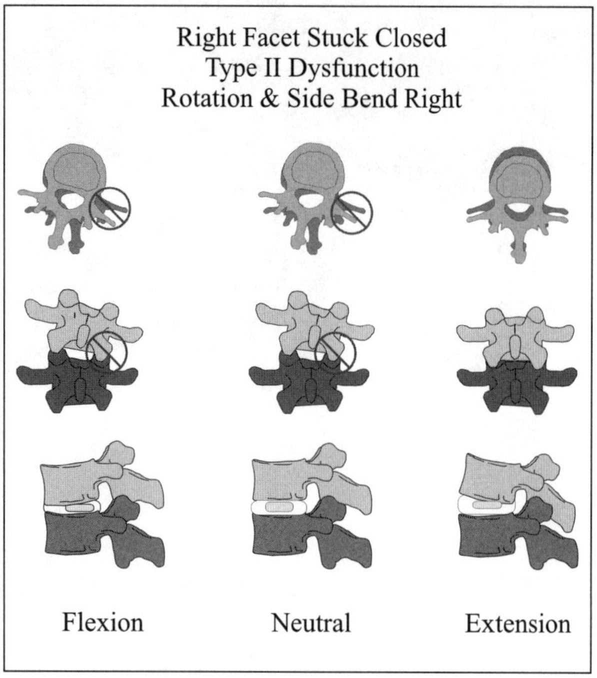

Type II Extended Lesions

Example

The right facet does not open. It is stuck closed. The right facet is stuck extended. During flexion the right transverse process will move posterior and prominent. During extension the right transverse process will move evenly, and there will not be any posterior prominence. The segment is stuck extended, and therefore cannot flex. This is an extended lesion, which will cause limitations of physiologic flexion of the spine. The vertebra is side bent to the right because the right facet is stuck closed. It cannot sidebend to the left, and it will cause limitation of physiologic left side bending whenever the spine is flexed. The vertebra is rotated to the right. (Rotation and side bending occur to the same side when extended.) The segment cannot rotate to the left, and it will cause limitation of physiologic left rotation whenever the spine is flexed. What sign indicates a Type II extended lesion? There is one sole transverse process, which is posteriorly prominent when the spine is flexed. This is a unisegmental (one vertebra only) dysfunction. (Two or more consecutive and identical rotation dysfunctions found in spinal flexion would not be a Type II dysfunction.)

Positional Diagnosis

If the right facet is stuck closed (extended), the posteriorly prominent transverse process is on the right.

1. Palpate transverse processes in flexion.
2. Discover one individual posteriorly prominent transverse process. (Press lightly over the paravertebral muscles. Palpate over the hypertonic muscle. It is not necessary to palpate bony prominence.)
3. Posterior prominence indicates rotation dysfunction of that vertebra.
4. The rotation of this vertebra is to the side of the posterior prominence.
5. Side bending is to the same side of the rotation (to the side of the posterior prominence). Extended, rotated right, side bent right = ERRSR.

Note: The dot is over the posteriorly prominent transverse process of L3. Therefore L3 is stuck rotated right. Therefore L3 is stuck side bent right. L3 is stuck extended, because the right transverse process becomes prominent in flexion. Therefore the positional diagnosis is ERSR.

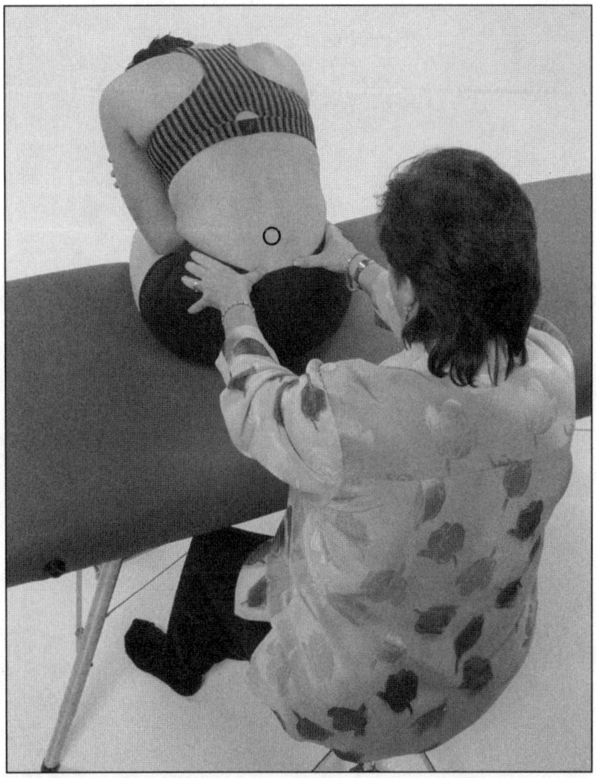

Extended, rotated, and side bent right = ERSR

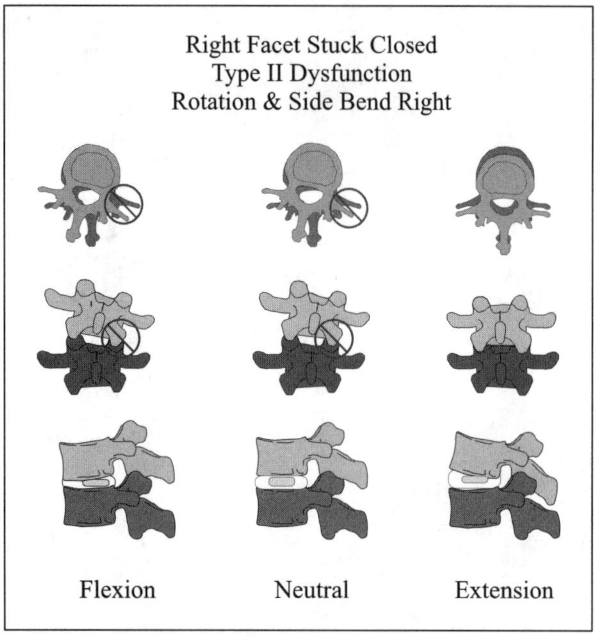

Flexion, rotation and side bending left = FRSL

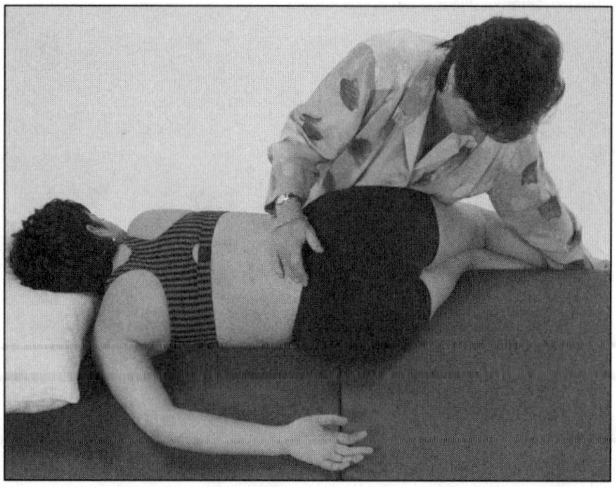

Flexion, rotation and side bending left = FRSL

Movement Barrier

The movement barrier is the opposite direction of the positional diagnosis. That vertebra is stuck extended, rotated and side bent right because the right facet is stuck closed. That segment cannot flex, rotate left, or side bend left. The movement barrier is: flexion, left rotation, left side bending = FRLSL.

Position of Treatment

The position of treatment is the same as the movement barrier, but opposite the positional diagnosis. If the movement barrier is FRSL, then that vertebra must be positioned in flexion, left rotation, and left side bending. The position of treatment is: flexion, rotation left, side bending left = FRSL.

Note: Sagittal plane (flexion) barrier is attained by bringing knees to chest and flexing up the spinal kinetic chain; transverse plane (rotation) barrier is attained by reaching to the floor with the upper hand and rotating down the kinetic chain; coronal plane (side bending) barrier is attained by bringing both feet to the floor and side bending up the spinal kinetic chain.

1. In Type II dysfunctions the hypomobile vertebra is positioned in a 3-planar interbarrier zone: sagittal plane (flexion); transverse plane (rotation to the movement barrier); coronal plane (side bending to the movement barrier). The barrier is palpated at the segment.
2. The patient lies on the side opposite the prominent transverse process.

Find extended Type II dysfunctions in physiologic flexion.

3. The sagittal plane flexion barrier is attained by flexing both hips (with knees bent), flexing up the kinetic chain to the barrier.
4. The transverse plane rotation barrier is attained by rotating the trunk anteriorly, rotating down the kinetic chain to the barrier.

5. The coronal plane side bending barrier is attained by dropping both feet to the floor (the upper leg is on top of the bottom leg), side bending up the kinetic chain to the barrier.

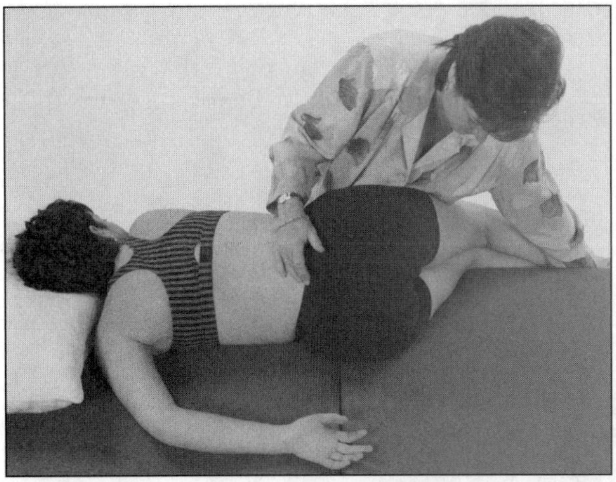

Isometric Resistance

The patient is side lying with one leg on top of the other, the knees just off the edge of the bed. The feet are lowered toward the floor to side bend up the kinetic chain. The resistance is to both feet: the patient pushes both feet up or both feet down.

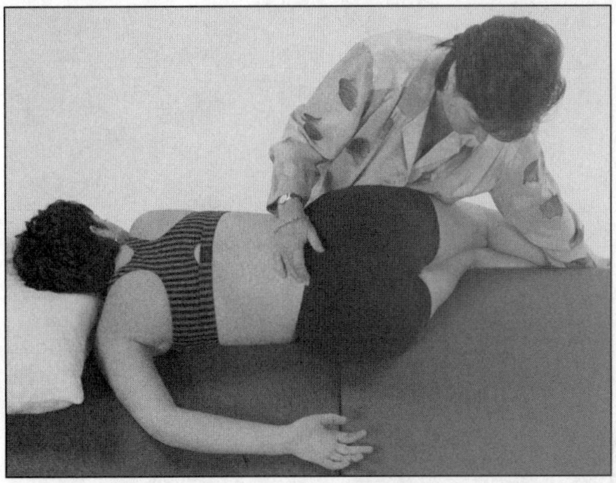

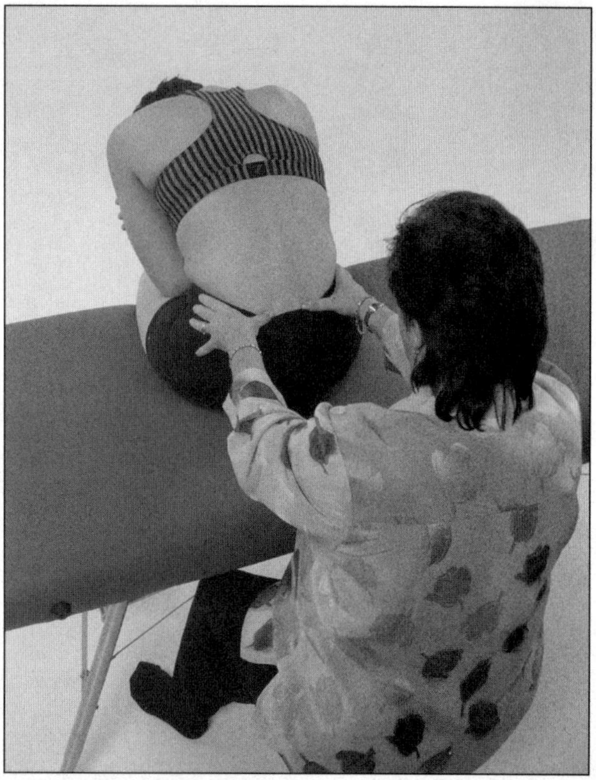

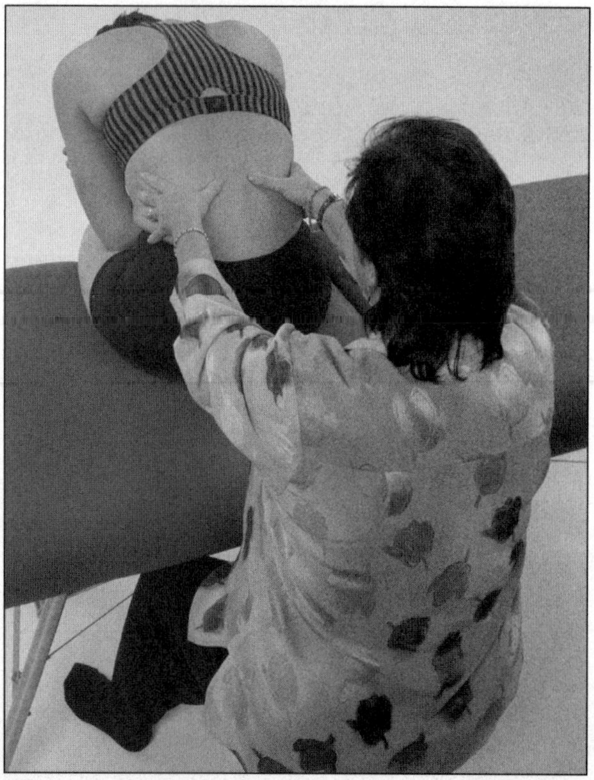

Extended Lesions

Diagnosis

1. Diagnose in sitting flexion.
2. Limitations in lumbar flexion.
3. Patient fully flexes.
4. Palpate transverse processes bilaterally from L5 up spine.
5. Find one sole posteriorly transverse process.
6. Ask patient to flex into full flexion and palpate segment during dynamic movement.

Rationale

Extended lumbar lesions are treated in a position similar to that for anterior sacral torsions. Position the patient with the posteriorly prominent transverse process up. In extended lesions, elevate the prominent transverse process. (The patient lies on the side opposite the prominent transverse process.)

Treatment

Patient side lying with posterior transverse process up and lower arm positioned behind back. (Position patient similar to the anterior sacral torsion treatment position.) Palpate barrier at segment (hypomobile vertebra).

First Barrier:
    Flex both hips to chest with knees flexed until palpation of barrier. Return to the interbarrier zone.

Second Barrier:
    Upper arm off bed reaches to floor as chest rotates anteriorly. Rotate down the kinetic chain until palpation of barrier. Return to the interbarrier zone.

Third Barrier:
    Lower both feet to the floor to sidebend up the kinetic chain until palpation of barrier. Return to interbarrier zone.

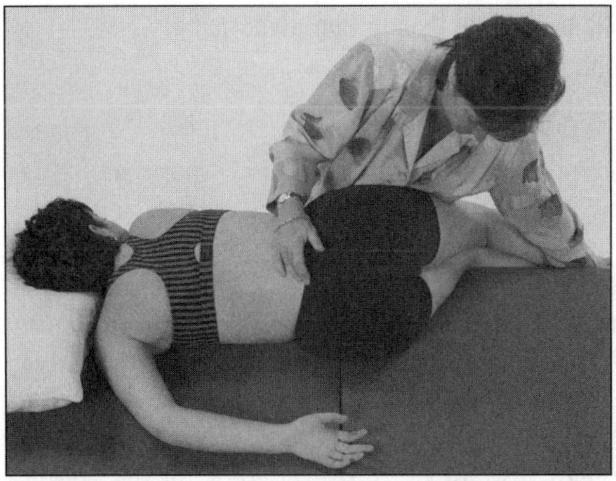

Resist: Isometric resistance: Push both feet up to the ceiling, or push both feet down to the floor, 6 to 10 seconds. Repeat 3 to 5 times.

Note: Type II extended lesions between L5 and T9 are treated in the side lying position. Extended lesions superior to T9 are treated in sitting.

Start here

EXTENDED TYPE II	FLEXED TYPE II
LEFT	RIGHT
• Facet stuck closed	• Facet stuck open
• Extended lesion	• Flexed lesion
• Rotated left	• Rotated left
• Side bent left	• Side bent left
ERLSL =	FRLSL =
• ERSL positional diagnosis	• FRSL positional diagnosis
• FRSR movement barrier	• ERSR movement barrier
• FRSR position of treatment	• ERSR position of treatment
• found and diagnosed in flexion	• found and diagnosed in extension
• Treat in flexion	• Treat in extension
• Prominent transverse process on same side of dysfunctional facet	• Prominent transverse process opposite side of dysfunctional facet

Biomechanical Dysfunction, Hypertonicity, and the Disc

Hypothesis

Hypertonicity (protective muscle spasm and spasticity) results in biomechanical dysfunction of the vertebra(e). Biomechanical dysfunction of the vertebra(e) compromises the disc(s) by causing abnormal hydrostatic pressure. Elimination of hypertonicity can diminish biomechanical dysfunction, normalizing the articular balance between the vertebral bodies, thereby equalizing the hydrostatic pressure exerted on the disc, resulting in spontaneous reduction of the disc(s) and correction of the discopathy. The disc can be viewed as a sign and a symptom.

Disc Behavior During Simple Movements

When the spine flexes, the vertebrae flex. Each segment flexes in isolation as Type II motion. During flexion, the facets of those segments open. The hydrostatic pressure on the disc will cause the disc to move somewhat in a posterior direction.

When the spine extends, the vertebrae extend. Each segment extends in isolation as Type II motion. During extension, the facets of those segments come together. The hydrostatic pressure on the disc will cause the disc to move somewhat in an anterior direction.

Apparently during flexion and extension this pressure influences the nucleus pulposus, rather than the annulus fibrosus.

When the vertebrae flex during spinal flexion, they respond in a Type II movement pattern. This means that when the spine rotates to one side during spinal flexion, those vertebrae will side bend to the same side. Unless there is spinal dysfunction, the rotation and side bending are pro-rated, which means there is an equal percentage of rotation and side bending of each vertebra.

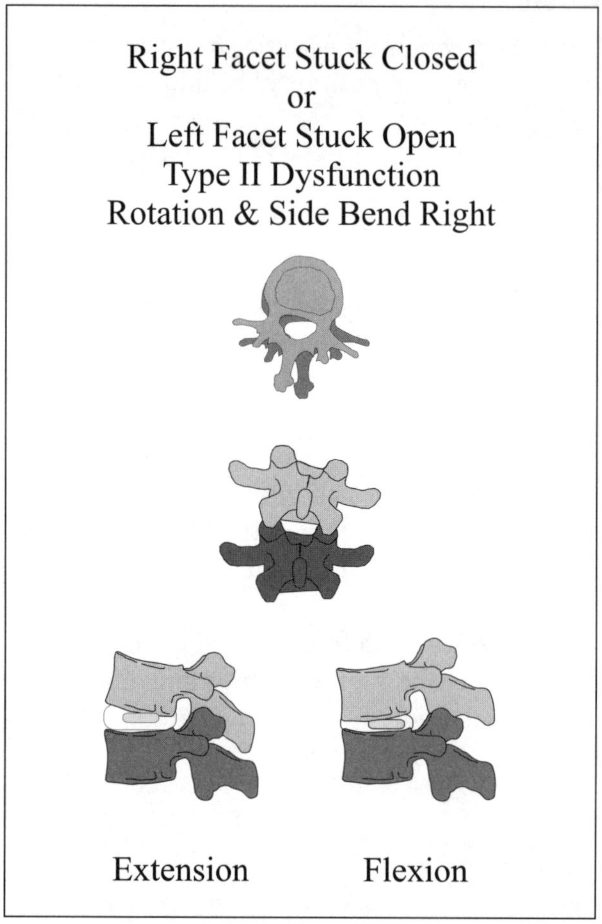

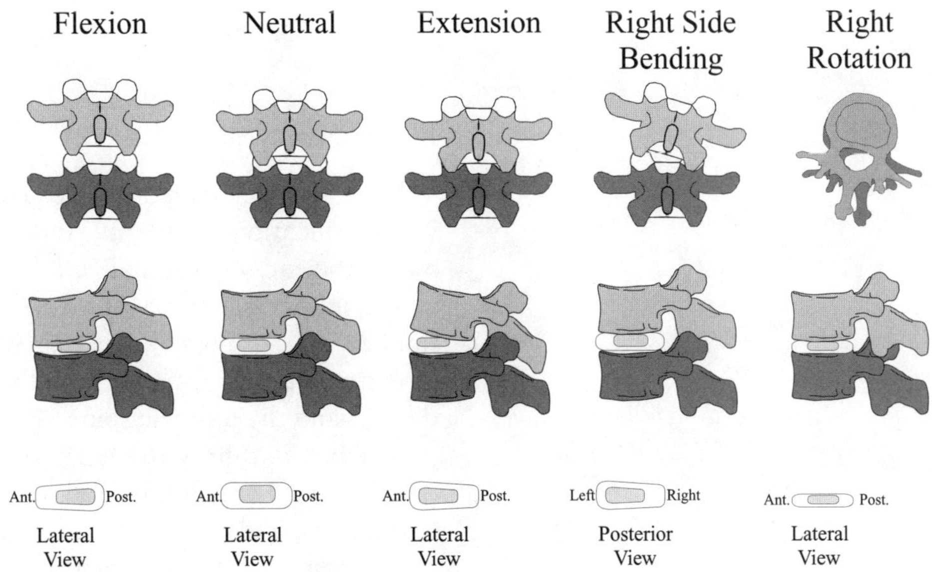

During flexion: The facets open. The upper vertebra moves anteriorly; the anterior interspace is reduced; the nuclear material is driven posteriorly.
During side bending: The facet closes on the side of side bending, and opens on the opposite side. The upper vertebra tilts toward the side of side bending. The nuclear material is driven to the opposite side.

During extension: The facets close. The upper vertebra moves posteriorly; the posterior interspace is reduced; the nuclear material is driven anteriorly.
During rotation: The oblique fibers, running counter to the direction of movement, are stretched. The nuclear material is strongly compressed, and internal intradiscal pressure rises in proportion to the angle of rotation.

The disc has contractile tissue. It responds to the Strain and Counterstrain approach. There is a remarkable body of knowledge today for therapy on the disc. The author refers the practitioner to: *Advanced Strain and Counterstrain for the Autonomic Nervous System and Related Disorders*, Giammatteo and (Weiselfish) Giammatteo (North Atlantic Books).

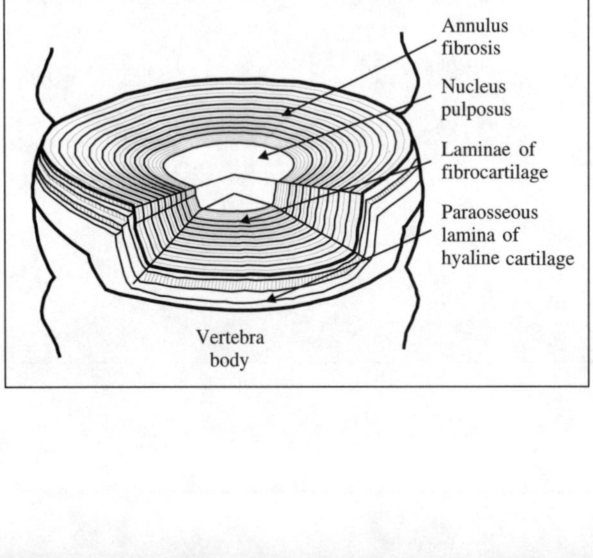

CHAPTER 23

BIOMECHANICAL DYSFUCNTION OF THE THORACIC SPINE

The Chest Cavity

The thoracic spine cannot be adequately presented without mentioning the ribs. Often treatment is more efficient when spinal vertebrae of the thoracic spine are addressed before the ribs. Yet occasionally, it is not possible to attain biomechanical integrity without going back and forth between the ribs and the thoracic spine.

Pleura, diaphragms, lung tissue, bronchi, breast tissue, lymph nodes and tissue, blood vessels, nerves, and other soft tissues greatly affect biomechanics of the rib cage (thoracic spine and rib joints). More often than not, when there are significant problems with these tissues, organs, and vessels, biomechanics of the rib cage joints cannot be optimally addressed.

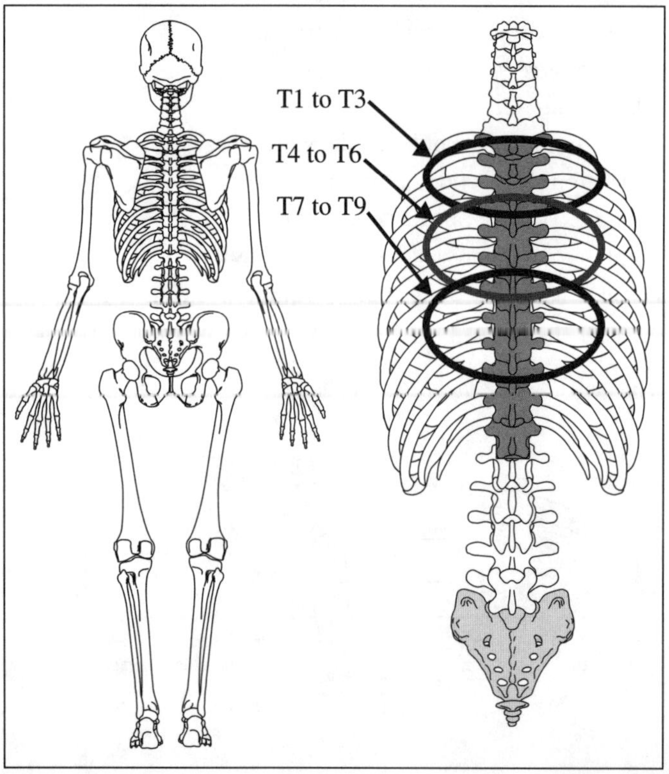

Thoracic Spine Type I Dysfunction

Biomechanics

The first thoracic vertebra acts like the cervical spine. The last thoracic vertebra acts like the lumbar spine. It is important, when working with the thoracic spine, to note the thoracic "Law of Threes":

1. T1 to T3: The transverse processes are one-half ($\frac{1}{2}$) vertebral segment cephalad to their respective spinous process.
2. T4 to T6: The transverse processes are one (1) vertebral segment cephalad to their respective spinous process.
3. T7 to T9: The transverse processes are one and one-half ($1\frac{1}{2}$) vertebral segments cephalad to their respective spinous process.

Diagnosis

Evaluate and diagnose the spine above T9 in sitting. Evaluate up to T9 in the recumbent position, like the lumbar spine. Above T9, the thoracic spine is more easily evaluated in the sitting position. There are crossovers of Type II dysfunction within a Type I curve (when one vertebra is rotated and side bent right, while the consecutive segment is rotated and side bent left).

Palpate transverse processes bilaterally. Observe two or more adjacent transverse processes prominent posteriorly.

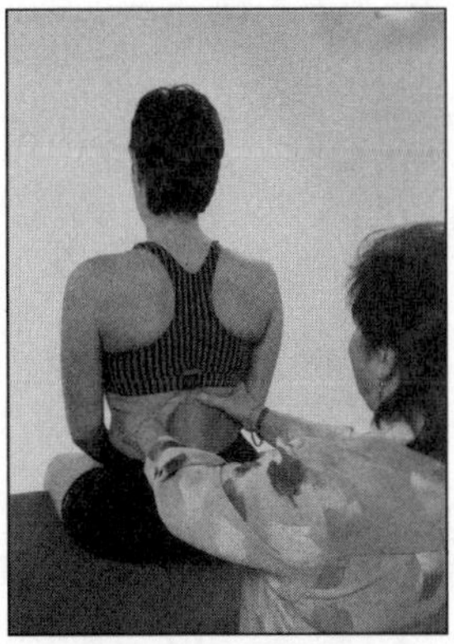

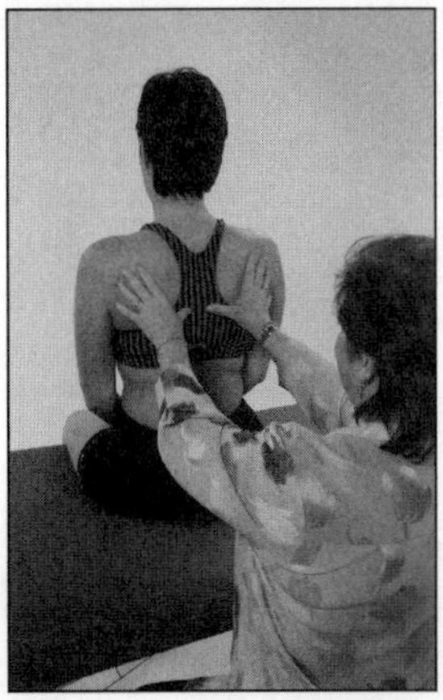

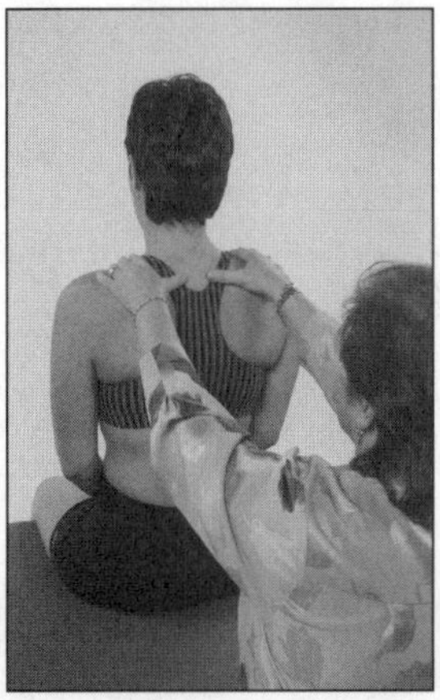

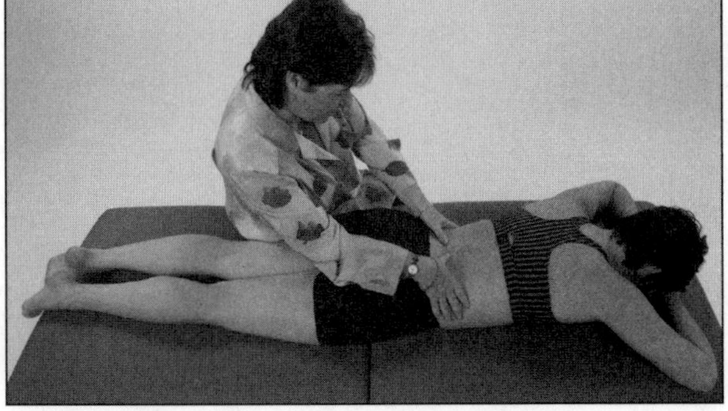

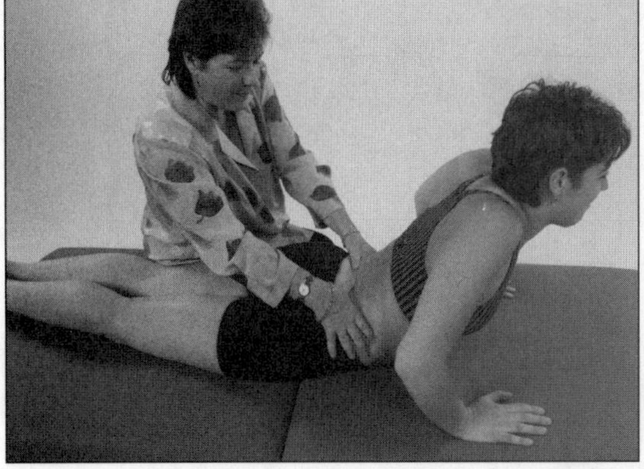

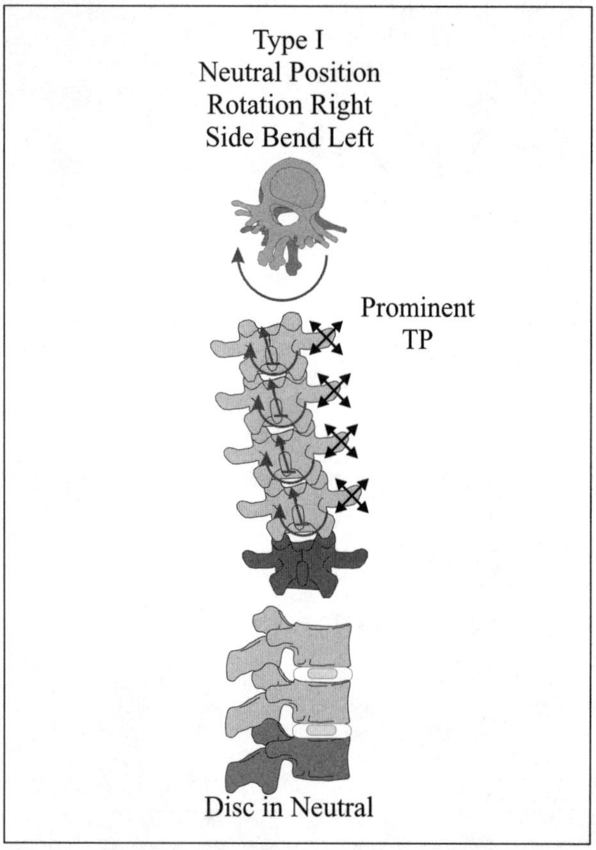

Neutral, side bent left = NSL

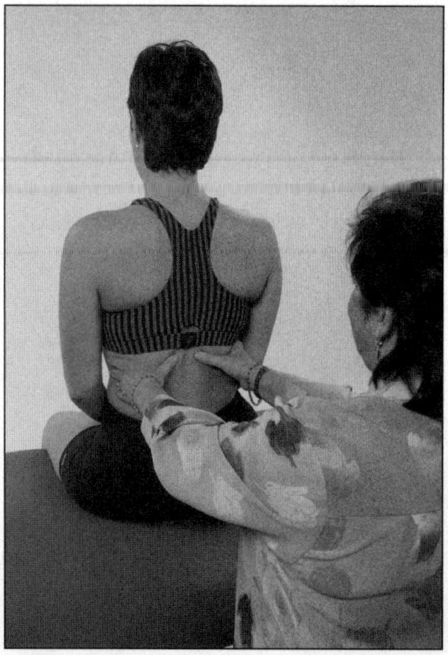

Positional Diagnosis

If the transverse processes of T4 to T9 are prominent posteriorly to the right, then those segments are stuck rotated right. Fryette's law I states that these segments will be side bent left because rotation and side bending occur to opposite sides in neutral. This dysfunction occurs only in AP neutral. Two or more consecutive vertebrae must be found with identical rotation dysfunction (posteriorly prominent transverse processes on the same side), because Type I dysfunctions are multisegmental. The positional diagnosis is: neutral, side bent left, rotated right = NSLRR.

Note: The dots are over 3 consecutive prominent transverse processes. Therefore, the positional diagnosis is: NSL.

Movement Barrier

The movement barrier is opposite the positional diagnosis. In neutral, the segments of T4 to T9 are limited in right side bending and left rotation. The movement barrier is: neutral, side bending right, rotation left = NSRRL.

Note: The patient with a positional diagnosis NSL will have limitations in physiologic right side bending and left rotation of that part of the spine.

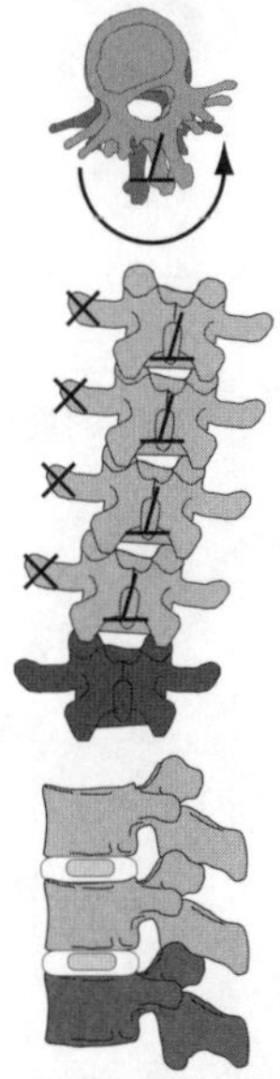

Neutral, side bending right = NSR

Position of Treatment

The position of treatment is the same as the movement barrier. In neutral, T4 to T9 are side bent to the right. The rotational component is not considered significant. Therefore side bending is to the barrier. The barrier is the vertebra at the apex of the curve.

Note: The therapist can stand closer to the patient for better contact and support.

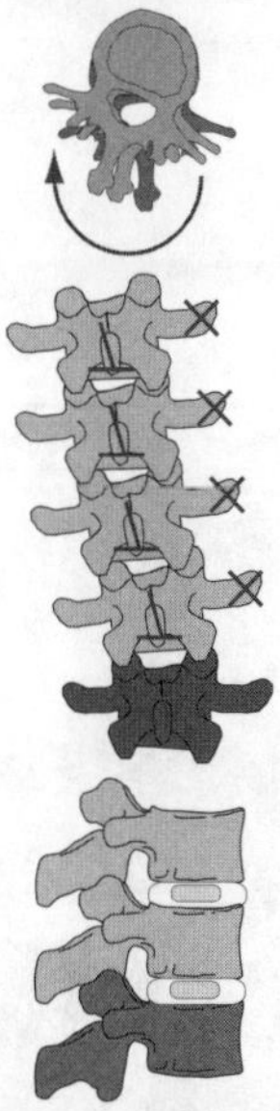

Neutral, side bending right = NSR

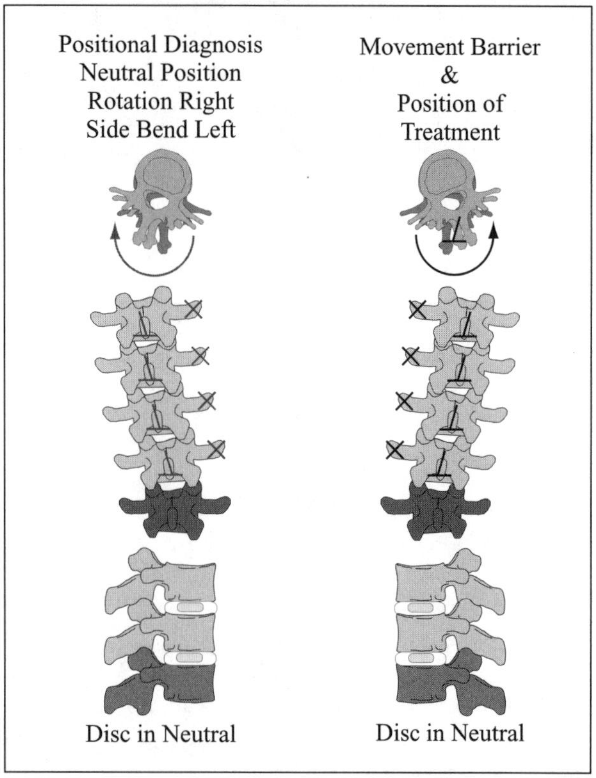

Treatment

If the Type I dysfunction is low thoracic (below T9) or thoracolumbar, then the patient lies on the side of the concavity, so that gravity assists in the correction. The hips are flexed 30 degrees to 45 degrees; the knees are flexed 90 degrees. One leg is on top of the other. Both feet are raised toward the ceiling, side bending up the kinetic chain to the barrier. There is no interbarrier zone in Type I treatment.

If the Type I dysfunction is above T9, then the patient is seated in neutral; often the patient is more comfortable with arms crossed and hands on shoulders.

Barrier: Side bend down the kinetic chain to the barrier to the side of the convexity. Palpate for the barrier at the segment (apex of curve).

Resist At the barrier, resist side bending isometrically 3 to 5 seconds. Repeat 5 to 6 times.

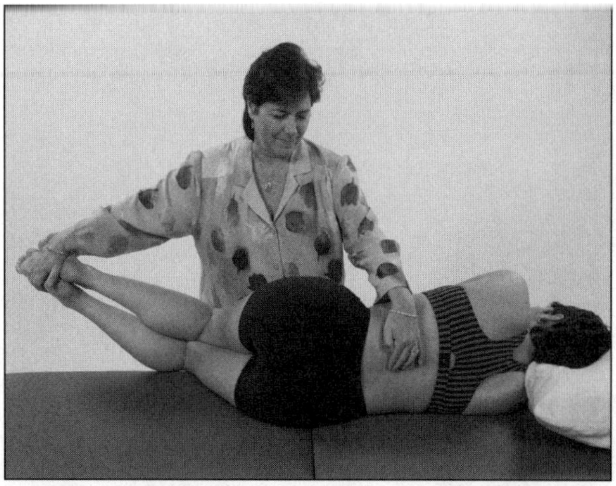

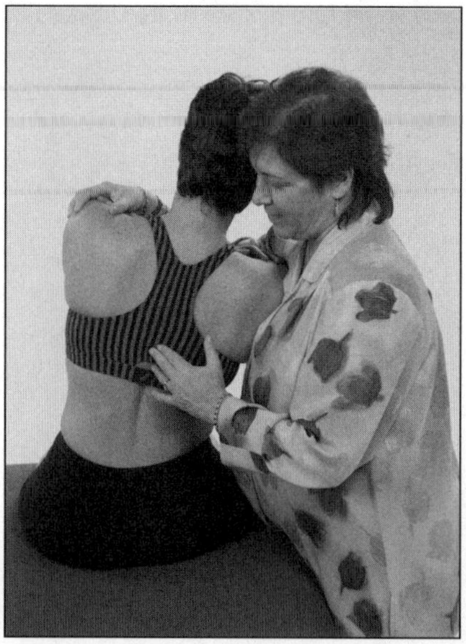

Isometric Resistance

For low thoracic and thoracolumbar Type I dysfunctions, the patient is side lying with one leg on top of the other. The feet are raised to the ceiling, side bent up the kinetic chain. The resistance is to both feet. The patient pushes both feet up or both feet down. For thoracic Type I dysfunctions treated in sitting, resistance can be applied to the shoulder girdles or to the head (gently).

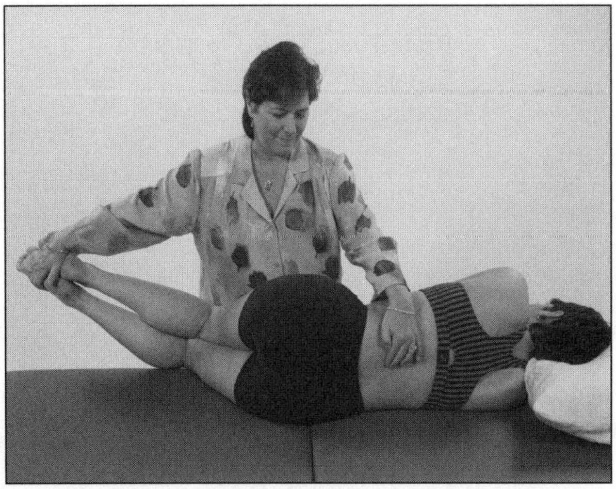

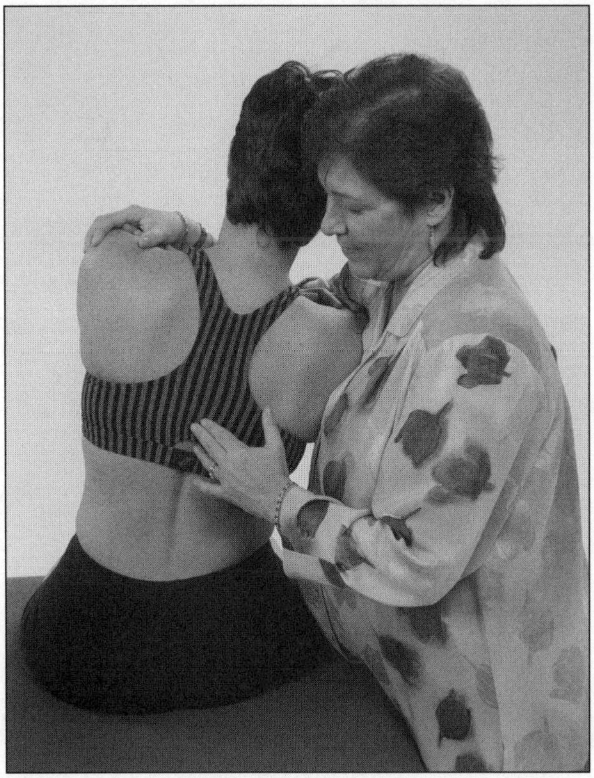

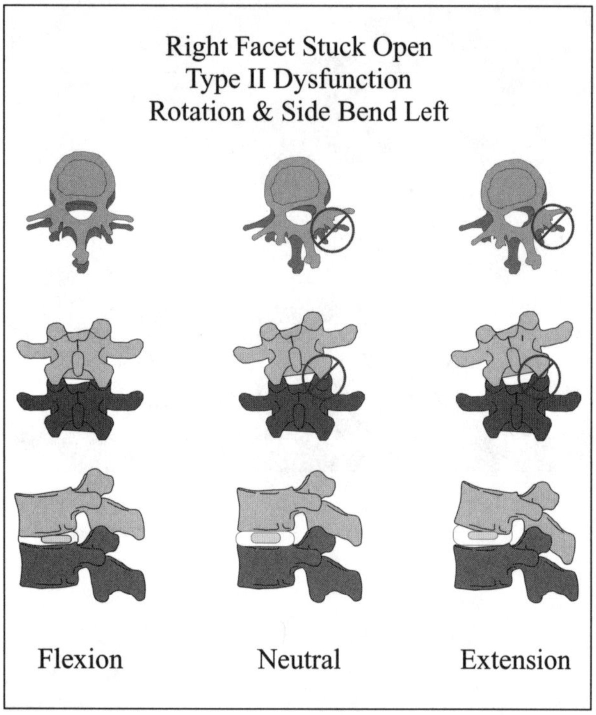

Thoracic Spine Type II Dysfunction: Flexed Lesions

Example

The right facet is stuck open. It does not close. The right facet is stuck flexed. During extension, the left transverse process will become posteriorly prominent. During flexion, the left transverse process will move more evenly, and there will not be any sign of posterior prominence. The segment is already flexed. This is a flexed lesion. Physiologic extension will be inhibited. The vertebra is stuck side bent left. It will not side bend right. Physiologic right side bending will be limited. The vertebra is stuck rotated left. The segment will not rotate right. Physiologic right rotation will be limited.

Positional Diagnosis

In seated extension, palpate one sole transverse process prominent posteriorly. A prominent transverse process reflects rotation dysfunction to the side of the prominence. If the prominent transverse process is on the left side, then the segment is rotated left. Fryette's law II states that the segment will be side bent left, because side bending and rotation occur to the same side. If the right facet is stuck open, the positional diagnosis is: flexed, rotated left, and side bent left = FRLSL.

Note: Only one sole prominent transverse process will be evident in a Type II dysfunction. (Two or more identical consecutive rotation dysfunctions in extension would not be a Type II dysfunction.)

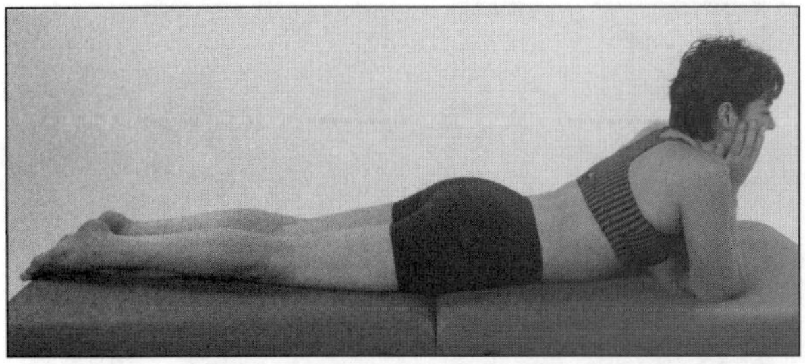

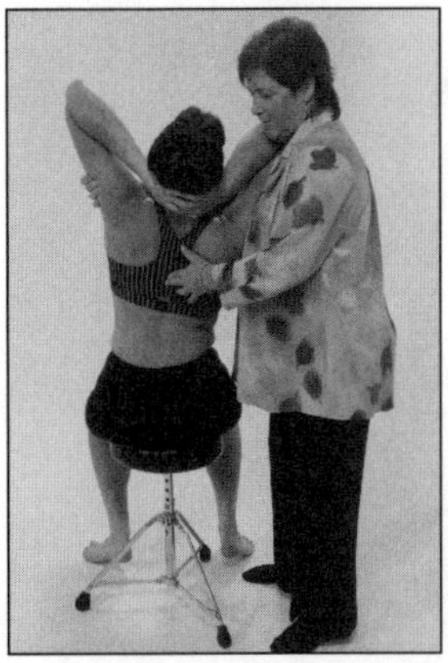

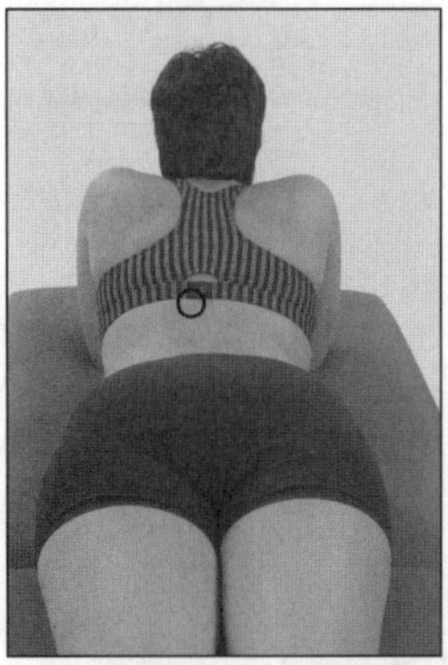

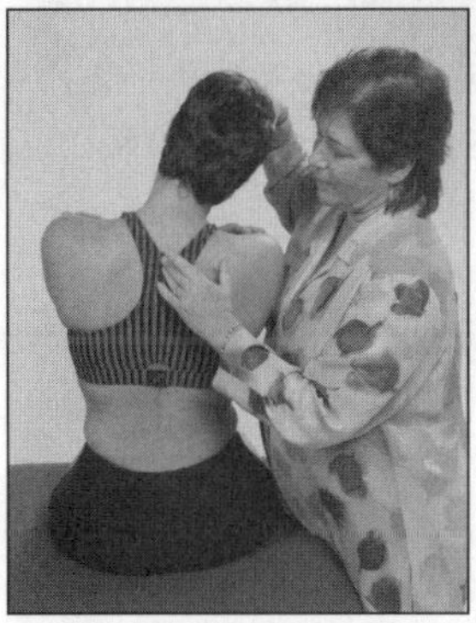

Flexed, rotated and side bent left = FRSL

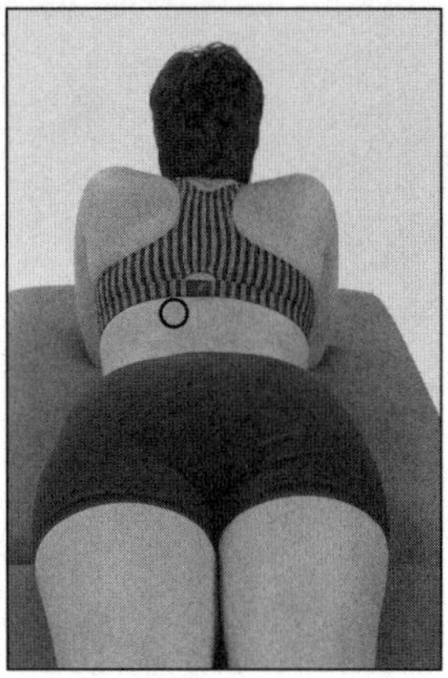

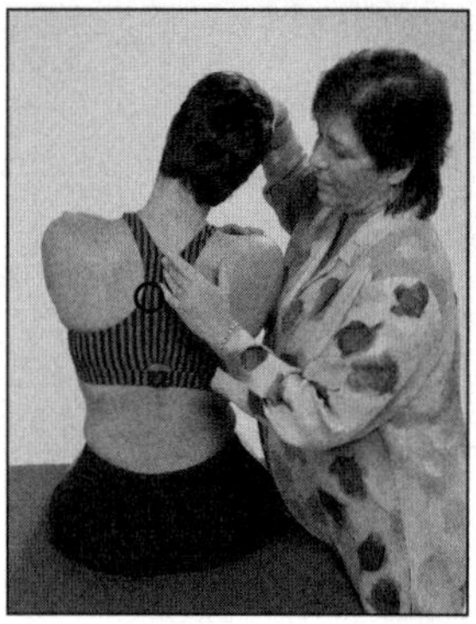

Extension, rotation and side bending right = ERSR

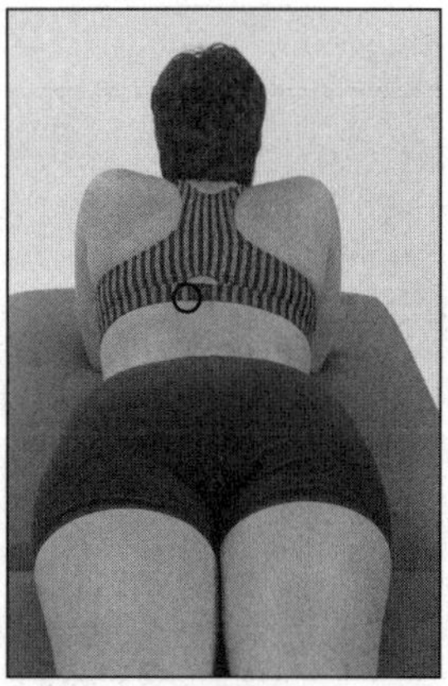

Extension, rotation and side bending right = ERSR

Movement Barrier

The movement barrier is opposite the positional diagnosis. If the facet is flexed, the segment is flexed; it cannot extend. If the segment is rotated left, it cannot rotate right. If the segment is side bent left, it cannot side bend right. If the right facet is stuck open, the movement barrier is: extension, rotation right, side bending right = ERRSR.

Note: Transverse processes can also be palpated with two thumbs. Ask the patient to bring the abdomen forward, and bring the head, neck, and arms backward.

Position of Treatment

The position of treatment is the same as the movement barrier. The vertebra must be positioned in extension, right rotation, and right side bending. The vertebra is positioned in a 3-planar interbarrier zone. The position of treatment is:

Treatment

If the Type II dysfunction is low thoracic (below T9) or thoracolumbar, then the patient is positioned side lying. If the Type II dysfunction is above T9, then the patient is positioned sitting; often the patient is more comfortable with arms crossed and hands on shoulders.

The patient is seated. The therapist stands on the side opposite the prominent transverse process. The patient's hands are clasped behind the head. The therapist palpates for the barrier at the vertebra. The patient extends: (1) brings the abdomen forward to extend up the kinetic chain; (2) extends the head and neck to extend down the kinetic chain to the sagittal plane barrier (interbarrier zone). The patient side bends down the kinetic chain to the right, to the transverse plane barrier (interbarrier zone). The patient rotates down the kinetic chain to the barrier (interbarrier zone). At the 3-planar interbarrier zone an isometric resistance is applied, to any direction, for 6 to 10 seconds. After the relaxation, this process is repeated 3 to 5 times.

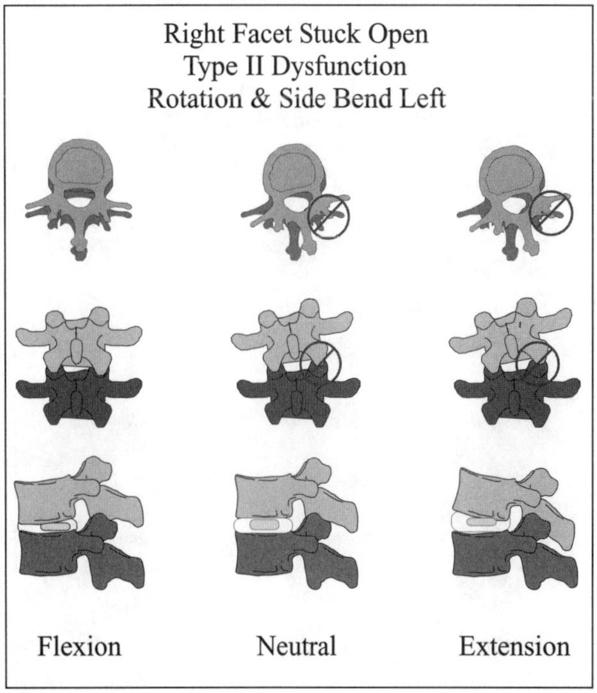

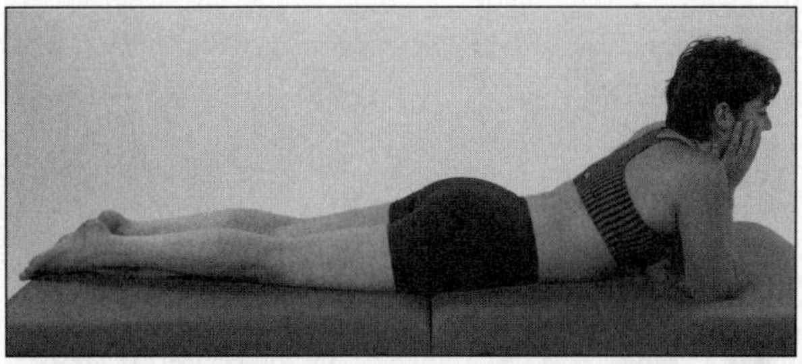

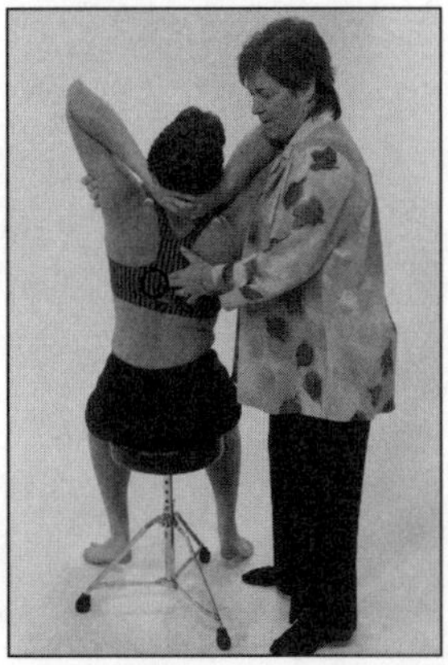

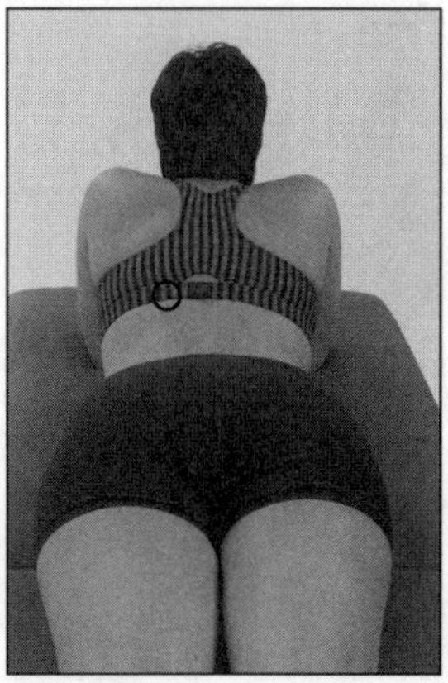

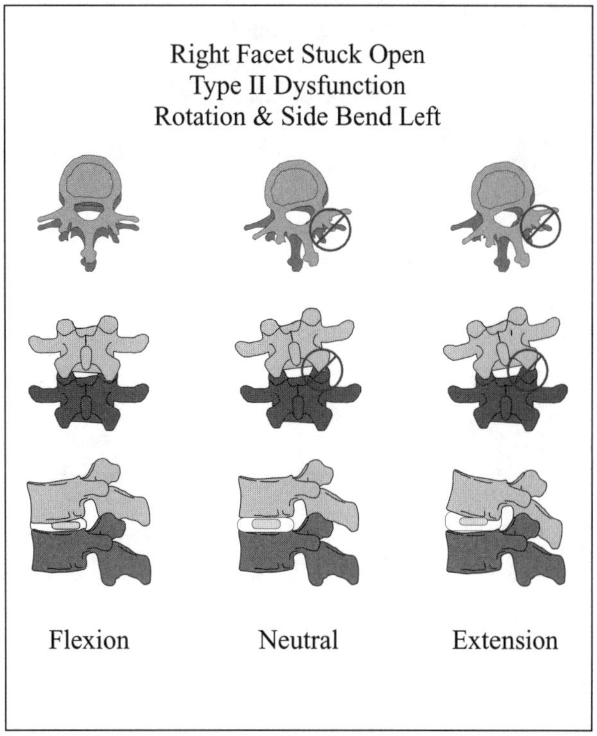

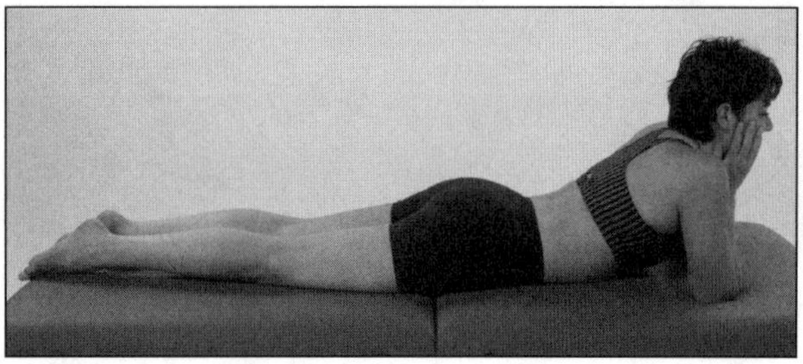

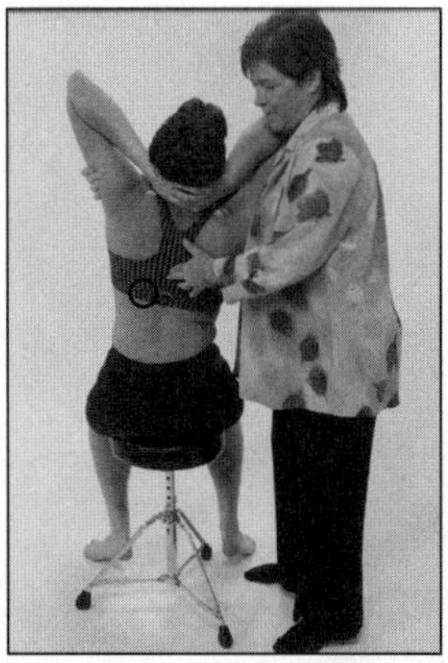

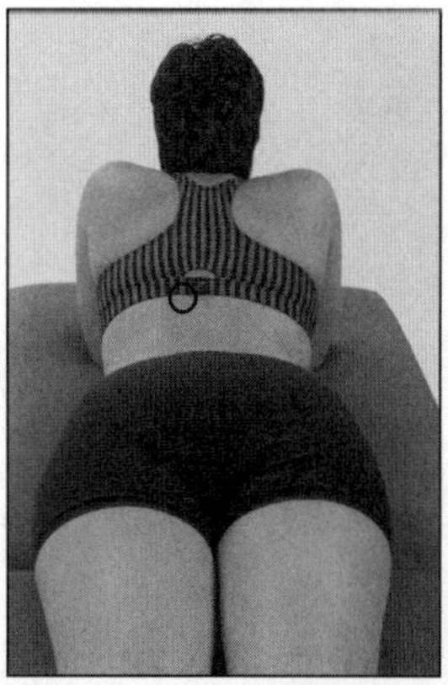

Muscle Energy and "Beyond" Technique for Flexed Lesions

Diagnosis

1. Diagnose in seated extension.
2. Limitations in thoracic extension.
3. Palpate one sole posteriorly prominent transverse process.
4. Ask patient to keep hands behind head for comfort and to inhibit the chin protrusion that elicits anterior shearing of the cervical spine.
5. Palpate segment dynamically into full extension.

Treatment

Seated extension. Hands behind head. Palpate barrier at segment:

First Barrier:

Protrude abdomen. Extend up the kinetic chain to barrier (interbarrier zone). Extend down the kinetic chain to barrier (interbarrier zone).

Second Barrier:

Rotate the head, neck, and thorax down the kinetic chain to barrier (interbarrier zone).

Third Barrier

Side bend head, neck, and thorax down the kinetic chain to barrier.

Resist

Isometric resistance in any direction 6 to 10 seconds. Relaxation. Repeat 3 to 5 times.

Extended Lesions

Example

The left facet is stuck closed. It does not open. The left facet is stuck extended. During flexion the left transverse process will become posteriorly prominent. During extension the left transverse process will move more evenly, and there will not be any posterior prominence. The segment is already extended. This is an extended lesion. This will cause limitation of physiologic flexion. The vertebra is stuck side bent left. It will not side bend right. Physiologic right side bending will be limited. The vertebra is stuck rotated left. The segment will not rotate right. Physiologic right rotation will be inhibited.

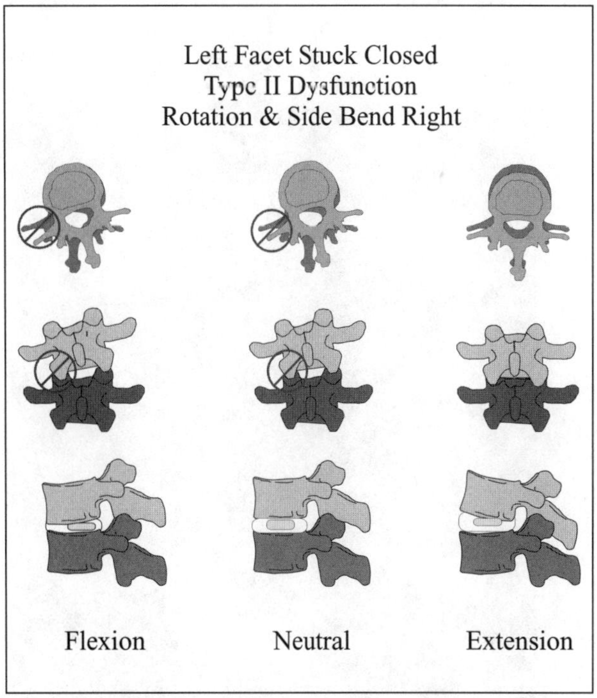

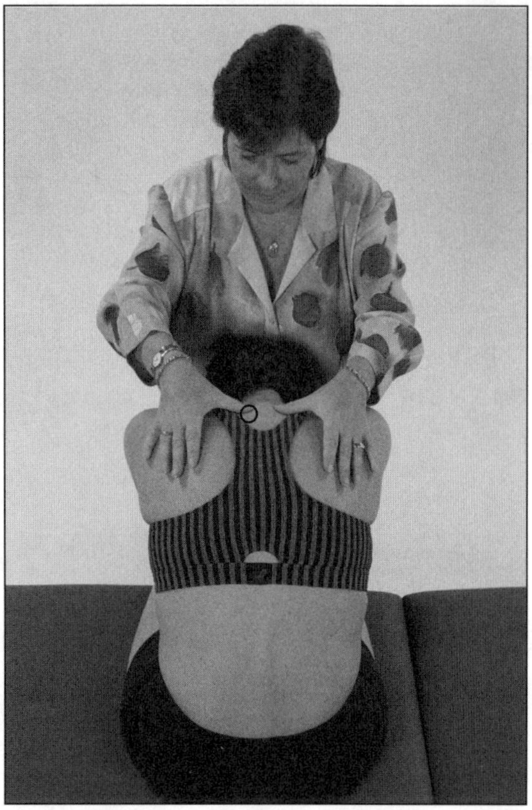

Positional Diagnosis

In seated flexion, palpate one sole transverse process prominent posteriorly. A prominent transverse process reflects rotation dysfunction to the side of the prominent transverse process. If the prominent transverse process is on the left side, then the segment is rotated left. Fryette's law II states that the segment will also be side bent left. If the left facet is stuck closed, the positional diagnosis is: extended, rotated left, and side bent left = ERLSL.

Note: Palpate one segment at a time down the kinetic chain, looking for one sole prominent transverse process. (Two or more consecutive and identical rotation dysfunctions would not be a Type II dysfunction.)

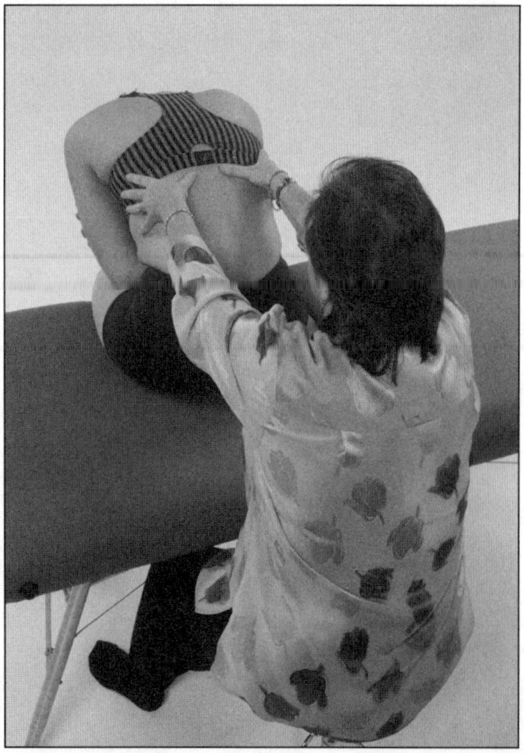

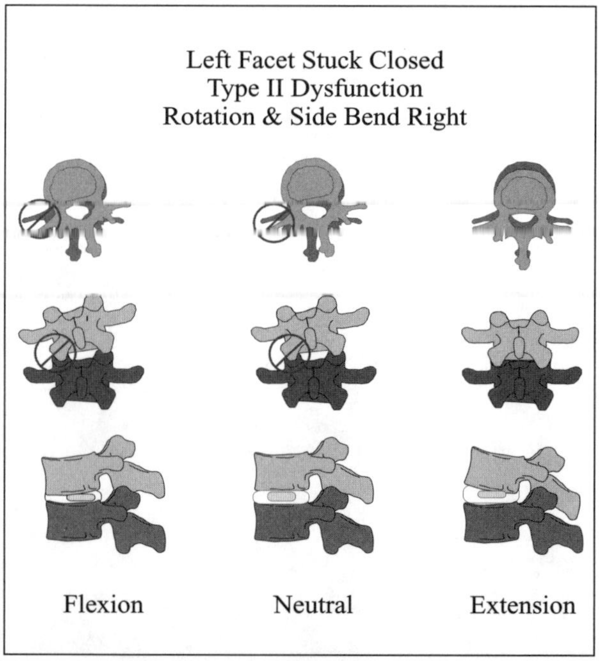

Extended, rotated and side bent left = ERSL

Movement Barrier

The movement barrier is opposite the positional diagnosis. If the facet is extended, the segment is extended; it cannot flex. If the segment is rotated left, it cannot rotate right. If the segment is side bent left, it cannot side bend right. If the left facet is stuck closed, the movement barrier is: flexion, rotation right, side bending right = FRRSR.

Note: The position of movement barrier for the vertebra is the same as the position of treatment, except for the interbarrier zone variation.

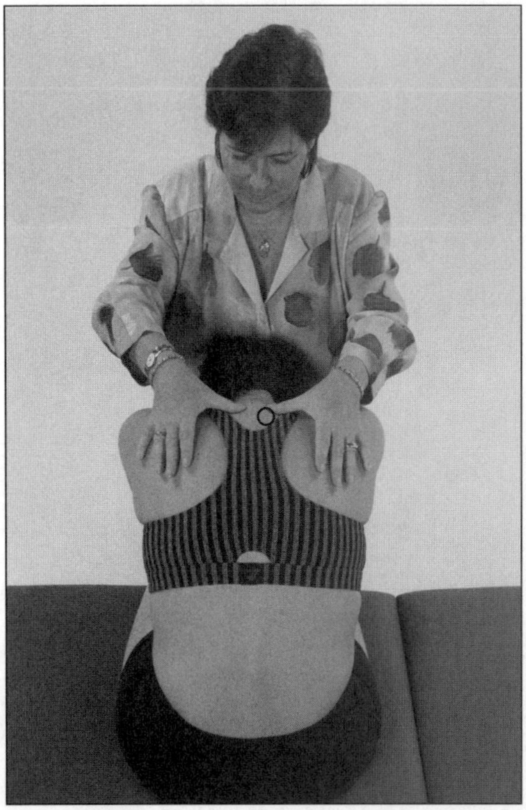

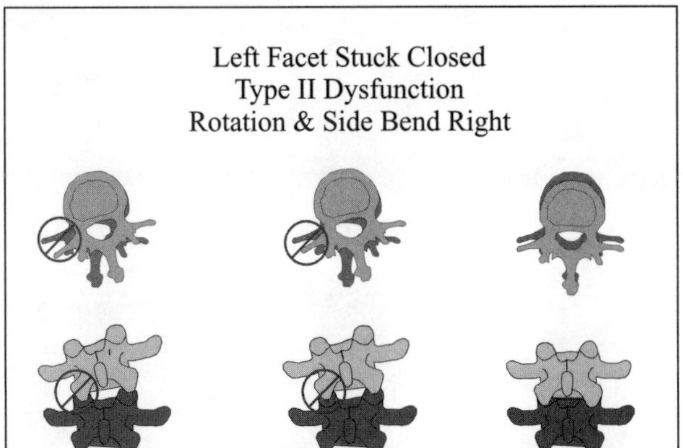

Flexion, rotation and side bending right = FRSR

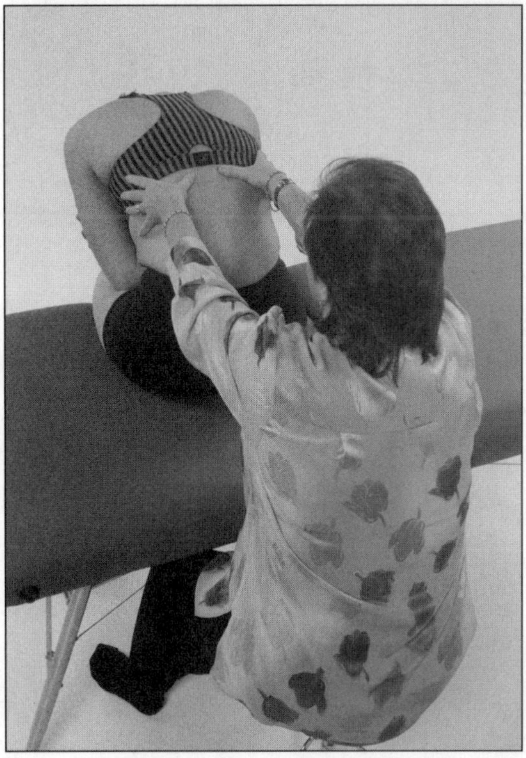

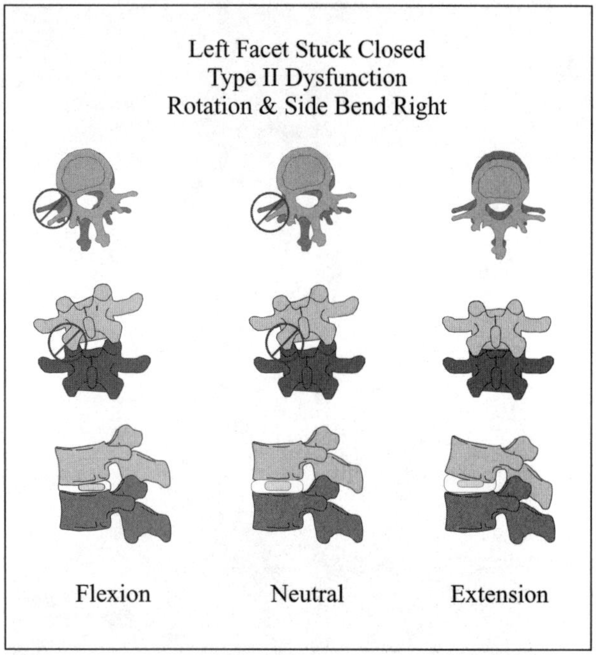

Flexion, rotation and side bending right = FRSR

Position of Treatment

Utilize respiration whenever treating the thoracics and rib cage. Ask the patient to inhale deeply, maintain inspiration for several seconds, then exhale. The position of treatment is the same as the movement barrier. The vertebra must be positioned in flexion, right rotation, and right side bending. The vertebra is positioned in a 3-planar interbarrier zone. The position of treatment is: flexion, rotation right, side bending right = FRRSR.

Treatment

The patient is seated. The therapist stands on the side opposite the prominent transverse process. The patient's arms are crossed in front, hands on shoulders. The therapist palpates for the barrier at the vertebra. The patient flexes: (1) the patient slouches, to flex up the kinetic chain to the sagittal plane barrier; (2) the patient flexes head and neck, flexing down the kinetic chain to the sagittal plane barrier (the interbarrier zone). The patient rotates to the right, down the kinetic chain, to the transverse plane barrier (interbarrier zone). The patient side bends to the right, down the kinetic chain, to the coronal plane barrier (interbarrier zone). At the 3-planar interbarrier zone, an isometric resistance is applied, to any direction, for 6 to 10 seconds. After the relaxation, this process is repeated 3 to 5 times.

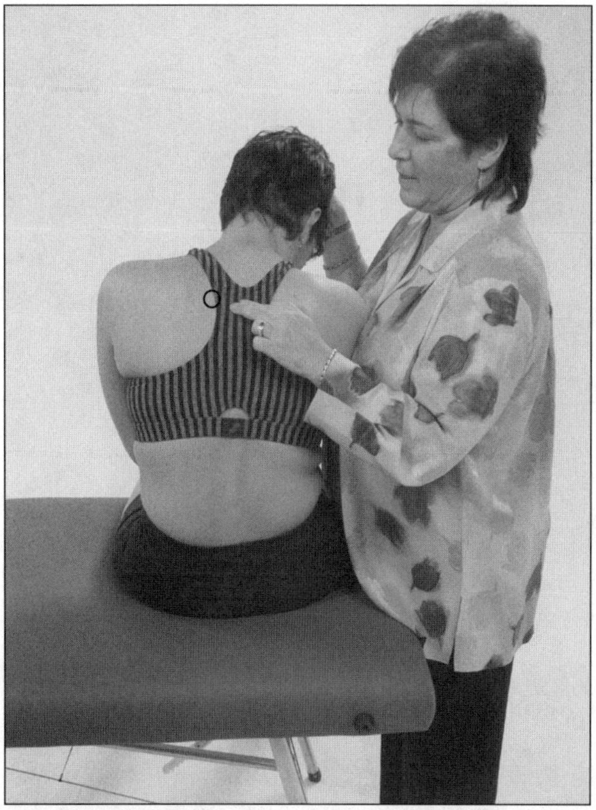

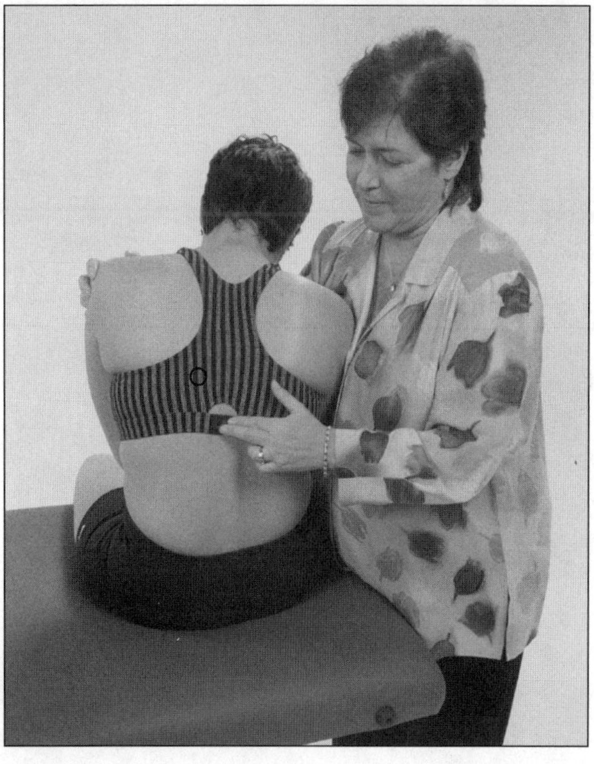

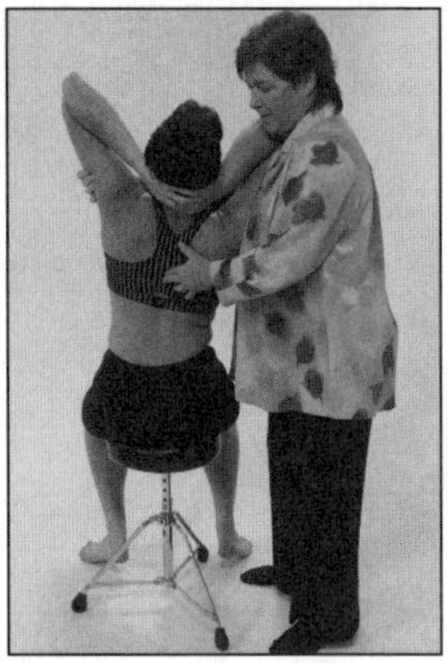

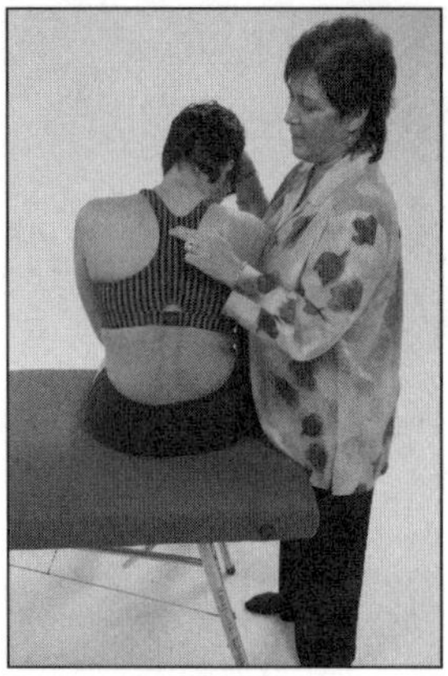

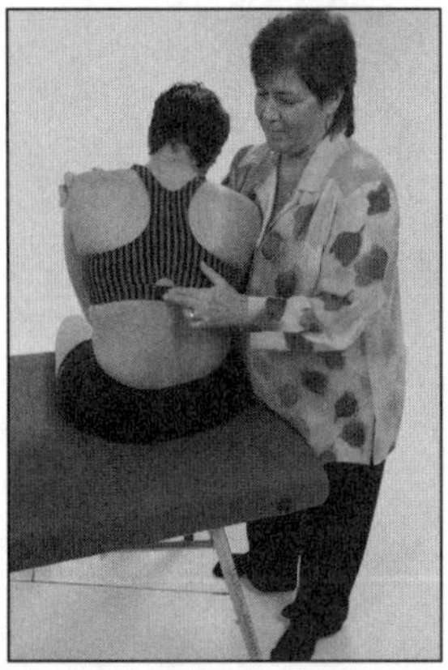

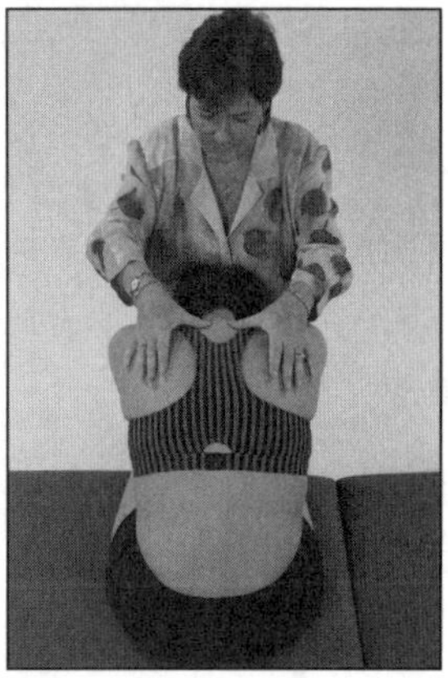

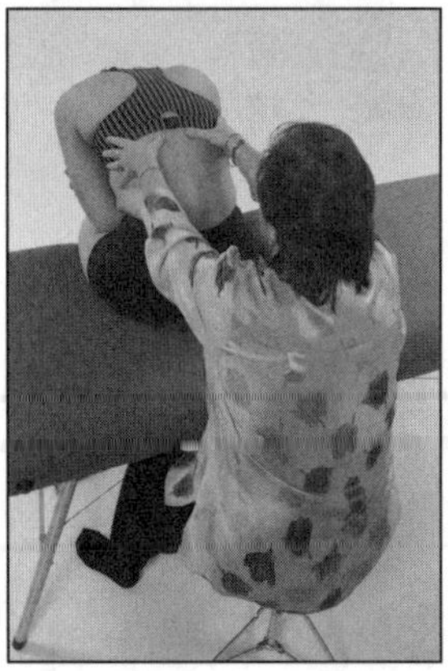

Muscle Energy and "Beyond" Technique for Flexed Lesions

Diagnosis

1. Diagnose in seated flexion.
2. Limitations in thoracic flexion.
3. Palpate one sole posteriorly prominent transverse process.
4. Ask patient to fully flex down the kinetic chain, one segment at a time, to palpate dysfunctional vertebrae dynamically.

Treatment

Seated flexion. Cross patient's arms, hands on shoulders. Palpate barrier at segment.

First Barrier:

Slouch: flex up the kinetic chain until palpation of barrier (interbarrier zone). Flex down the kinetic chain until palpation of barrier (interbarrier zone).

Second Barrier:

Rotate head, neck, and thorax down the kinetic chain to barrier (interbarrier zone).

Third Barrier:

Side bend head, neck, and thorax down the kinetic chain to barrier (interbarrier zone).

Resist:

Isometric resistance in any direction 6 to 10 seconds. Relaxation, then repeat 3 to 5 times.

CHAPTER 24

THE THORACOLUMBAR JUNCTION

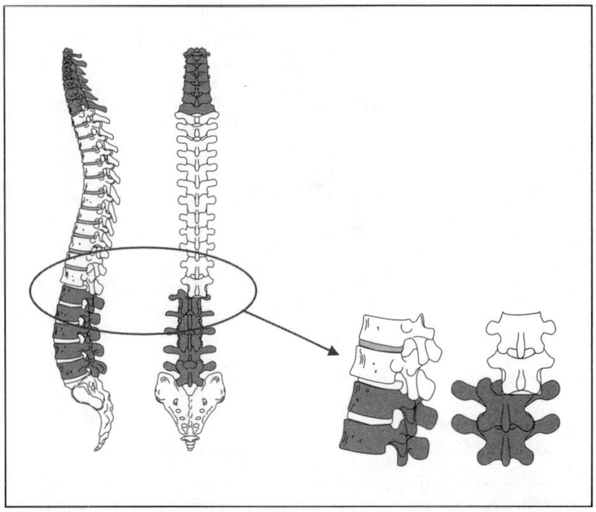

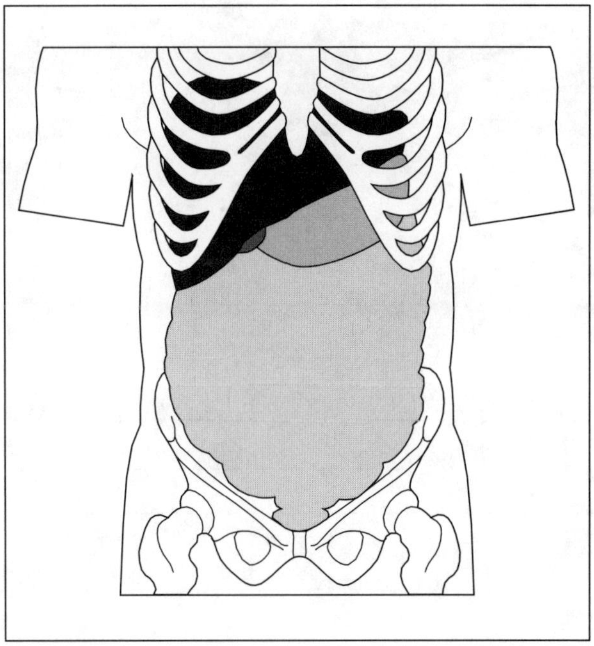

The thoracolumbar junction has often been described as difficult. Yet the difficulty is not because of biomechanical dysfunction. The following are issues that affect biomechanical integrity at the T12/ L1 junction:

1. The thoracolumbar junction is affected by the reciprocal movement forces that impact the spine at the lumbosacral junction which is typically restricted.
2. The thoracolumbar junction is not built to withstand reciprocal movement forces; they are meant for the L5/S1 junction.
3. Reciprocal motion forces at T12/L1 cause shears. Anterior and lateral shears are common at both T12 and L1!

It is common to have severe protective muscle spasm and fascial dysfunction of the diaphragms, especially the respiratory abdominal diaphragm, which will cause biomechanical dysfunction of the thoracolumbar junction.

Discogenic problems are common at the T12/L1 junction. Typical problems include shears of T12 and L1, anterior and lateral. The direction of lateral shears of T12 and L1 will usually be opposite. T12 disc is compressed. The crura of the respiratory abdominal diaphragm are in spasm, with a superior and anterior pull on L1, L2, and L3 vertebrae and discs.

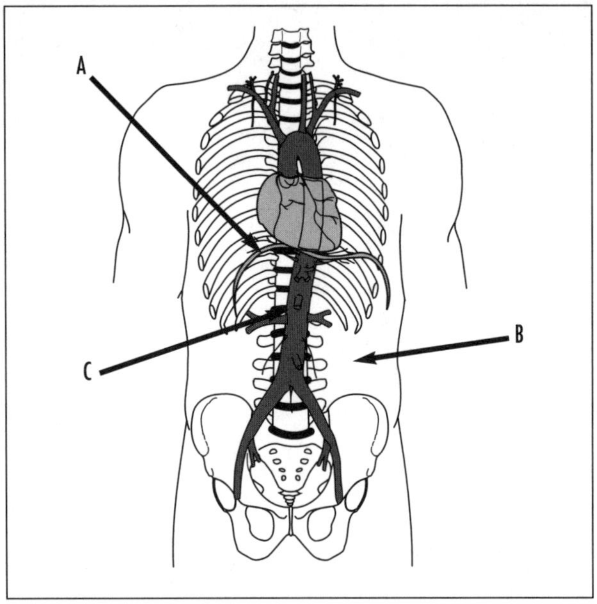

Usually there is considerable fascial dysfunction of the highly specialized (collagen specialty as compared to elastin) thoracolumbar fascia.

Often problems of the abdominal aorta and inferior vena cava affect the biomechanical integrity of the thoracolumbar junction.

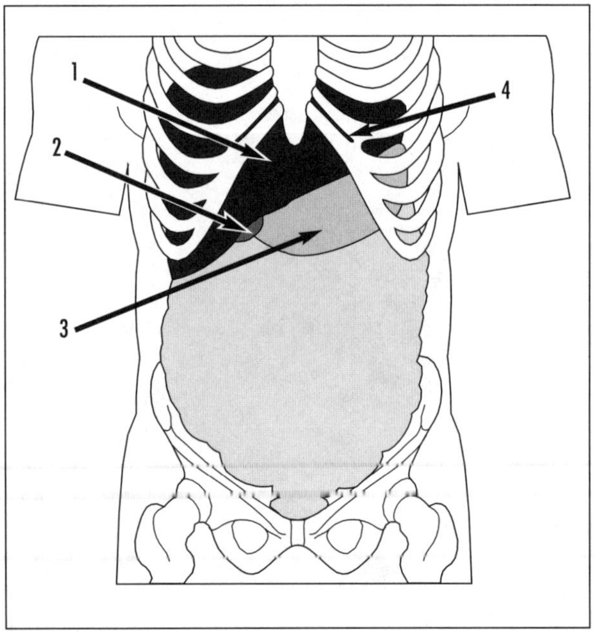

Common problems that cause Type II dysfunctions of T12 and L1 include:

1. Liver tension
2. Gall bladder tension
3. Stomach and cardiac sphincter tension

Whenever it is not easy and straightforward to correct T12 and L1 Type I and Type II dysfunctions, look at the conditions reflected by the Integrated Systems Approach. Often, inexperienced practitioners believe that Type I dysfunctions are common at the thoracolumbar junction. Indeed, primary Type I dysfunctions of the thoracolumbar junction are not common. Rather, these biomechanical problems will occur secondary to other soft tissue problems.

CHAPTER 25

BIOMECHANICAL DYSFUNCTION OF THE CERVICAL SPINE

Locating Cervical Spine Landmarks

C1: Palpate transverse process between mastoid process and ear lobe.
C2: Locate mastoid process bilaterally. Spinous process is located 1 to 2 cm inferior to midline of occiput. This is a wide bifid spinous process.
C3 to C5: Spinous processes are not easily palpated. Facets are found posteriorly one finger breadth lateral to the midline. C4 has the shortest spinous process. Extend neck to relax paravertebral muscles to facilitate palpation.
C6: Spinous process is palpable; 50 percent of the population has a bifid spinous process.
C7: Known as vertebra prominens due to the prominent protrusion of its nonbifid spinous process. Occasionally, when T1 protrudes further than C7, it is mistaken for C7. To locate C7, place your thumb on the spinous process of what you believe to be C7 and hyperextend the patient's head. If it is C7, the vertebra will slide forward. There is less extension of T1.

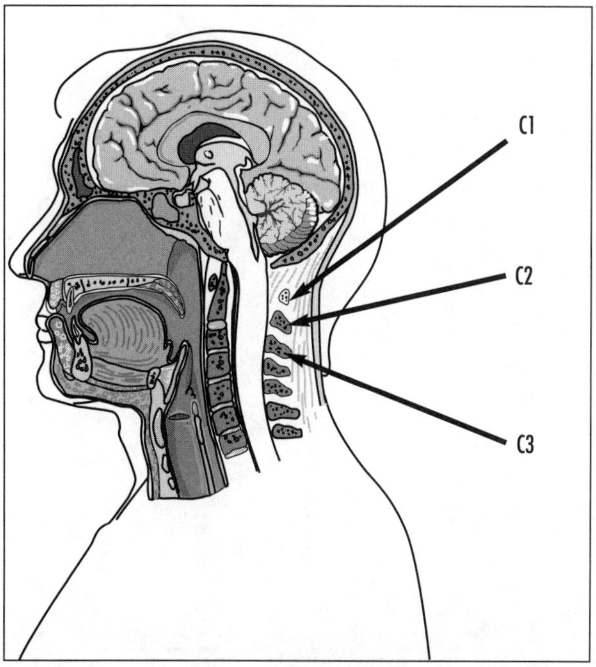

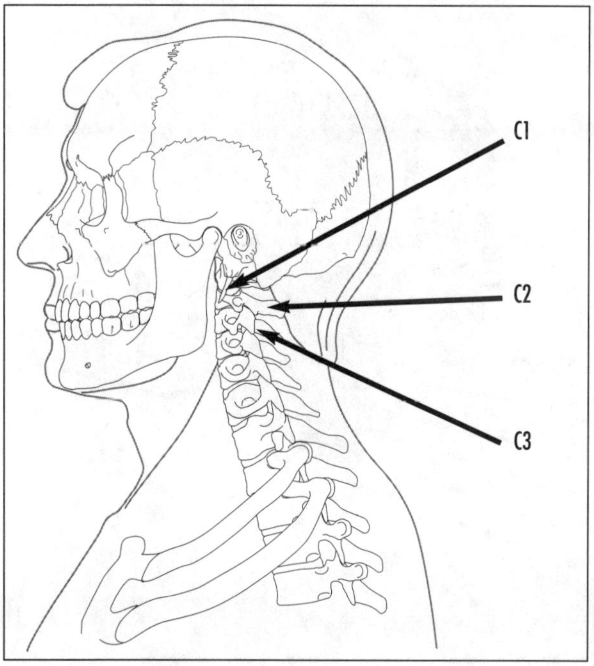

About the Neck

The neck is the most vulnerable region of the spine because it is not impacted between two major bony structures (i.e., the lumbar spine is impacted between the pelvis and rib cage), and because it is not contained between other major bony structures (i.e., the sacrum is contained by the pelvis, and the thoracic spine is contained by the rib cage). What are the parameters that must be understood prior to working on the cervical spine?

1. Correct biomechanics of the pelvic joints.
2. Correct biomechanics of the sacral joints.
3. Observe the anterior soft tissues and

structures of the neck, including: anterior cervical flexor muscles, which might be in spasm; connective tissue, which might be in a state of fascial dysfunction; integrity of the hyoid bone and joints; integrity of the thyroid cartilage and joints; integrity of the cricoid bone and joints; the arytenoid processes; the vocal cords.

4. Observe the lateral soft tissues of the neck, including: middle and anterior scalene muscles, which might be in spasm; connective tissue flexibility of the lateral neck fascia; integrity of the common carotid, external carotid, and especially the internal carotid arteries; integrity of the vertebral artery.

Often, the neck is defined as a cervical complex, rather than the cervical spine. Important tissues to assess, prior to working on the cervical spine include: the cervical plexus; the brachial plexus; nerve roots; discs, especially the C5 disc; and more.

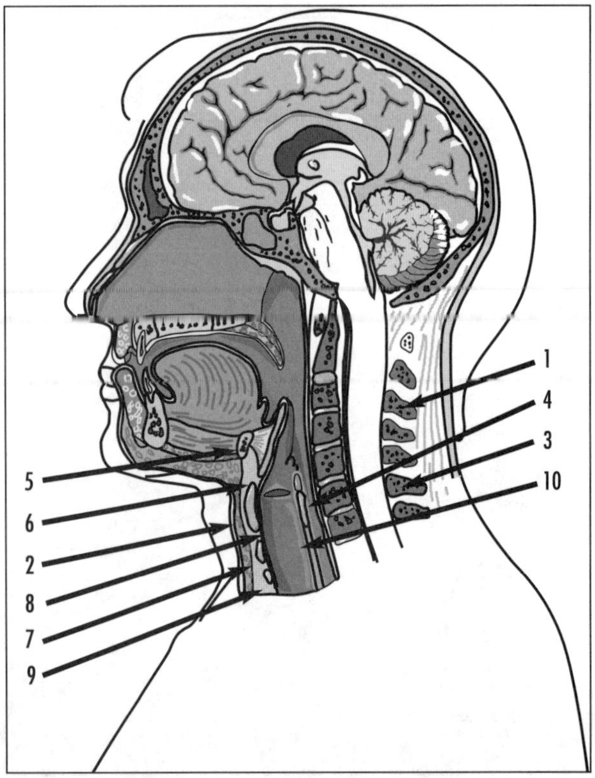

Anterior Palpations

1. C3
2. Thyroid cartilage: C4-superior border; C5-inferior border
3. Cricoid-C6
4. Epiglottis
5. Hyoid Bone
6. Thyrohyoid Membrane
7. Thyroid Cartilage
8. Cricothyroid Ligament
9. Cricoid Cartilage
10. Trachea

Type II Dysfunction of C2 to C7

In Muscle Energy philosophy, there exists only Type II motion in the cervical spine (C2 to C7). There is no cervical Type I movement.

Flexed Lesions: Cervicals 2 to 7

Diagnosis

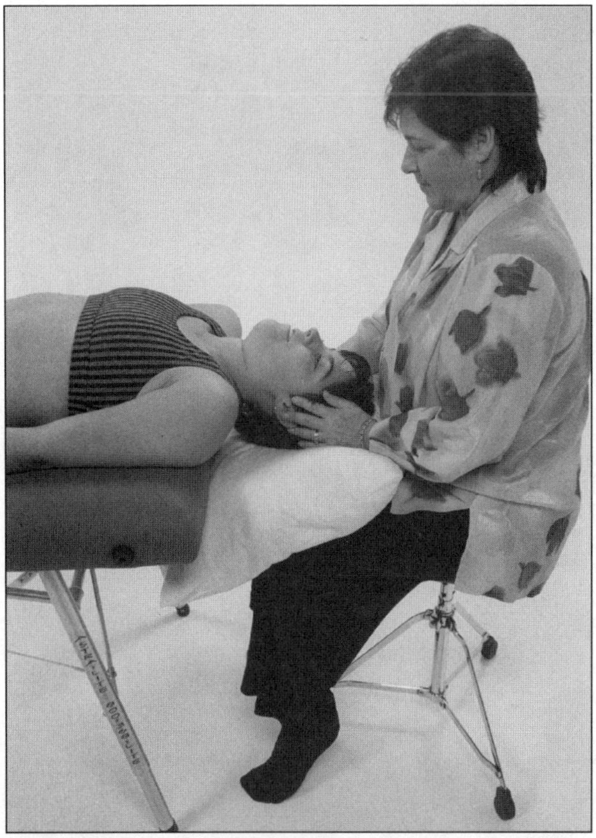

The patient is supine, with cervical extension, with head, neck and shoulder girdles off the edge of the bed almost to mid-scapulae. Therapist sits at the head of the table with a pillow on his or her lap and patient's head resting on the pillow. This position will inhibit chin protrusion and anterior shearing of cervical vertebrae. Maintain neutral position initially. Palpate articular facets bilaterally, one segment at a time, palpating for one sole posteriorly prominent transverse process.

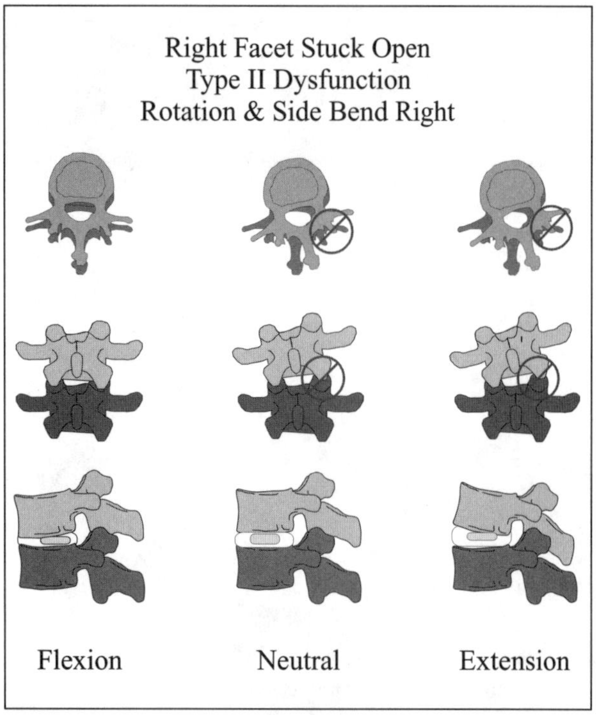

Positional Diagnosis

The right facet is stuck open. The vertebra is rotated and side bent to the side opposite the stuck open facet. If the open facet is on the right side, the positional diagnosis is: Flexed, rotated and side bent left = FRSL

Note: Flexed Type II lesions can be palpated in sitting or in supine. In supine, the neck should be completely off the bed. Be aware: Inhibit excessive chin protrusion, which will cause anterior shearing of the cervical vertebrae that will appear to be cervical extension.

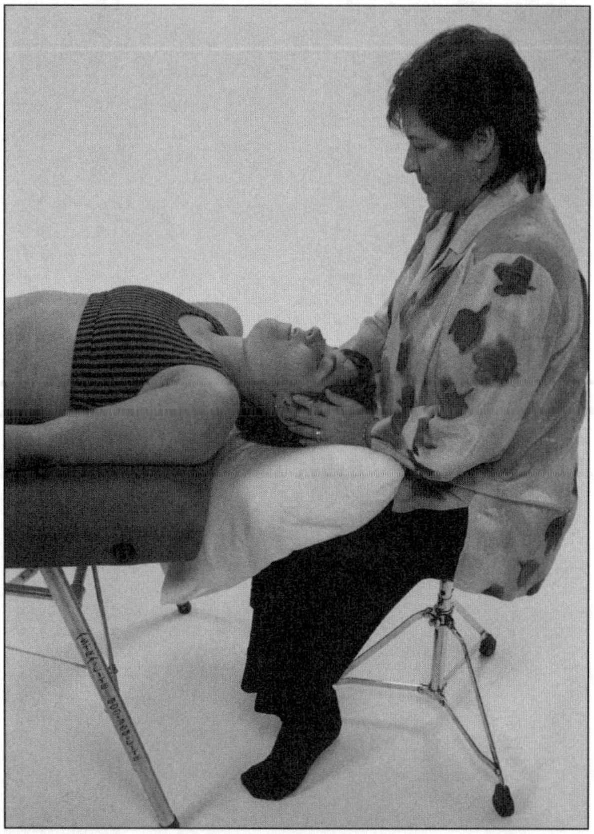

Flexed, rotated and side bent left = FRSL

Movement Barrier

The movement barrier is opposite the positional diagnosis. The vertebra is stuck flexed, rotated left, and side bent left. Therefore that vertebra cannot extend, rotate right, and side bend right. The movement barrier is: Extension, rotation and side bending right = ERSR

Note: Type II cervical flexed lesions can be palpated and treated in sitting and in supine cervical extension.

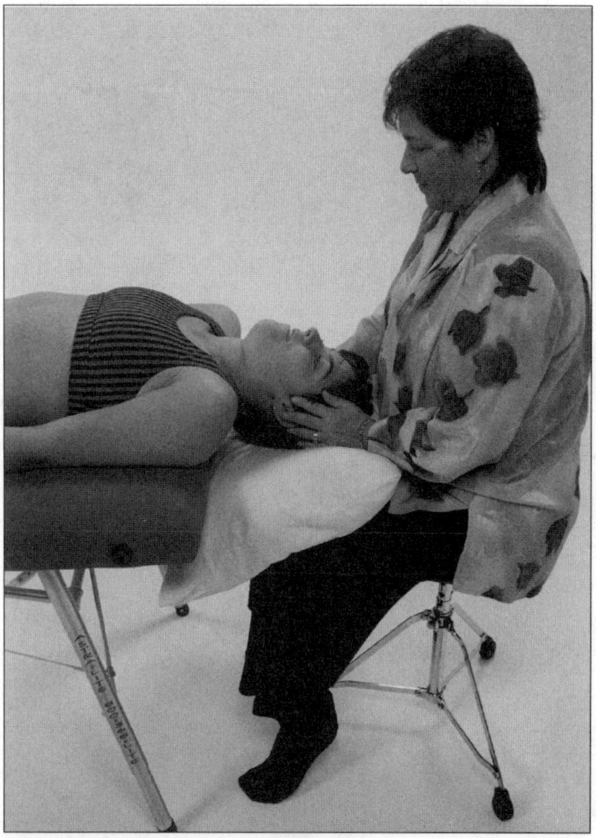

Extension, rotation and side bending right = ERSR

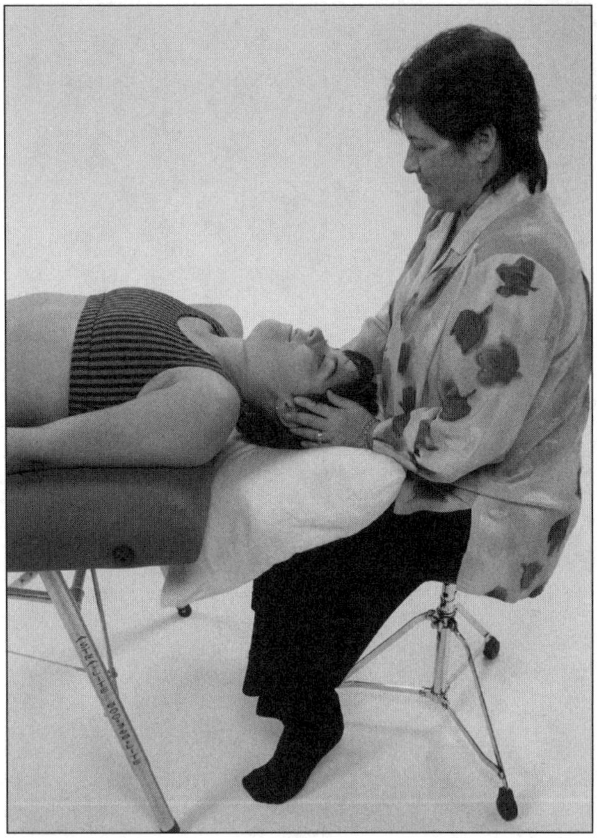

Extension, rotation and side bending right = ERSR

Position of Treatment

The position of treatment is the same as the movement barrier, opposite the positional diagnosis. The vertebra is positioned in extension, right rotation, and right side bending. The position of treatment is: Extension, rotation and side bending right = ERSR

Treatment

Position the patient in supine cervical extension with head, neck, and upper shoulder girdles off the edge of the bed. Therapist sits at the head of the bed with a pillow on his or her lap and the patient's head on the pillow. Palpate barrier at segment.

First Barrier:
: Extend down the kinetic chain until palpation of barrier (interbarrier zone).

Second Barrier:
: Rotate head and neck down the kinetic chain until palpation of barrier (interbarrier zone).

Third Barrier:
: Side bend head and neck down the kinetic chain until palpation of barrier (interbarrier zone).

Resist
: Isometric resistance of any direction for 6 to 10 seconds. Repeat 3 to 5 times.

Note: The common reason for failure to correct Cervical Type II flexed lesions: The chin is protruded and true physiologic cervical extension is not performed.

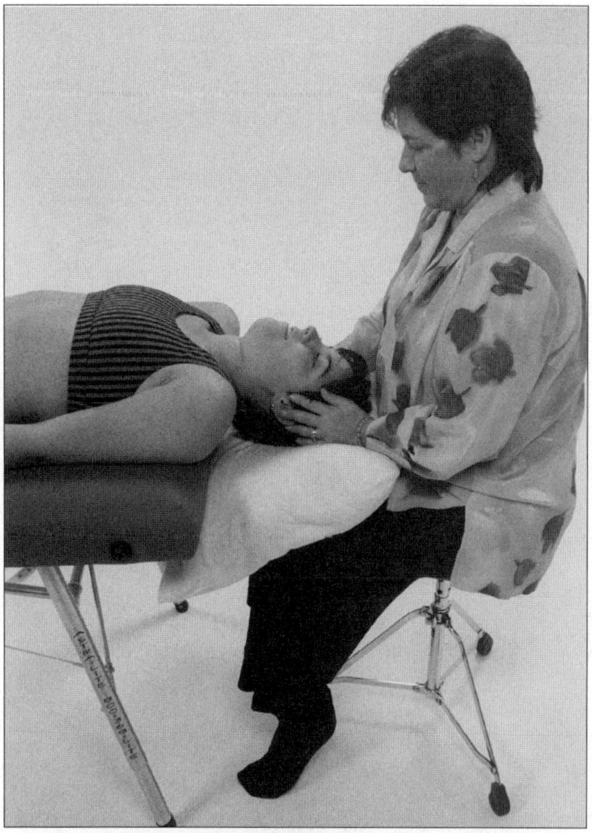

Extension, rotation and side bending right = ERSR

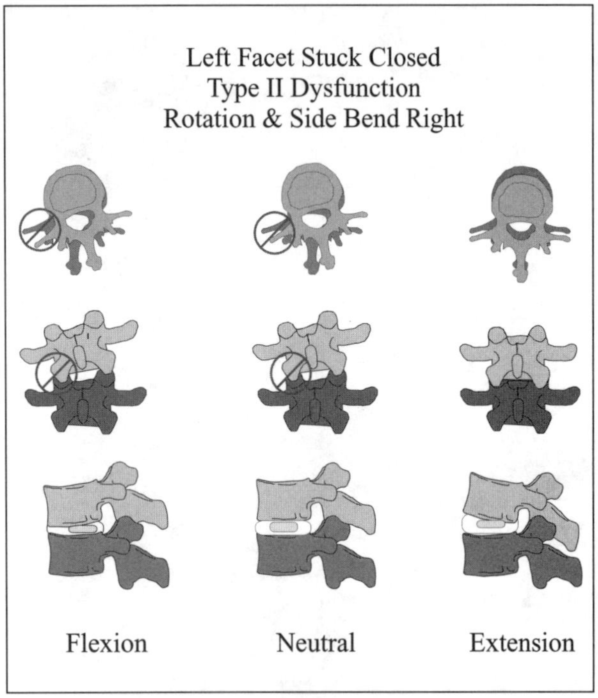

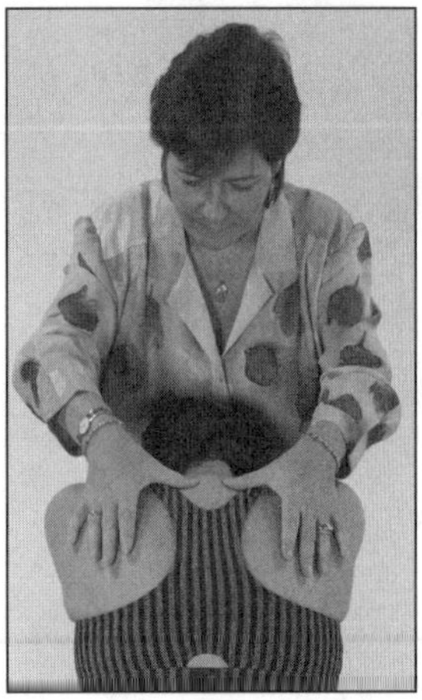

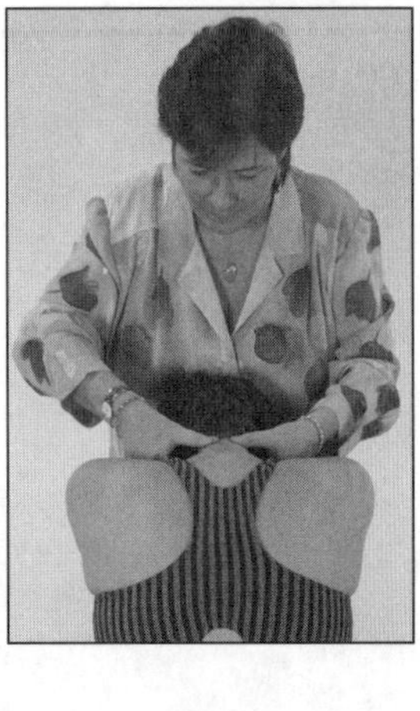

Extended Lesions: Cervicals 2 to 7

Diagnosis

1. Seated flexion.
2. Limitation of cervical flexion.
3. Palpate bilateral articular pillars of each vertebral segment individually.
4. Palpate one sole posteriorly prominent transverse process.
5. Palpate segment dynamically through full cervical flexion.

Positional Diagnosis

The left facet is stuck closed. The vertebra is rotated and side bent left toward the side of the stuck-closed facet. If the closed facet is on the left side, the positional diagnosis is: Extended, rotated and side bent left = ERSL

Note: Transverse processes can be palpated with two thumbs as patient maintains flexed posture.

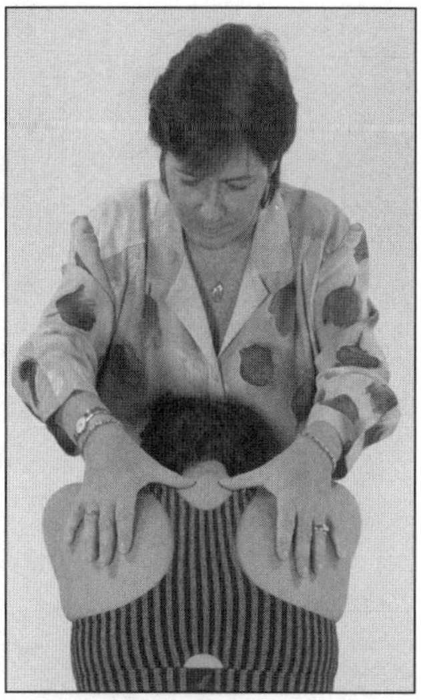

Extended, rotated and side bent left = ERSL

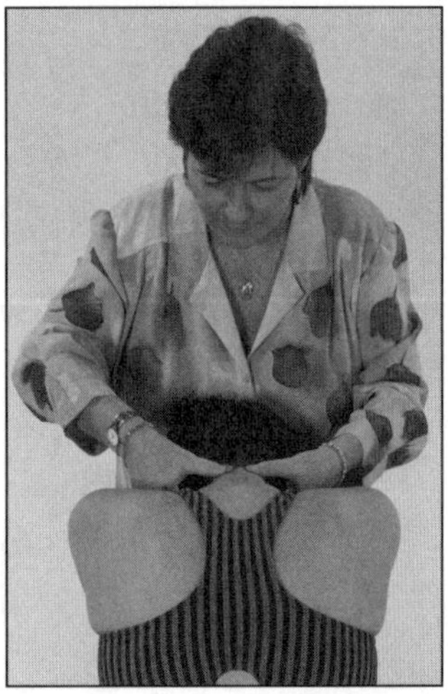

Flexion, rotation, and side bending right = FRSR

Movement Barrier

The movement barrier is opposite the positional diagnosis. The vertebra is stuck extended, rotated left, and side bent left. Therefore, the vertebra cannot flex, rotate right, and side bend right. The movement barrier is: Flexion, rotation, and side bending right = FRSR

Note: The only difference between the movement barrier position and the position of treatment is the interbarrier zone.

Position of Treatment

The position of treatment is the same as the movement barrier, opposite the positional diagnosis. The vertebra is positioned in flexion, right rotation, and right side bending. The position of treatment is: Flexion, rotation and side bending right = FRSR

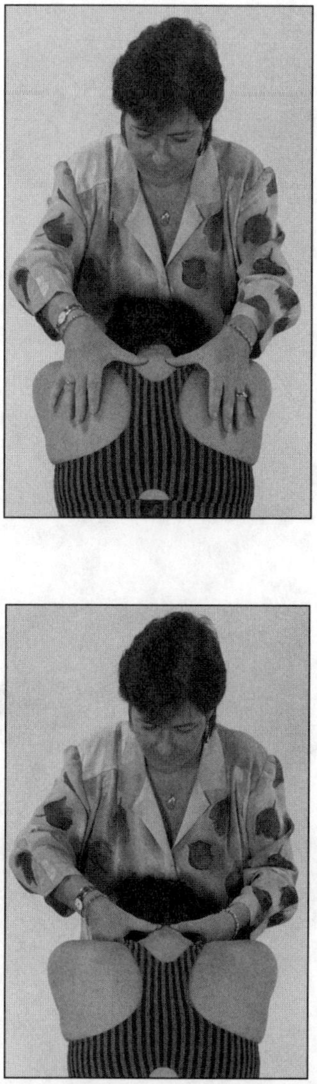

Flexion, rotation and side bending right = FRSR

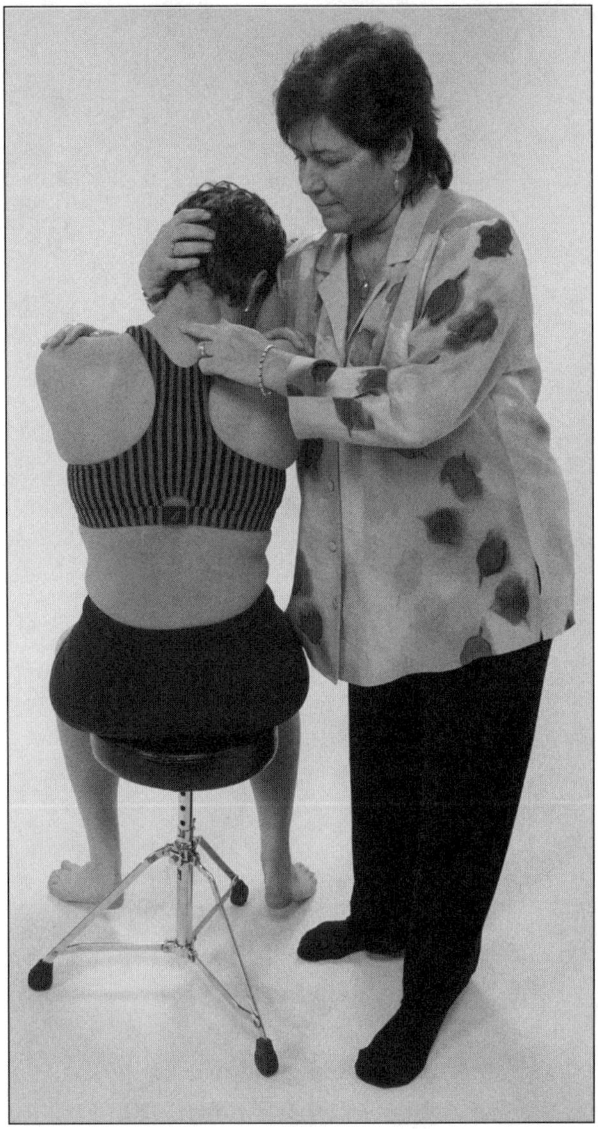

Cervical Spine Superior Segments: Atlantoaxial Joint (A/A: C1/C2)

Treatment

Seated flexion. Control head and neck for security and comfort. Palpate barrier at segment.

First Barrier:
Flex down the kinetic chain to barrier (interbarrier zone). Maintain chin tuck.

Second Barrier:
Rotate head and neck down the kinetic chain into movement barrier (interbarrier zone).

Third Barrier:
Side bend head and neck down the kinetic chain into movement barrier (interbarrier zone).

Resist:
Isometric resistance in any direction for 6 to 10 seconds. Repeat 3 to 5 times.

Note: The common reason for failure to correct cervical Type II extended lesions: The patient protrudes the chin, instead of maintaining a chin tuck; the barrier is not correct because the patient does not have true physiologic cervical flexion.

The Craniocervical Junction

This region is often called the suboccipital space. Yet the term appears to be limiting. Problems of the craniocervical junction are considered by the author to include: the occiput and the foramen magnum; all soft tissues of the suboccipital region, including cranial nerves 9 (glossopharyngeal), 10 (vagus), 11 (spinal accessory), and 12 (hypoglossal); the atlas (C1); the axis (C2); the odontoid process (dens); the alar ligaments; and more.

Typical problems that affect the craniocervical junction and that must be addressed prior to successful intervention with Muscle Energy and "Beyond" Technique to restore biomechanical integrity for the occipitoatlantal joints and atlantoaxial joints include:

1. Bone bruises (microtrabecular fractures) of the dens, the anterior arch of the atlas, and the occiput
2. Tension of the tentorium cerebellum
3. Muscle spasm of the suboccipital musculature
4. Fascial dysfunction of the suboccipital space and transverse sinus
5. Neural tissue tension of cranial nerves numbers 9, 10, 11, and 12
6. Vascular insufficiency, especially of the circle of Willis (a) and the basilar arteries (b).

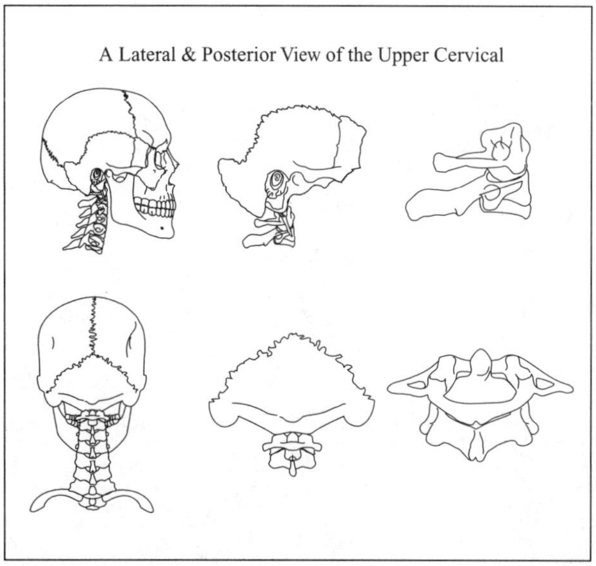

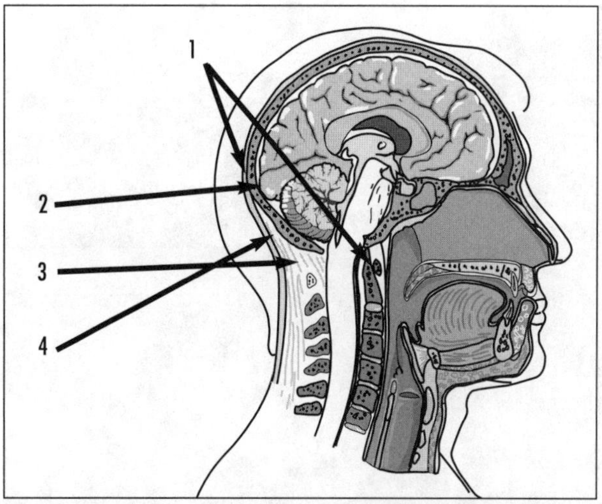

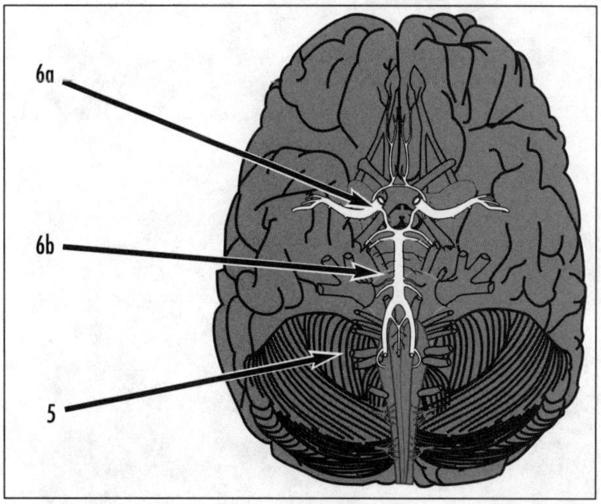

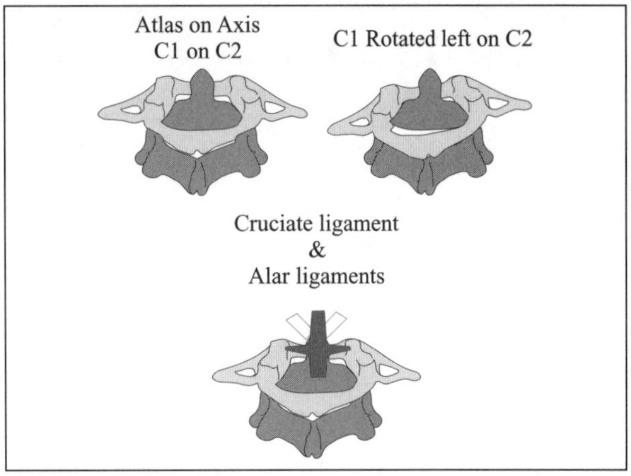

Dysfunction

The A/A joint contributes rotation up to 45 percent of cervical rotation. There is minimal flexion and extension (less than 5 degrees); there is minimal side bending (less than 5 degrees). Open-mouth x-rays are needed to assess the integrity of this joint. There are three A/A joints: two joints with the articular masses of C1 and C2; one joint with the odontoid process (dens) and the anterior arch of C1. The dens is maintained in its position by the cruciate and alar ligaments.

Diagnosis

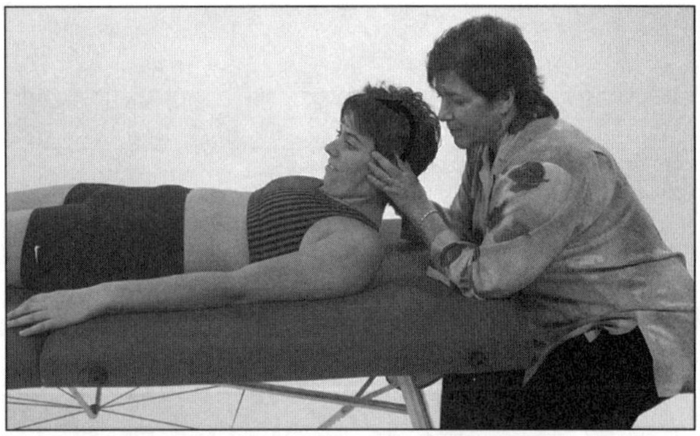

1. Fully flex patient's head and neck, without overpressure, to lock the lower cervical vertebrae (C3 to C7).

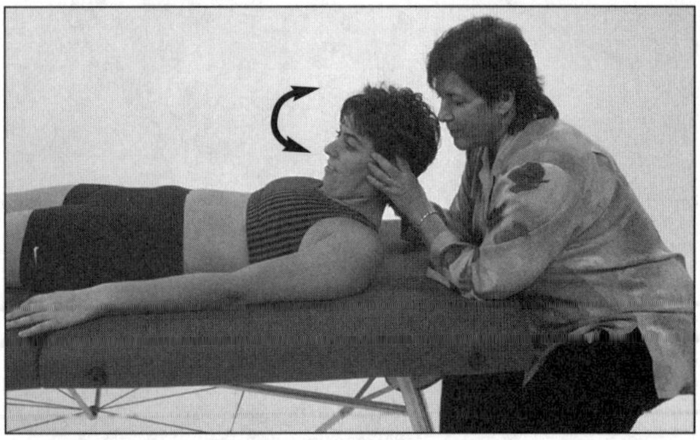

2. Passively rotate patient's head to both sides.

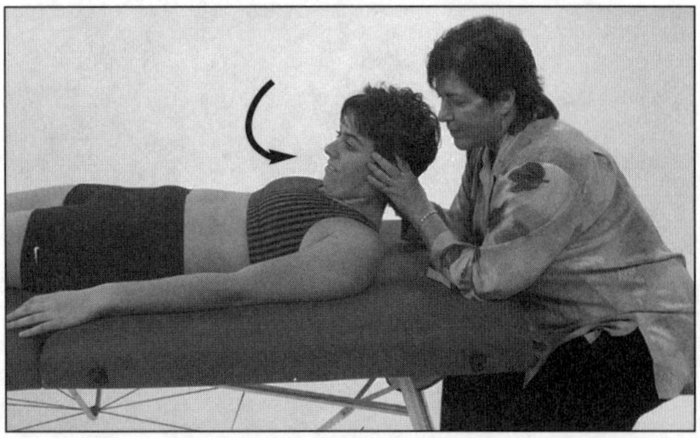

3. Compare degree of restriction in rotation.
4. Movement barrier is to the restricted side.

Rationale

This is almost a pure rotational dysfunction. The treatment involves only rotation.

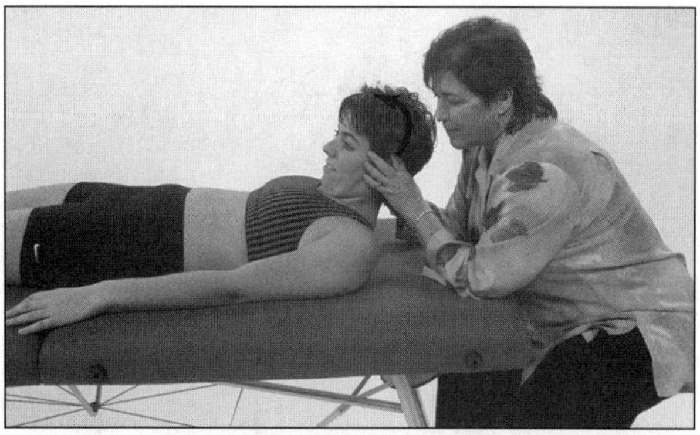

Treatment

Supine or seated flexion. Palpate barrier at suboccipital space, palpating the suboccipital musculature. Fully flex cervical spine without overpressure.

Barrier:

Rotate head into movement barrier until barrier is palpated (interbarrier zone).

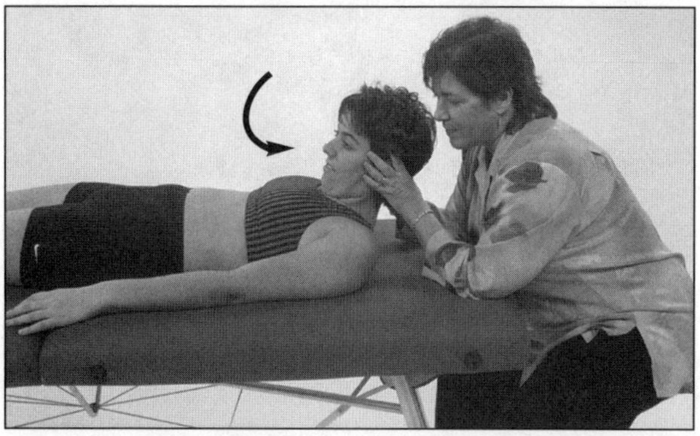

Resist:

Head rotation to opposite side 6 to 10 seconds. Repeat 3 to 5 times.

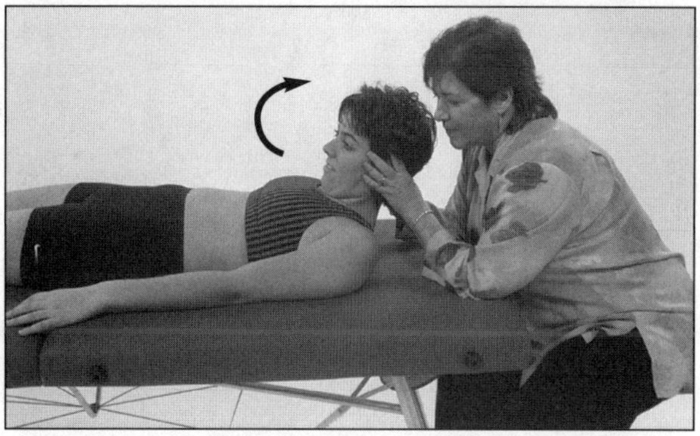

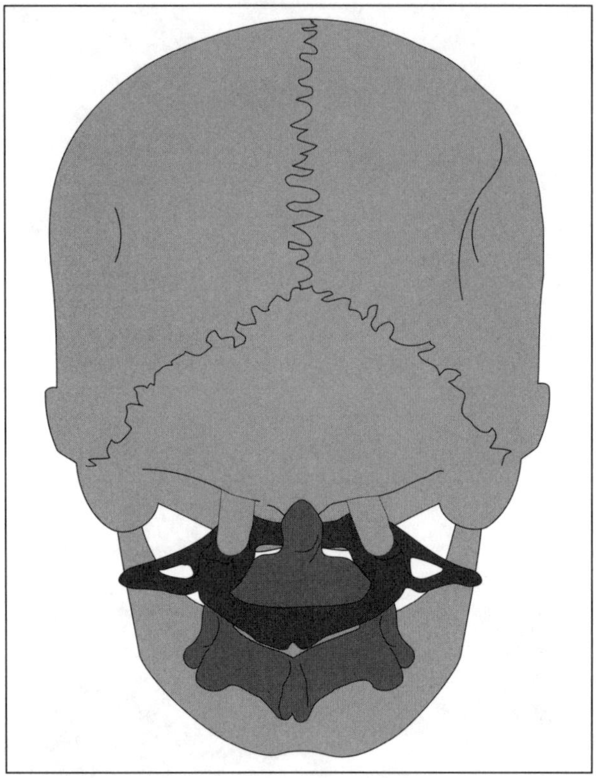

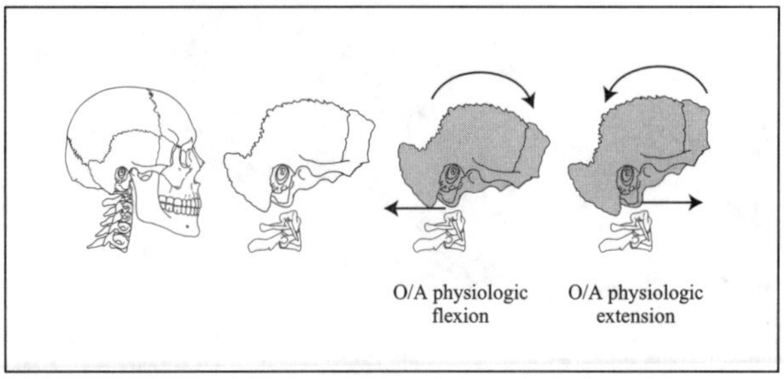

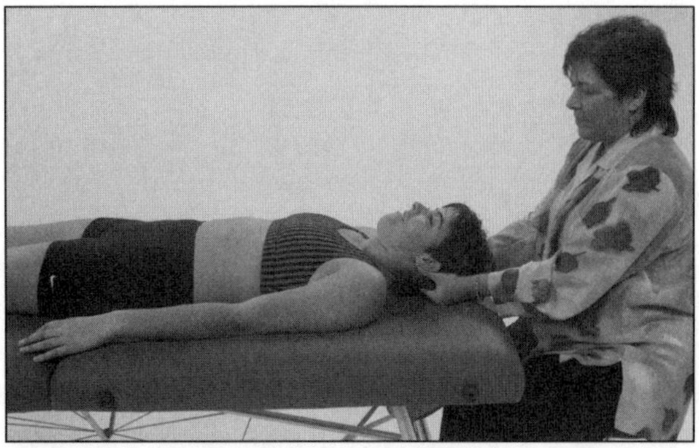

Cervical Spine Superior Segments: Occipitoatlantal Joint (O/A)

Biomechanics

There is Type I movement only at the O/A joint, which means that rotation and side bending are to opposite sides. There are 15 degrees of flexion (chin tuck) and 15 degrees of extension (chin protrusion). There is minimal side bending and only slight rotation (less than 5 degrees). Dysfunction of the O/A joint can compromise the suboccipital space.

Convex-Concave Theory

Because biomechanical dysfunction of the occipitoatlantal joint can cause compression of the suboccipital space, an important area to heal in neurorehabilitation, it is important to comprehensively correct this dysfunction. The treatment is a combination of mobilization philosophy (convex-concave theory) and Muscle Energy techniques.

Application Rationale

1. Convex moving on concave: Occiput is convex moving on concave atlas.
2. O/A physiologic flexion occurs with accessory movement of posterior glide of occiput.
3. O/A physiologic extension occurs with accessory movement of anterior glide of occiput.

Diagnosis and Treatment

Supine position. Therapist sits at patient's head. Therapist's forearms are supported on table. Patient's head rests in palms of therapist's hands while fingers palpate suboccipital musculature for barrier.

Treatment of the Occipitoatlantal Joints to Decompress the Craniocervical Junction

This treatment approach, developed by the author, recognizes the significance and difficulty in eliminating dysfunction of a compressed suboccipital space. To correct biomechanical dysfunction of the O/A joints, it is understood that there are two joints, a right O/A joint and a left O/A joint. Each joint can potentially have two lesions: Type I anterior lesion and a Type I posterior lesion. Each joint may need two corrections to improve physiologic flexion and physiologic extension. If there is good physiologic flexion of occiput, there is good posterior glide of occiput on atlas.

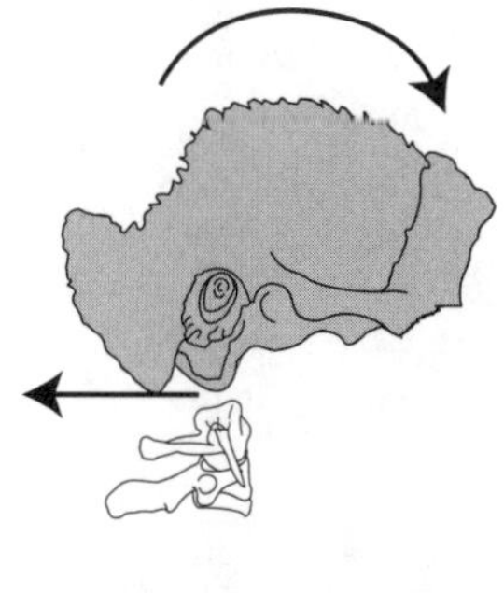

O/A physiologic flexion & Posterior glide

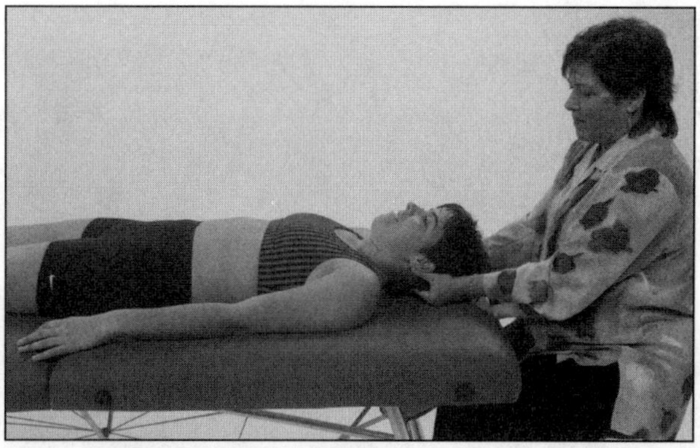

If there is good physiologic extension of occiput, there is good anterior glide of occiput on atlas.

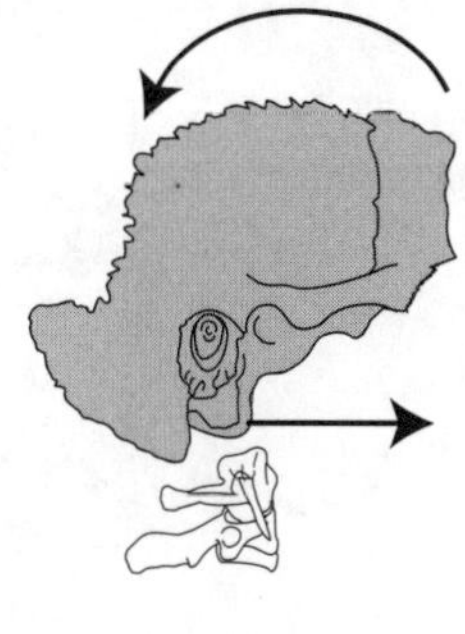

O/A physiologic extension & Anterior glide

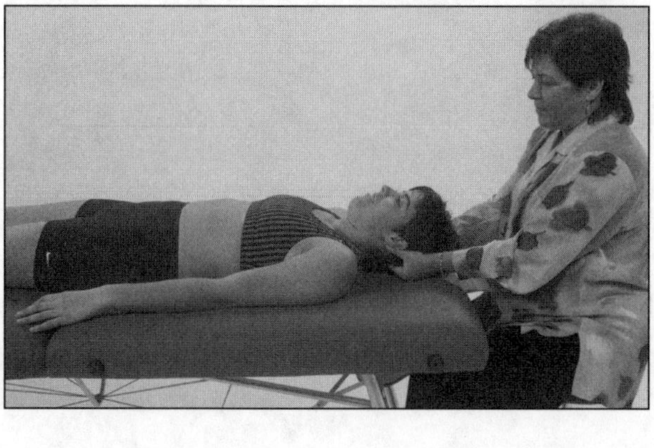

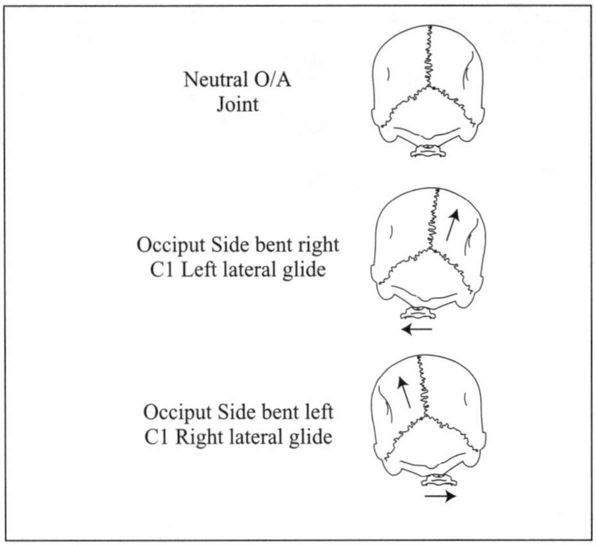

With physiologic right side bending of occiput, there is left lateral glide of occiput on atlas.

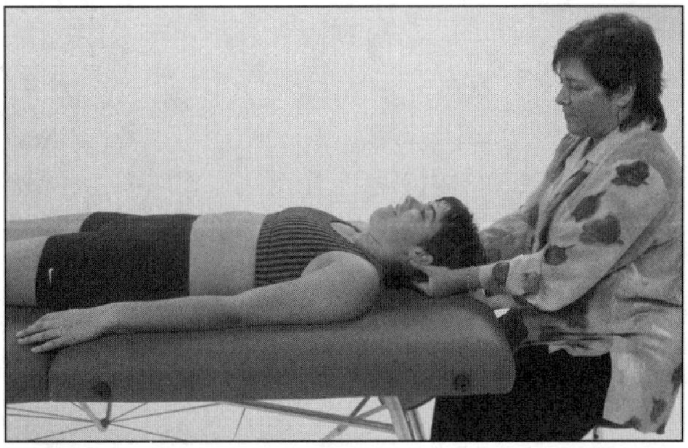

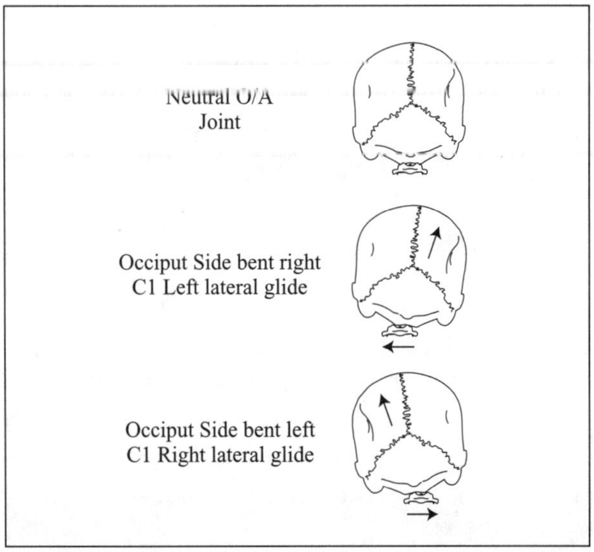

With physiologic left side bending of occiput, there is right lateral glide of occiput on atlas.

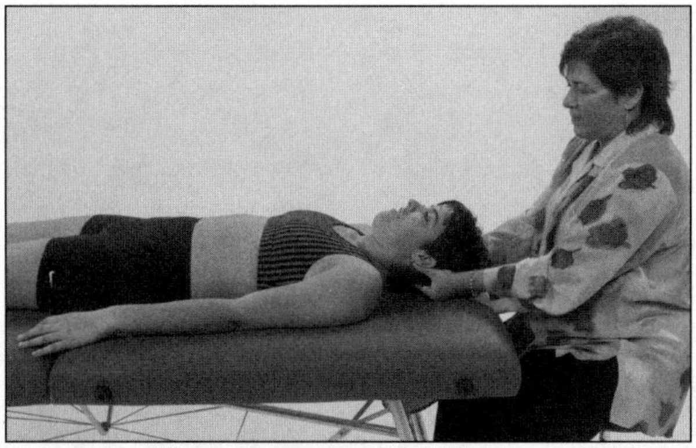

Rationale of Treatment for Biomechanical Dysfunction of the O/A Joints

When physiologic flexion of O/A is more limited than physiologic extension of O/A, then posterior glide will be more limited than anterior glide. Because Muscle Energy and "Beyond" Technique is a direct technique, bring occiput into posterior glide (the direction of hypomobility) with 5 grams of force (the weight of a nickel). Palpate the suboccipital musculature for the muscle barrier. Then add (stack) physiologic flexion, just slightly, so as not to surpass the barrier. Palpate the suboccipital musculature for the muscle barrier. Then test which direction of accessory joint movement is more hypomobile: left lateral glide or right lateral glide. Add (stack) the more hypomobile direction. If left lateral glide of occiput is more hypomobile than right lateral glide, add (stack) right side bending, just slightly. Palpate the suboccipital musculature for the muscle barrier. When all movements, physiologic and accessory, have been stacked (posterior glide plus physiologic flexion plus left lateral glide plus right side bending), add physiologic rotation of occiput. There is Type I movement of the O/A joint in rotation and side bending because of the auricular-shaped surface. Therefore physiologic rotation and side bending of occiput are to opposite sides. Add (stack) physiologic left rotation of occiput, just slightly. Palpate the suboccipital musculature for the muscle barrier. Now all movements, physiologic plus accessory, have been stacked: posterior glide plus physiologic flexion plus left lateral glide plus right side bending plus left rotation. Provide isometric resistance (any direction, flexion or extension or right/left side bending or right/left rotation). Maintain 6 to 10 seconds' isometric resistance. Relax. Repeat the total process: posterior glide plus physiologic flexion plus left lateral glide plus right side bending plus left rotation, all movements just slightly. Palpate the suboccipital musculature for the muscle barrier. Resist. Relax. Repeat the total process. Do 3 repetitions.

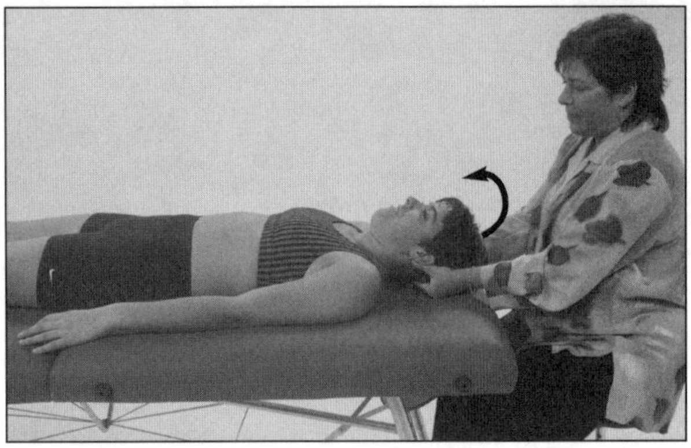

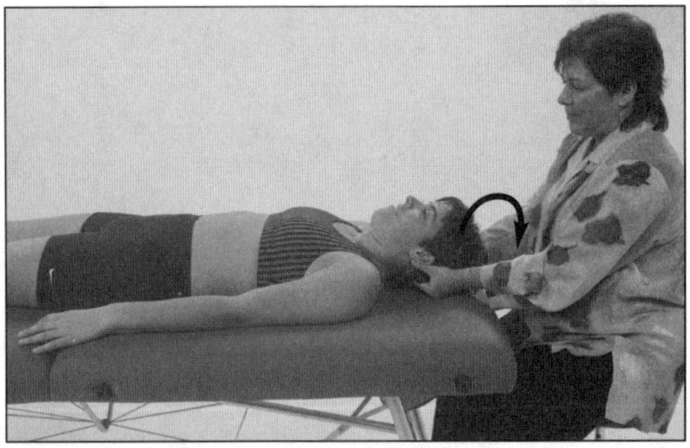

Then retest anterior and posterior glide of occiput. Palpate hypomobility. Continue to treat the O/A joints until all flexed and extended dysfunctions have been treated.

Treatment of Biomechanical Dysfunction of the Occipitoatlantal Joint

Always move into barrier of hypomobility. Remember convex on concave theory. Remember O/A presents Type I movement.

Note: This patient is treated in physiologic flexion; therefore, there was greater in posterior glide of occiput.

Note: This patient is treated in physiologic hypomobility extension; therefore, there was greater hypomobility in anterior glide of occiput.

PHASE 1

STEP 1. COMPARE
1ST BARRIER

Anterior glide of occiput (accessory movement) If more hypomobile, ▼	Posterior glide of occiput (accessory movement) If more hypomobile, ▼

STEP 2. STACK
2ND BARRIER

Extend head on neck (physiologic movement) (Protract chin) ▼	Flex head on neck (physiologic movement) (Chin tuck) ▼

STEP 3. COMPARE
3RD BARRIER

Right lateral glide (accessory movement) If more hypomobile, ▼	Left lateral glide (accessory movement) If more hypomobile, ▼	Right lateral glide (accessory movement) If more hypomobile, ▼	Left lateral glide (accessory movement) If more hypomobile, ▼

STEP 4. STACK
4TH BARRIER

Left side bend head (physiologic movement) ▼	Right side bend head (physiologic movement) ▼	Left side bend head (physiologic movement) ▼	Right side bend head (physiologic movement) ▼

STEP 5: STACK
5TH BARRIER

Rotate right head (physiologic movement)	Rotate left head (physiologic movement)	Rotate right head (physiologic movement)	Rotate left head (physiologic movement)

STEP 6: RESIST
(1ST REPETITION)

Resist: Isometric resistance any direction, 6 to 10 seconds. Repeat: 5 to 6 times.

Treatment is: Step 1 plus Step 2 plus Step 3 plus Step 4 plus Step 5 plus Step 6. Phase 2, Phase 3, and Phase 4 may be needed. Repeat the process from Step 1 through Step 6 until each direction is mobile.

CHAPTER 26

THE RIB CAGE

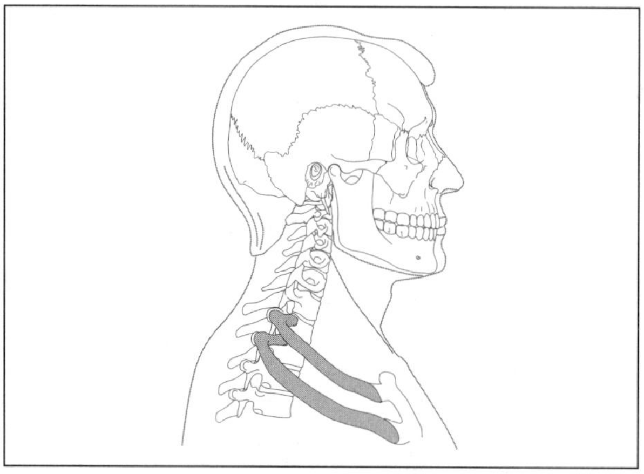

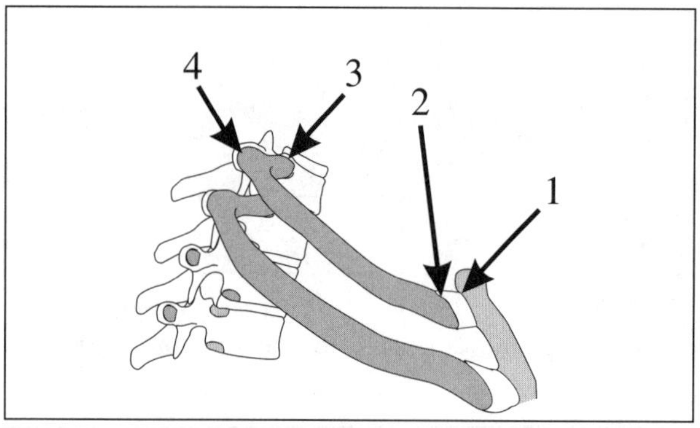

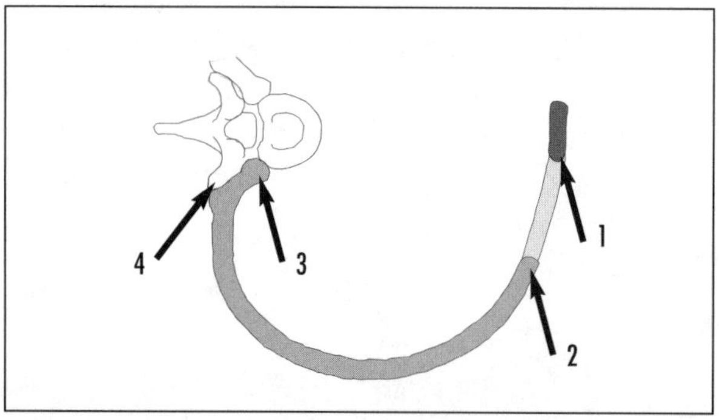

The author developed these techniques for a more "projectile" effect. When the rib cage joints (costovertebral, costotransverse, sternochondral, costochondral) are addressed with Muscle Energy and "Beyond" Technique there is a profound effect throughout the rib cage, diaphragms, peripheral joints, and cranium.

The techniques included in this chapter are appropriate for all clients, whether acute or chronic, mild or severe. Coughing, breathing difficulties, asthma, COPD (chronic obstructive pulmonary disease), emphysema, atelectasis, cystic fibrosis, and any other pulmonary problem can be assisted with these techniques. Certainly pain in the rib cage, mobility problems, and limitations of motion are helped with this approach.

Determination of appropriateness to utilize these rib techniques:

1. Postural assessment of chest and rib cage
2. Mobility testing of rib joints
3. Weakness and functional limitations of thoracic spine movements, upper extremity motions, and rib cage movements

The rib cage has been divided into four (4) major sections. There is an anterior category and a posterior category:

Anterior Category

1. The Sternochondral Joints
2. The Costochondral Joints

Posterior Category

3. The Costovertebral Joints
4. The Costotransverse Joints

The Sternochondral Joints

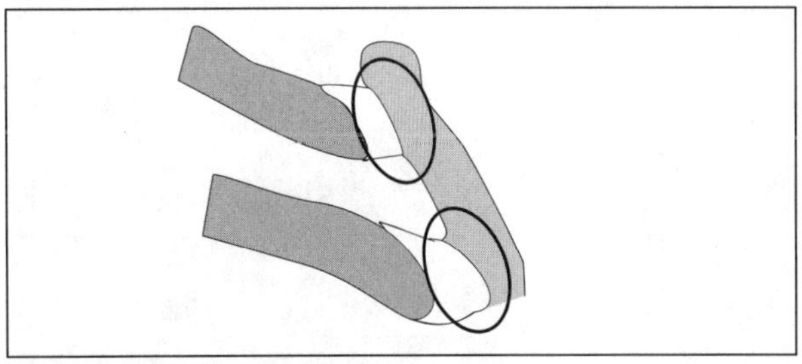

The sternochondral joints do not have a great deal of joint mobility. Typically, the articular balance is compromised; the chondral articular surface is compressed or distracted. Multiple problems occur secondary to this joint dysfunction. Sternochondral joint dysfunction is a modified Type II dysfunction.

Often restrictions of sternochondral joints cause severe pain, which resembles Tsi-Tsi syndrome and is an inflammation of the sternochondral region. Breathing may appear like asthma, because there is also spasm and dysfunction of the bronchi. Pain on palpation may be equal to pain in prone.

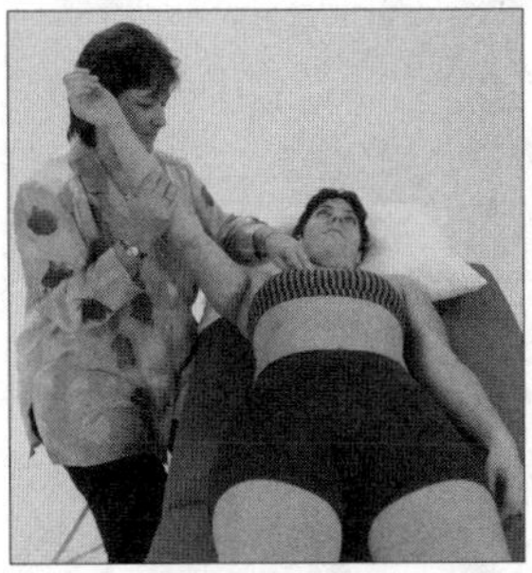

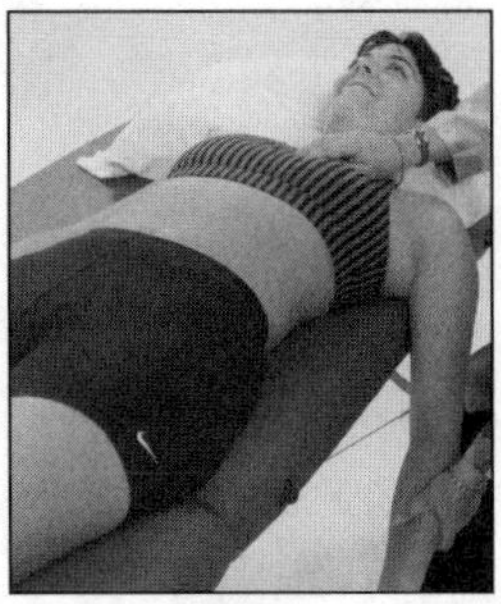

Positional Diagnosis

The Sternochondral joint can be flexed or extended (sagittal plane); or compressed or distracted. The four possible positional diagnoses are:

1. Flexed, compressed
2. Flexed, distracted
3. Extended, compressed
4. Extended, distracted

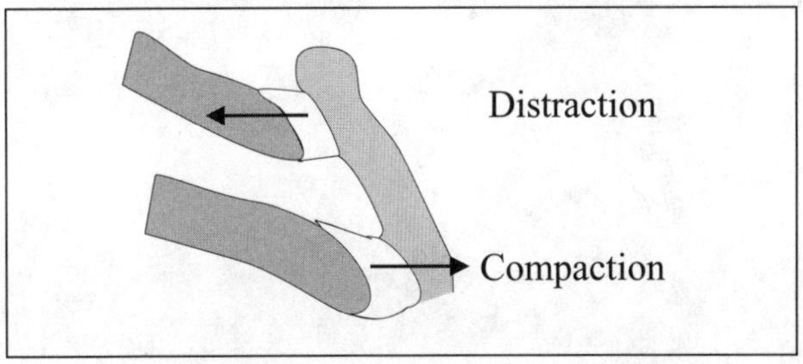

Palpation of the proximal chondral articular surface. Does the proximal chondral articular surface compress (approximate) or distract during respiration (inhalation and exhalation)? Inhalation is flexion. Exhalation is extension. If the compression or distraction is palpated during inhalation (flexion), the dysfunction is extended, because the rib cage can't flex.

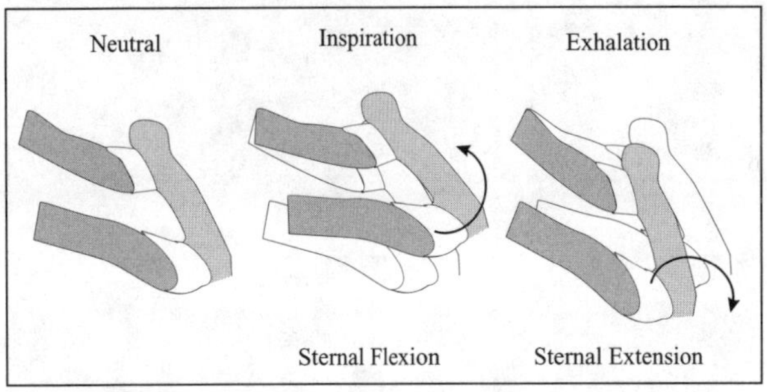

If the compression or distraction is palpated during exhalation (extension), the dysfunction is flexed, because the rib cage can't extend.

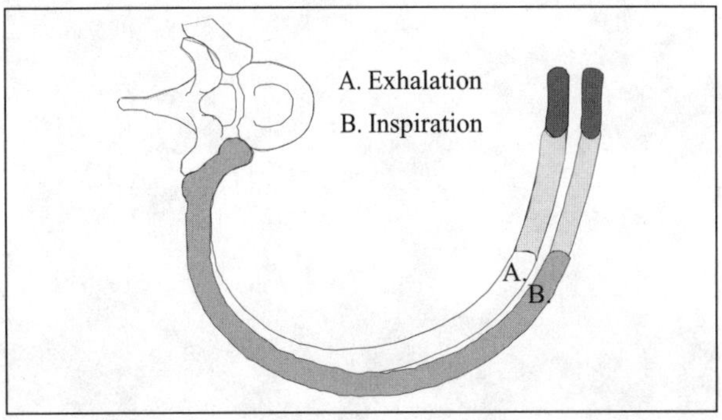

Flexed, compressed:

Find during exhalation. The proximal chondral articular surface approximates.

Flexed, distracted:

Find during exhalation. The proximal chondral articular surface distracts.

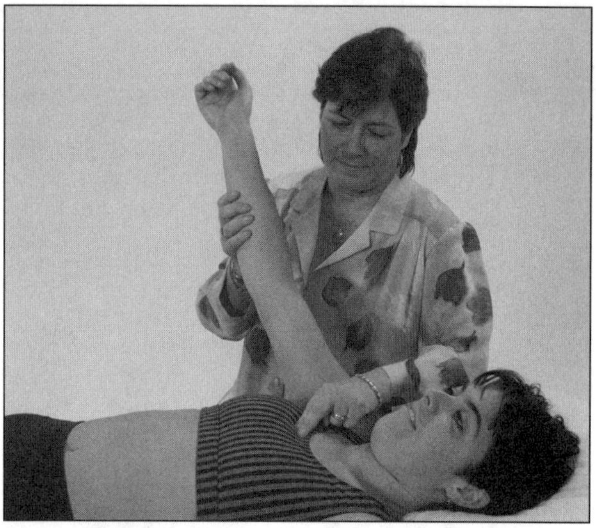

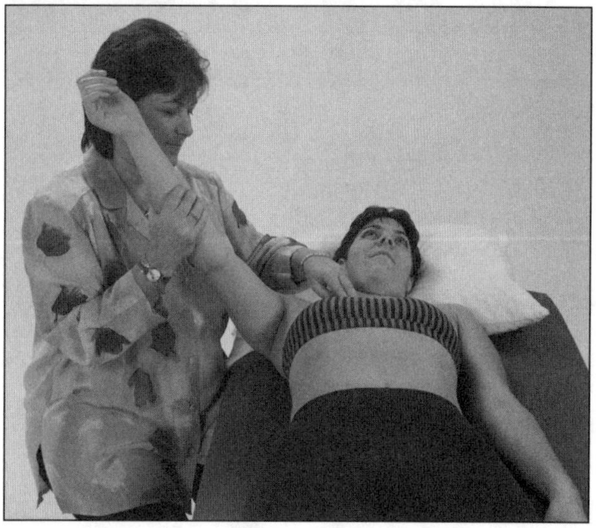

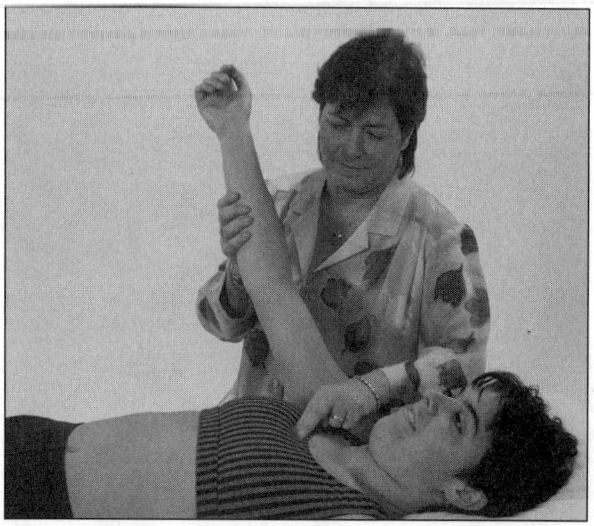

Extended, compressed:

Find during inhalation. The proximal chondral articular surface approximates.

Extended, distracted:

Find during inhalation. The proximal chondral articular surface distracts.

Movement Barrier

1. When the joint is **flexed, compressed:** The movement barrier is extended, distracted. Exhalation is inhibited.
2. When the joint is **flexed, distracted:** The movement barrier is extended, compressed. Exhalation is inhibited.
3. When the Joint is **extended, compressed:** The movement barrier is flexed, distracted. Inhalation is inhibited.
4. When the joint is **extended, distracted:** The movement barrier is flexed, compressed. Inhalation is inhibited.

Treatment

*When the joint is **flexed, compressed:***

1. The patient is supine. The shoulder and arm is off the edge of the table.
2. Palpate the proximal chondral articular surface.
3. Extend the arm until the proximal chondral articular surface begins to approximate. (This is the muscle barrier.) Return a few degrees to the interbarrier zone.
4. Abduct the arm until the proximal chondral articular surface begins to approximate. (This is the muscle barrier.) Return a few degrees to the interbarrier zone.
5. Perform isometric resistance in flexion or extension for six (6) seconds. Exhale (blow out air) during the muscle contraction.
6. Relax. Extend the arm to the next barrier. Return to the interbarrier zone.
7. Abduct the arm to the next barrier. Return to the interbarrier zone.
8. Perform isometric resistance in the same direction.

9. Exhale (blow out air) during the muscle contraction.
10. Relax. Move the arm for the third and last repetition. Extend the arm to the next barrier. Return to the interbarrier zone.
11. Abduct the arm to the next barrier. Return to the interbarrier zone.
12. Perform isometric resistance in the same direction. Exhale (blow out air) during the muscle contraction.
13. Relax.

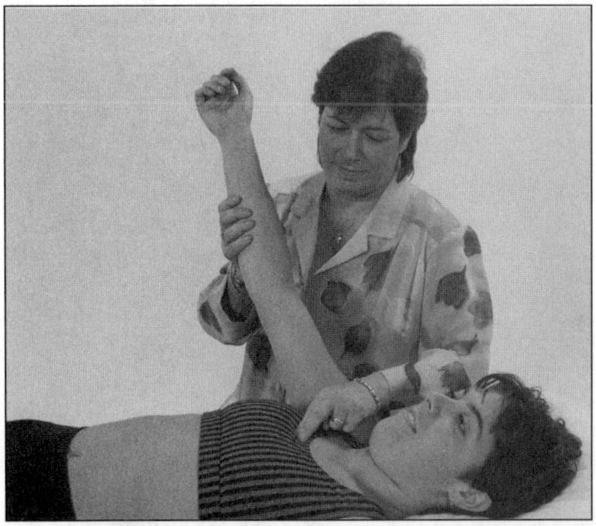

*When the joint is **flexed, distracted:***

1. The patient is supine. The shoulder and arm is off the edge of the table.
2. Palpate the proximal chondral articular surface.
3. Extend the arm until the proximal chondral articular surface begins to distract. (This is the muscle barrier.) Return a few degrees to the interbarrier zone.
4. Adduct the arm until the proximal chondral articular surface begins to distract. (This is the muscle barrier.) Return a few degrees to the interbarrier zone.
5. Perform isometric resistance in flexion or extension for six (6) seconds. Exhale (blow out air) during the muscle contraction.
6. Relax. Extend the arm to the next barrier. Return to the interbarrier zone.
7. Adduct the arm to the next barrier. Return to the interbarrier zone.
8. Perform isometric resistance in the same direction. Exhale (blow out air) during the muscle contraction.
9. Relax. Move the arm for the third and last repetition. Extend the arm to the next barrier. Return to the interbarrier zone.
10. Adduct the arm to the next barrier. Return to the interbarrier zone.
11. Perform isometric resistance in the same direction. Exhale (blow out air) during the muscle contraction.
12. Relax.

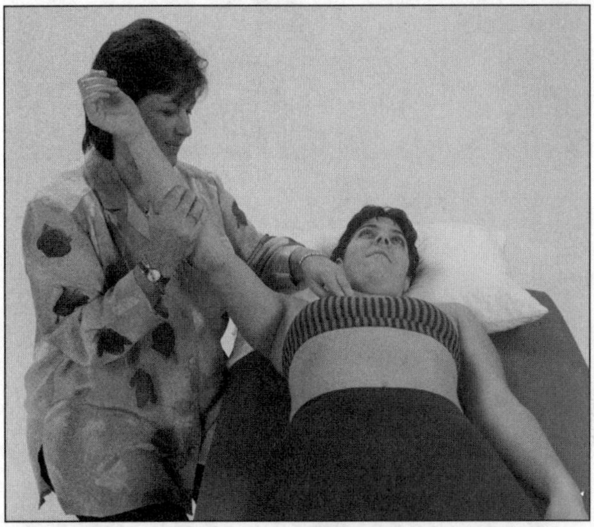

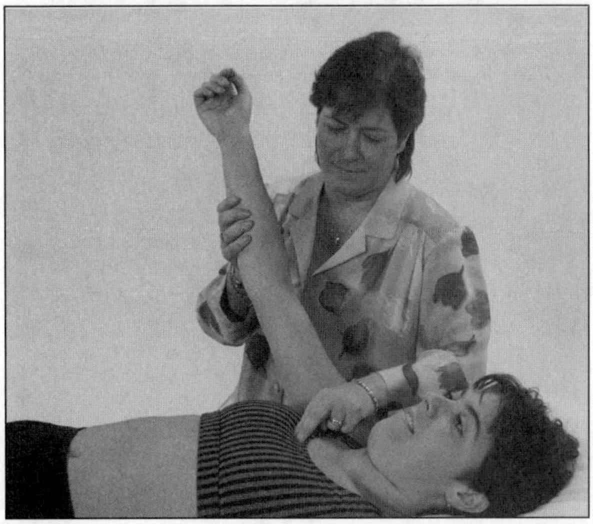

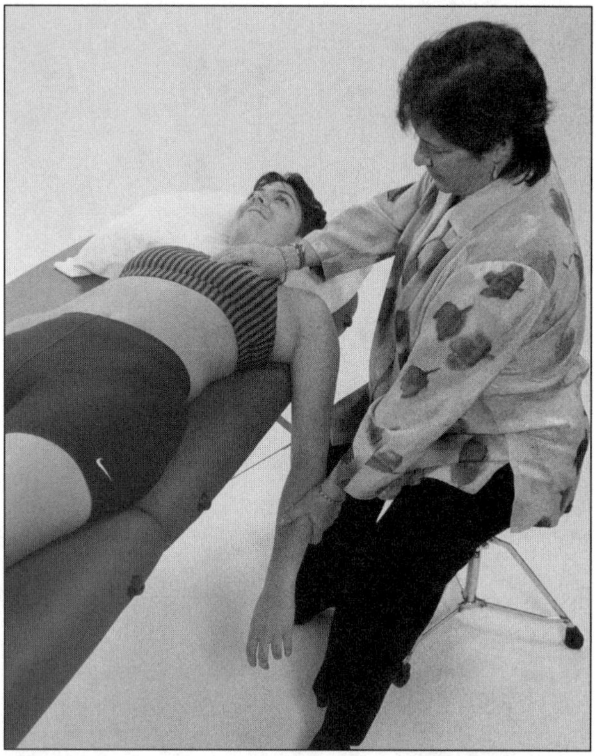

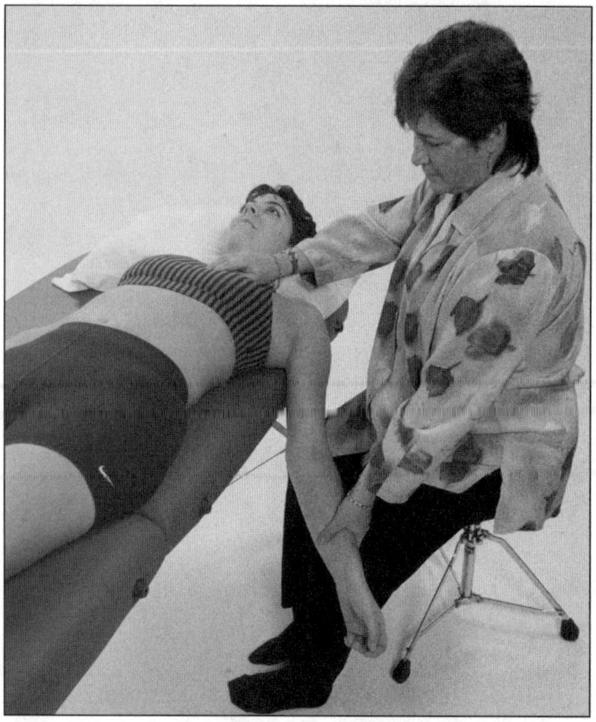

When the joint is ***extended, compressed:***

1. The patient is supine. The shoulder and arm is on the table.
2. Palpate the proximal chondral articular surface.
3. Flex the arm until the proximal chondral articular surface begins to approximate. (This is the muscle barrier.) Return a few degrees to the interbarrier zone.
4. Abduct the arm until the proximal chondral articular surface begins to approximate. (This is the muscle barrier.) Return a few degrees to the interbarrier zone.
5. Perform isometric resistance in flexion or extension for six (6) seconds. Inhale (take in air) during the muscle contraction.
6. Relax. Flex the arm to the next barrier. Return to the interbarrier zone.
7. Abduct the arm to the next barrier. Return to the interbarrier zone.
8. Perform isometric resistance in the same direction. Inhale (take in air) during the muscle contraction.
9. Relax. Move the arm for the third and last repetition. Flex the arm to the next barrier. Return to the interbarrier zone.
10. Abduct the arm to the next barrier. Return to the interbarrier zone.
11. Perform isometric resistance in the same direction. Inhale (take in air) during the muscle contraction.
12. Relax.

*When the joint is **extended, distracted:***

1. The patient is supine. The shoulder and arm is on the table.
2. Palpate the proximal chondral articular surface.
3. Flex the arm until the proximal chondral articular surface begins to distract. (This is the muscle barrier.) Return a few degrees to the interbarrier zone.
4. Adduct the arm until the proximal chondral articular surface begins to distract. (This is the muscle barrier.) Return a few degrees to the interbarrier zone.
5. Perform isometric resistance in flexion or extension for six (6) seconds. Inhale (take in air) during the muscle contraction.
6. Relax. Flex the arm to the next barrier. Return to the interbarrier zone.
7. Adduct the arm to the next barrier. Return to the interbarrier zone.
8. Perform isometric resistance in the same direction. Inhale (take in air) during the muscle contraction.
9. Relax. Move the arm for the third and last repetition. Flex the arm to the next barrier. Return to the interbarrier zone.
10. Adduct the arm to the next barrier. Return to the interbarrier zone.
11. Perform isometric resistance in the same direction. Inhale (take in air) during the muscle contraction.
12. Relax.

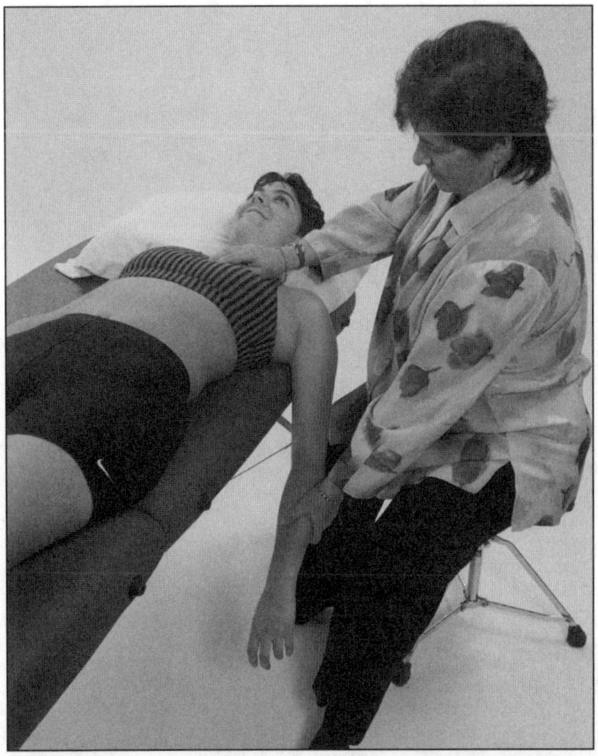

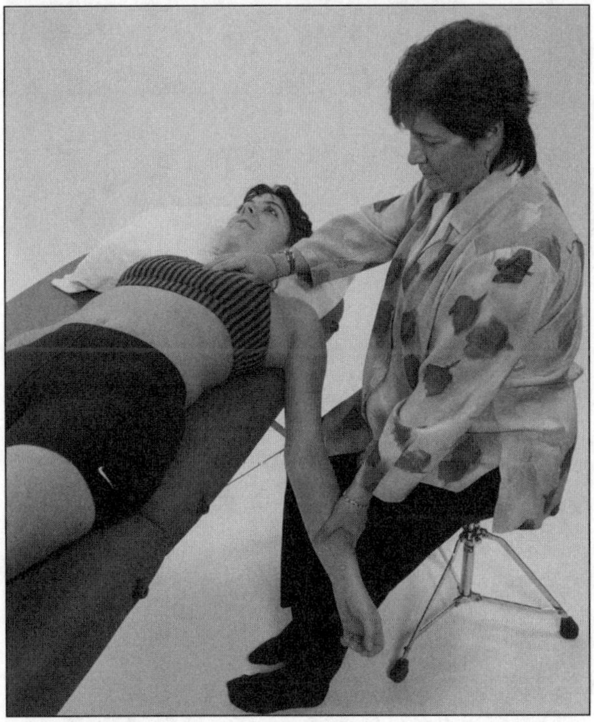

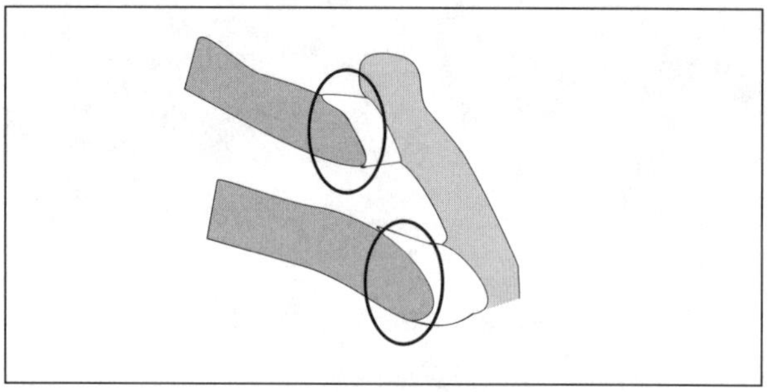

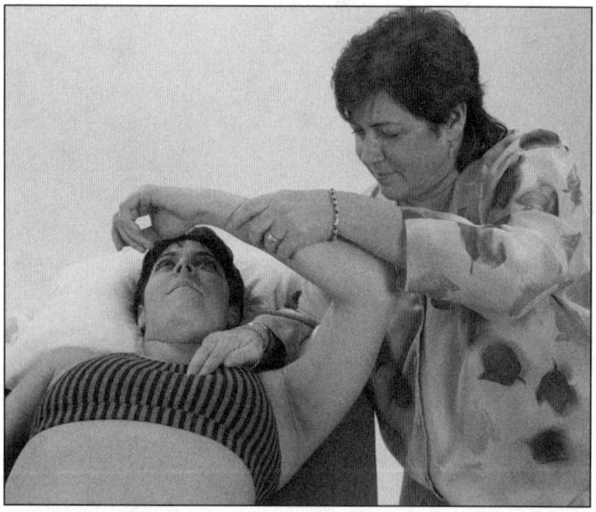

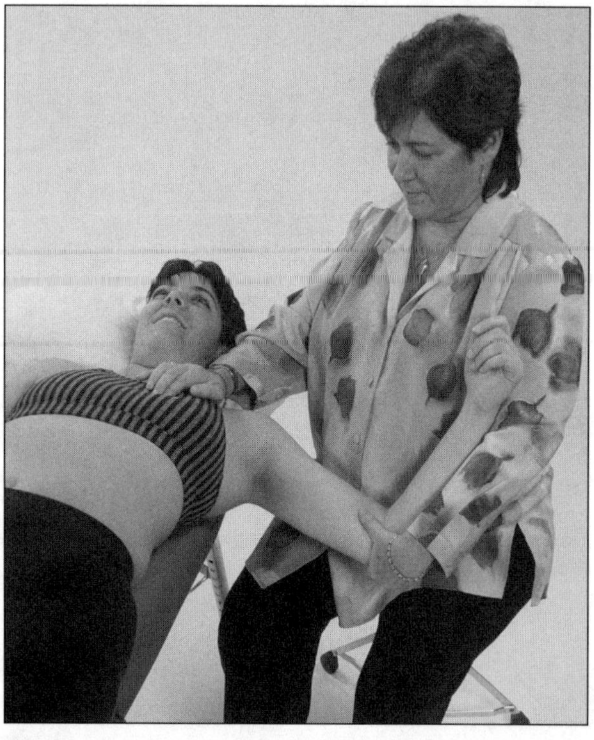

The Costoochondral Joints

The costochondral joint has modified Type I behavior. Type I motion means that in neutral, side bending and rotation occur to opposite sides. Although there is only Type II movement at the sternochondral joints, there is only Type I movement at the costochondral joints.

Costochondral joint dysfunction has multiple secondary problems: Bronchial spasm and bronchitis are the most common result of prolonged costochondral joint dysfunction. Bronchiectasis may be eliminated when costochondral joint dysfunction is corrected. Pleurisy may also be corrected (i.e., signs and symptoms eradicated) when costochondral joint dysfunction is treated with Muscle Energy and "Beyond" Technique.

Motion Palpation for the Muscle Barrier

Palpation for the muscle barrier is at the lateral aspect of the costochondral joint, immediately lateral to the rib bone articular surface.

Palpate for one of two possible motions:

1. Horizontal abduction, which is coronal plane side bending.
2. Horizontal adduction, which is coronal plane side bending.

A costochondral joint with biomechanical integrity will show evidence of either horizontal abduction or horizontal adduction during movement. Evidence of compromised horizontal abduction and/or horizontal adduction is evidence of Type I movement dysfunction.

Type I Dysfunction of the Costochondral Joint

There are two (2) possible Type I dysfunctions of the costochondral joint:

1. All compressed
2. All distracted

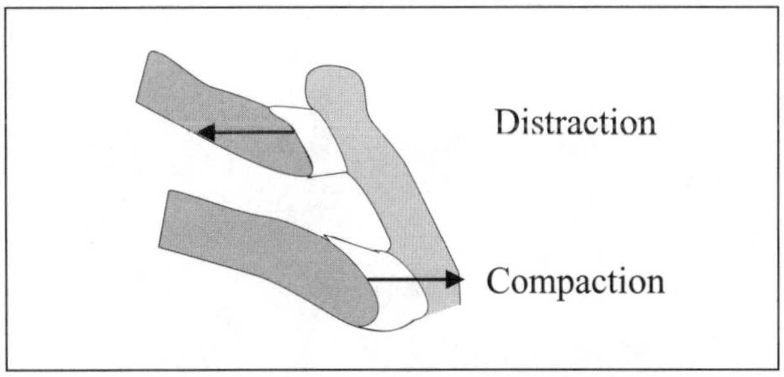

The coronal plane portion of Type I dysfunction of the costochondral joint:

1. All compressed: The rib is stuck in horizontal adduction; there is a compression of the anterior bony rib articular surface into the distal chondral portion of the rib.
2. All distracted: The rib is stuck in horizontal abduction; there is a lateral distraction of the anterior bony rib articular surface away from the distal chondral portion of the rib.

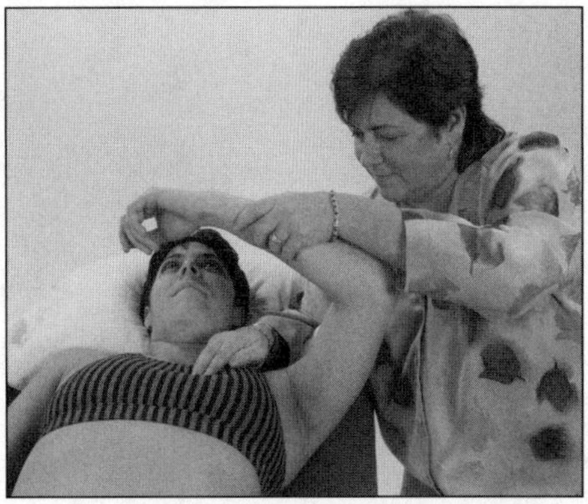

The transverse plane portion of Type I dysfunction of the costochondral joint:

1. All compressed: When the rib is stuck in horizontal adduction, there is an inferior rotation of the total bony rib complex, which affects the costochondral joint, but also affects the costovertebral joints and the costotransverse joint.
2. All distracted: When the rib is stuck in horizontal abduction, there is a superior rotation of the total bony rib complex, which affects the costochondral joint, but also affects the costovertebal joint and the costotransverse joint.

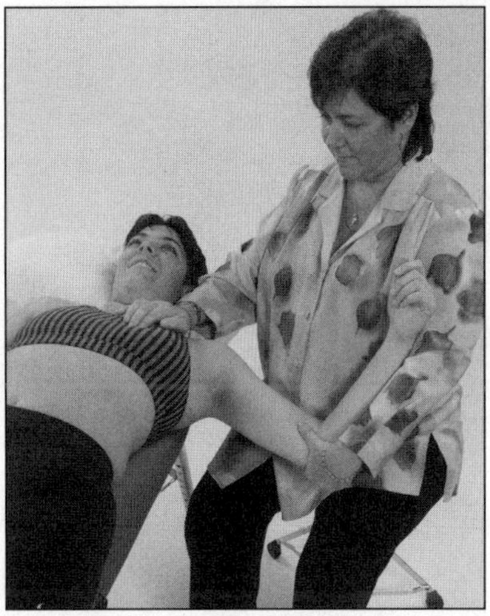

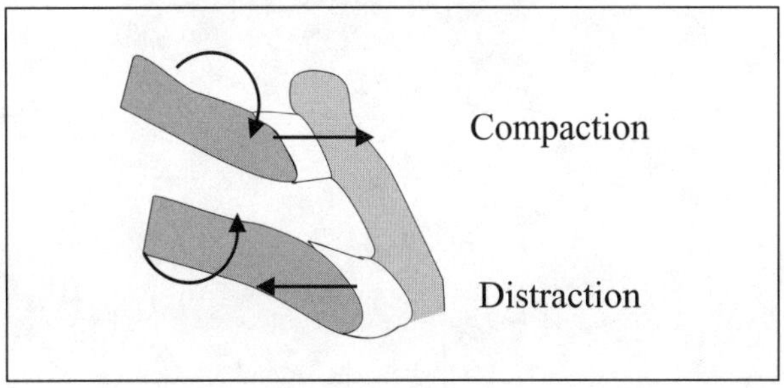

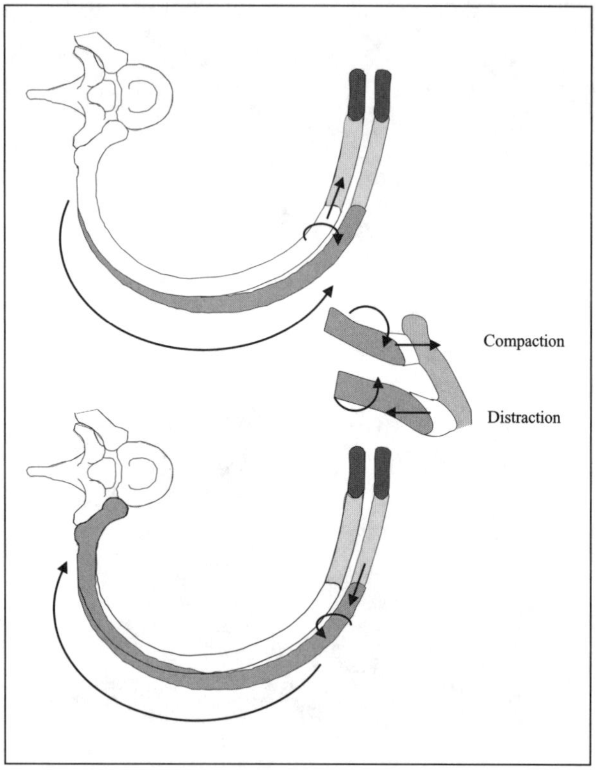

Considering the coronal and transverse plane portions listed above, the Type I dysfunctions of the costochondral joint are as follows:

1. All compressed: The rib is stuck in horizontal adduction and inferior rotation.
2. All distracted: The rib is stuck in horizontal abduction and superior rotation.

Movement Barrier

1. All compressed: The rib is stuck in horizontal adduction and Inferior Rotation. The Movement Barrier is Horizontal Abduction and Superior Rotation.
2. All Distracted: The rib is stuck in Horizontal Abduction and Superior Rotation. The Movement Barrier is Horizontal Adduction and Inferior Rotation.

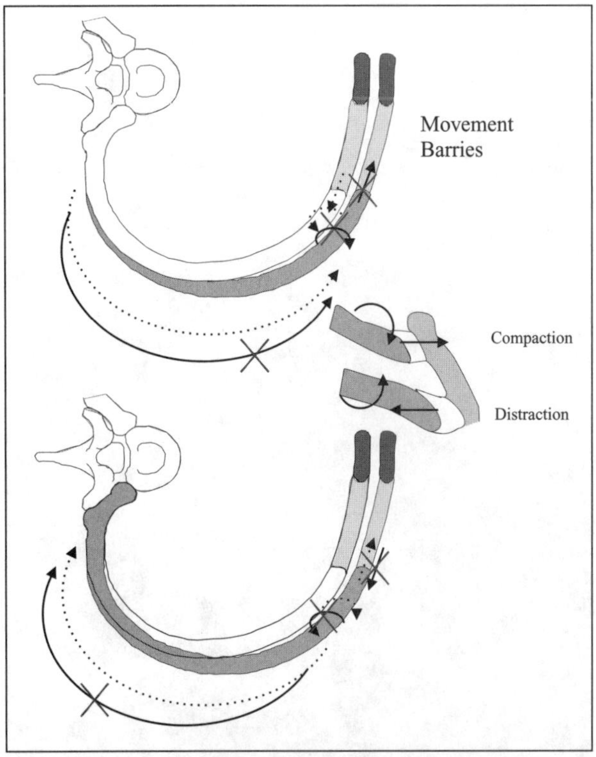

Correlation with Respiratory Difficulty

1. All compressed: There is limited expansion of the rib cage during respiration. Inhalation will be inhibited. Side bending of the trunk to the opposite side of the Type I dysfunction will be limited. Rotation of the trunk will be limited to both sides. Abduction and elevation of the ipsilateral arm will be limited.

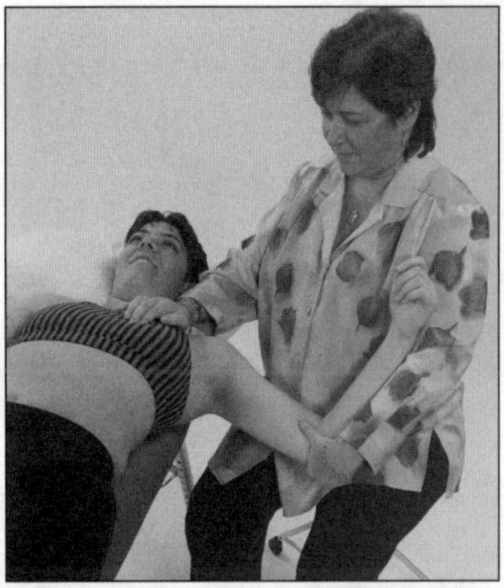

2. All distracted: There is limited compression of the rib cage during respiration. Exhalation will be inhibited. Side bending of the trunk to the side of the Type I dysfunction will be limited. Rotation of the trunk will be limited to both sides. Full adduction and extension of the ipsilateral arm will be limited.

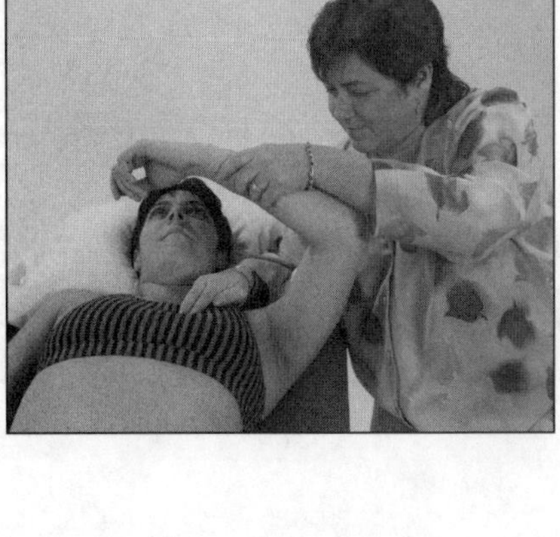

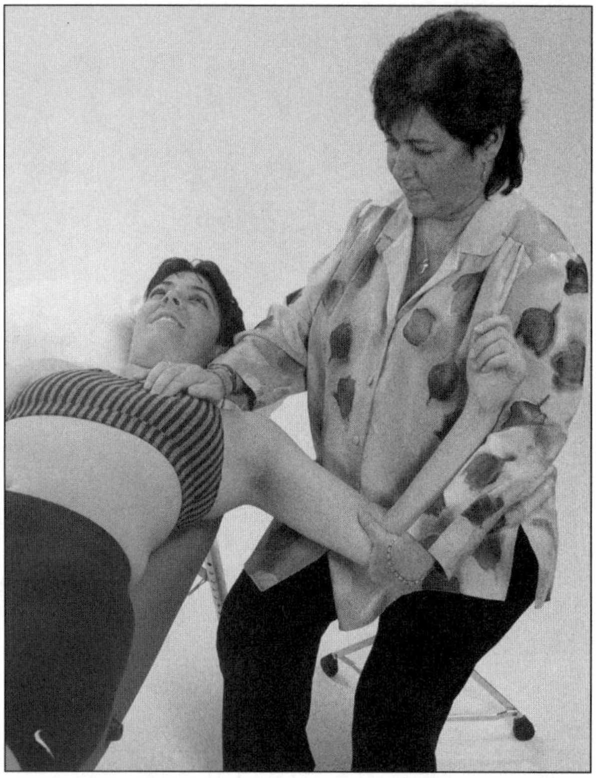

Treatment

All compressed:

1. The patient can be in supine or sitting. Ask the patient to inhale (take in air). Place the ipsilateral arm in 90 degrees elbow flexion.
2. Abduct the arm to the movement barrier.
3. Palpate the movement barrier at the proximal bony rib articular surface. When the articular surface begins to approximate (i.e., the proximal bony rib articular surface begins to move medial), the movement barrier has been reached. Return a few degrees to the interbarrier zone.
4. Perform an isometric resistance (5 grams resistance). Resist horizontal abduction or adduction for 6 seconds.
5. Relax. Then move the arm further into horizontal abduction until the second movement barrier is palpated. Return a few degrees to the interbarrier zone.
6. Repeat the same direction of isometric resistance as before for 6 seconds.
7. Relax. For the third and last repetition, move the arm into horizontal abduction to the barrier. Return to the interbarrier zone.
8. Perform a similar isometric resistance.
9. Relax.

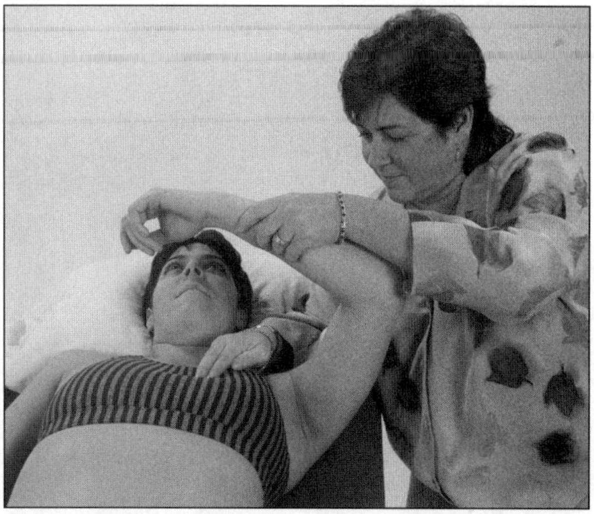

All distracted:

1. The patient can be in supine or sitting. Ask the patient to exhale (blow out air). Place the ipsilateral arm in 90 degrees elbow flexion.
2. Start at 90 degrees abduction.
3. Bring the arm into horizontal adduction to the movement barrier. Palpate the movement barrier at the proximal bony rib articular surface. When the articular surface begins to distract lateral, the movement barrier has been reached. Return a few degrees to the interbarrier zone.
4. Perform an isometric resistance (5 grams resistance). Resist horizontal abduction or adduction for 6 seconds.

5. Relax. Then move the arm further into horizontal adduction until the second movement barrier is palpated. Return a few degrees to the interbarrier zone.
6. Repeat the same direction of isometric resistance as before for 6 seconds.
7. Relax. For the third and last repetition, move the arm into horizontal adduction to the barrier. Return to the interbarrier zone.
8. Perform a similar isometric resistance.
9. Relax.

The Costovertebral Joints

There is modified Type I movement at the costovertebral joint. The joint does not flex or extend. There is flexion and extension at the thoracic spine facets, but the articulation between the proximal rib articular surface and the vertebral body has no flexion or extension. The costovertebral joint remains in neutral at all times, whether the thoracic vertebra is in neutral, in flexion, or in extension. Side bending and rotation occur to opposite sides in Type I movement. This is true of the costovertebral joint. The majority of rib cage rotation occurs due to the movement at the costovertebral joints, rather than at any other joint within the rib cage complex. Restore motion at the costovertebral joints, and there will be a significant increase in rib cage rotations.

Pain because of costovertebral joint dysfunction can remain sharp and acute, even when the dysfunction is chronic. Rotations and side bendings of the trunk can cause referred pain along the length of the rib. Lifting, bending, and carrying can be intensely painful even when there is only one costovertebral joint dysfunction. Breathing can be markedly improved, with significant increased rib excursion, when costovertebral joint dysfunction is corrected.

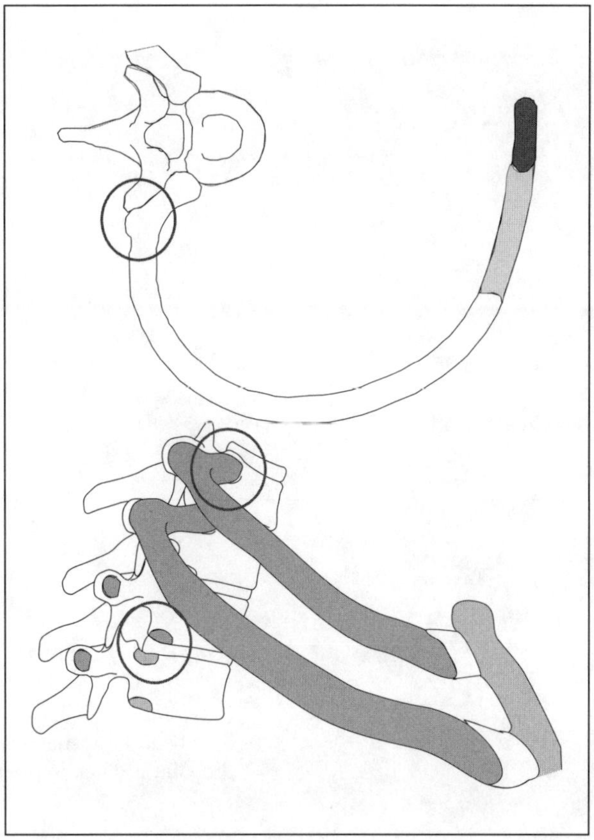

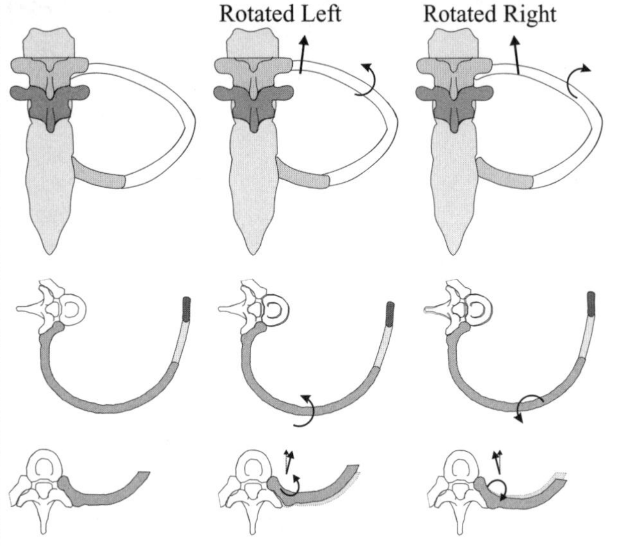

Type I Dysfunction of the Costovertebral Joints

The coronal plane motion at the costovertebral joint is side bending to the right and to the left. The transverse plane movement at the costovertebral joint is medial rotation and lateral rotation. There is no sagittal plane motion at the costovertebral joint. Rotation and side bending occur to opposite sides. When the right costovertebral joint side bends to the right, it rotates to the left. This is medial rotation. When the right costovertebral joint side bends to the left, it rotate to the right. This is lateral rotation. Side bending to the ipsilateral side occurs with medial rotation.

Side bending to the contralateral side occurs with lateral rotation.

When the costovertebral joint is stuck in side bending to the ipsilateral (same) side of the dysfunctional joint, the joint is stuck in medial rotation.

When the costovertebral joint is stuck in side bending to the contralateral (opposite) side of the dysfunctional joint, the joint is stuck in lateral rotation.

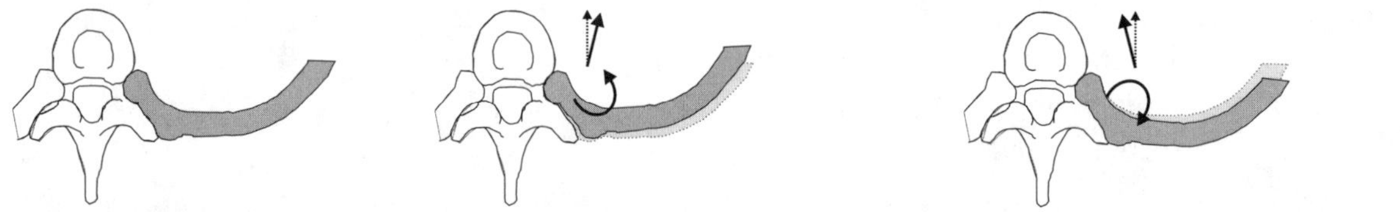

Muscle Barrier Palpation

Palpate the muscle barrier lateral to the transverse process of the involved rib joint. The dysfunctional articulation is between the vertebral body and the proximal rib head, but the muscle barrier is palpated just lateral to the transverse process. Palpate for medial rotation and lateral rotation of the proximal rib. Medial rotation of the rib feels like an anterior motion. Lateral rotation of the rib feels like a posterior motion.

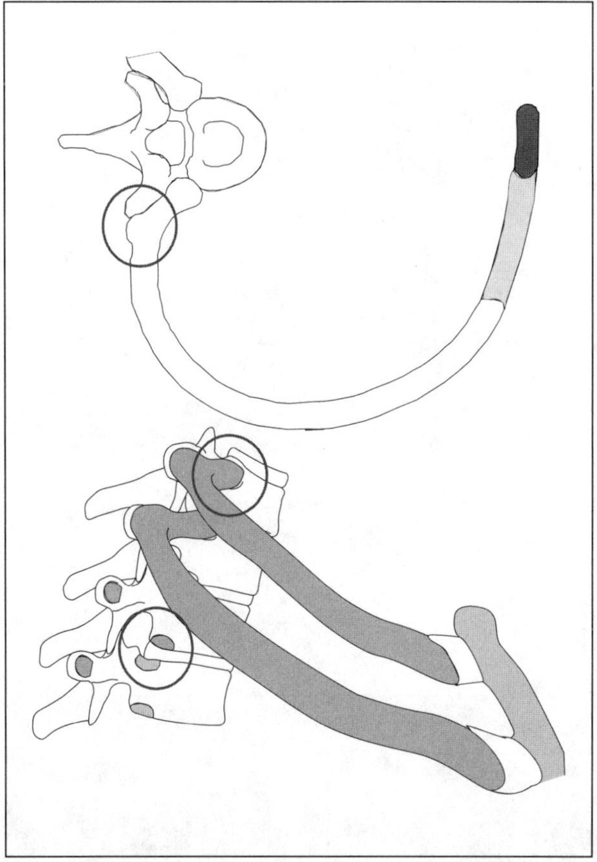

Positional Diagnosis

There are two possible positional diagnoses when assessing for costovertebral joint dysfunction:

1. Lateral rotated, side bent: The costovertebal joint is stuck side bent to the opposite side and lateral rotated.
2. Medial rotated, side bent: The costovertebral joint is stuck side bent to the same side and medial rotated.

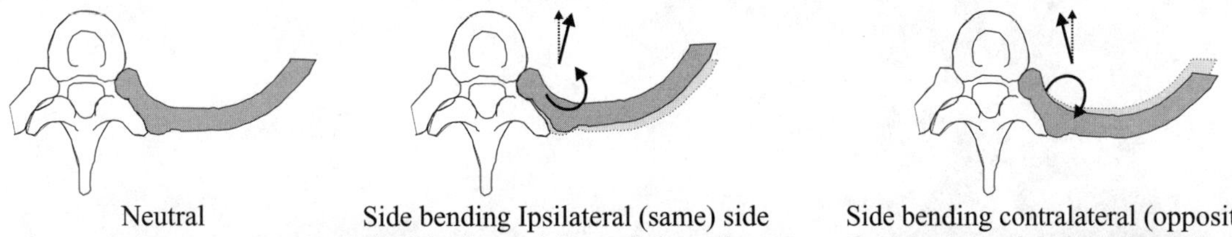

Neutral

Side bending Ipsilateral (same) side the joint is stuck in medial rotation.

Side bending contralateral (opposite) side, the joint is stuck in lateral rotation

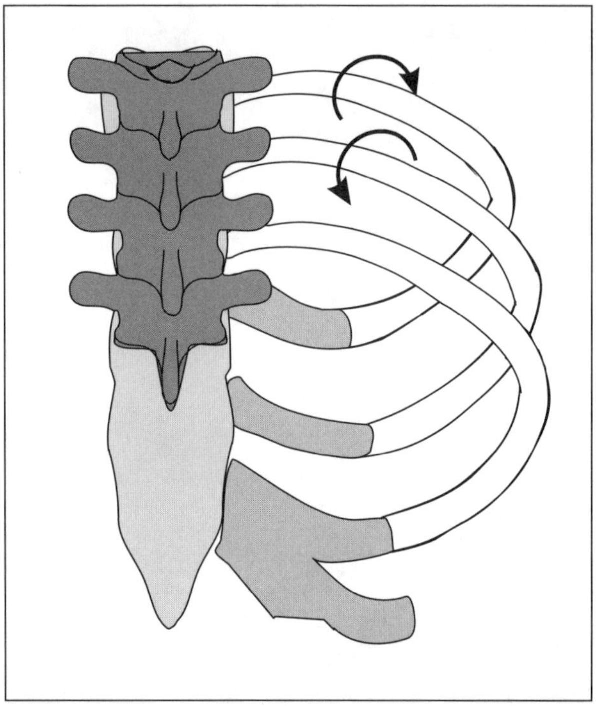

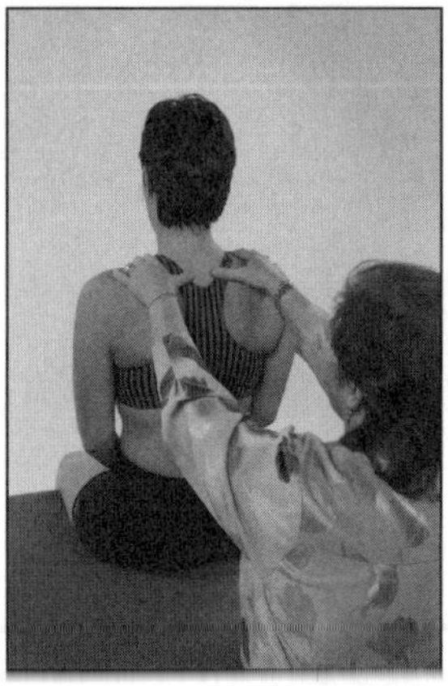

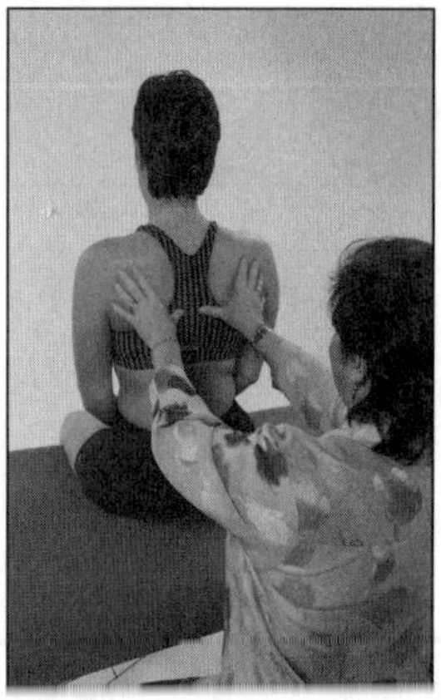

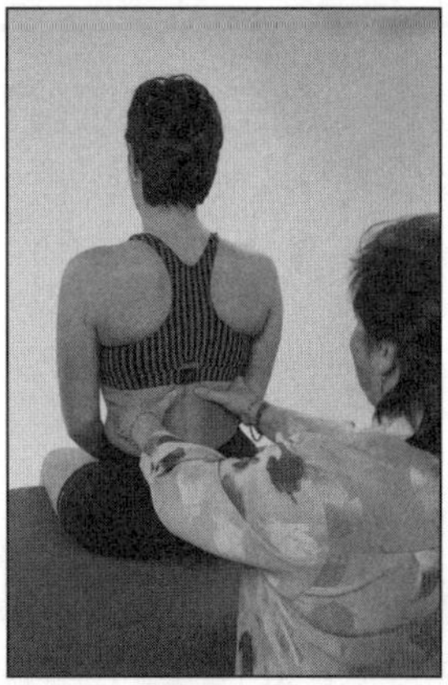

Movement Barrier

Palpate the muscle barrier lateral to the transverse process of the involved rib joint. The dysfunctional articulation is between the vertebral body and the proximal rib head, but the muscle barrier is palpated just lateral to the transverse process. Palpate for medial rotation and lateral rotation of the proximal rib. Medial rotation of the rib feels like an anterior motion. Lateral rotation of the rib feels like a posterior motion.

1. When the muscle barrier is palpated as a posterior motion of the rib, the costovertebral joint is lateral rotated, side bent. The movement barrier is medial rotation, side bending to the same side.
2. When the muscle barrier is palpated as an anterior motion of the rib, the costoverterbal joint is medial rotated, side bent. The movement barrier is lateral rotation, side bending to the opposite side.

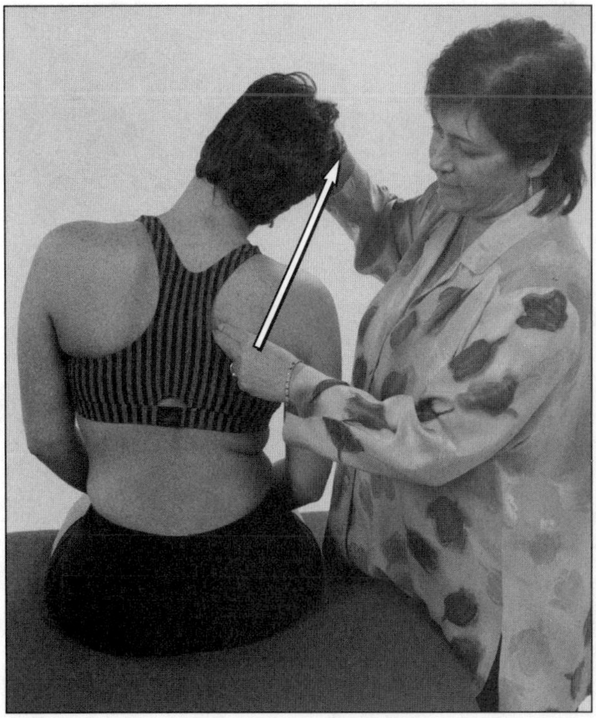

Treatment

Lateral rotated, side bent:

1. The patient is sitting. Supine is possible for the more involved patient (i.e., the neurologically involved patient).
2. Side bend the patient down the kinetic chain towards the side of the dysfunctional rib until the muscle barrier is palpated lateral to the transverse process as a posterior motion of the rib. Return a few degrees to the interbarrier zone.
3. Push the rib gently anterior, just lateral to the transverse process.
4. Perform an isometric resistance, 5 grams of force, for 6 seconds. Resist side bending of the trunk (rib cage). Either right or left side bending can be resisted.
5. Relax. Then side bend further down the kinetic chain towards the side of the dysfunctional rib until the muscle barrier is palpated again for the second repetition. Return a few degrees to the interbarrier zone.
6. Push the rib gently anterior. Perform an isometric resistance, 5 grams of force, for 6 seconds. Resist side bending of the trunk (rib cage). Either right or left side bending can be resisted.
7. Relax. Then side bend further down the kinetic chain towards the side of the dysfunctional rib until the muscle barrier is palpated again for the third and last repetition. Return a few degrees to the interbarrier zone.
8. Push the rib gently anterior. Perform an isometric resistance, 5 grams of force, for 6 seconds. Resist side bending of the trunk (rib cage). Either right or left side bending can be resisted.
9. Relax.

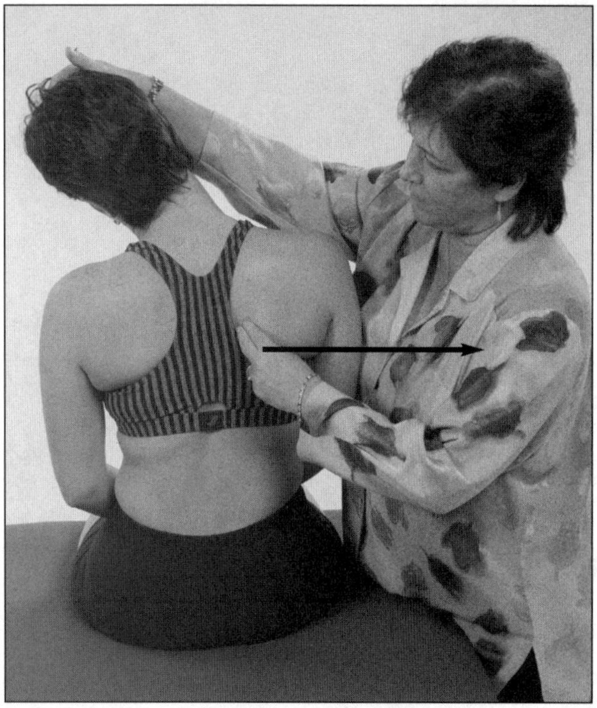

Medial rotated, side bent:

1. The patient is sitting. Supine is possible for the more involved patient (i.e., the neurologically involved patient).
2. Side bend the patient down the kinetic chain away from the side of the dysfunctional rib until the muscle barrier is palpated lateral to the transverse process as an anterior motion of the rib. Return a few degrees to the interbarrier zone.
3. Push the rib gently posterior. Perform an isometric resistance, 5 grams of force, for 6 seconds. Resist side bending of the trunk (rib cage). Either right or left side bending can be resisted.
4. Relax. Then side bend further down the kinetic chain away from the side of the dysfunctional rib until the muscle barrier is palpated again for the second repetition. Return a few degrees to the interbarrier zone.
5. Push the rib gently posterior. Perform an isometric resistance, 5 grams of force, for 6 seconds. Resist side bending of the trunk (rib cage). Either right or left side bending can be resisted.
6. Relax. Then side bend further down the kinetic chain away from the side of the dysfunctional rib until the muscle barrier is palpated again for the third and last repetition. Return a few degrees to the interbarrier zone.
7. Push the rib gently posterior. Perform an isometric resistance, 5 grams of force, for 6 seconds. Resist side bending of the trunk (rib cage). Either right or left side bending can be resisted.
8. Relax.

The Costotransverse Joint

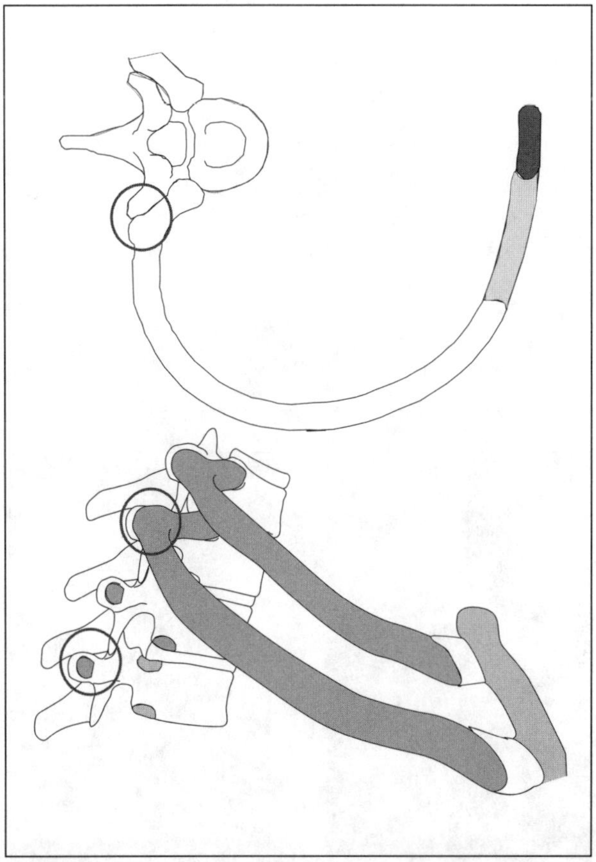

The costotransverse joint has a modified Type I movement. There is no sagittal plane flexion or extension at this joint. There are glides, medial and lateral, which are modified coronal plane side bending. Transverse plane motions are not palpated because there is a superior rotation with the external rotation and an inferior rotation with the medial rotation. The functional motions are the medial and lateral glides. The medial glide and lateral glide is palpated lateral to the transverse process of the dysfunctional costotransverse joint. The medial glide is similar to a side bending to the same side. The lateral glide is similar to a side bending to the opposite side.

The costotransverse joint dyfunction is perhaps the most debilitating of all biomechanical dysfunctions of the rib cage. This is because of the location of the sympathetic ganglia, anterior to the costotransverse joints.

Sympathetic dysautonomia is common with costotransverse joint dysfunction. Often, somatovisceral dysfunction, which has been severe and chronic, will be eliminated with correction of the costotransverse joint dysfunction. It is definitely worthwhile to assess the spinal level of innervation with all visceral problems. Then treat all biomechanical dysfunction of thoracic and rib cage associated levels. Focus on the costotransverse joints of that same level of spinal innervation relevant to that organ.

Pain with breathing, trunk motions, rib movement, and all ranges of upper quadrant movements can occur secondary to costotransverse joint dysfunction.

Even herpes zoster has been eliminated after therapy for correction of costotransverse joint dysfunction. This indicates that herpes zoster may be (at least) augmented because of sympathetic ganglion facilitation.

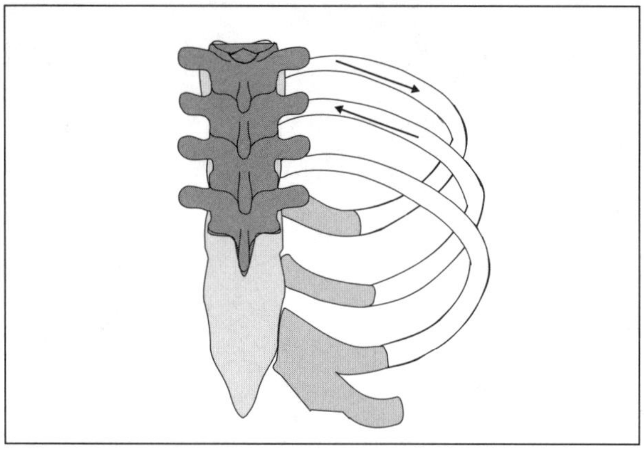

Palpation of the Muscle Barrier

Palpate the muscle barrier just lateral to the transverse process of the dysfunctional rib. Palpate for a medial or a lateral glide of the bony proximal rib.

The medial glide is side bending towards the ipsilateral (same) side.

The lateral glide is side bending towards the contralateral (opposite) side.

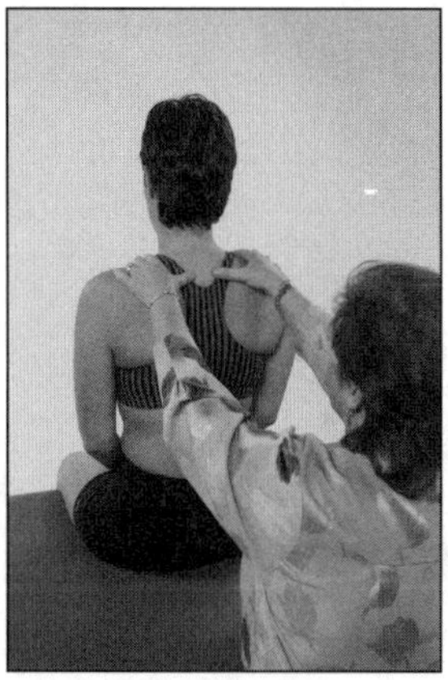

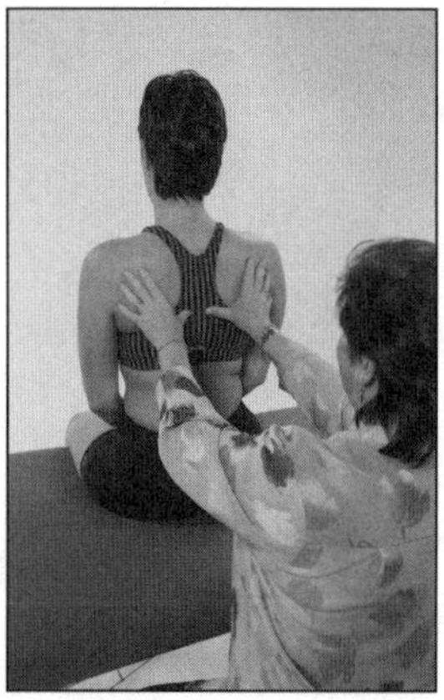

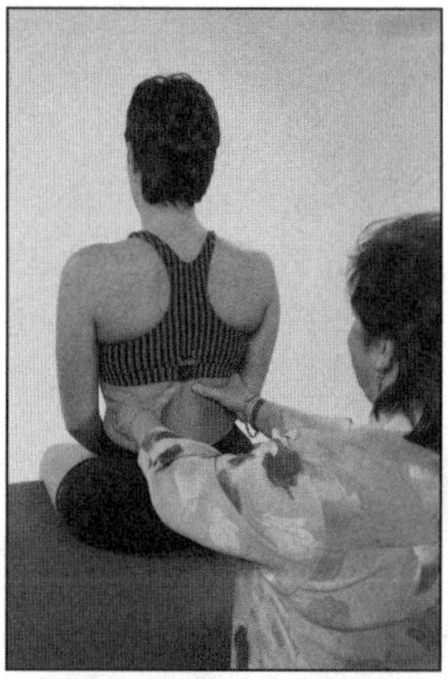

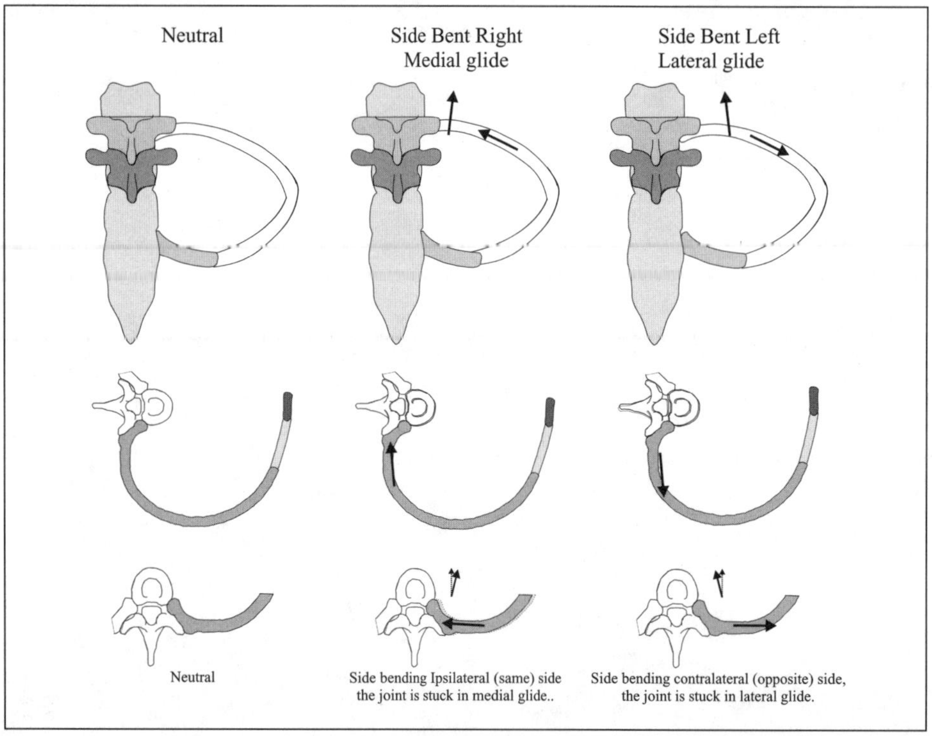

Positional Diagnosis

There are two options for Type I dysfunction of the costotransverse joint:

1. Medial glide: The costotransverse joint is stuck side bent to the same side and stuck in medial rotation and inferior rotation.
2. Lateral glide: The costotransverse joint is stuck side bent to the opposite side and stuck in external rotation and superior rotation.

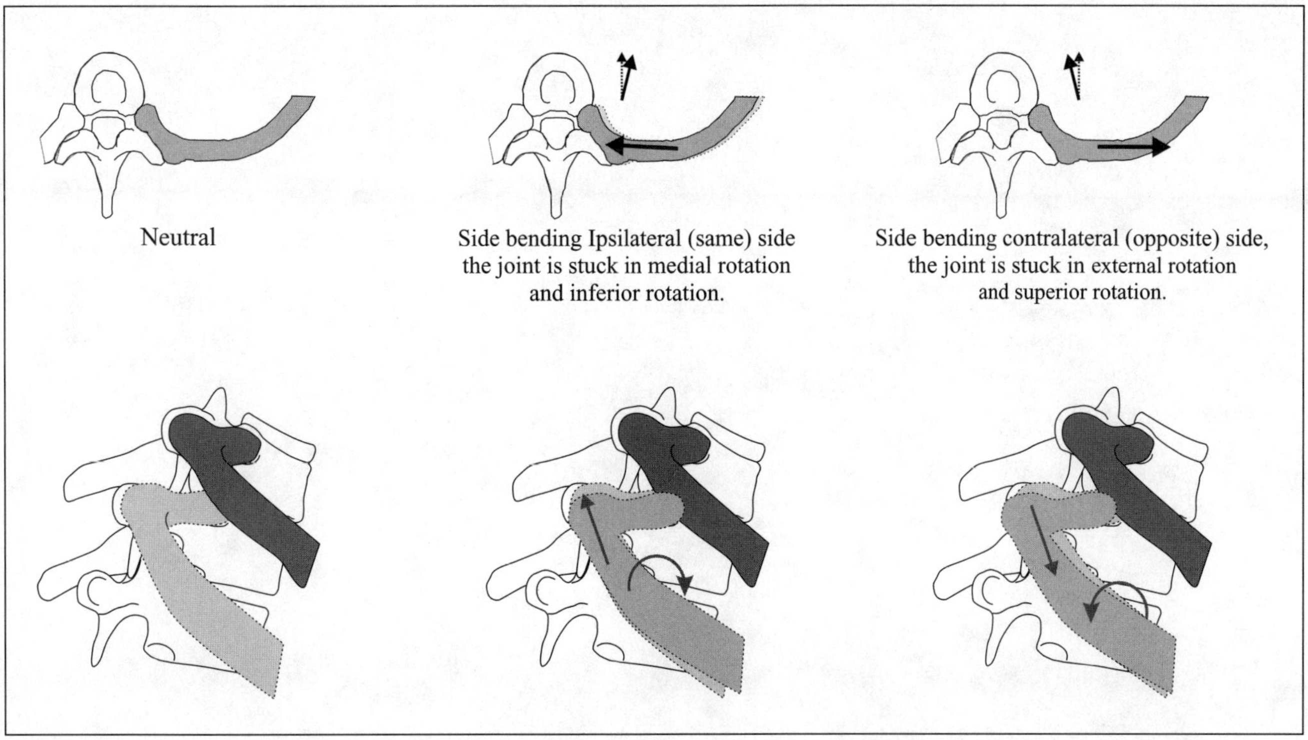

Movement Barrier

1. Medial glide: The costotransverse joint is stuck side bent to the same side and stuck in internal rotation and inferior rotation. The movement barrier is side bending to the opposite side, external rotation and superior rotation.
2. Lateral glide: The costotransverse joint is stuck side bent to the opposite side and stuck in external rotation and superior rotation. The movement barrier is side bending to the same side, internal rotation and inferior rotation.

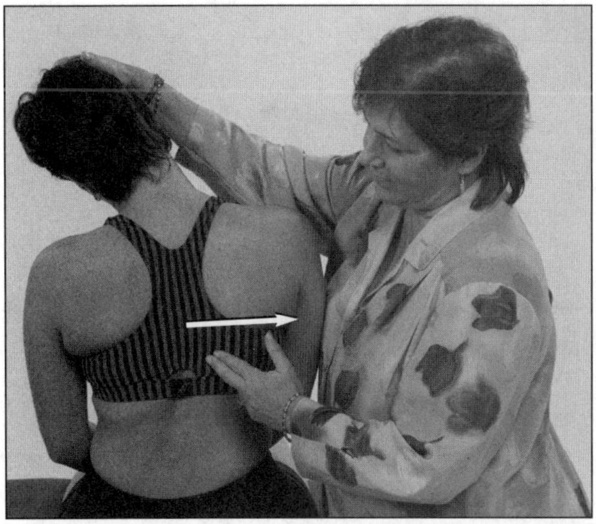

Treatment

Medial glide:

1. The patient is sitting. Supine position is possible for the more involved patient (e.g., the neurologic patient).
2. Side bend the patient one segment at a time down the kinetic chain, to locate the precise level, away from the side of the dysfunctional rib until the movement barrier is palpated as a medial glide, lateral to the transverse process. Return to the interbarrier zone. (The costotransverse joint is stuck side bent to the same side and stuck in internal rotation and inferior rotation. The movement barrier is side bending to the opposite side, external rotation and superior rotation.)
3. Push the rib gently into lateral glide.
4. Resist side bending of the trunk (rib cage) with 5 grams of force for 6 seconds. Resist either direction of side bending.
5. Relax. Side bend the patient down the kinetic chain away from the side of the dysfunctional rib for the second repetition until the movement barrier is palpated as a lateral glide, lateral to the transverse process. Return to the interbarrier zone.
6. Push the rib gently into lateral glide.
7. Resist side bending of the trunk (rib cage) with 5 grams of force for 6 seconds. Resist the same direction of side bending.
8. Relax. Side bend the patient down the kinetic chain away from the side of the dysfunctional rib until the movement barrier is palpated as a lateral glide, lateral to the transverse process for the third and last repetition. Return to the interbarrier zone.
9. Push the rib gently into lateral glide.
10. Resist side bending of the trunk (rib cage) with 5 grams of force for 6 seconds.
11. Relax.

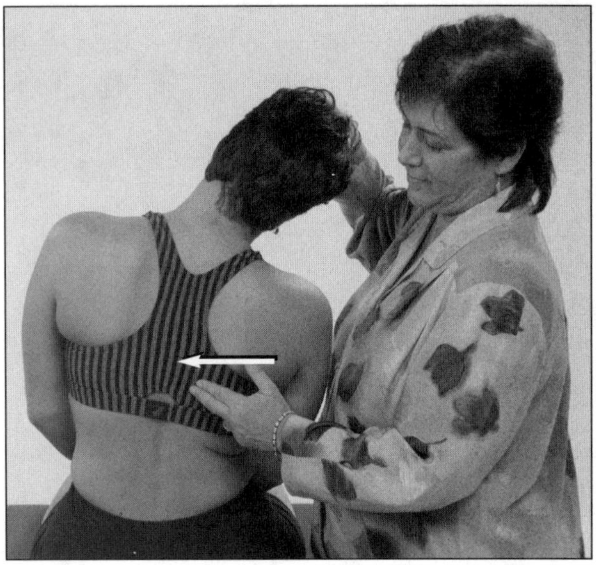

Lateral glide:

1. The patient is sitting. Supine position is possible for the more involved patient (e.g., the neurologic patient).
2. Side bend the patient down the kinetic chain towards the side of the dysfunctional rib until the movement barrier is palpated as a medial glide, lateral to the transverse process. Return to the interbarrier zone. (The costotransverse joint is stuck side bent to the opposite side and stuck in external rotation and superior rotation. The movement barrier is side bending to the same side, internal rotation and inferior rotation.)
3. Push the rib gently into medial glide.
4. Side bend the patient down the kinetic chain towards the side of the dysfunctional rib for the second repetition until the movement barrier is palpated as a medial glide, lateral to the transverse process. Return to the interbarrier zone.
5. Push the rib gently into medial glide.
6. Resist side bending of the trunk (rib cage) with 5 grams of force for 6 seconds. Resist the same direction of side bending.
7. Relax. Side bend the patient down the kinetic chain towards the side of the dysfunctional rib until the movement barrier is palpated as a medial glide, lateral to the transverse process for the third and last repetition. Return to the interbarrier zone.
8. Push the rib gently into medial glide.
9. Resist side bending of the trunk (rib cage) with 5 grams of force for 6 seconds.
10. Relax.

CHAPTER 27

THE HIP JOINT

The hip joints are a source of constant, chronic pain, which can be diagnosed as osteoarthrosis and/ or osteoarthritis. This Type II dysfunction causes friction of the femoral joint surface and acetabular joint surface, with a resultant strain of ligaments and capsule, compromise of the vertical dimension of the intra-articular joint space, and inflammation. Total hip replacement surgery can be required by this dysfunction, i.e., the pain, defibrillation of hyaline cartilage lining, and inflammation are caused by this dysfunction. When this technique is utilized and possibly repeated multiple times, total hip replacement surgery can often be prevented.

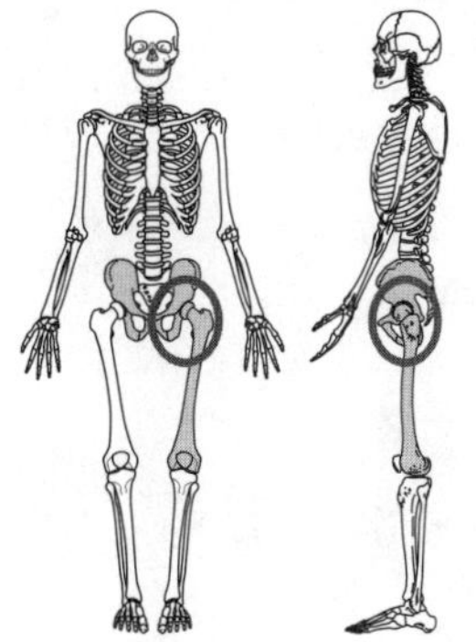

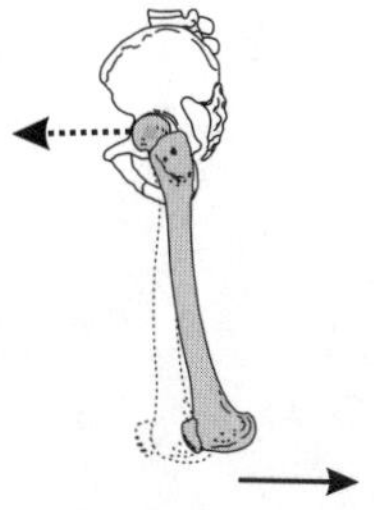

Anterior glide
with
Hip extension

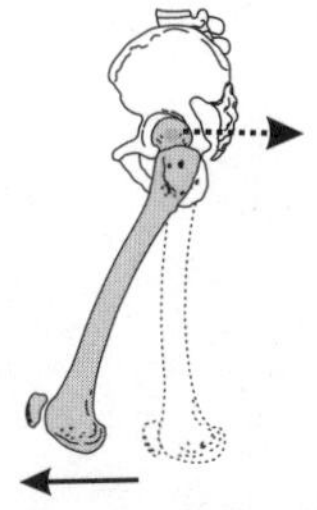

Posterior glide
with
Hip flexion

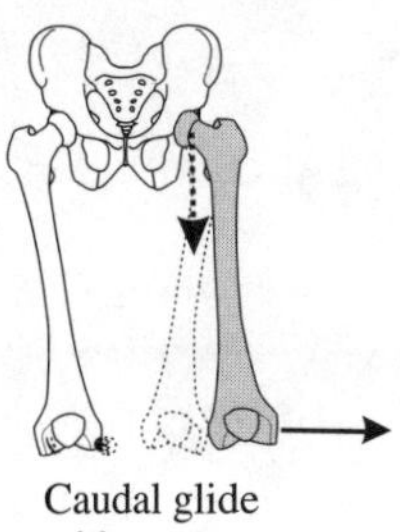

Caudal glide
with
Hip abduction

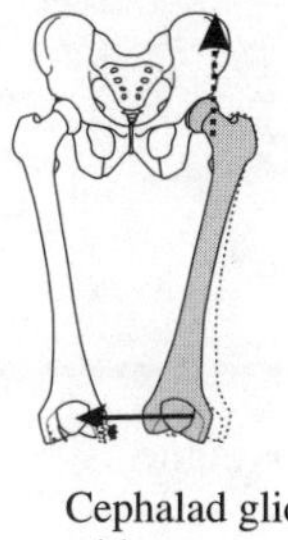

Cephalad glide
with
Hip adduction

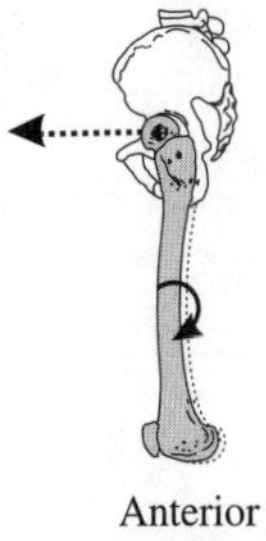

Anterior glide
with
Hip internal
rotation

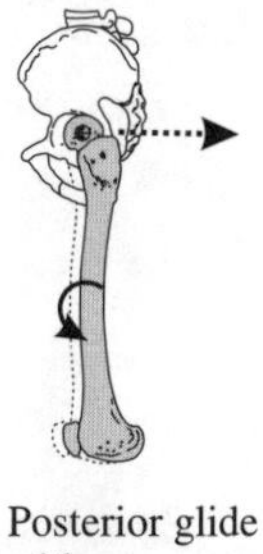

Posterior glide
with
Hip external
rotation

Type II Movement of the Hip

Hip joints are femoral/acetabular joints. Physiologic movements of the hips are flexion, extension (sagittal plane); abduction, adduction (coronal plane); external rotation, internal rotation (transverse plane). Direct accessory movements of the femoral head are:

1. Anterior glide for extension
2. Posterior glide for flexion
3. Caudal glide for abduction
4. Cephalad glide for adduction
5. Anterior glide plus cephalad glide for internal rotation
6. Posterior glide plus caudal glide for external rotation.

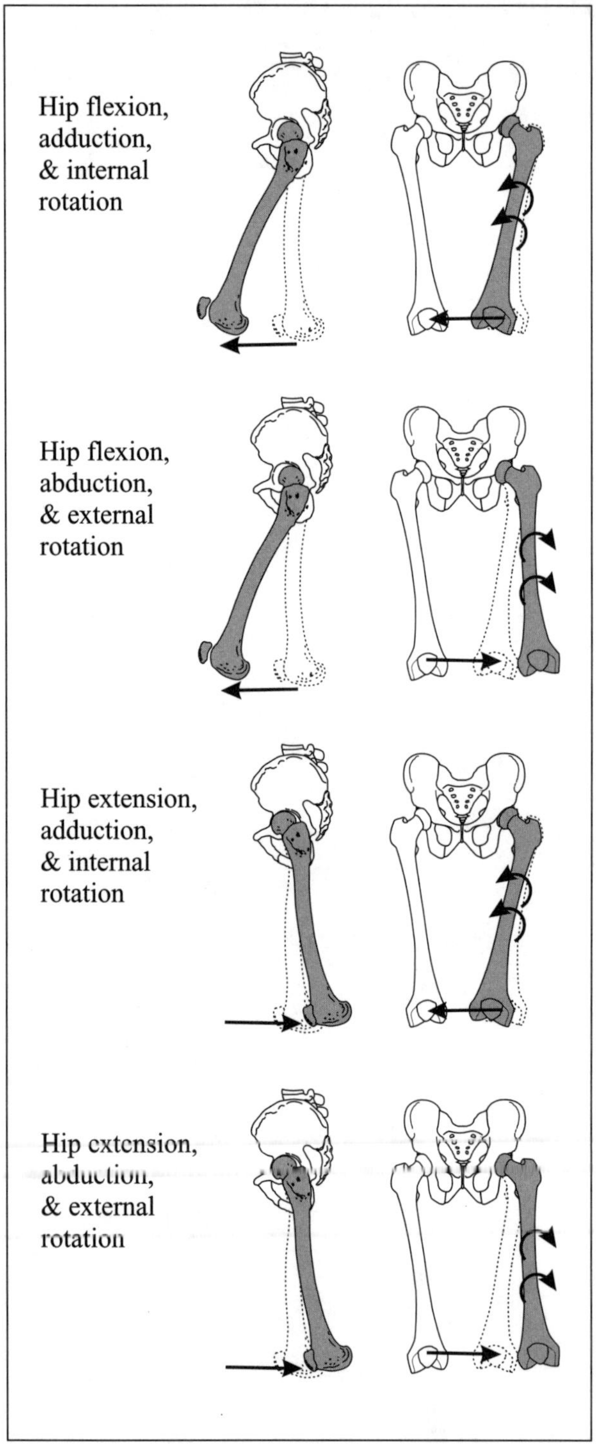

Type II movements occur whenever the hip is flexed or extended. The Type II dysfunctions are: flexed and extended. Type II movement dysfunctions are 3-planar torsion dysfunctions. Four possible Type II dysfunctions can occur at the hip (femoral-acetabular) joint:

1. Flexed, adducted, internal rotated
2. Flexed, abducted, external rotated
3. Extended, adducted, internal rotated
4. Extended, abducted, external rotated

Internal rotation occurs with adduction in Type II movement. External rotation occurs with abduction in Type II movement.

Assessment

Palpate the glides of the femoral head at the greater trochanter during physiologic hip flexion and extension.

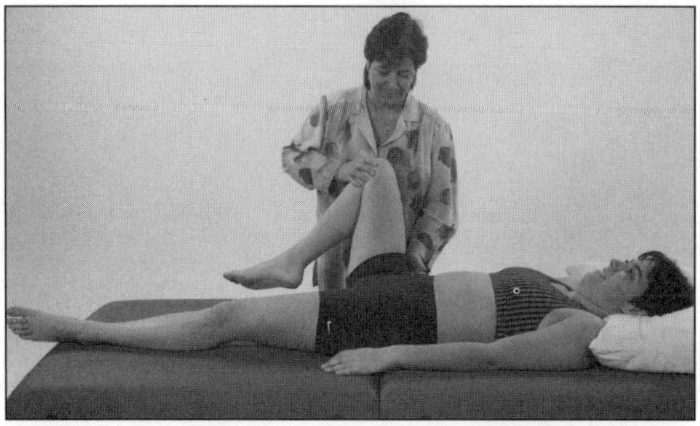

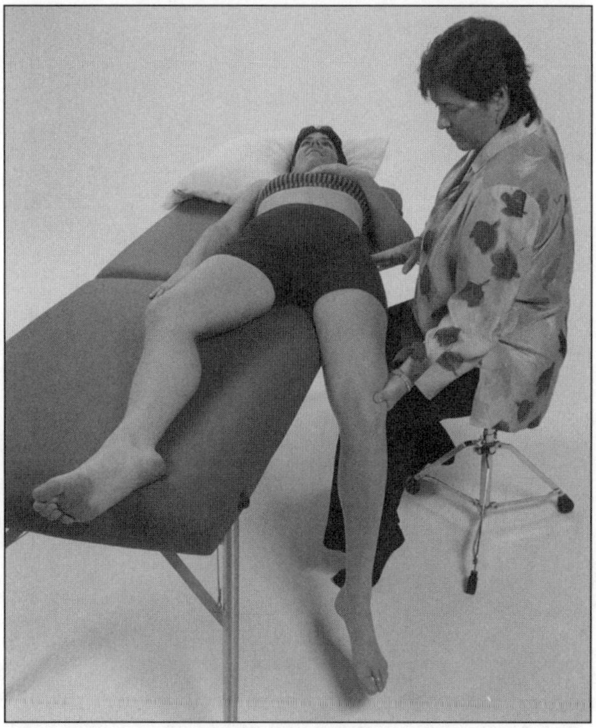

There should be no approximation or distraction of the femoral head during hip flexion and extension.

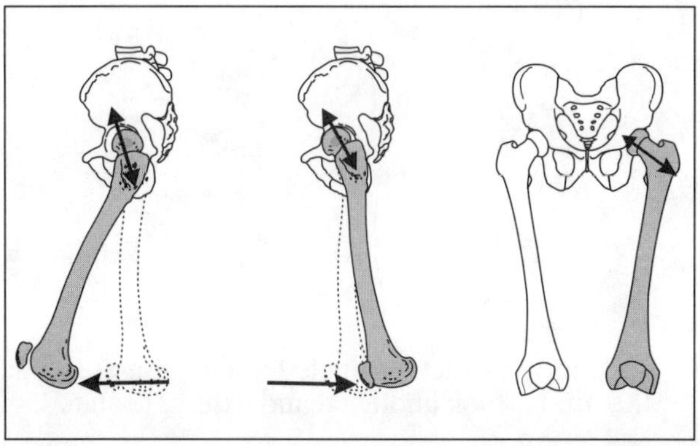

If the femoral head approximates, there is a medial pull palpated at the greater trochanter. If the femoral head distracts, a lateral distraction is palpated at the greater trochanter. If the femoral head approximates, palpated at the greater trochanter as a medial pull, the hip is stuck adducted and internal rotated.

If the femoral head laterally distracts, palpated at the greater trochanter as a lateral distraction, the hip is stuck abducted and external rotated.

If the movement dysfunction is palpated at the greater trochanter during flexion, the hip cannot flex, therefore the hip is stuck extended. If the movement dysfunction is palpated at the greater trochanter during extension, the hip cannot extend, therefore the hip is stuck flexed.

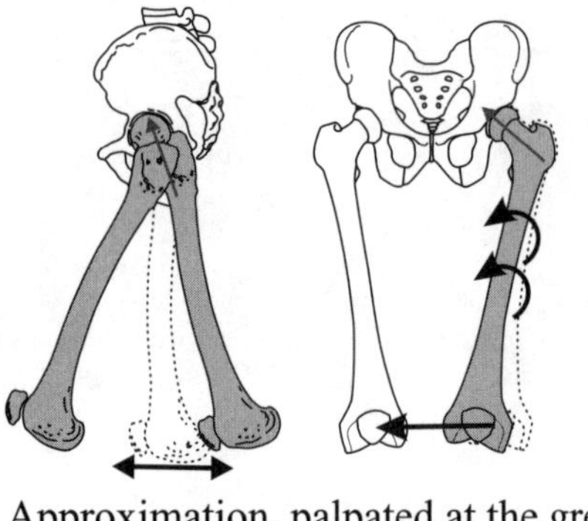

Approximation, palpated at the greater trochanter as a medial pull, the hip is stuck adducted and internal rotated.

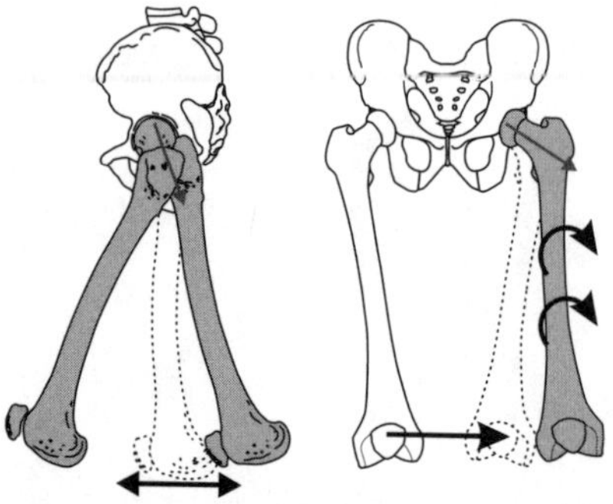

Laterally distracts, palpated at the greater trochanter as a lateral distraction, the hip is stuck abducted and external rotated.

Positional Diagnosis

EADIR:

Find in flexion: extended. Approximation of femoral head: adducted, internally rotated

EABER:

Find in flexion: extended. Lateral distraction of femoral head: abducted, externally rotated

FADIR:

Find in extension: flexed. Approximation of femoral head: adducted, internally rotated

FABER:

Find in extension: flexed. Lateral distraction of femoral head: abducted, externally rotated

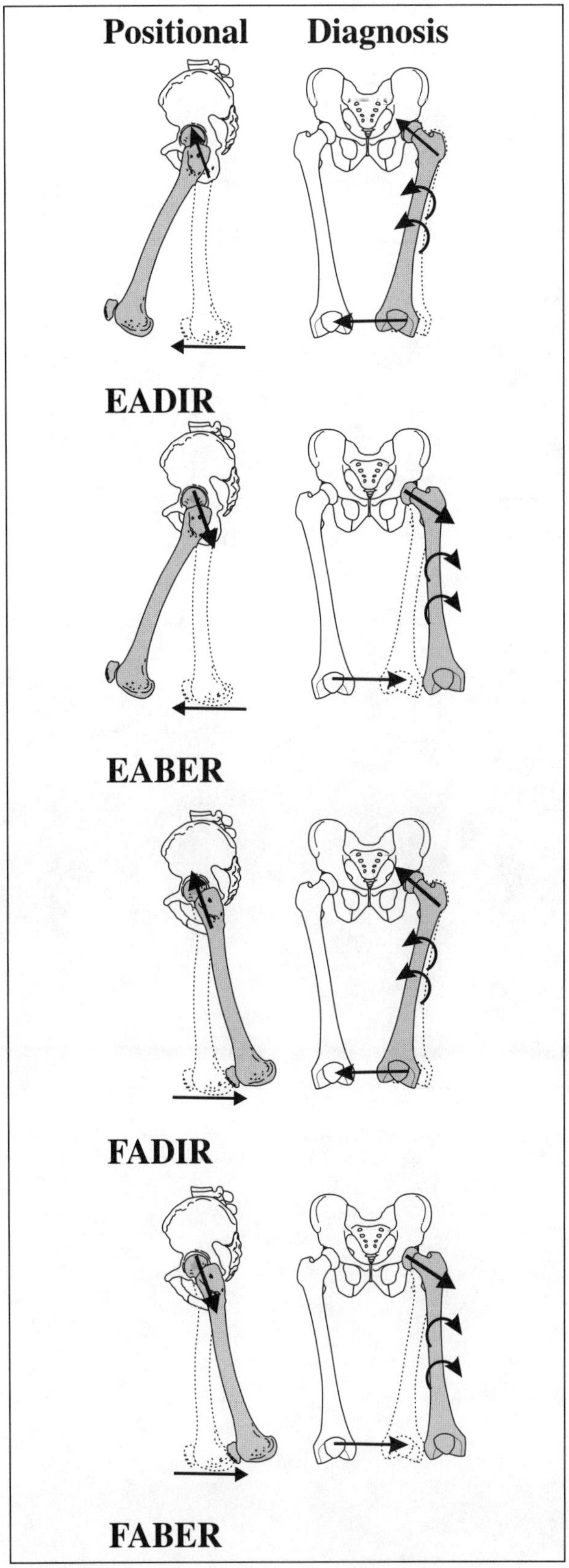

Movement Barrier

Check the physiologic limitations of motion. The limitations should correlate with the movement barriers.

EADIR:

> The positional diagnosis: extended, adducted, and internally rotated. The movement barriers are: flexion, abducted, and external rotation: FABER

EABER:

> The positional diagnosis is: extended, abducted and externally rotated. The movement barriers are: flexion, adduction, and internal rotation: FADIR

FADIR:

> The positional diagnosis is: flexed, adducted, and internally rotated. The movement barriers are: extension, abduction, external rotation: EABER

FABER:

> The positional diagnosis: flexed, abducted, and externally rotated. The movement barriers are: extension, adduction, and internal rotation: EADIR

Position of Treatment

EADIR

The positional diagnosis is: extended, adducted and internally rotated.
The movement barrier is: FABER.
The position of treatment is the same as the movement barrier. FABER: Flex, abduct, and externally rotate the hip to the interbarrier zones.

EABER

The positional diagnosis is: extended, abducted, and externally rotated.
The movement barrier is FADIR.
The position of treatment is the same as the movement barrier. FADIR: Flex, adduct, and internally rotate the hip to the interbarrier zones.

FADIR

The positional diagnosis is: flexed, adducted, and internally rotated.
The movement barrier is: EABER.
The position of treatment is the same as the movement barrier. EABER: Extend, abduct, and externally rotate the hip to the interbarrier zones.

FABER

The positional diagnosis is: flexed, abducted, and externally rotated.
The movement barrier is EADIR.
The position of treatment is the same as the movement barrier. EADIR: Extend, adduct and internally rotate the hip to the interbarrier zones.

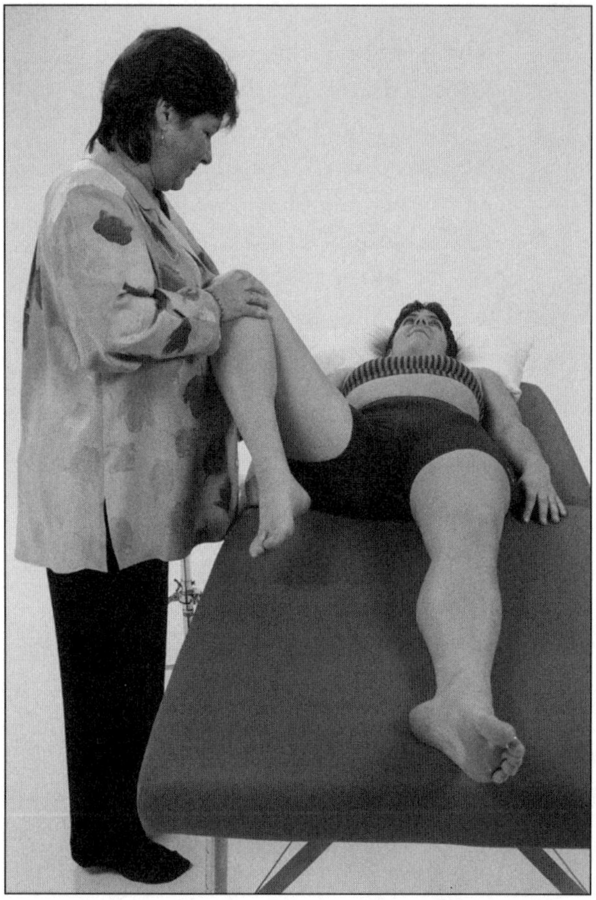

Treatment

Extended:

1. Palpate the muscle barrier at the opposite ASIS in supine. Move the hip and leg passively to the interbarrier zones on all 3 planes to a 3-planar interbarrier zone. For extended dysfunctions, move the leg into flexion.
2. Resist any movement of the hip. Resistance: isometric, 5 grams resistance, 6 seconds resistance, unidirectional/uniplanar resistance.
3. Relax. Progress to next 3-planar interbarrier zone.
4. Repetitions: 3.
5. Reassess.

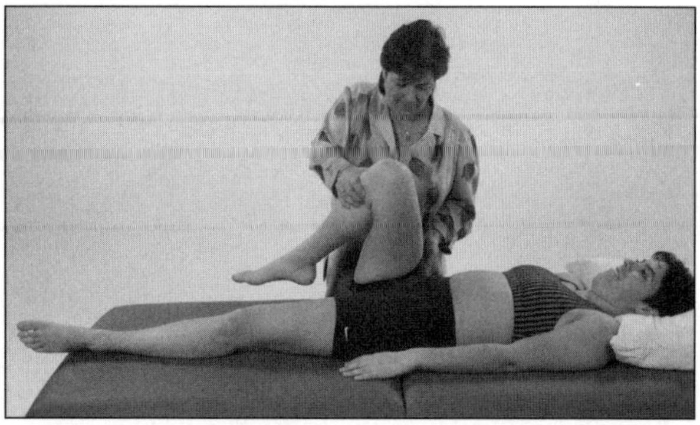

Flexed:

1. Palpate the muscle barrier at the opposite PSIS in prone. Move the hip and leg passively to the interbarrier zones on all 3 planes to a 3-planar interbarrier zone. For flexed dysfunctions, move the leg into extension.
2. Resist any movement of the hip. Resistance: isometric, 5 grams resistance, 6 seconds resistance, unidirectional/uniplanar resistance.
3. Relax. Progress to next 3-planar interbarrier zone.
4. Repetitions: 3.
5. Reassess.

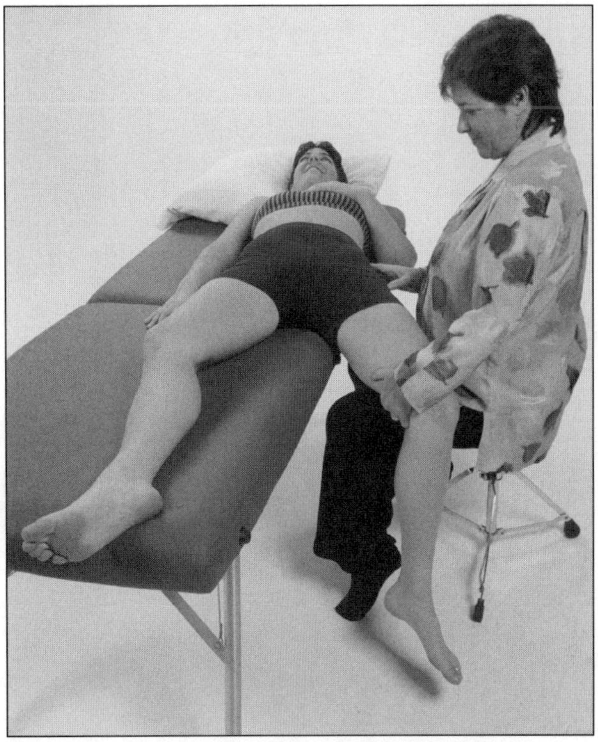

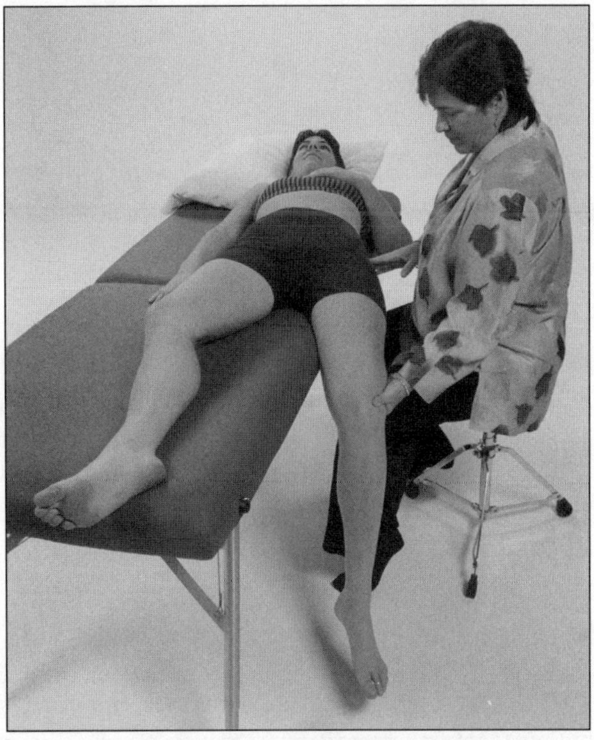

CHAPTER 28

THE KNEE JOINT

This knee joint dysfunction is as significant for pain and disability as the hip joint Type II dysfunction. There will be strain of ligaments and capsule, inflammation, and defibrillation of hyaline cartilage lining, secondary to the compromise of vertical dimension of the intra-articular space between the distal femoral surface and the proximal tibial surface.

It is possible to reverse the dysfunction with multiple repetitions of these techniques over a long period, possibly eliminating the need for total knee surgery.

Type II Knee Joint Dysfunction

There are flexed and extended Type II knee joint dysfunctions. Physiologic movements of the knee joint include flexion and extension (sagittal plane). Direct accessory movements of the proximal tibial articular surface are:

1. Anterior glide for extension
2. Posterior glide for flexion

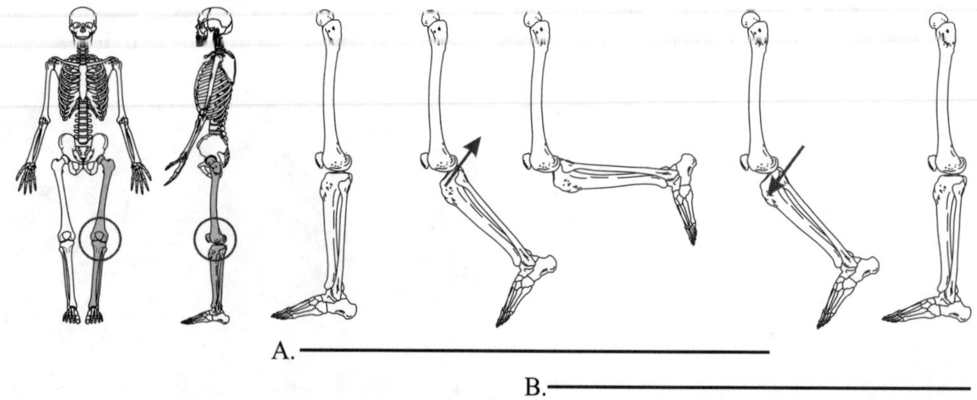

A. Knee movement into flexion and posterior glide
B. Knee movement into extension and anterior glide

Lateral glides are dysfunctional if they occur before the last 5 degrees of flexion and extension. Rotations are dysfunctional if they occur before the last 5 degrees of flexion and extension.

Type II movements occur whenever the knee joint is flexed or hyperextended. It is dysfunctional for the knee joint to be hyperextended (recurvatum). Normal Type II movements of the knee occur in flexion.

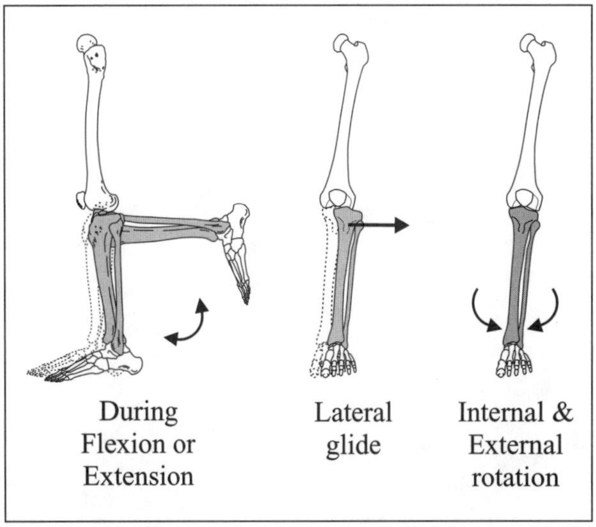

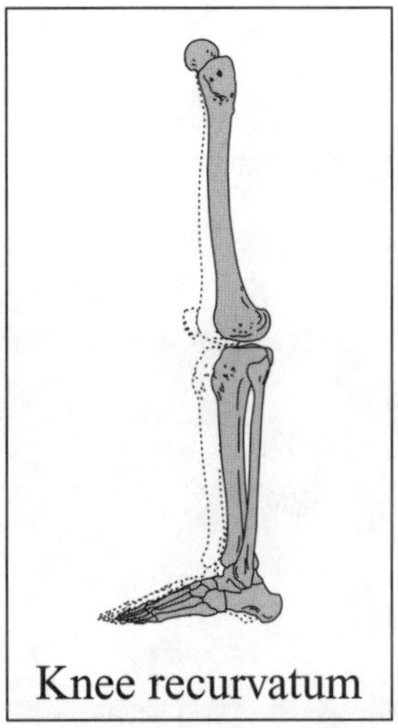

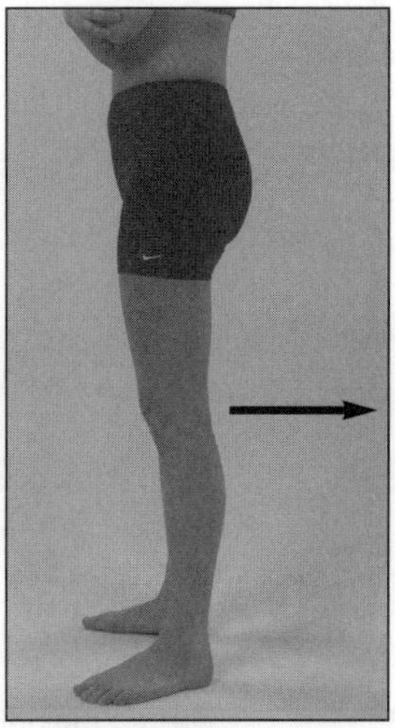

Muscle Barrier

The muscle barrier for assessment of Type II knee dysfunctions can be palpated at the proximal tibial head on the lateral surface.

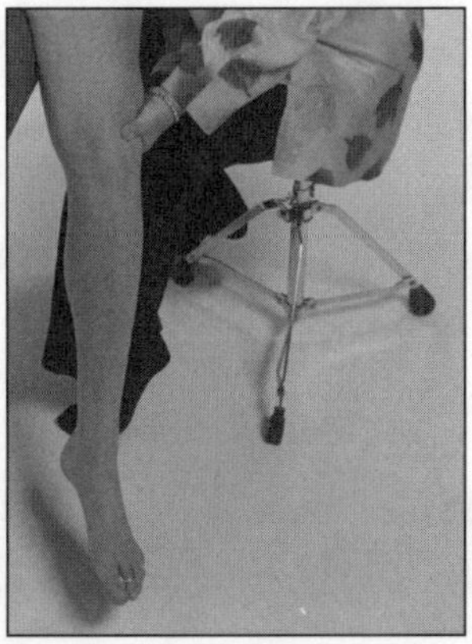

Type II Knee Joint Movements

During flexion, if the proximal tibial head glides laterally, it also externally rotates.

During flexion, if the proximal tibial head glides medial, it also internally rotates.

During hyperextension (in genu recurvatum), if the proximal tibial head glides lateral, it also externally rotates.

During hyperextension (in genu recurvatum), if the proximal tibial head glides medial, it also internally rotates.

External rotation occurs with lateral glide.

Internal rotation occurs with medial glide.

Glides (medial and lateral) are the modified and natural coronal plane side bending/abduction/adduction motion for the knee. The most common Type II knee dysfunction is flexed, lateral glide, external rotated.

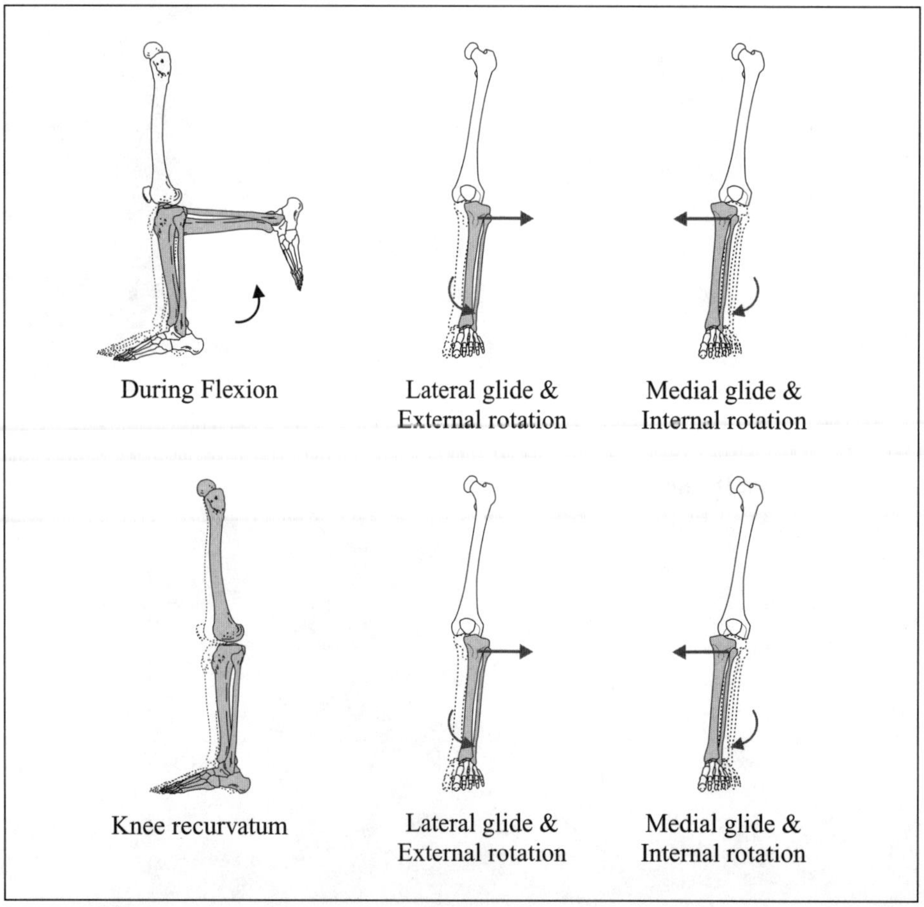

Assessment

In prone, assess flexion of the knee joint. Assess knee joint movement with palpation of the proximal tibial head. Observe:

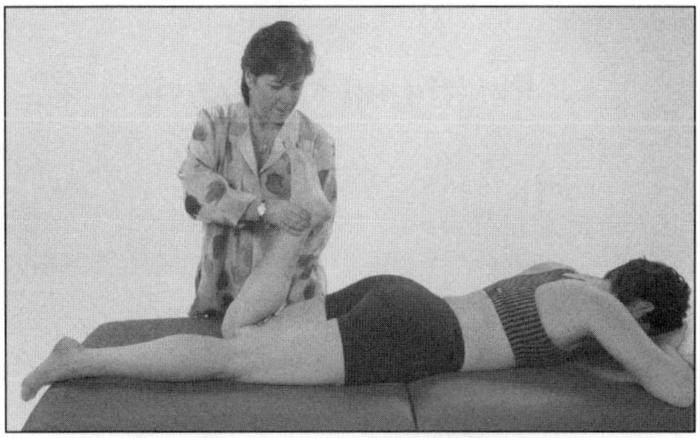

1. Is there lateral or medial glide of the proximal tibial head before the last 5 degrees of end-range knee flexion?
2. Lateral glide is modified knee abduction. Medial glide is modified knee adduction.
3. If there is lateral glide of the proximal tibial head, there is also external rotation of the proximal tibial head. If there is medial glide of the proximal tibial head, there is also internal rotation of the proximal tibial head.

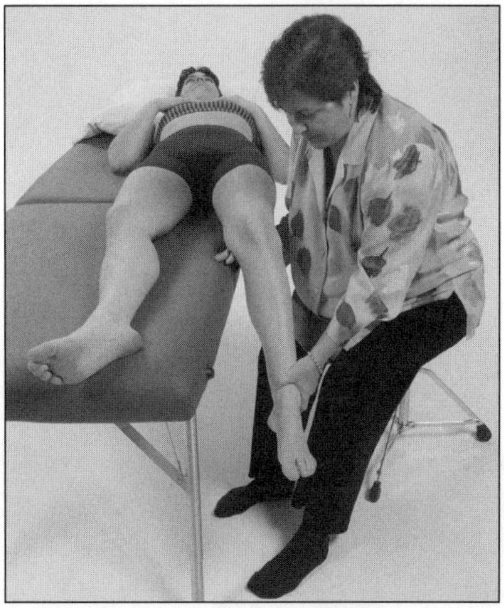

In supine, assess extension (hyperextension) of the knee joint. Assess knee joint movement with palpation of the proximal tibial head. Observe:

1. Is there lateral or medial glide of the proximal tibial head before the last 5 degrees of end-range knee flexion?
2. Lateral glide is modified knee abduction. Medial glide is modified knee adduction.
3. If there is lateral glide of the proximal tibial head, there is also external rotation of the proximal tibial head. If there is medial glide of the proximal tibial head, there is also internal rotation of the proximal tibial head.

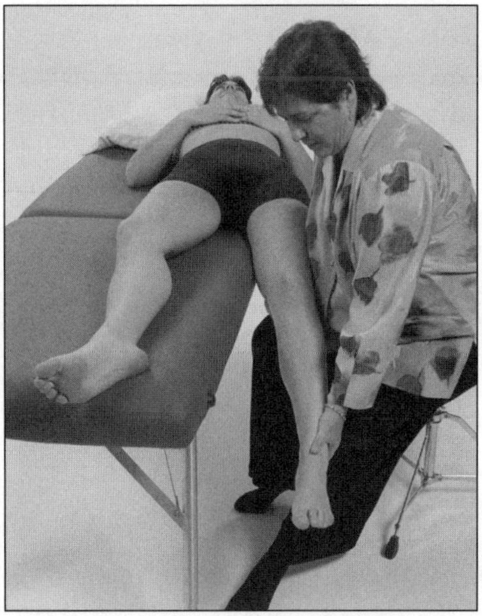

Positional Diagnosis

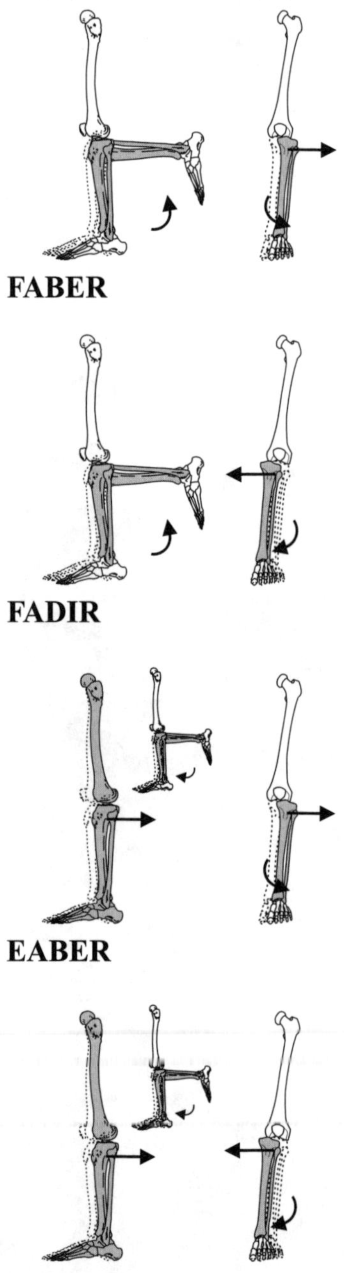

Positional Diagnosis

If dysfunctional glides and rotations occur in knee flexion:

1. The knee cannot flex.
2. The knee is stuck in extension.
3. There is an Extended Type II knee joint dysfunction.

If dysfunctional glides and rotations occur in knee hyperextension:

1. The knee cannot extend.
2. The knee is stuck flexed.
3. There is a flexed Type II knee joint dysfunction.

If a lateral glide/external rotation occurs before the last 5 degrees of end-range knee flexion and/or during hyperextension, the knee is stuck abducted and externally rotated. If a medial glide/internal rotation occurs before the last 5 degrees of end-range knee flexion and/or during hyperextension, the knee is stuck adducted and internally rotated: EADIR.

There are four possible Type II knee joint dysfunctions:

1. Flexed, abducted and externally rotated: FABER
2. Flexed, adducted and internally rotated: FADIR
3. Extended, abducted and externally rotated: EABER
4. Extended, adducted and internally rotated: EADIR

The most common Type II knee joint dysfunction is FABER.

Movement Barrier

FABER:

The knee is stuck flexed, abducted (in lateral glide), and externally rotated. The movement barrier is: EADIR: The knee cannot extend, adduct (medial glide) and internally rotate.

FADIR:

The knee is stuck flexed, adducted (in medial glide) and internally rotated. The movement barrier is: EABER: The knee cannot extend, abduct (lateral glide), and externally rotate.

EABER:

The knee is stuck extended, abducted (in lateral glide) and externally rotated. The movement barrier is: FADIR: The knee cannot flex, adduct (medial glide), and internally rotate.

EADIR:

The knee is stuck extended, adducted (in medial glide) and internally rotated. The movement barrier is: FABER: The knee cannot flex, abduct (lateral glide), and externally rotate.

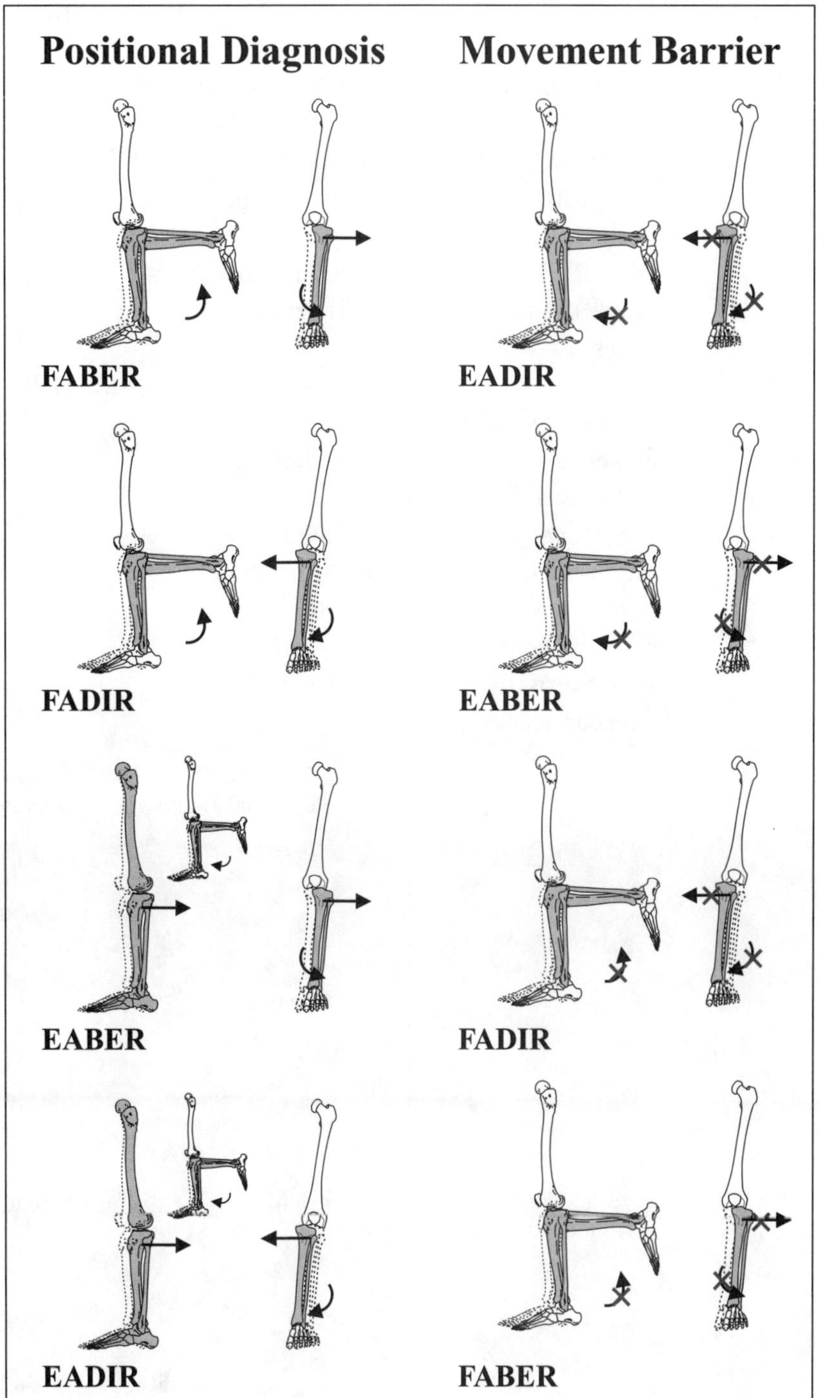

Position of Treatment

FABER:

The positional diagnosis is:
Flexed, abducted (in lateral glide) and externally rotated.
The movement barrier is: EADIR.
The position of treatment is the same as the movement barrier:
EADIR: Extend, adduct, and internally rotate the knee to the interbarrier zones.

FADIR:

The positional diagnosis is:
Flexed, adducted (put in medial glide), and internally rotated.
The movement barrier is: EABER.
The position of treatment is the same as the movement barrier:
EABER: Extend, abduct (lateral glide), and externally rotate the knee to the interbarrier zones.

EABER:

The positional diagnosis is:
Extended, abducted (lateral glide) and externally rotated.
The movement barrier is: FADIR.
The position of treatment is the same as the movement barrier:
FADIR: Flex, adduct (medial glide) and internally rotate the knee to the interbarrier zones.

EADIR:

The positional diagnosis is:
Extended, adducted, internally rotated.
The movement barrier is: FABER.
The position of treatment is the same as the movement barrier.
FABER: Flex, abduct (lateral glide), and externally rotate the knee to the interbarrier zones.

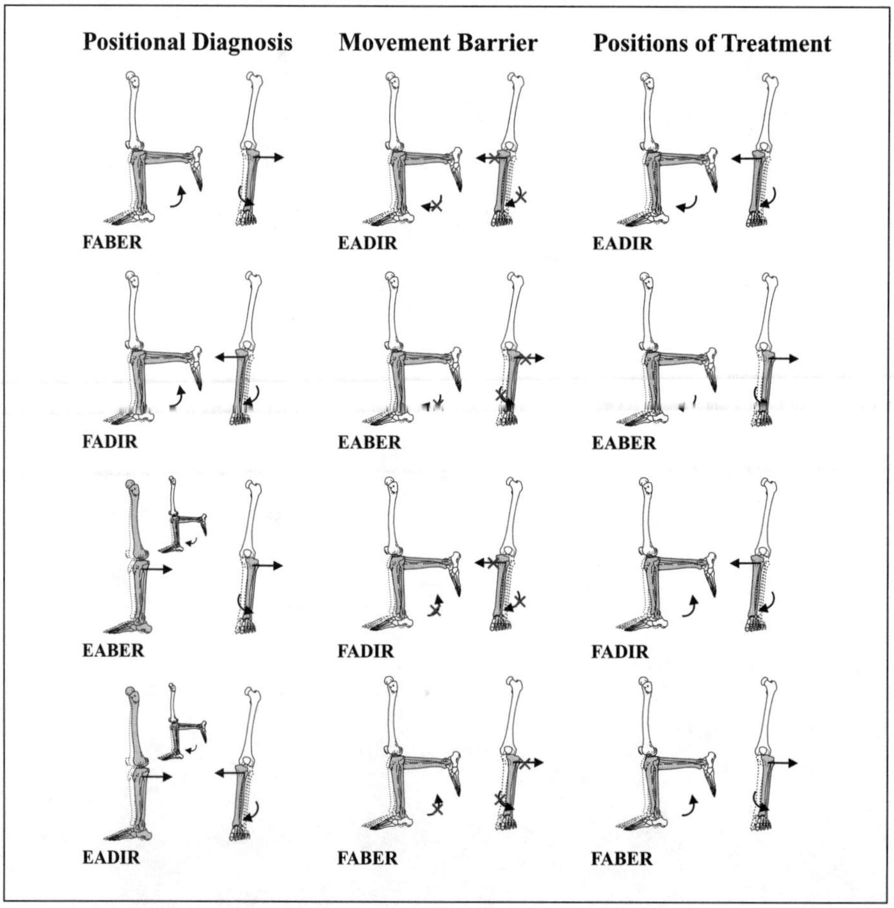

Treatment

Flexed:

1. In prone or in supine with knee off the bed, keep hip aligned in neutral (not abducted, not rotated).
2. Palpate for the muscle barrier at the proximal tibial head. Focus on the lateral aspect of the tibial head.
3. Move the lower leg passively on all 3 planes to a 3-planar interbarrier zone. Do lateral glide instead of abduction. Do medial glide instead of adduction.
4. For flexed dysfunctions, move into extension.

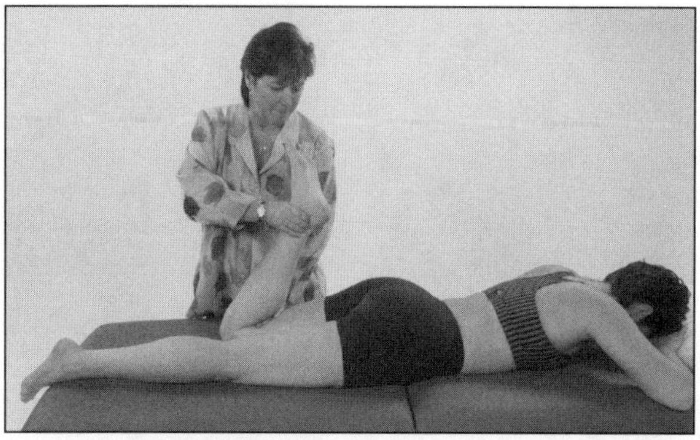

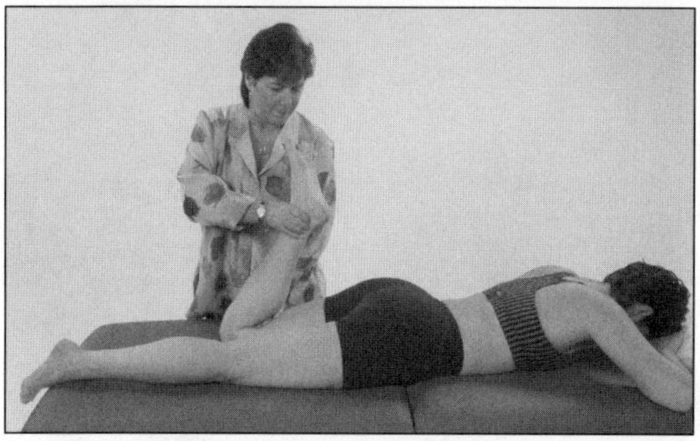

5. Resist knee extension or knee flexion.

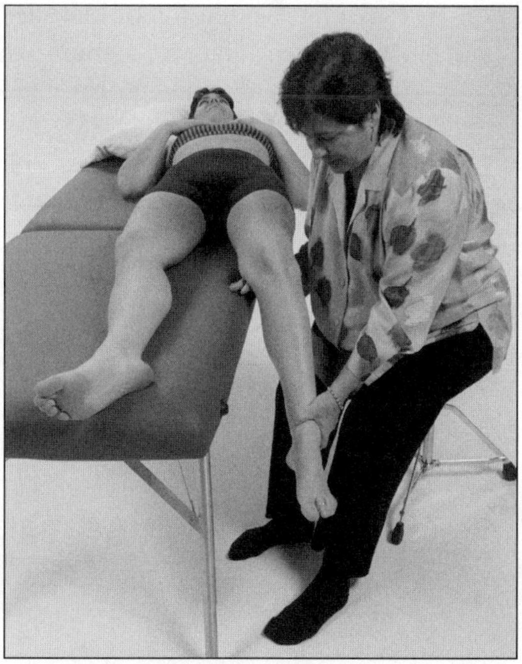

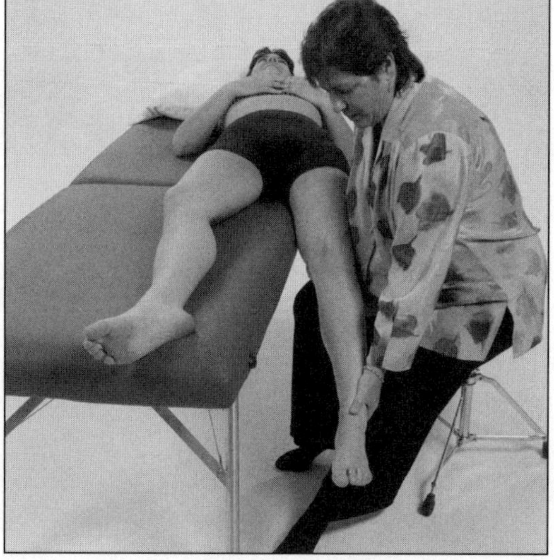

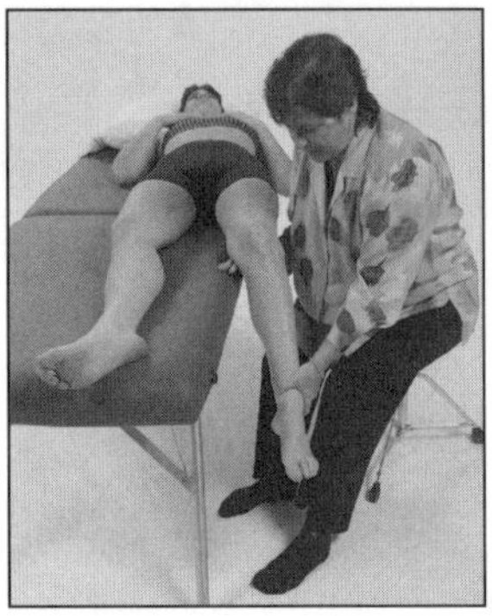

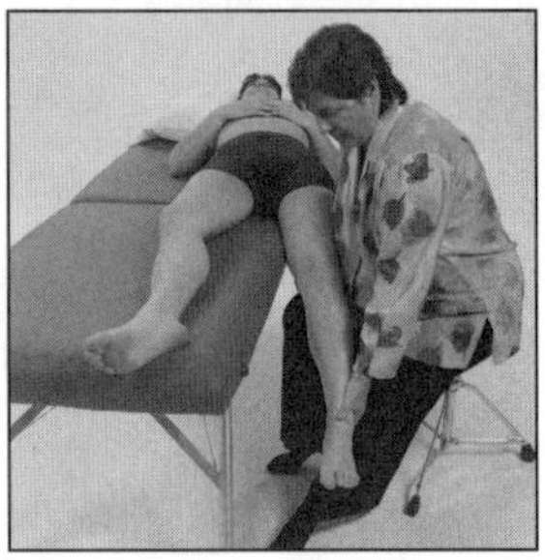

6. Resistance: isometric, 5 grams resistance, unidirectional/uniplanar resistance, 6 seconds resistance.
7. Relaxation: Progress to next 3-planar interbarrier zone.
8. Repetitions: 3.
9. Reassess.

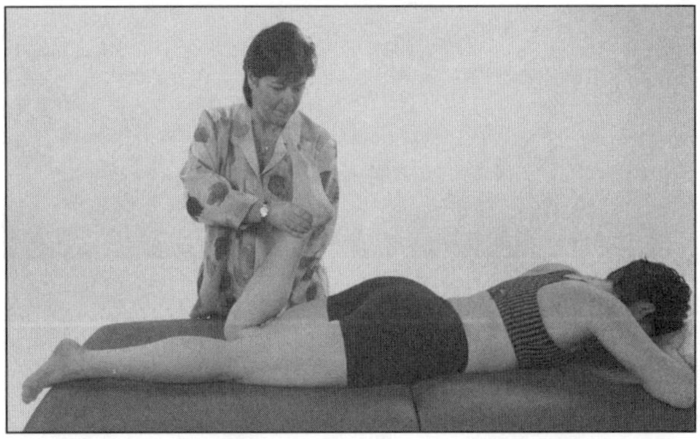

Extended:

1. In prone or supine with knee off the bed, keep hip aligned in neutral (not abducted, not rotated).
2. Palpate for the muscle barrier at the proximal tibial head. Focus on the lateral aspect of the tibial head.
3. Move the lower leg passively on all 3 planes to a 3-planar interbarrier zone. Do lateral glide instead of abduction. Do medial glide instead of adduction.
4. For extended dysfunctions, move into flexion.
5. Resist knee extension or knee flexion.
6. Resistance: isometric, 5 grams resistance, unidirectional/uniplanar resistance, 6 seconds resistance.
7. Relaxation: Progress to next 3-planar interbarrier zone.
8. Repetitions: 3.
9. Reassess.

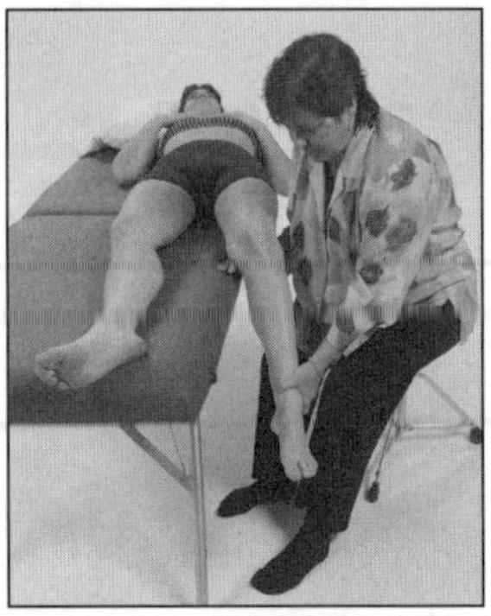

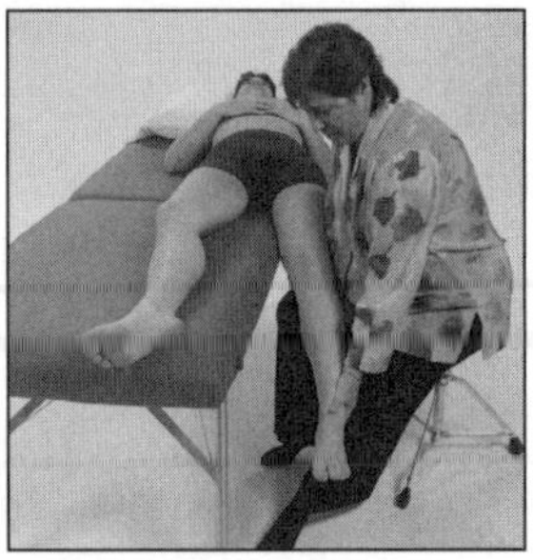

CHAPTER 29

THE ANKLE JOINT

In clinical research the author has found that dorsiflexion is the most commonly limited movement in the body after limitations of motion in the lumbosacral joint. The result of limited dorsiflexion is far greater than expected by most health care professionals. In standing, there is a need for zero (0) degrees dorsiflexion. Otherwise the tibia cannot articulate with the talus as required. Standing will then be in plantar flexion; the distal tibial surface will be positioned too posterior relative to the superior surface of the talus.

During ambulation, the tibia needs to glide anterior on the talus.

When there is a limitation of dorsiflexion, less than ten (10) degrees, there is insufficient anterior tibial glide on the talus.

Extension forces occur at the ankle joint proper (tibiotalar joint). The gastrocnemius will be in muscle spasm, contributing to knee joint problems and/or recurvatum at the knee.

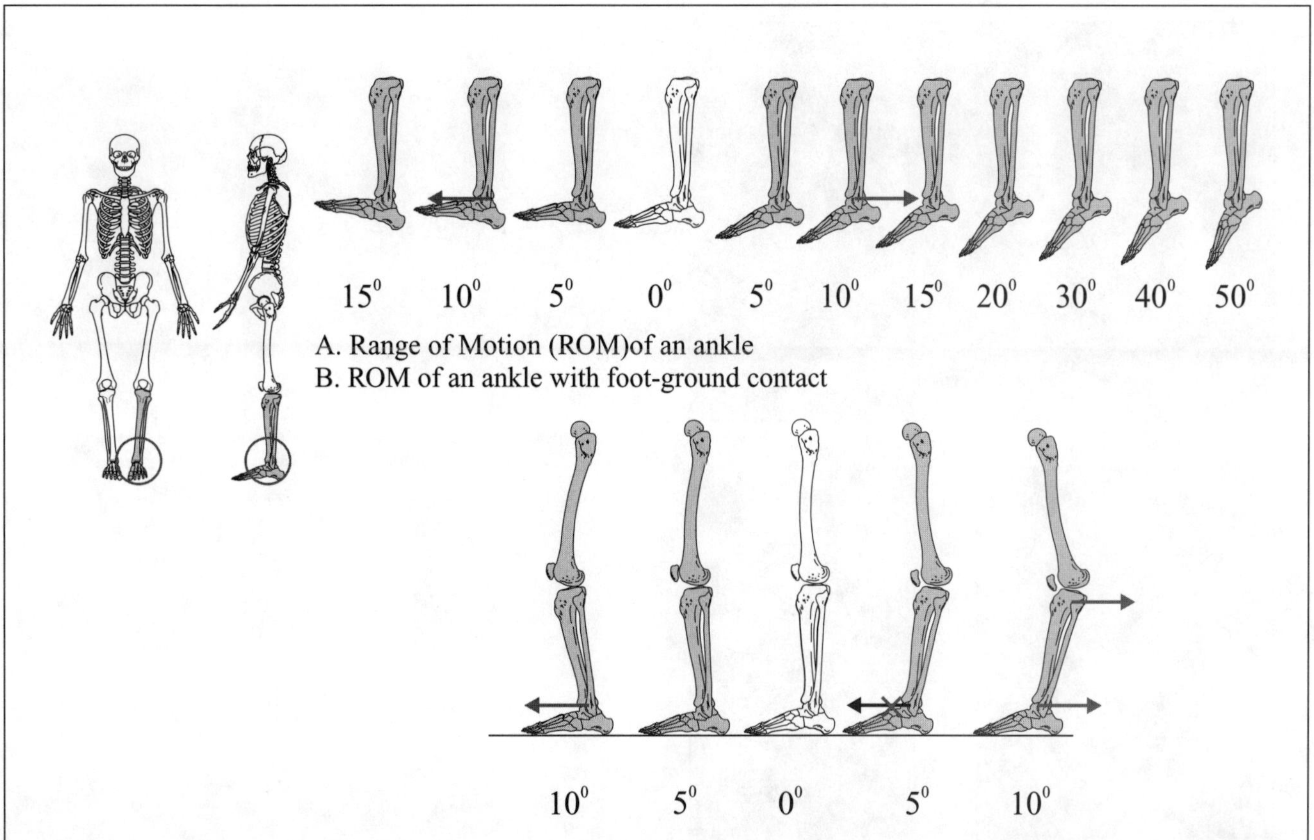

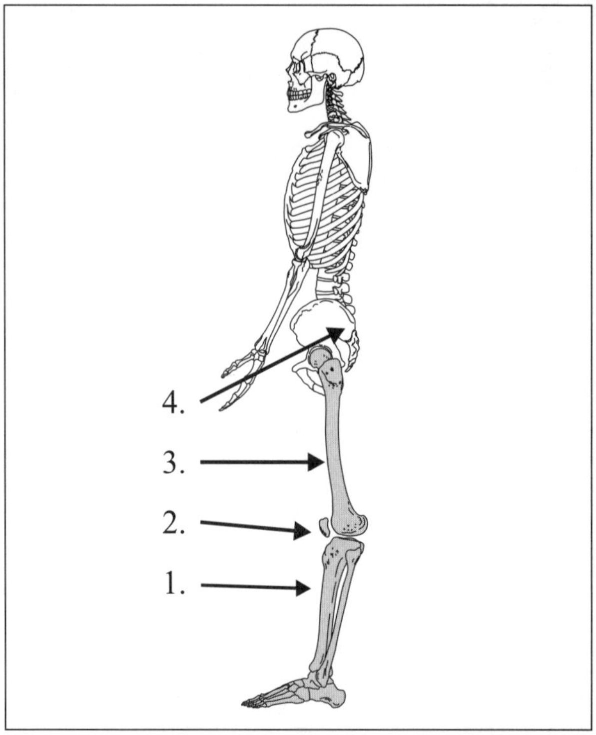

Extension forces that occur from foot flat (stance phase of ambulation) through toe off (stance phase of ambulation) will cause tension at the following areas:

1. Anterior tibia, causing shin splints
2. Posterior patella, causing chondromalasia
3. Anterior thigh, causing quadriceps spasm and secondary patella problems
4. S2, causing lumbosacral junction and sacroiliac joint problems, especially extended and posterior biomechanical dysfunctions.

Correction of ankle joint biomechanical dysfunctions will result in increased dorsiflexion; multiple improvements up the kinetic chain may occur!

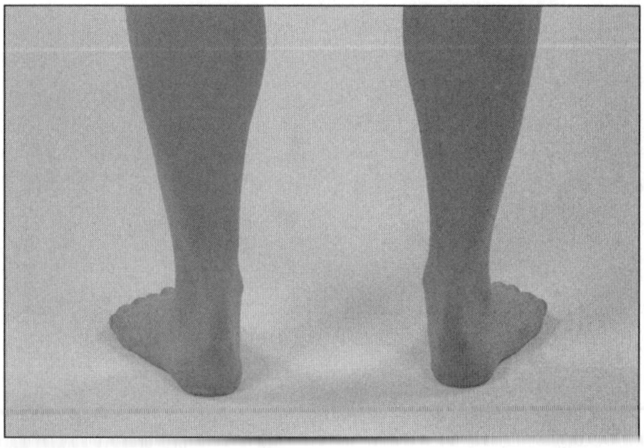

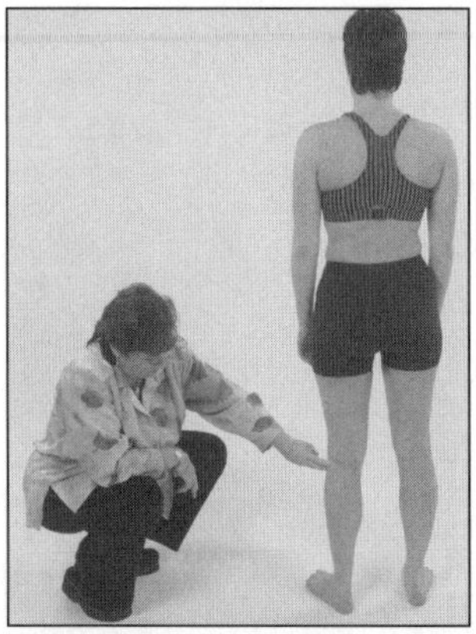

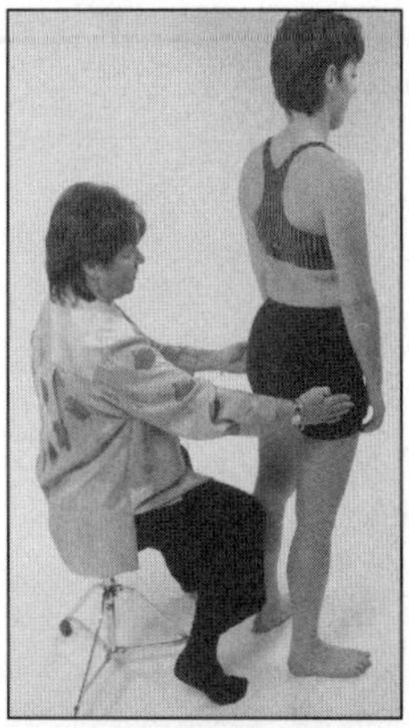

Type II Ankle Joint Dysfunctions (Tibiotalar Joint)

This joint requires slightly different orientation, since it is dependent on the subtalar (talocalcaneal) joint for mobility and articular balance. The subtalar joint frequently has intra-articular adhesions. There is no effective Muscle Energy technique for the subtalar (talocalcaneal) joint. Manipulation and Strain and Counterstrain and mobilization are more effective approaches for the subtalar joint. Prior to, or after, treatment of Type II tibiotalar joint dysfunction, treat the subtalar joint to restore mobility and articular balance.

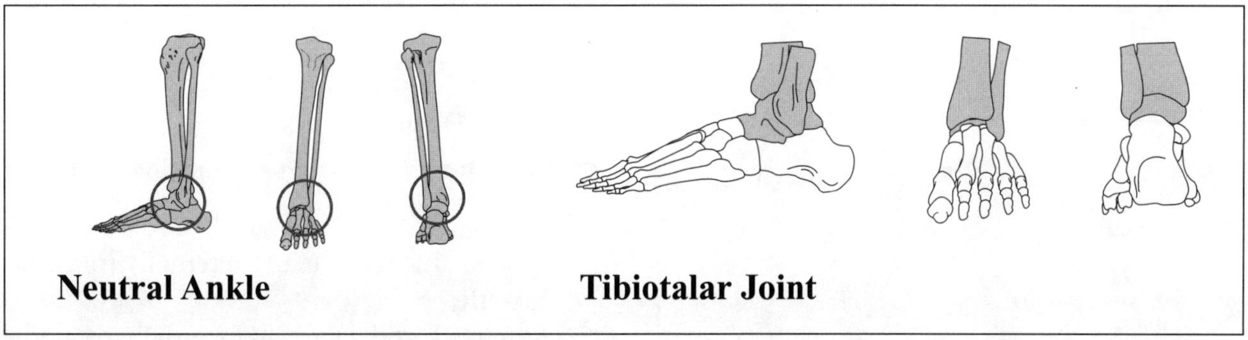

When treating the ankle, it is important to note a few factors: Dorsiflexion is always limited; there is never limitation of plantar flexion. Dorsiflexion limitations are extended Type II dysfunctions of the tibiotalar joint. The ankle is stuck plantar flexed (extended). The ankle cannot flex (dorsiflex). The positional diagnosis is always extended.

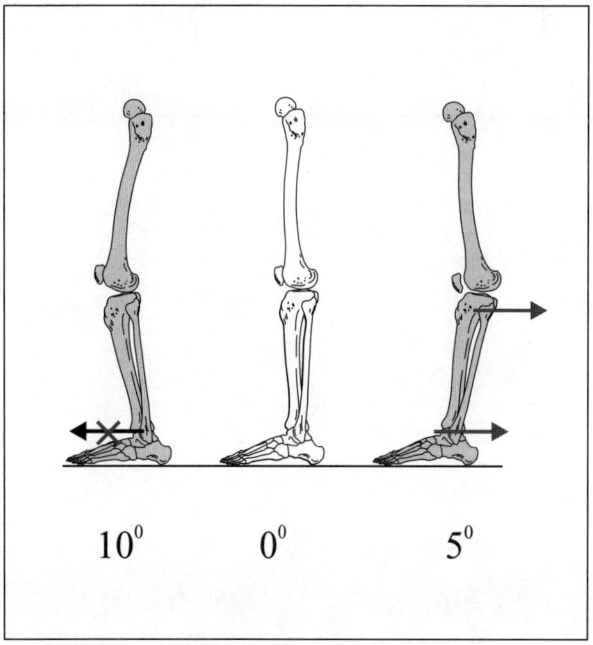

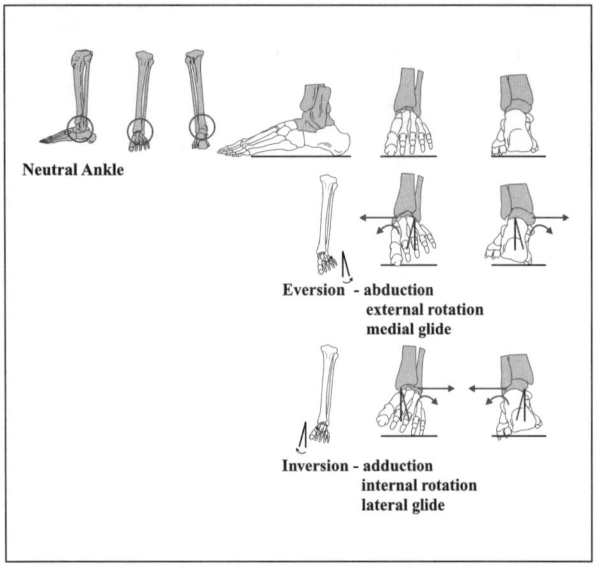

Inversion is adduction at the ankle.
Eversion is abducted.
Inversion occurs with internal rotation.
Eversion occurs with external rotation.

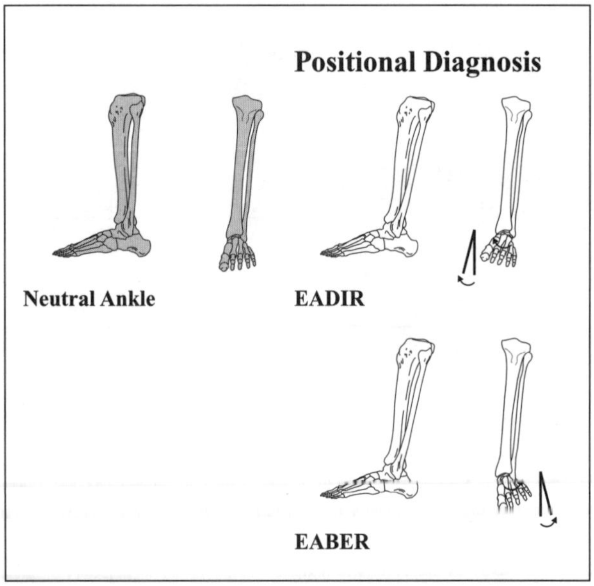

Positional Diagnosis

The Type II torsion patterns of the tibiotalar joint will be:

1. Extended, adducted and internally rotated: EADIR; or
2. Extended, abducted and externally rotated: EABER.

The most common ankle joint dysfunction is EADIR.

Movement Barrier

EADIR:

The tibiotalar joint is stuck: extended (plantar flexed), adducted, and internally rotated (inverted). The movement barrier is: FABER. The ankle joint cannot flex (dorsiflexion), abduct, and externally rotate (evert).

EABER:

The tibiotalar joint is stuck extended (plantar flexed), abducted, and externally rotated (everted). The movement barrier is: FADIR. The ankle joint cannot flex (dorsiflex), adduct, and internally rotate (invert).

Assess the movement barrier at the superior surface of the talus at the ankle mortise.

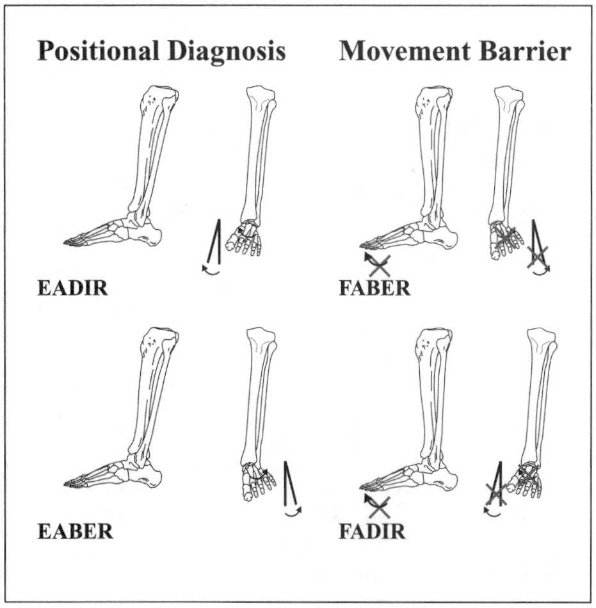

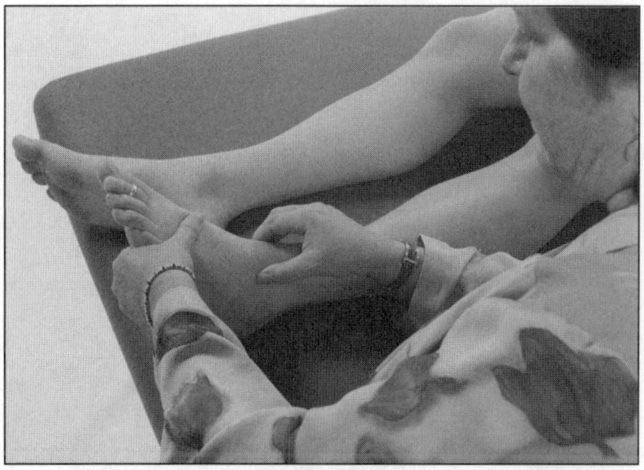

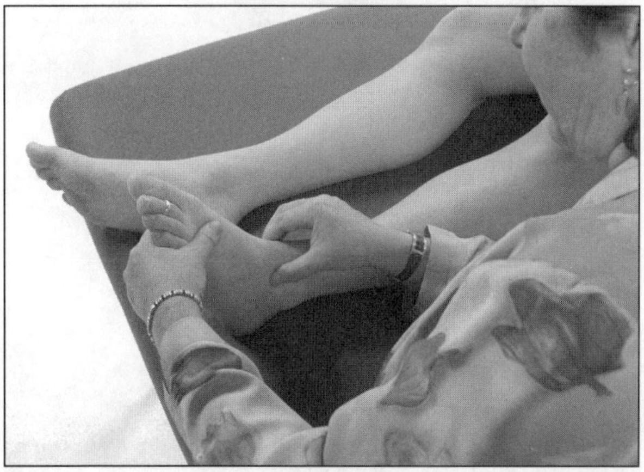

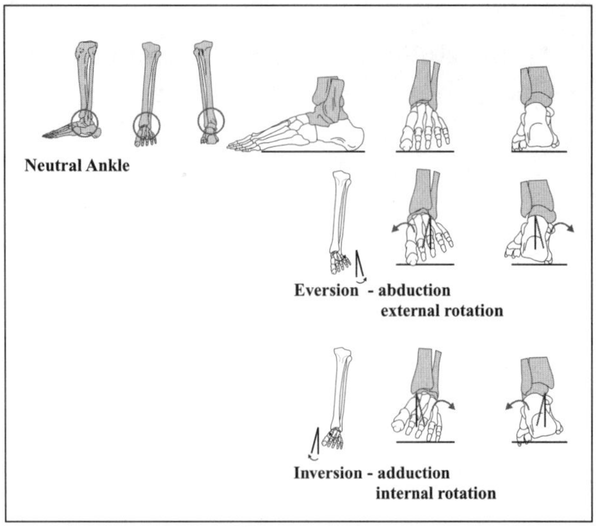

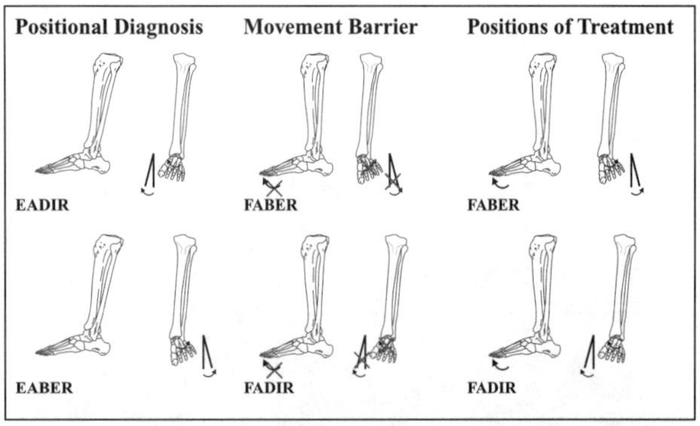

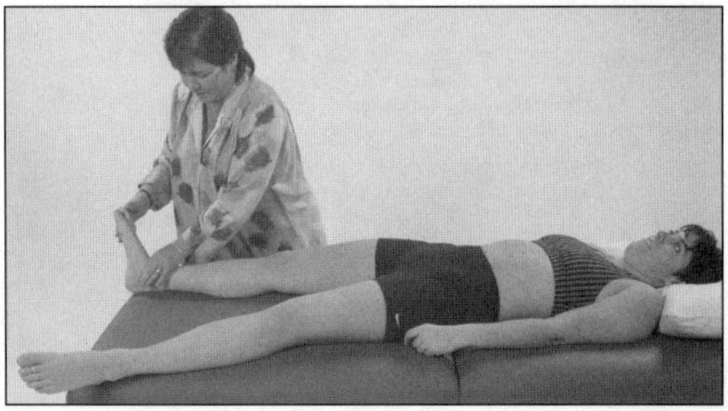

A lateral glide of the talus occurs during internal rotation and adduction.

A medial glide of the talus occurs during external rotation and abduction.

Position of Treatment

EADIR:

> The positional diagnosis is: extended, (plantar flexed), adducted, and internally rotated. The movement barrier is FABIR. The position of treatment is the same as the movement barrier. FABER: Flex (dorsiflex), abduct, and externally rotate the tibiotalar joint to the interbarrier zones.

EABER:

> The positional diagnosis is: extended (plantar flexed), abducted and externally rotated. The movement barrier is FADIR. The position of treatment is the same as the movement barrier. FADIR: Flex (dorsiflex), adduct, and internally rotate the tibiotalar joint to the interbarrier zones.

Treatment

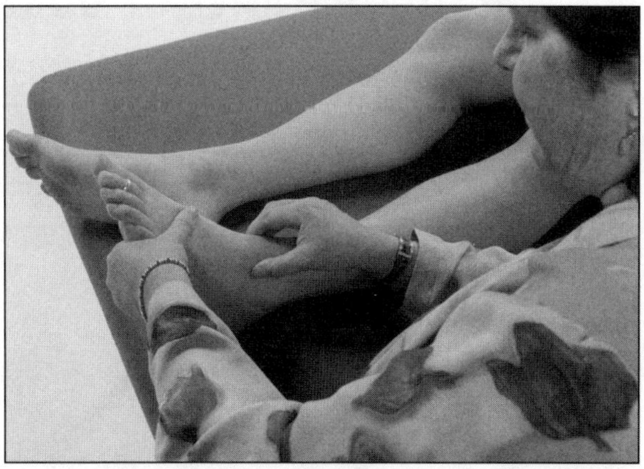

1. Treat in supine. Dorsiflexion is the sagittal plane movement. Palpate this movement barrier at the mortise. Observe lateral and medial glides of the superior surface of the talus.
   A lateral glide occurs during adduction and internal rotation.
   A medial glide occurs during abduction and external rotation.
2. Move the ankle passively in all 3 planes to the 3-planar interbarrier zone. Resist dorsiflexion or plantar flexion.
3. Resistance: isometric, 5 grams force, unidirectional/uniplanar resistance; 6 seconds resistance. Relaxation: Progress to new 3-planar interbarrier zone. Repetitions: 3. Reassess.

CHAPTER 30

THE SIGNIFICANCE OF DORSIFLEXION FOR SPINAL MOTION

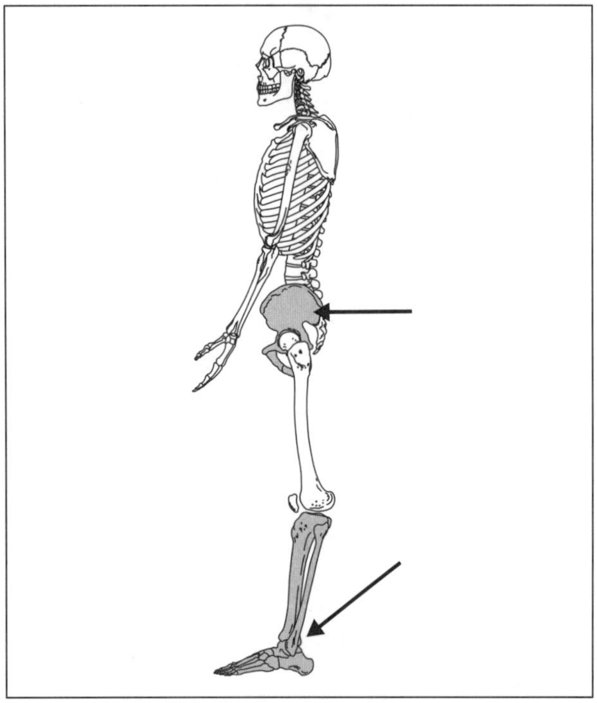

There is an essential loss of motion at the lumbosacral junction when there is a limitation of passive dorsiflexion, unilateral or bilateral. The tibia will glide anterior over the talus (when there is dorsiflexion) when the lumbar spine flexes and the sacral base glides posteriorly.

When there is hypomobility of anterior glide of tibia on talus, there is an extension force transcribed up the legs because of sustained plantar flexion during standing and stance phase of gait. This force will cause extension and posterior forces on the sacrum, contributing to lumbosacral junction dysfunctional movement.

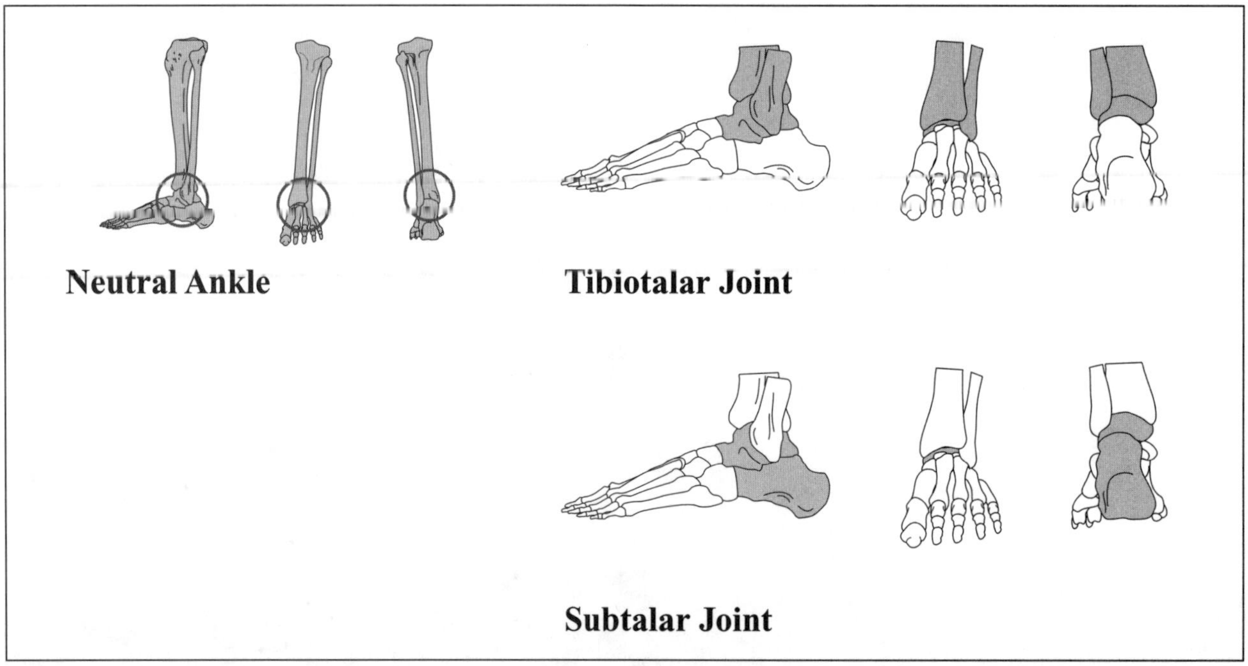

Manipulation of Tibiotalar Joint

Indication for treatment: Limitation in passive dorsiflexion.

Treatment

1. Grasp foot with good finger contact on talus. Subtalar neutral and loose-packed foot.
2. Distract talus longitudinally off tibia. Take up the slack.
3. Manipulate high velocity, low amplitude, low force, talus off tibia.

   Result: Increased passive dorsiflexion.

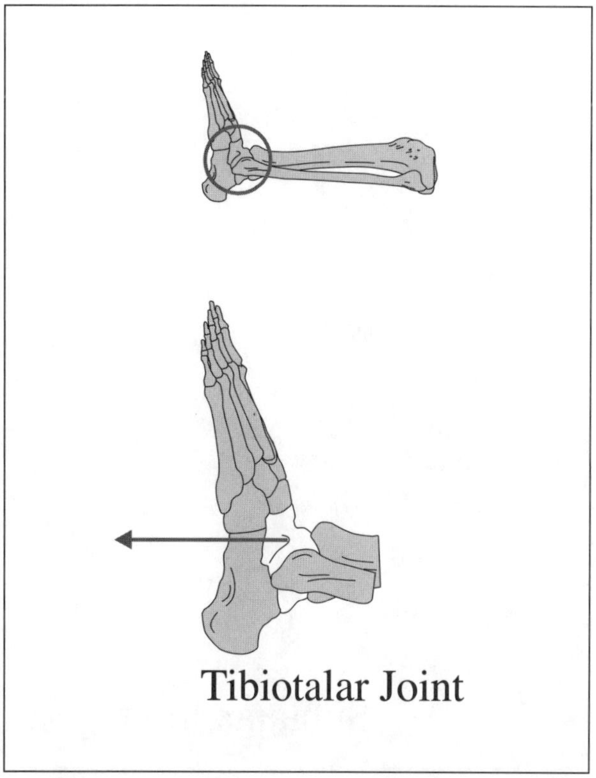

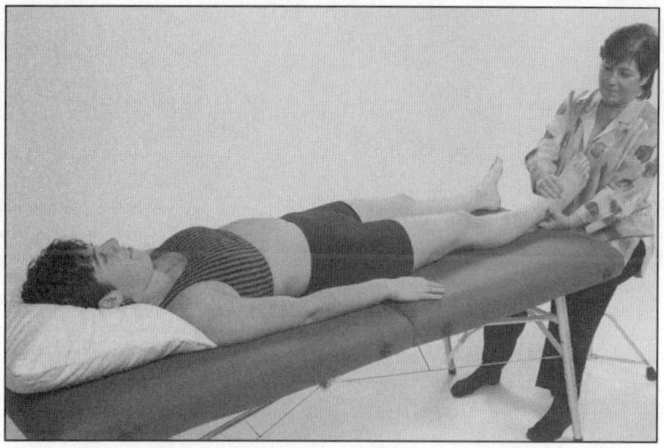

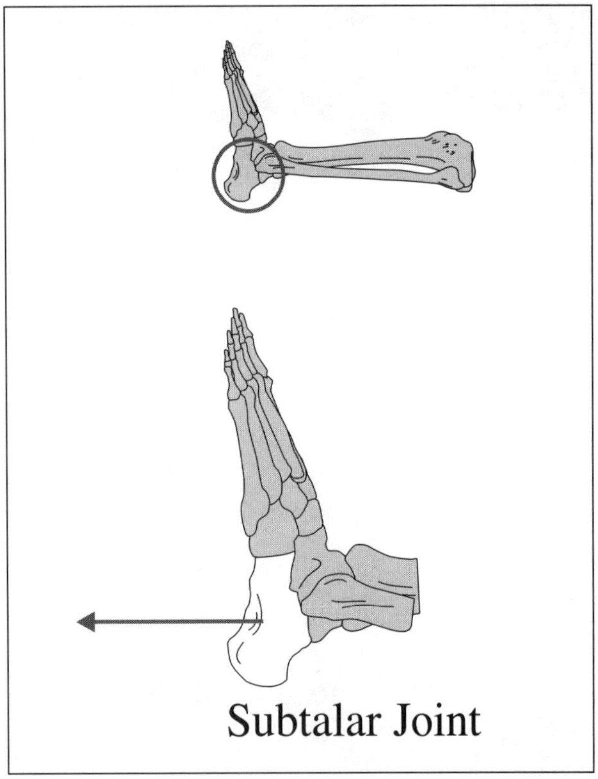

Manipulation of Subtalar Joint

Indication for treatment: Limitations in passive dorsiflexion.

Treatment

1. Grasp calcaneus. Maintain subtalar neutral and loose-packed foot.
2. Distract calcaneus longitudinally off talus. Take up the slack.
3. Manipulate high velocity, low amplitude, calcaneus off talus.

Result: Functional 10 degrees of dorsiflexion is required during gait for lumbosacral junction mobility.

CHAPTER 31

THE SHOULDER GIRDLE JOINT

Shoulder girdle joint dysfunctions can be soft tissue and/or joint problems. Often, other reasons, rather than Type II shoulder joint dysfunctions, are the cause of problems in this region. For example, muscle spasm is common. Consider elimination of protective muscle spasm of the following muscles prior to utilizing these techniques. Use Strain and Counterstrain Technique for the following muscles: pectoralis minor (second depressed rib technique); subscapularis; latissimus dorsi; supraspinatus; infraspinatus. The author recommends using Jones' Strain and Counterstrain Technique to eliminate primary protective muscle spasm.

Fascial dysfunction of the clavipectoral fascia is also common and can cause secondary Type II biomechanical dysfunction of the shoulder girdle joints. The author recommends Myofascial Release, the 3-Planar Fascial Fulcrum Approach, for treatment of clavipectoral fascial dysfunction. Occasionally, the Myofascial Release must be utilized prior to Muscle Energy and "Beyond" Technique.

Primary neural tissue tension problems of the brachial plexus might be the problem that is causing Type II biomechanical dysfunction. Assess the following regions, which may be constricting the brachial plexus: scalenes, between the middle and anterior scalenes; the costoclavicular joint, between clavicle and first rib; underneath the pectoralis minor; within axilla, compressed between the humeral head (because the latissimus dorsi is

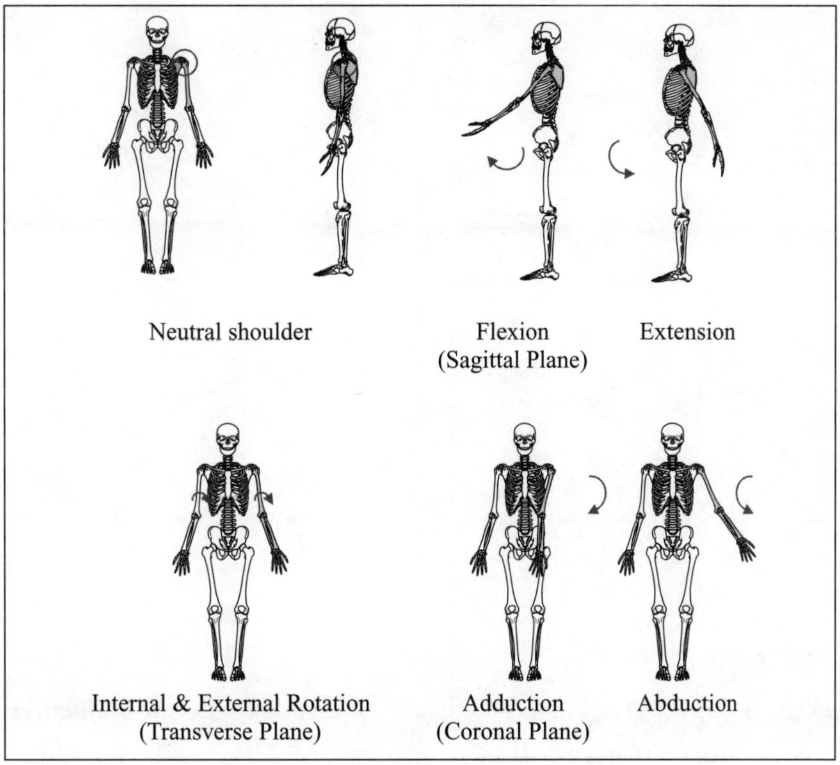

in spasm and causing a caudal subluxation of the humeral head) and inferior soft tissues of the armpit. First consider techniques to correct soft tissue and joint dysfunction. Then correct Type II dysfunctions of the shoulder joint.

The physiologic movements of the shoulder joint are: flexion, extension (sagittal plane); abduction; adduction (coronal plane); external rotation; internal rotation (transverse plane). Direct accessory movements of the humeral head are:

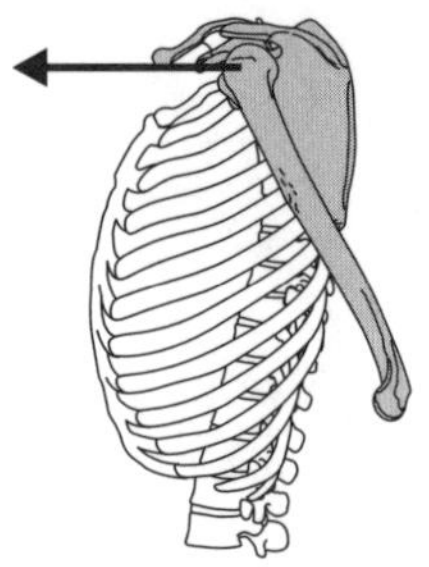

Extension
Anterior glide

1. Anterior glide for extension

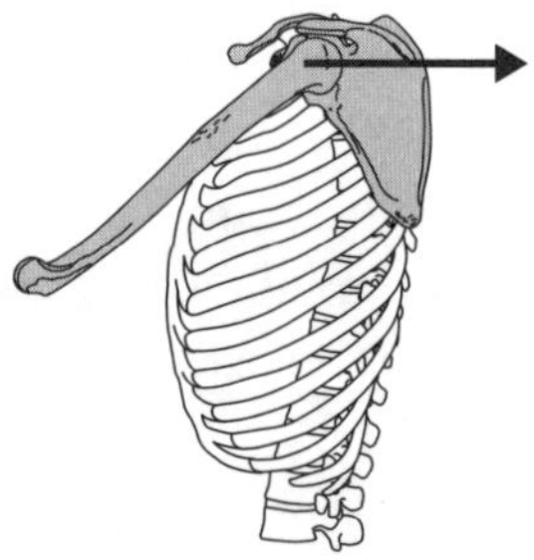

Flexion
Posterior glide

2. Posterior glide for flexion

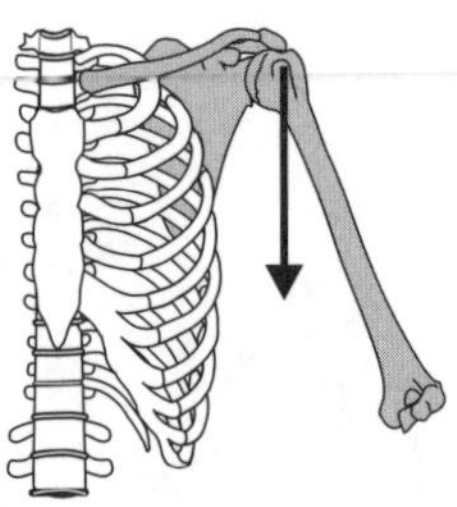

Abduction
Caudal glide

3. Caudal glide for abduction

4. Cephalad glide (to neutral) for adduction (from abduction)

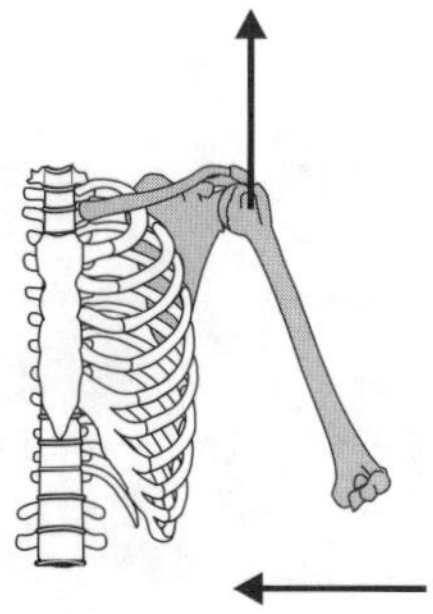

5. Anterior glide plus caudal glide for internal rotation

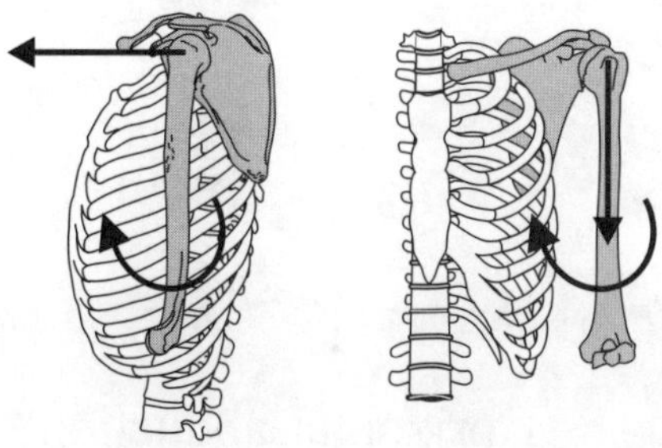

6. Posterior glide plus cephalad glide for external rotation

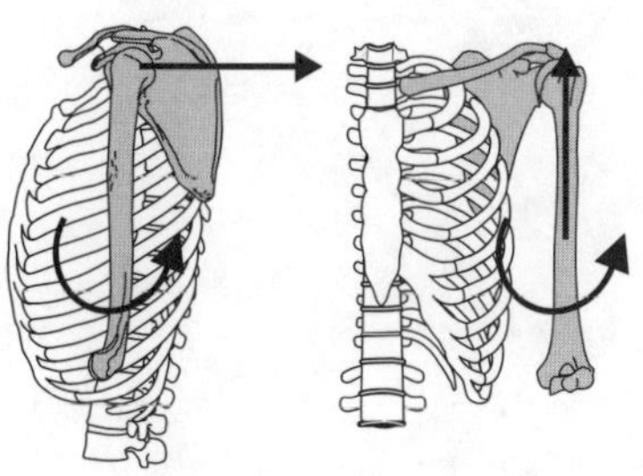

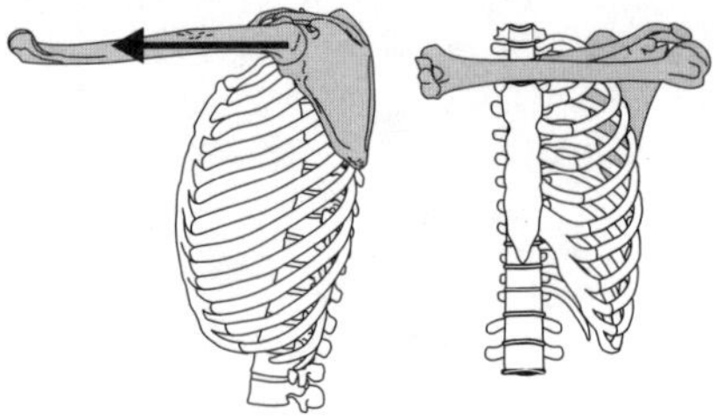

Horizontal adduction
Anterior glide

7. Anterior glide for horizontal adduction

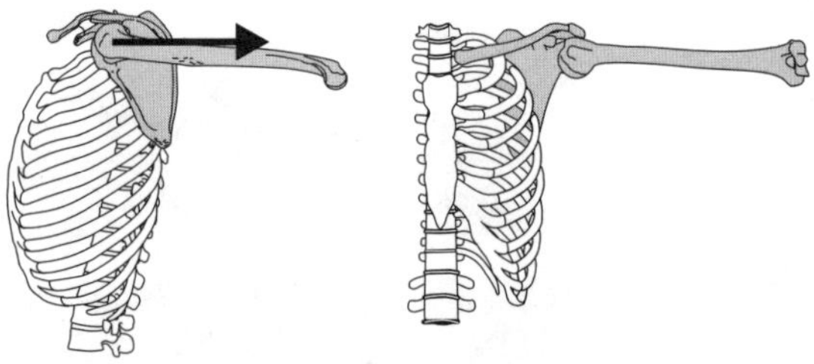

Horizontal abduction
Posterior glide

8. Posterior glide for horizontal abduction

Type II movements occur whenever the shoulder joint is flexed or extended. The Type II dysfunctions are flexed and extended. Type II movement dysfunctions are 3-planar torsion dysfunctions. Four possible Type II dysfunctions can occur at the shoulder (glenohumeral) joint:

1. Flexed, adducted, internal rotated
2. Flexed, abducted, external rotated
3. Extended, adducted, internal rotated
4. Extended, abducted, external rotated

Internal rotation occurs with adduction in Type II movement.

External rotation occurs with abduction in Type II movement.

Positional Diagnosis

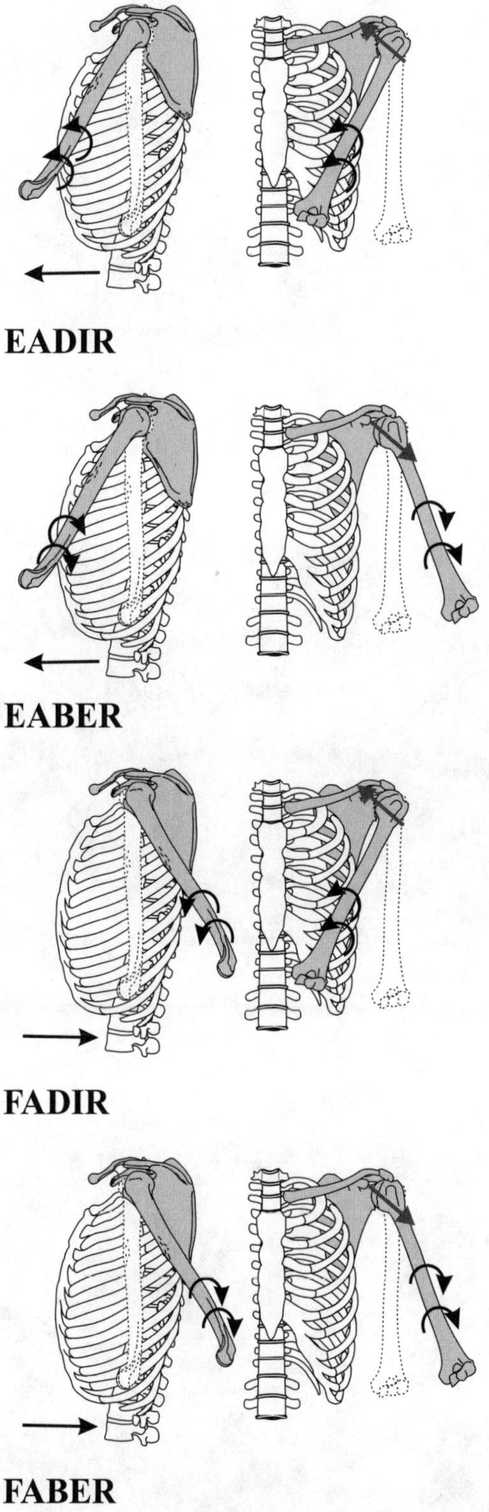

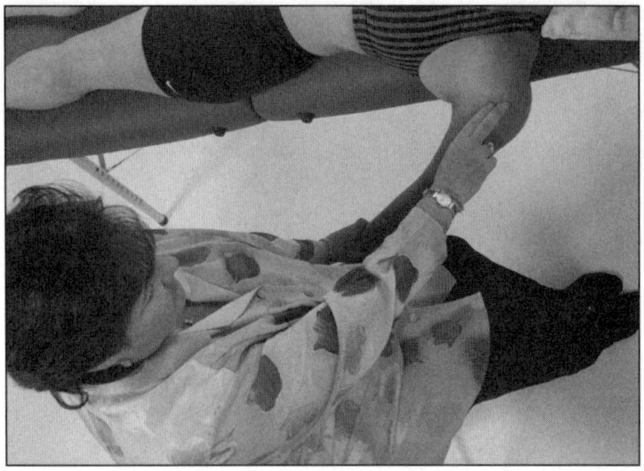

Assessment

Palpate the glides of the humeral head at the greater tuberosity during physiologic shoulder flexion and extension.

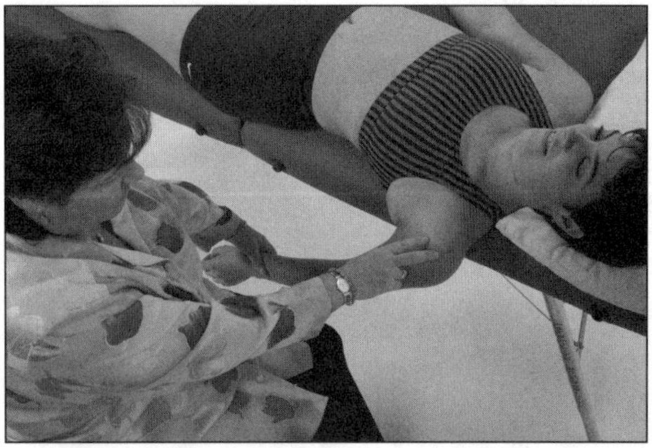

There should be no approximation or distraction of the humeral head during shoulder flexion and extension.

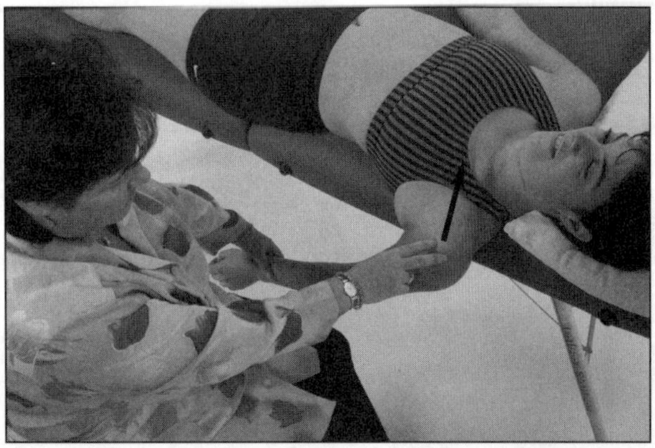

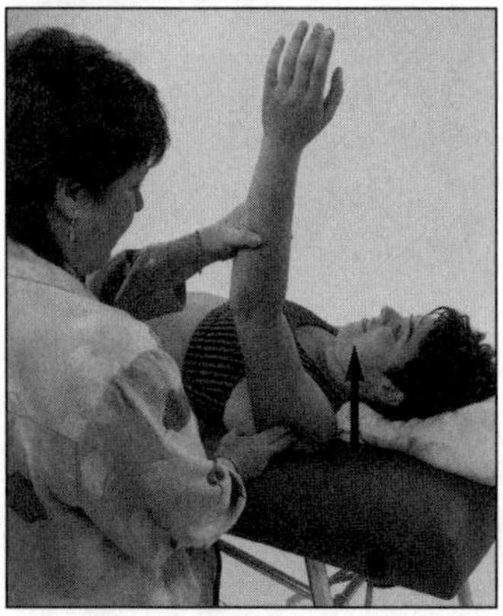

If the humeral head approximates, a medial pull is palpated at the greater tuberosity.

If the humeral head distracts, a lateral distraction is palpated at the greater tuberosity.

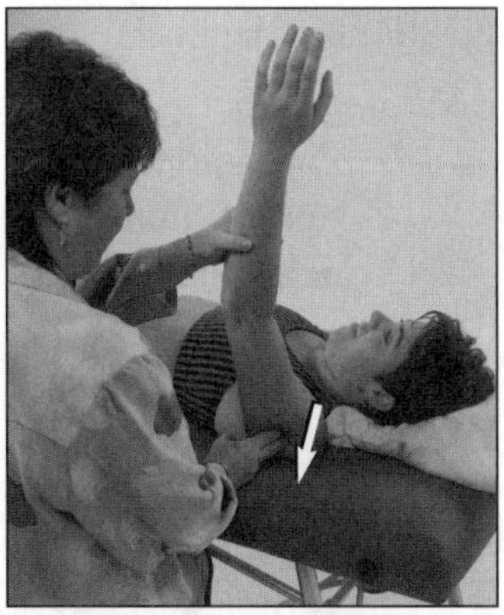

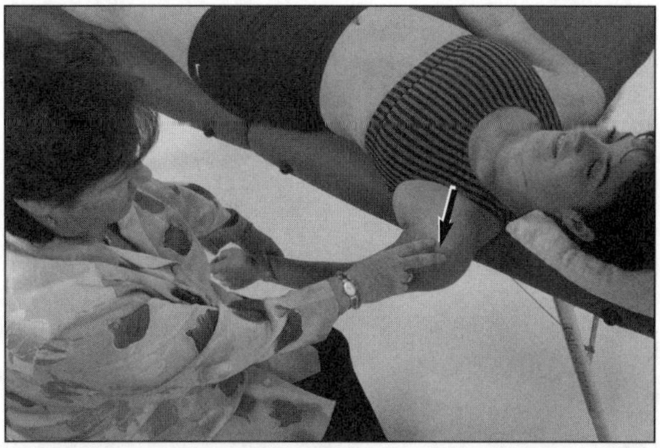

If the humeral head approximates, palpated at the greater tuberosity as a medial pull, the shoulder is stuck adducted and internally rotated.

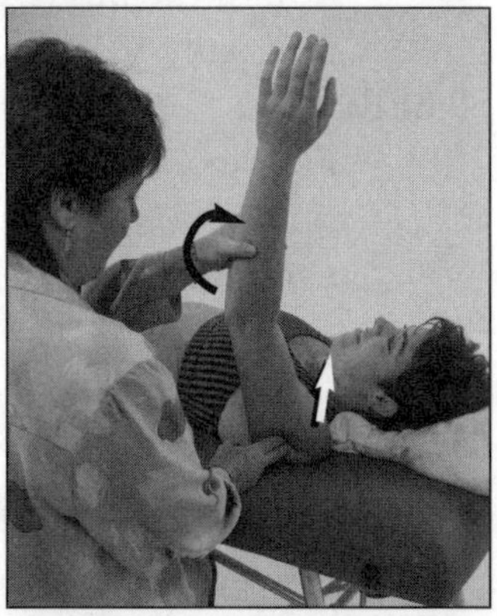

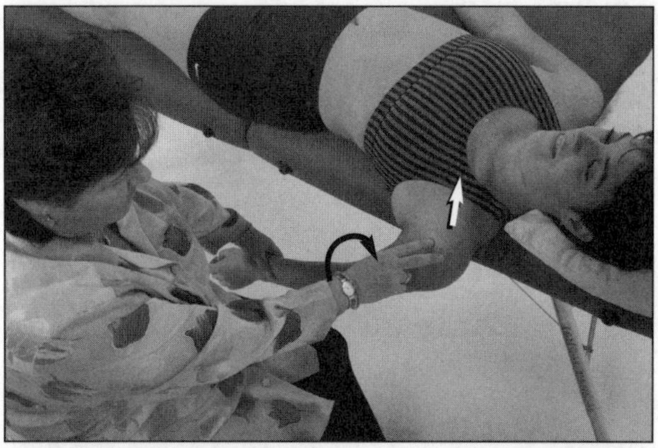

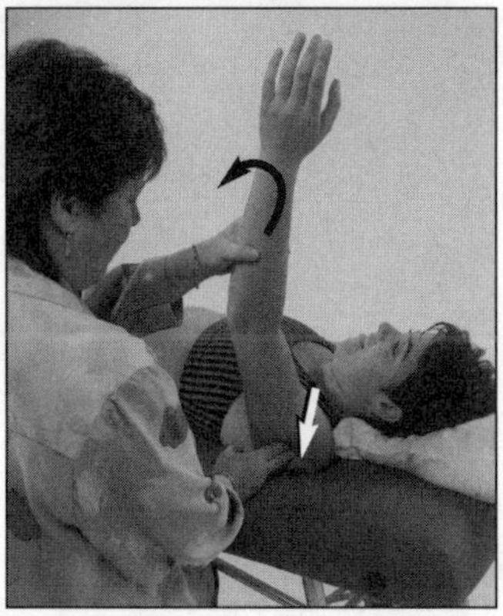

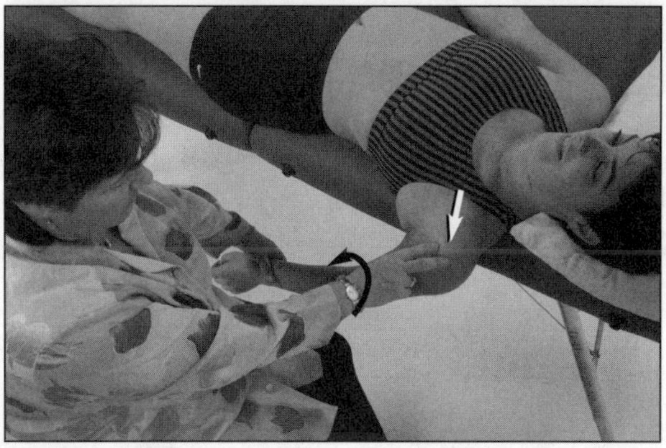

If the humeral head laterally distracts, palpated at the greater tuberosity as a lateral distraction, the shoulder is stuck abducted and external rotated.

If the movement dysfunction is palpated at the greater tuberosity during flexion, the shoulder can't flex, therefore the shoulder is stuck extended. If the movement dysfunction is palpated at the greater tuberosity during extension, the shoulder can't extend, therefore the shoulder is stuck flexed.

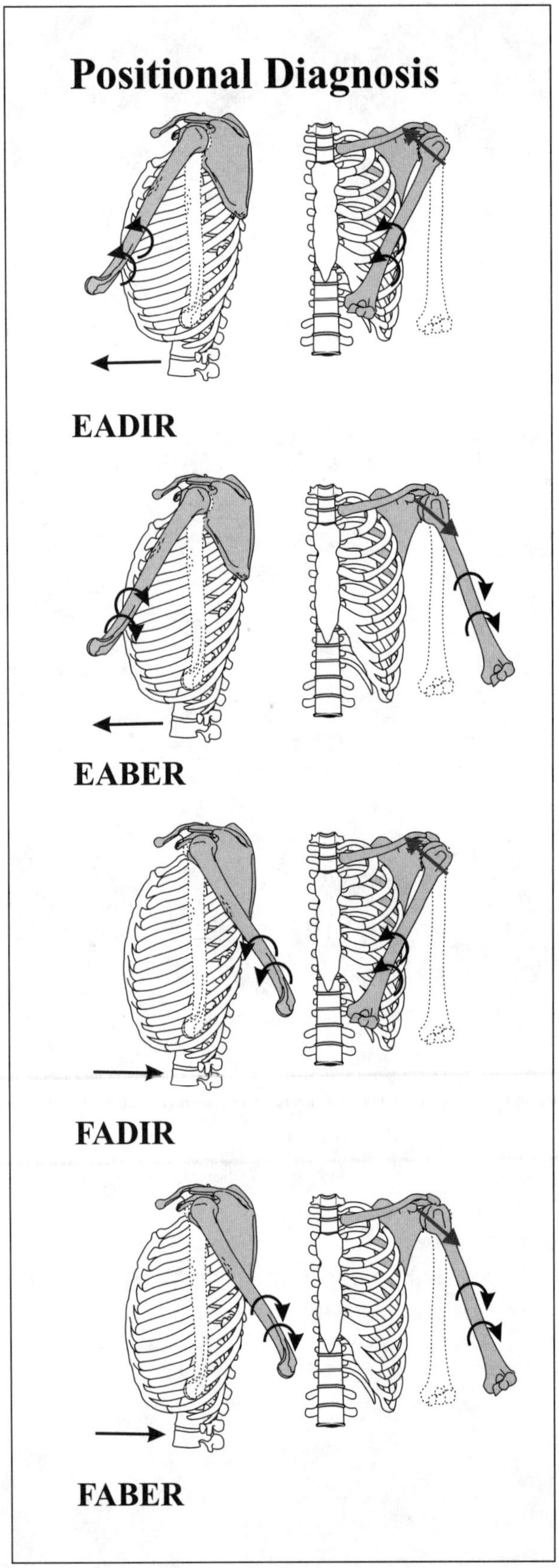

Positional Diagnosis

EADIR:

Find in flexion: extended. Approximation of humeral head: adducted, internally rotated.

EABER:

Find in flexion: extended. Lateral distraction of the humeral head: abducted, externally rotated.

FADIR:

Find in extension: flexed. Approximation of the humeral head: adducted, internally rotated.

FABER:

Find in extension: flexed. Lateral distraction of the humeral head: abducted, externally rotated.

Movement Barrier

Check the physiologic limitations of motion. Thcsc limitations should correlate with the following movement barriers.

EADIR:

The positional diagnosis is: extended, adducted, and internally rotated. The movement barriers are: flexion, abducted, and external rotation: FABER.

EABER:

The positional diagnosis is: extended, abducted, and externally rotated. The movement barriers are: flexion, adduction, and internal rotation: FADIR.

FADIR:

The positional diagnosis is: flexed, adducted, and internally rotated. The movement barriers are: extension, abduction, external rotation: EABER.

FABER:

The positional diagnosis is: flexed, abducted, and externally rotated. The movement barriers are: extension, adduction, and internal rotation: EADIR.

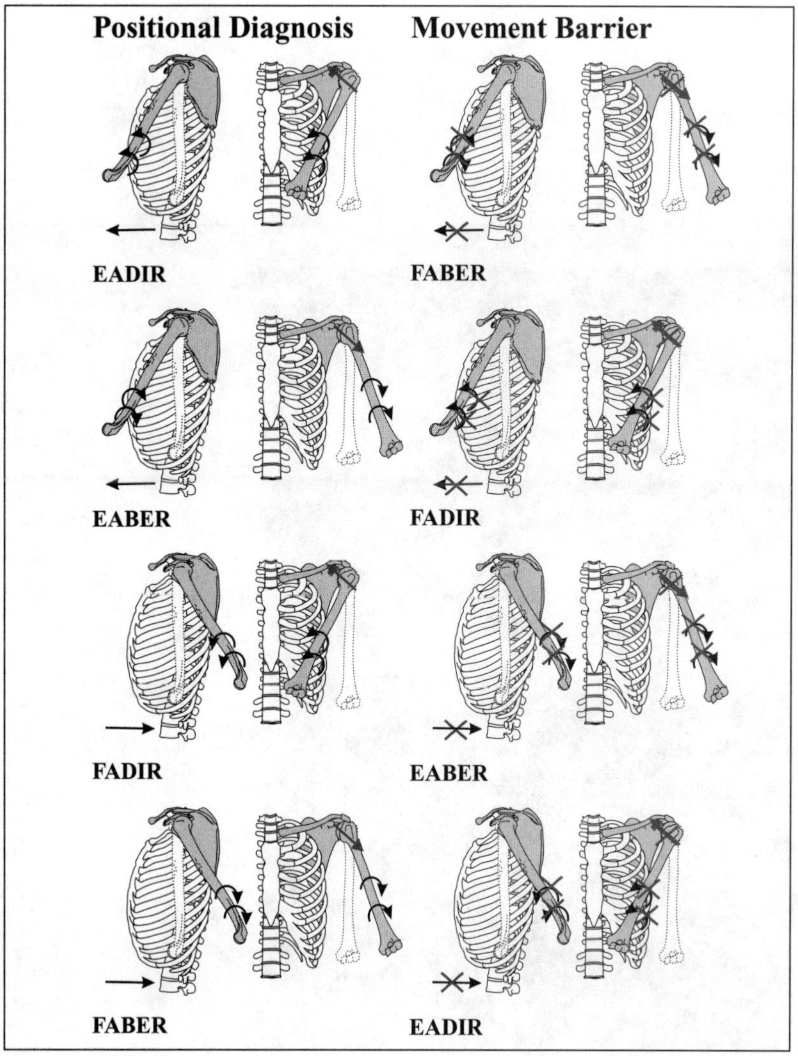

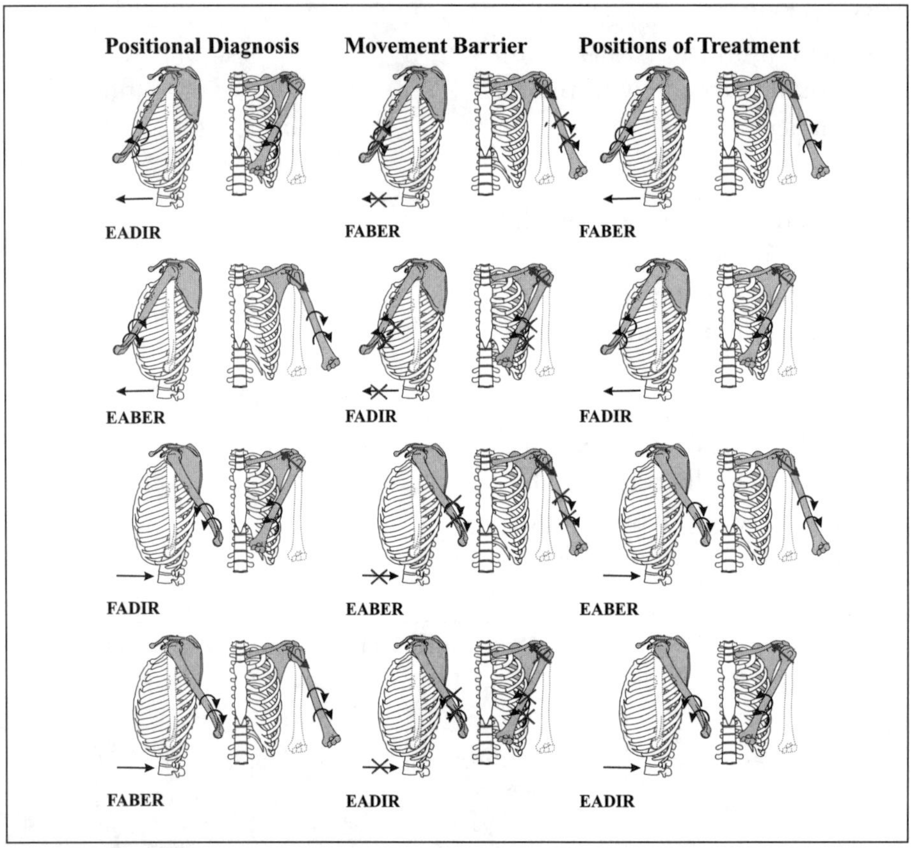

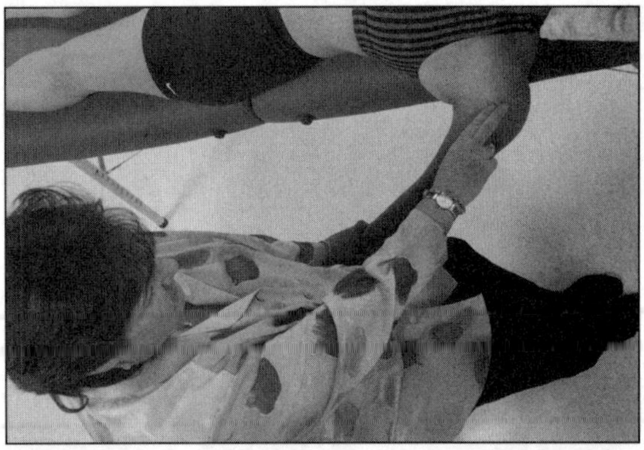

Position of Treatment

EADIR:

The positional diagnosis is: extended, adducted and internally rotated.
The movement barrier is: FABER.
The position of treatment is the same as the movement barrier: FABER: Flex, abduct, and externally rotate the shoulders to the interbarrier zones.

EABER:

The positional diagnosis is: extended, abducted, and externally rotated.
The movement barrier is: FADIR.
The position of treatment is the same as the movement barrier. FADIR: Flex, adduct, and internally rotate the shoulder to the interbarrier zones.

FADIR:

The positional diagnosis is: flexed, adducted, and internally rotated. The movement barrier is: EABER. The position of treatment is the same as the movement barrier. EABER: Extend, abduct, and externally rotate the shoulder to the interbarrier zones.

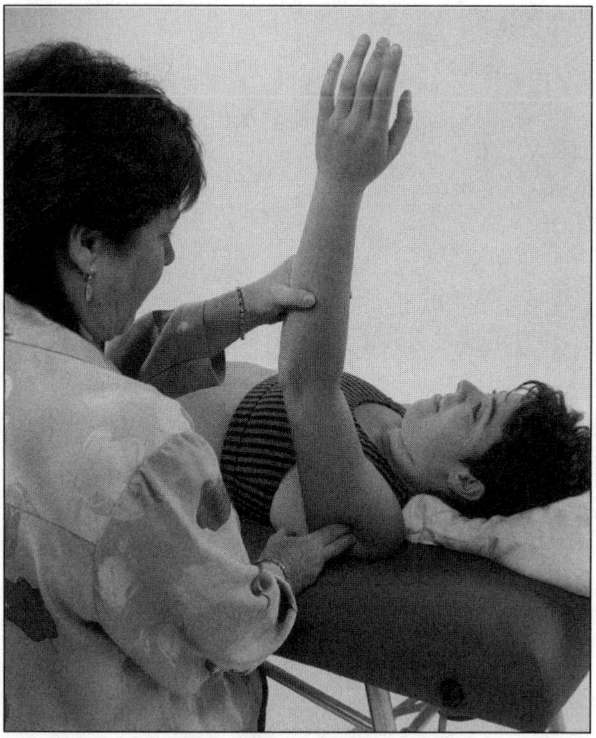

FABER:

The positional diagnosis is: flexed, abducted, and externally rotated. The movement barrier is EADIR. The position of treatment is the same as the movement barrier. EADIR: Extend, adduct, and internally rotate the shoulder to the interbarrier zones.

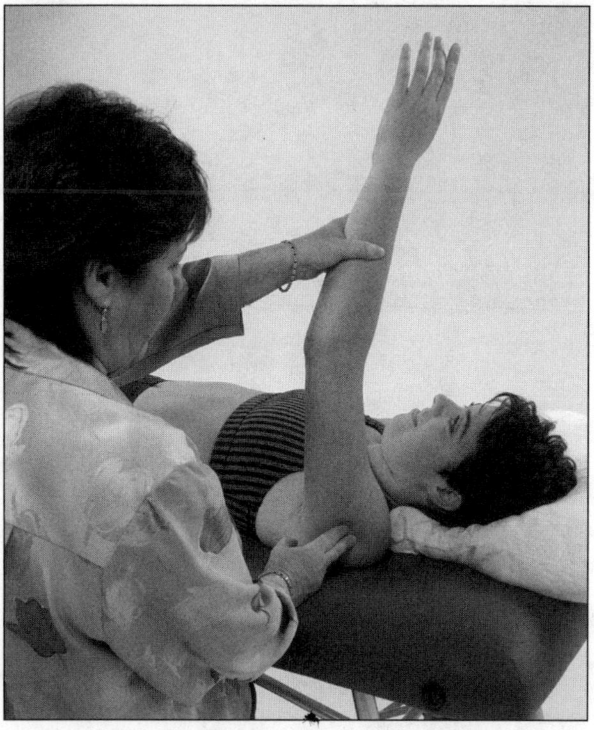

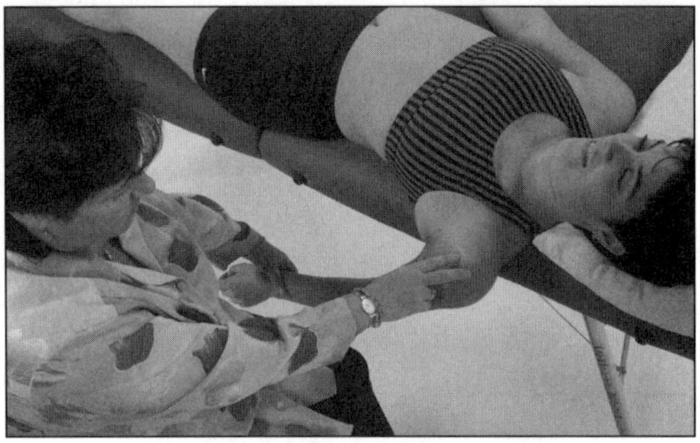

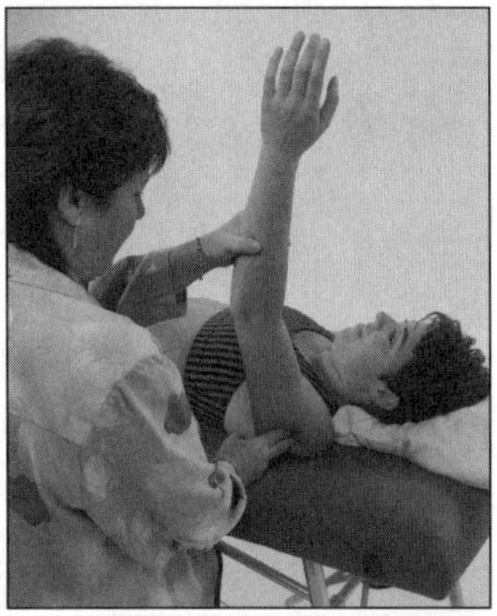

Treatment

Extended:

1. Palpate the muscle barrier at the greater tuberosity, in supine.
2. Move the shoulder and arm passively to the interbarrier zone on all 3 planes to a 3-planar interbarrier zone.
3. Resist any movement of the shoulder. Resistance: isometric; 5 grams resistance; 6 seconds resisting; unidirectional/uniplanar resistance.
4. Relax. Progress to next 3-planar interbarrier zone.
5. Repetitions: 3.
6. Reassess.

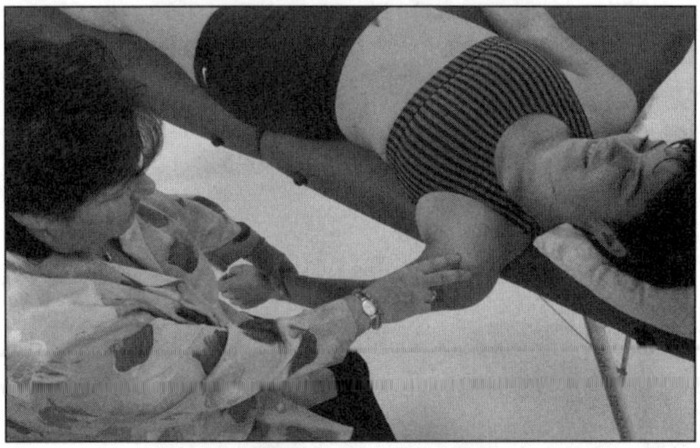

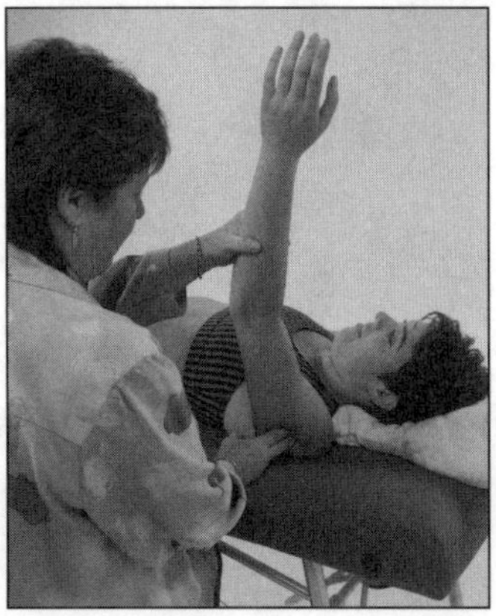

Flexed:

1. Palpate the muscle barrier at the medial aspect of the greater tuberosity, in prone, or at the opposite acromion head of scapula, in supine, with the shoulder off the bed.
2. Move the shoulder and arm passively to the interbarrier zone on all 3 planes to a 3-planar interbarrier zone.
3. Resist any movement of the shoulder. Resistance: isometric; 5 grams resistance; 6 seconds resisting; unidirectional/uniplanar resistance.
4. Relax. Progress to next 3-planar interbarrier zone.
5. Repetitions: 3.
6. Reassess.

The Scapulothoracic Joint

The scapulothoracic joint is a pseudo-joint without primary biomechanical dysfunction. Effective treatment for mobility and articular balance of the scapulothoracic joint is attained with Strain and Counterstrain; mobilization; Visceral Mobilization; Myofascial Release.

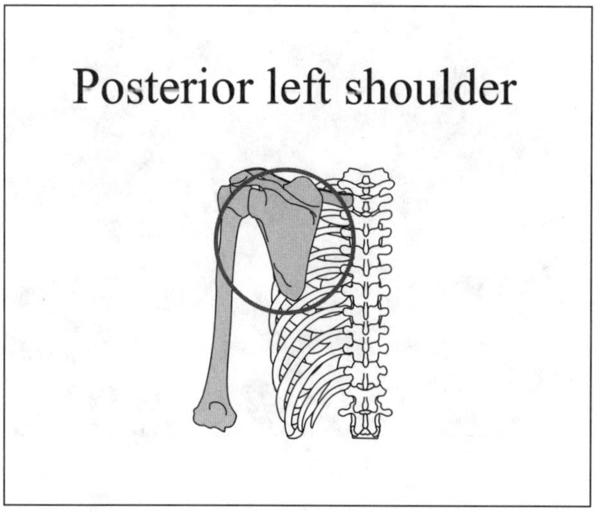

The Sternoclavicular Joint

Correction of biomechanical dysfunction of the sternoclavicular joint can cause exceptional changes, including decreases in pain and disability. Whenever there is a question regarding how treatment of such a small joint can cause so many shifts, consider the following: The sternoclavicular joint is the gateway for the thoracic inlet; lung and pleura tissue send fibers to the sternoclavicular joint capsule. This joint is a modified physiologic joint. Direct accessory movements are:

1. Caudal glide of the clavicular head during abduction of the shoulder

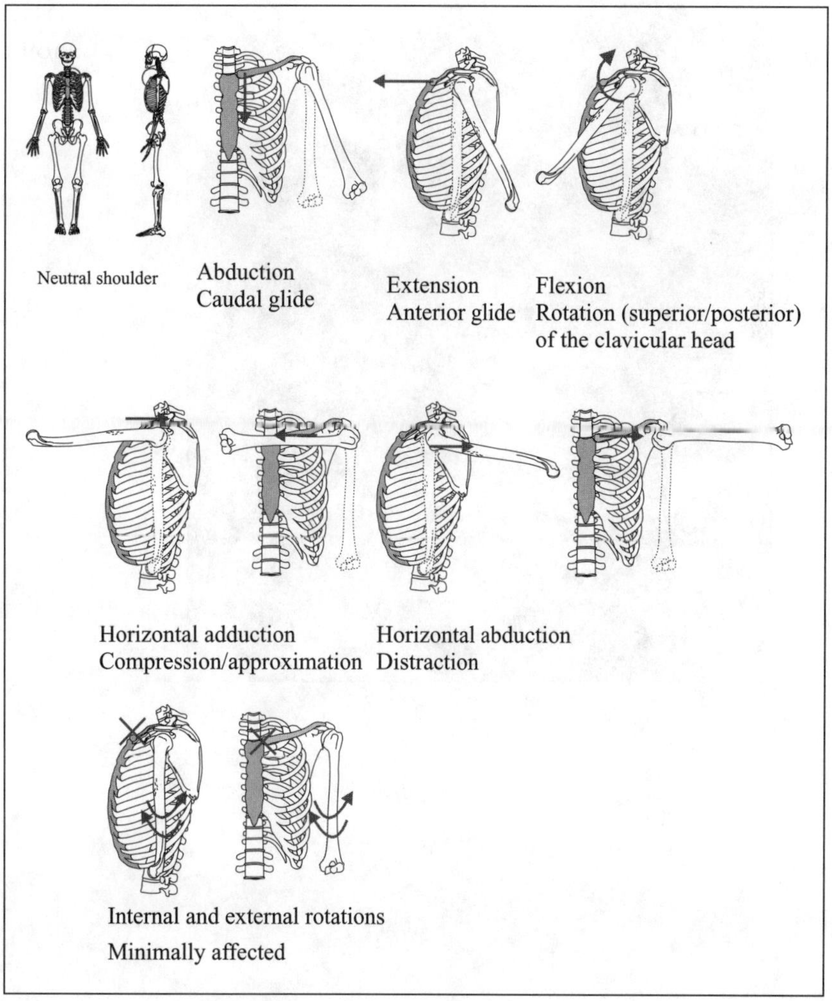

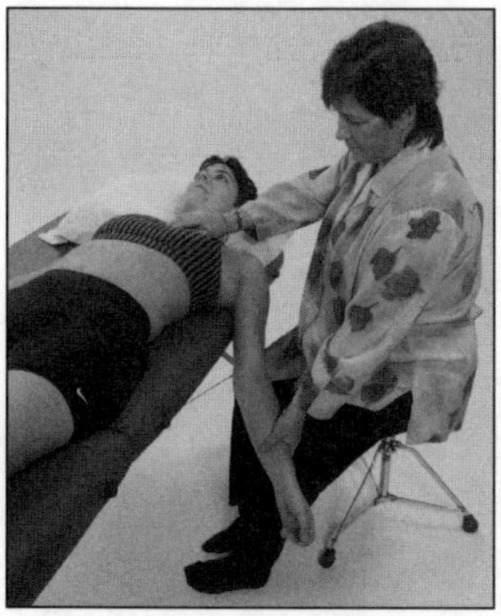

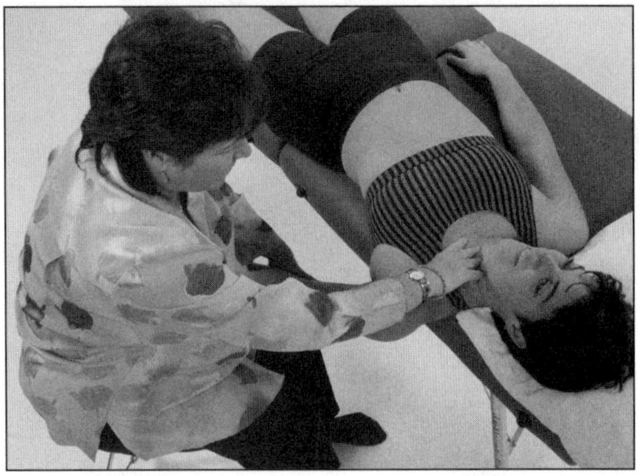

2. Anterior glide of the clavicular head during extension of the shoulder

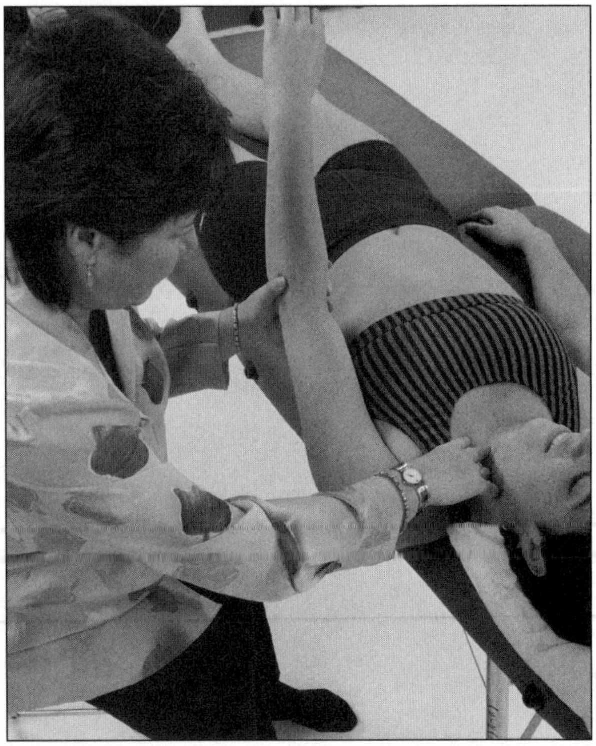

3. Rotation (superior/posterior) of the clavicular head during flexion of the shoulder

4. Compression/approximation of the clavicular head during horizontal adduction of the shoulder.

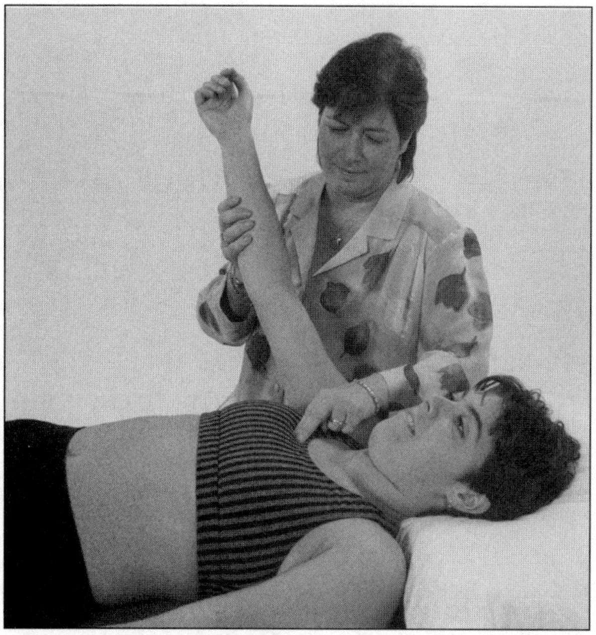

5. Distraction of the clavicular head during horizontal abduction of the shoulder
6. Internal and external rotations are minimally affected/affect the sternoclavicular joint; these movements are focused at the acromioclavicular joint.

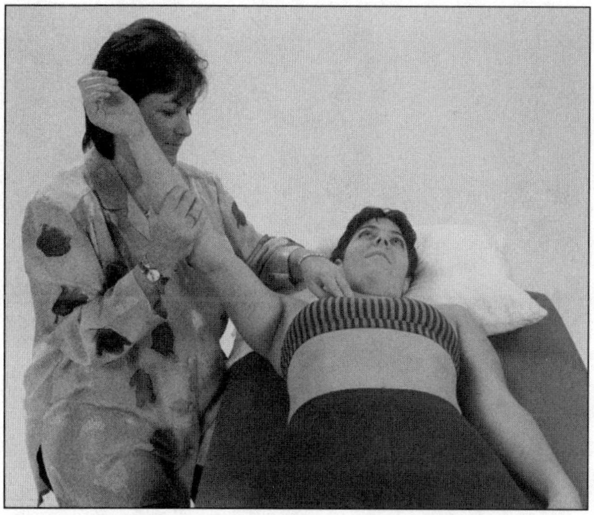

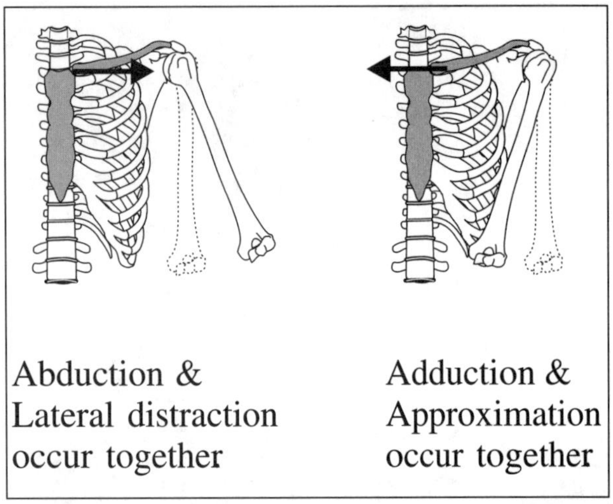

Type II Movements of the Sternoclavicular Joint

Abduction and lateral distraction occur together.
Adduction and approximation occur together.

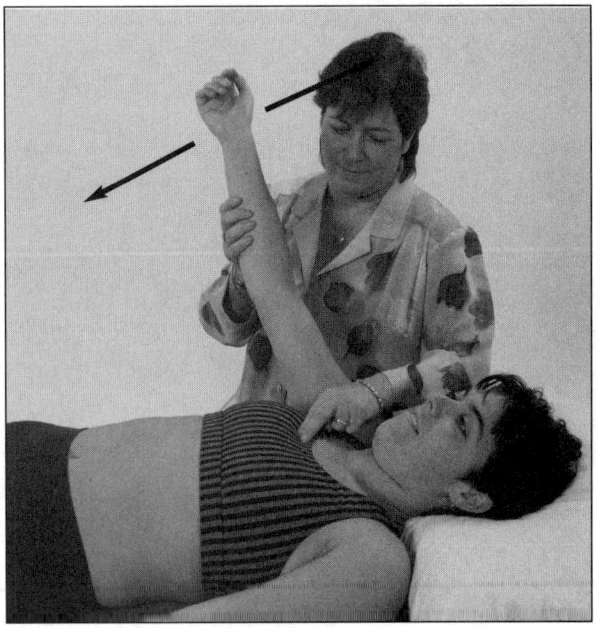

Flexed has two possible Type II variations:

1. Abduction and lateral distraction
2. Adduction and approximation.

Extended has two possible Type II variations:

1. Abduction and lateral distraction
2. Adduction and approximation

Positional Diagnosis

The four possible positional diagnoses of biomechanical dysfunction at this joint are:

1. Flexed, abducted, lateral distracted: FABLD
2. Flexed, adducted, approximation: FADA
3. Extended, abducted, lateral distracted: EABLD
4. Extended, adducted, approximated: EADA

FABLD:

Find in extension: The sternoclavicular joint is stuck flexed. Treat in shoulder extension.

FADA:

Find in extension: The sternoclavicular joint is stuck flexed. Treat in shoulder extension.

EABLD:

Find in flexion: The sternoclavicular joint is stuck extended. Treat in shoulder flexion.

EADA:

Find in flexion: The sternoclavicular joint is stuck extended. Treat in shoulder flexion.

Assess: lateral distraction and approximation. Lateral distraction occurs with abduction. Approximation occurs with adduction.

Positional Diagnosis

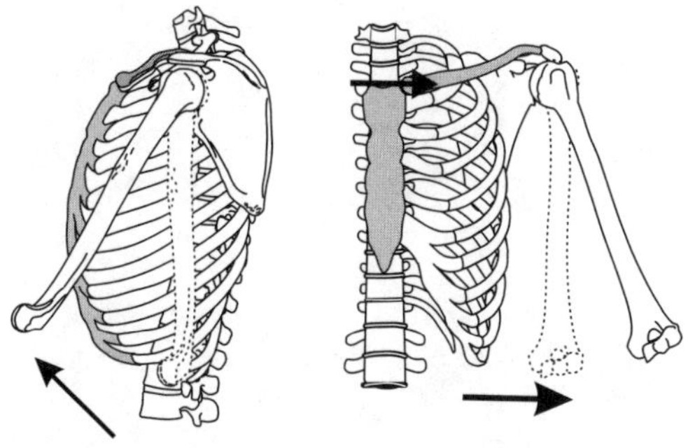

FABLD

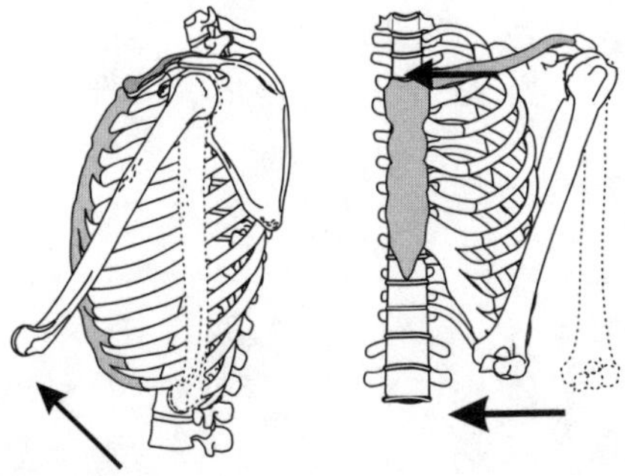

FADA

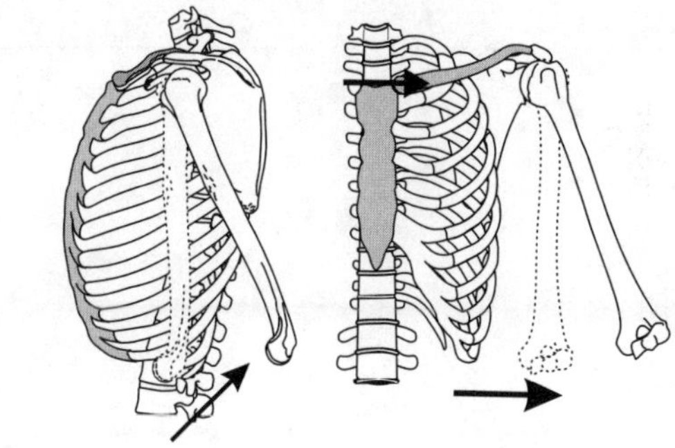

EABLD

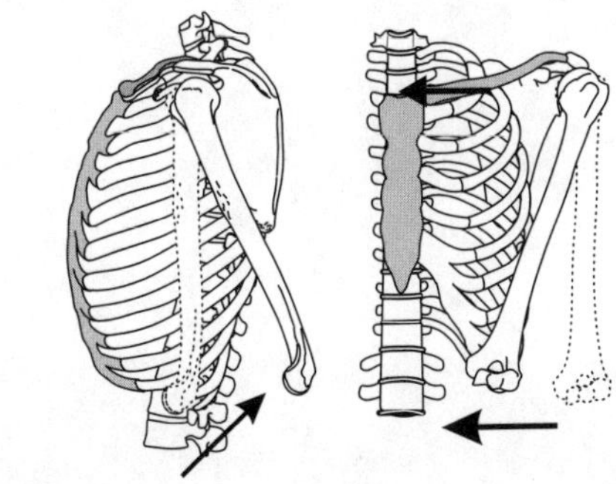

EADA

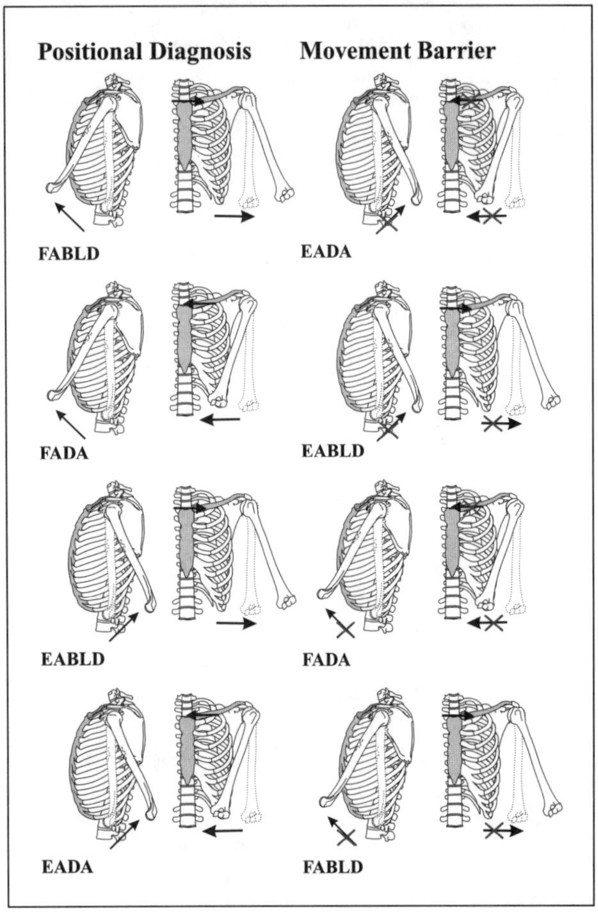

Movement Barrier

Palpate the movement barrier at the proximal head of clavicle.

FABLD:

The Sternoclavicular joint is stuck flexed (superior/posterior rotation), abducted, and lateral distracted. The movement barrier is: extension, adduction, and approximation: EADA.

FADA:

The Sternoclavicular joint is stuck flexed (superior/posterior rotation), adducted, and approximated. The movement barrier is: extension, abduction, and lateral distraction: EABLD.

EABLD:

The Sternoclavicular joint is stuck extended (anterior glide), abducted, and lateral distracted. The movement barrier is: flexion, adduction, and approximation: FADA.

EADA:

The Sternoclavicular joint is stuck extended, adducted, and approximated. The movement barrier is: flexion, abduction, and lateral distraction: FABLD.

Position of Treatment

To position the arm/shoulder to assess movement barriers, and for treatment, follow the following procedures:

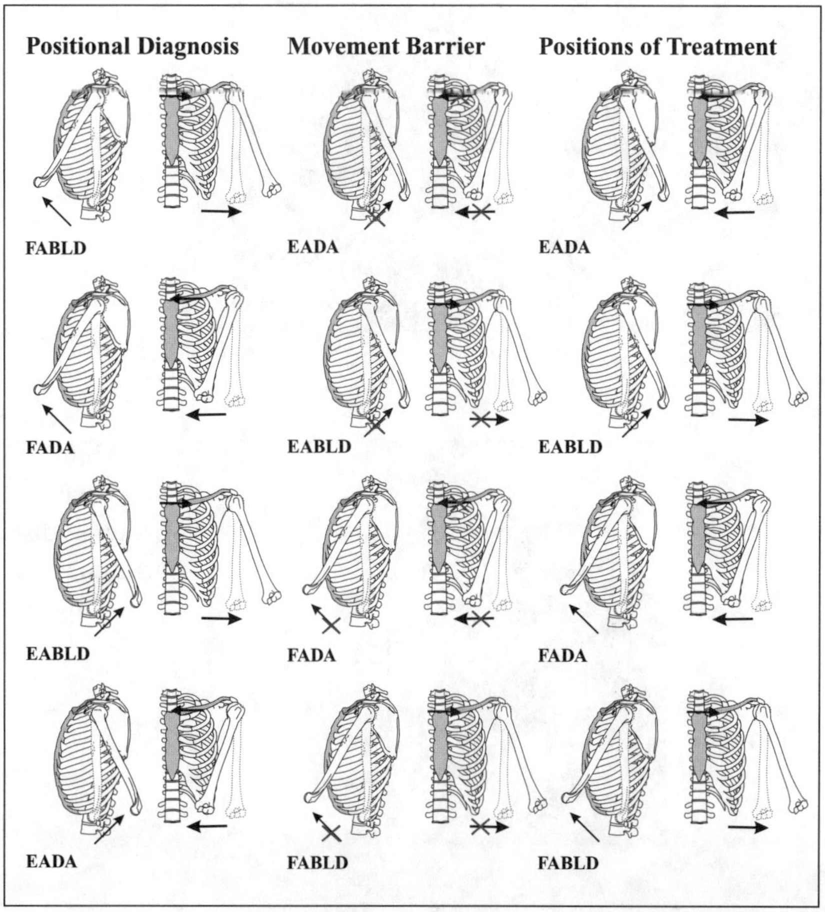

Flexed: In supine; shoulder off edge of the bed; straight arm; no abduction or rotation; passively move into extension.

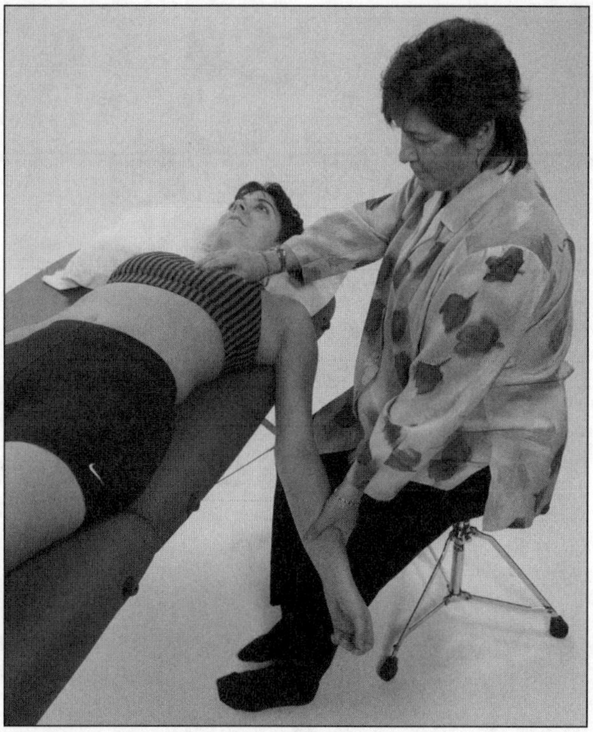

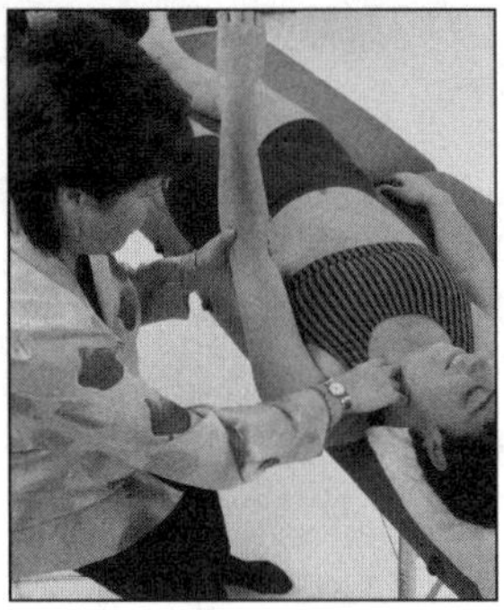

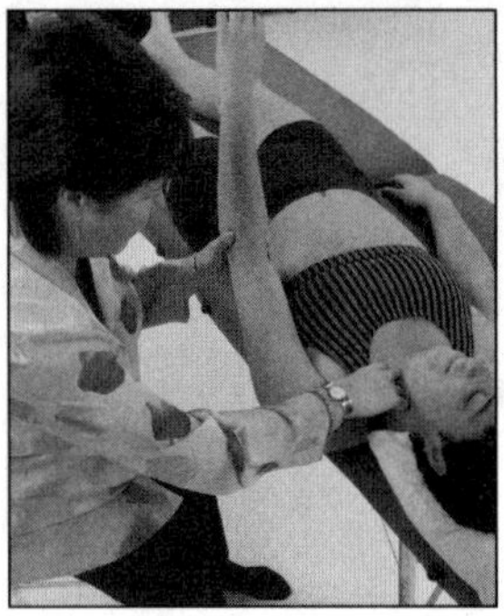

Extended: In supine; straight arm; no abduction/adduction or rotation; passively move into flexion.

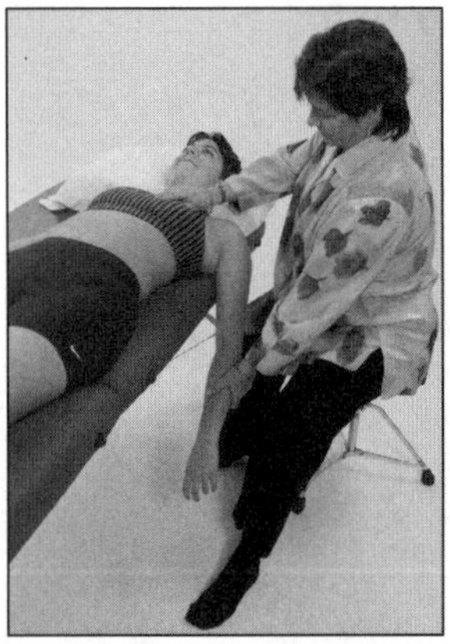

Abducted and lateral distracted: Use fingers on proximal clavicle head. Adduct the clavicle by bringing distal clavicle caudal and compress the proximal clavicle head towards the manubrium articular surface.

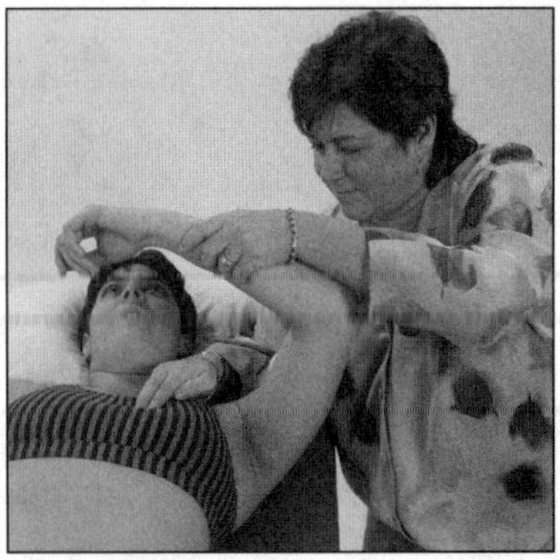

Adducted and approximated: Use fingers on proximal clavicle head. Abduct the clavicle by bringing distal clavicle superior and distract the proximal clavicle head away from the manubrium articular surface.

Treatment

FABLD:

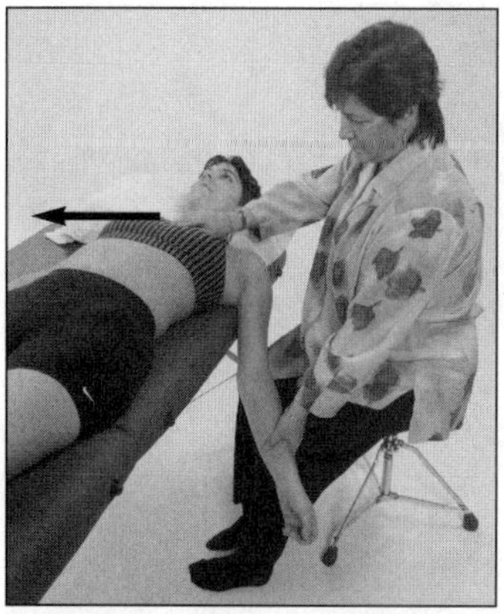

1. Extend the arm off the edge of the table until the movement barrier (lateral distraction) is palpated at the proximal clavicle head.
2. Then adduct the clavicle and approximate the proximal clavicle head.
3. Position at a 3-planar interbarrier zone.

FADA:

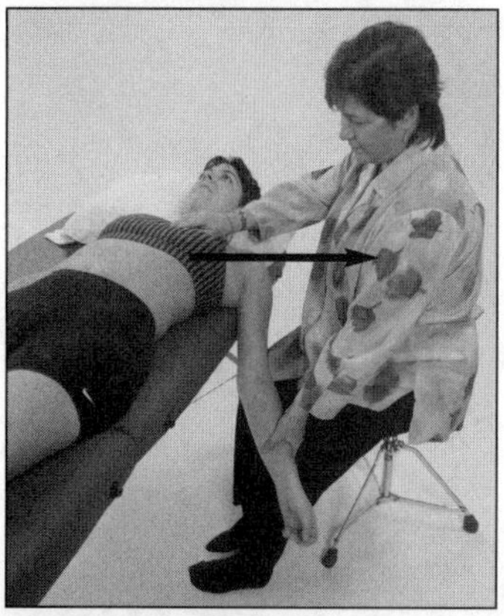

1. Extend the arm off the edge of the table until the movement barrier (lateral distraction) is palpated at the proximal clavicle head.
2. Then abduct the clavicle and laterally distract the proximal clavicle head.
3. Position at a 3-planar interbarrier zone.

EABLD:

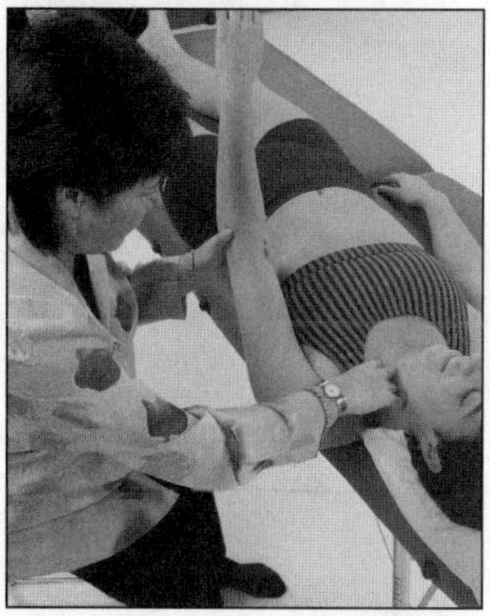

1. Flex the arm until the movement barrier (approximation) is palpated at the proximal clavicle head.
2. Then adduct the clavicle and approximate the proximal clavicle head.
3. Position at a 3-planar interbarrier zone.

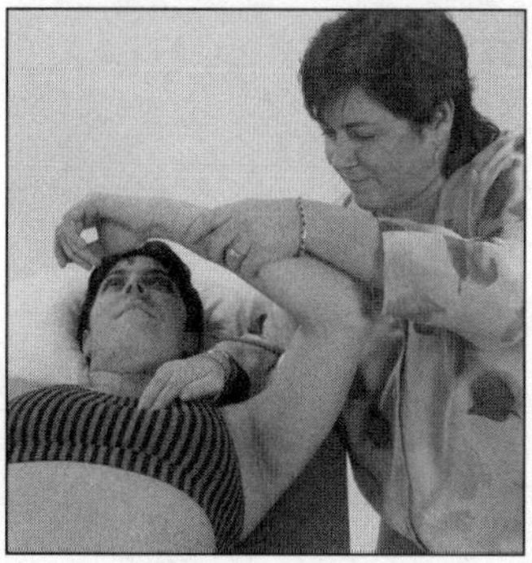

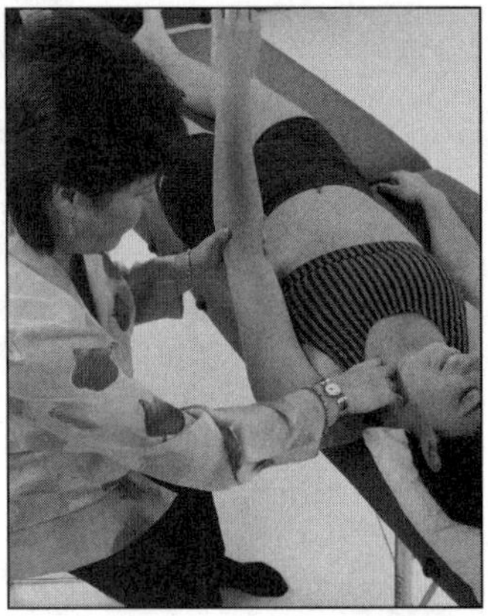

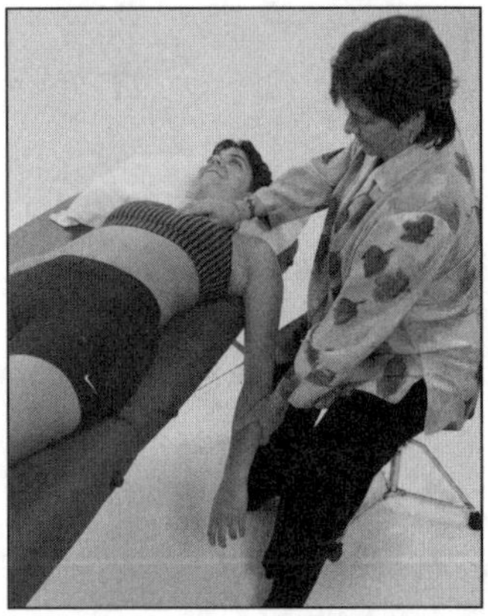

EADA:

1. Flex the arm until the movement barrier (approximation) is palpated at the proximal clavicle head.
2. Then abduct the clavicle and laterally distract the proximal clavicle head.
3. Position at a 3-planar interbarrier zone.

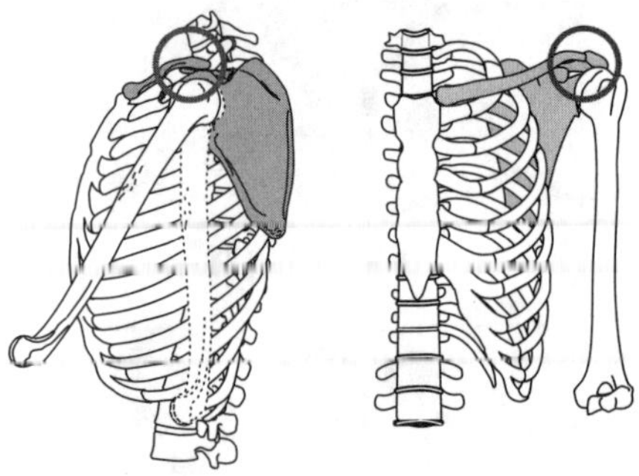

Acromioclavicular Joint

The Acromioclavicular Joint

This joint does not respond well to Muscle Energy and "Beyond" Technique philosophy. Excellent articular balance and mobility can be attained with Strain and Counterstrain. Use the anterior acromioclavicular joint Tender Point and the posterior acromioclavicular joint Tender Point.

CHAPTER 32

THE ELBOW JOINT

The elbow joint is often a surprise. Time and energy, effort and money, are often expended in treatment of elbow pain and disability. Often there is improvement; just as often there is residual dysfunction. Muscle Energy and "Beyond" Technique for the elbow can surprise the practitioner with superb results in severe and chronic elbow problems.

Inflammation of the soft tissue in the region of the elbow is common; neural tissue tension problems are common in the elbow region; vascular insufficiency is also typical. Yet in all those problems, and in other scenarios excellent results can usually be attained with these techniques.

The elbow has three joints:

1. The radioulnar joint
2. The radiohumeral joint
3. The humeroulnar joint

When the elbow joints are treated in the above sequence, the Muscle Energy and "Beyond" Technique approach is an extremely effective and efficient way to restore joint mobility and articular balance:

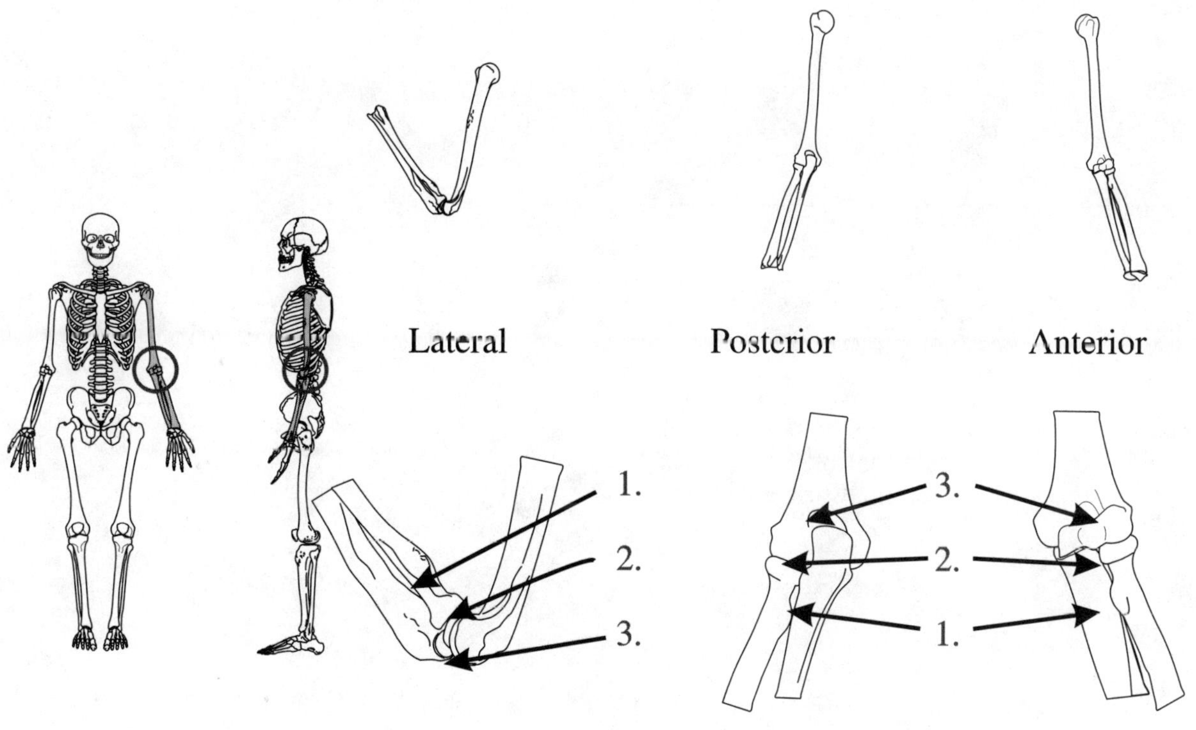

1. Radioulnar joint
2. Radiohumeral joint
3. Humeroulnar joint

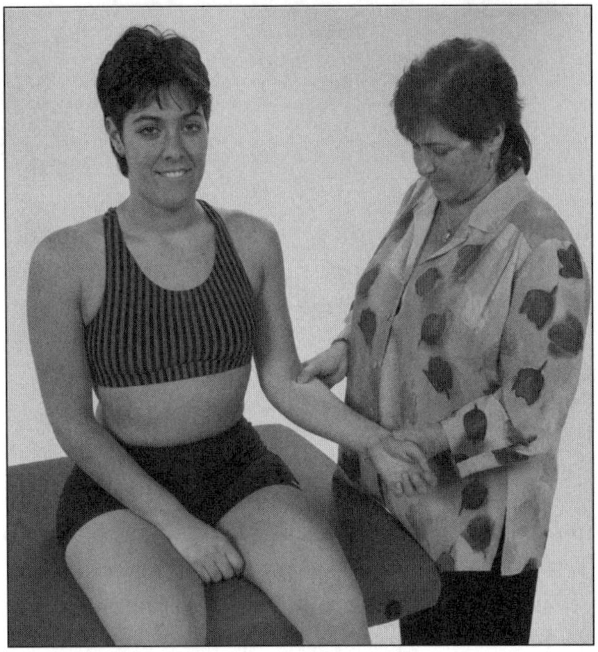

The Radioulnar Joint

Assessment

Simply assess pronation and supination. Palpate the muscle barrier at the radial head. During supination, the radial head glides anterior.

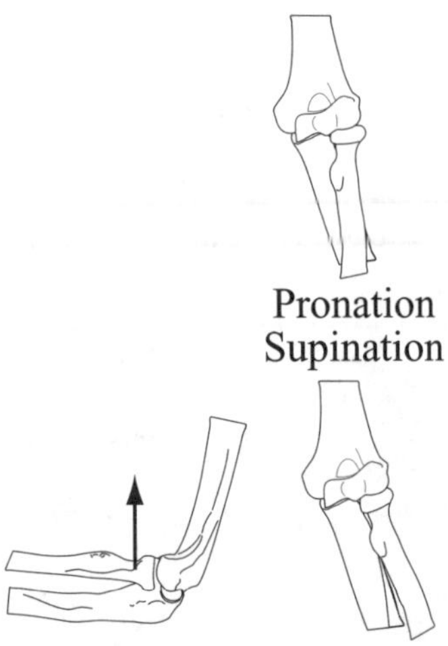

Pronation moving into supination, the radial head glides anterior.

During pronation, the radial head glides posterior.

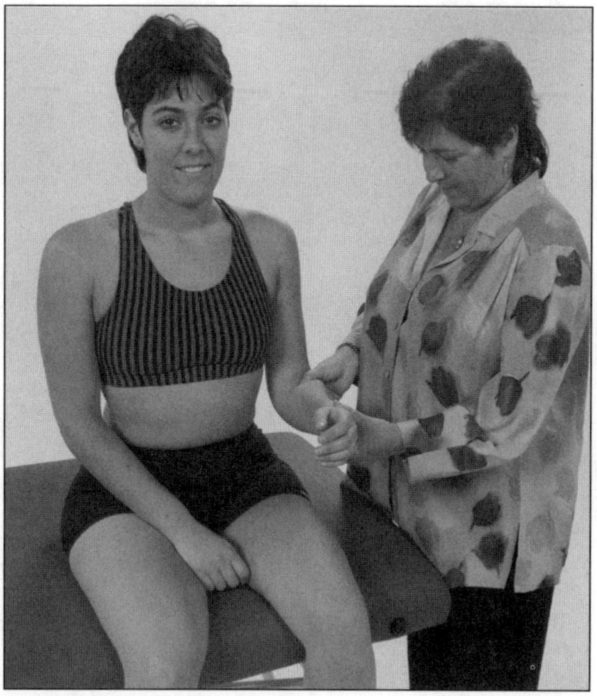

During supination and pronation there should be no medial or lateral motion of the radial head. Palpate for medial and lateral motion of the radial head during supination and pronation.

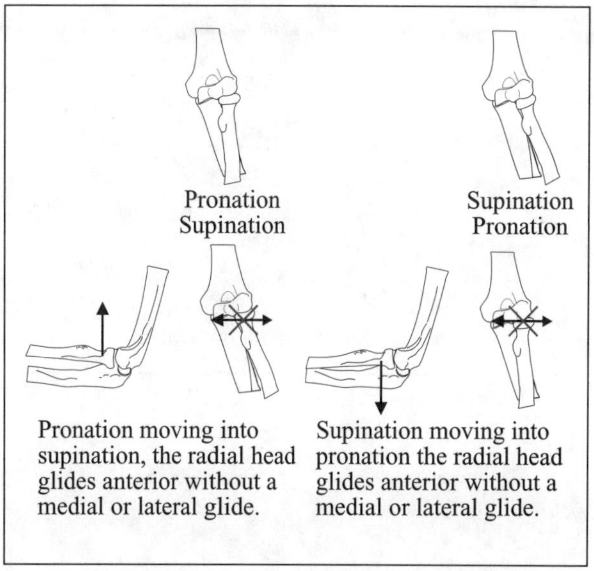

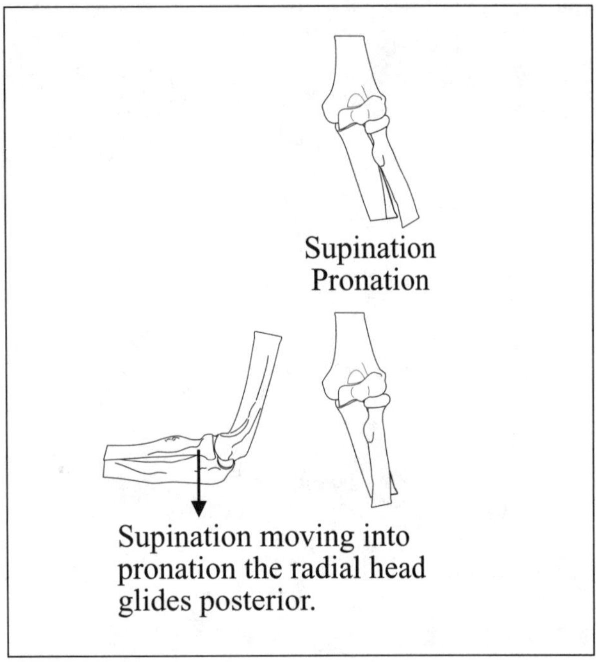

Positional Diagnosis

Supinated:

The radial head is stuck supinated in anterior glide and cannot pronate with posterior glide.

Pronated:

The radial head is stuck pronated in posterior glide and cannot supinate with anterior glide.

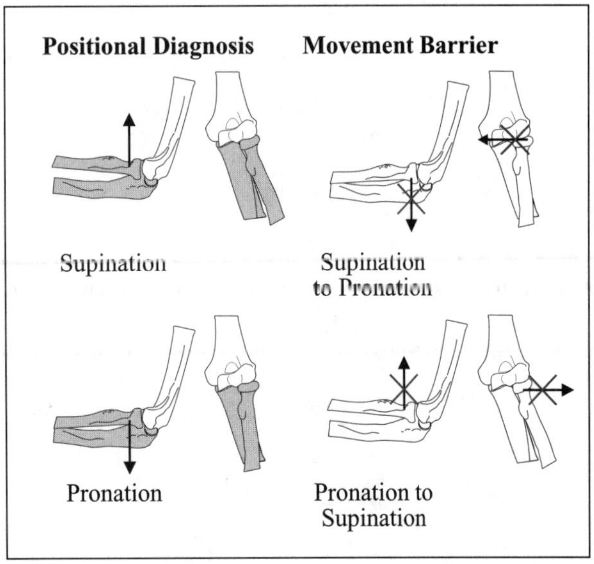

Movement Barrier

Supinated:

Pronation cannot occur passively without lateral glide of radial head. Palpate muscle barrier of lateral glide of radial head.

Pronated:

Supination cannot occur passively without medial gliding radial head. Palpate muscle barrier of medial glide of radial head.

Treatment

Supine or sitting with elbow flexion of 90 degrees, shoulder in 0 degree abduction, 0 degree flexion, and 0 degree rotation.

Supinated:

1. Passively pronate the forearm until the interbarrier zone. (The proximal radioulnar joint is stuck supinated. The movement barrier is pronation.)
2. Palpate for lateral glide of the radial head.
3. Resist supination or pronation. Isometric: 6 seconds resistance; 5 grams of force.
4. Relax. Repeat 3 repetitions.
5. Reassess.

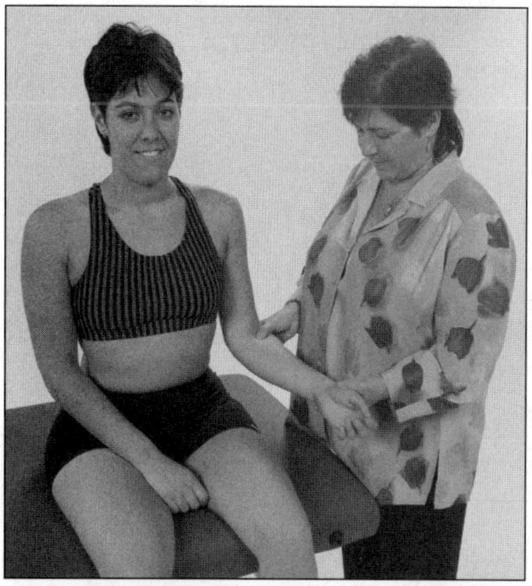

Pronated:

1. Passively supinate the forearm until the interbarrier zone. (The proximal radioulnar joint is stuck pronated. The movement barrier is supination.)
2. Palpate for medial glide of the radial head.
3. Resist supination or pronation. Isometric: 6 seconds resistance, 5 grams of force.
4. Relax. Repeat 3 repetitions.
5. Reassess.

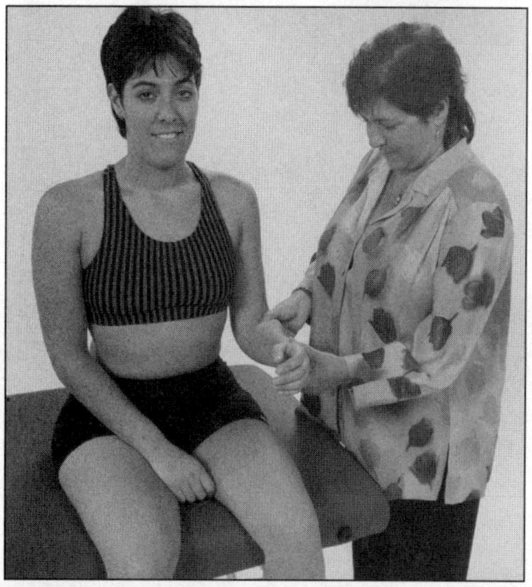

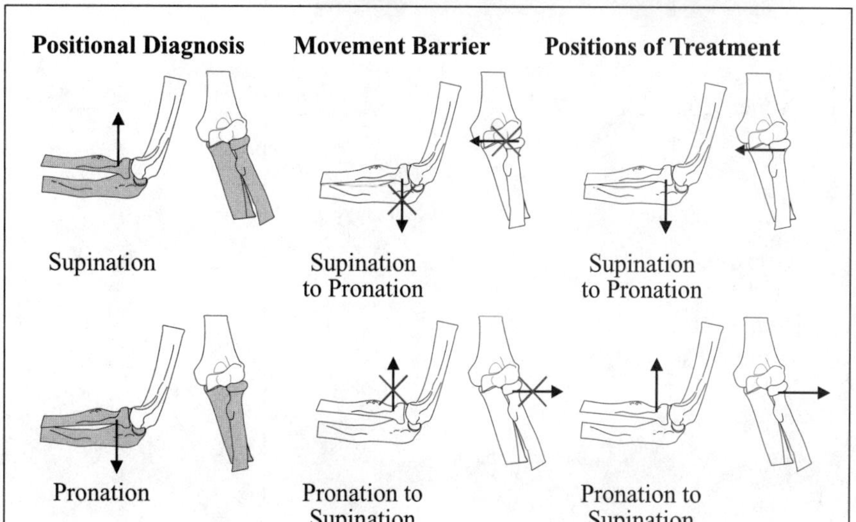

The Radiohumeral Joint

The radiohumeral joint is where elbow extension occurs. The humeroulnar joint is the joint where flexion occurs.

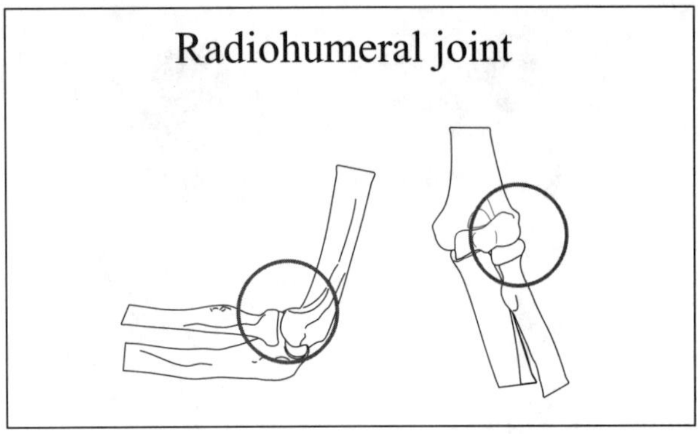

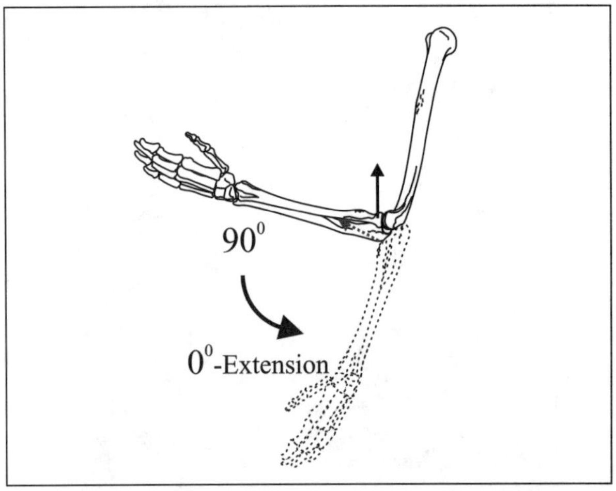

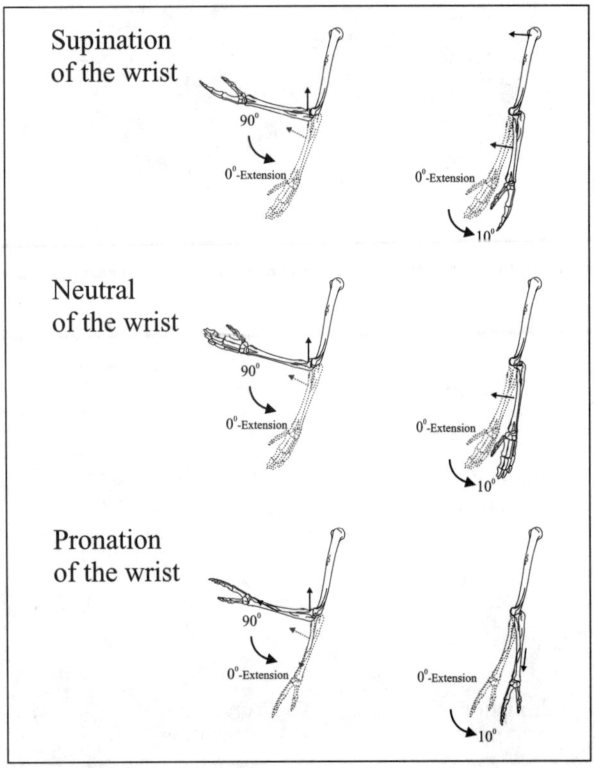

Positional Diagnosis

There are two possible Type II dysfunctions that occur at the radiohumeral joint: flexed, pronated and flexed, supinated. This radiohumeral joint gets stuck flexed; the movement barrier is extension. The radiohumeral joint cannot extend.

There are accessory movements that occur at this joint as well. During elbow extension, which primarily occurs at the radiohumeral joint, there is anterior glide of the radial head.

During elbow extension with supination (after 0 degree neutral through 90 degrees supination) there is anterior glide of the radial head.

During elbow extension with pronation (after 0 degree neutral through 90 degrees pronation) there is anterior glide of the radial head with longitudinal distraction of the radial head (rather than posterior glide, which occurs with pronation in flexion).

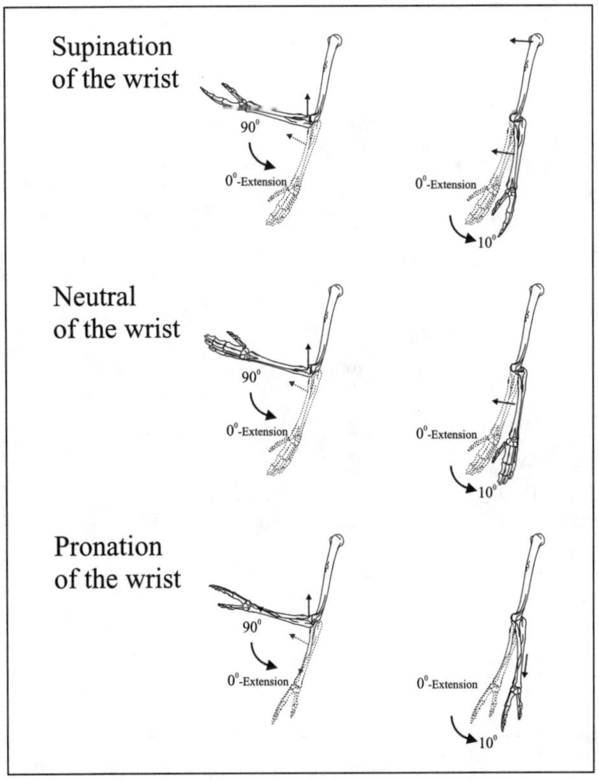

Extension is considered to be 0 degree.

Flexion is any elbow flexion after 0 degree.

During elbow extension from flexion, in supination and in neutral (0 degree supination/pronation), the radial head is gliding anterior from a posterior glide position.

During elbow extension after 0 degree, which is hyperextension, there is anterior glide of the radial head.

When supination is present, there is anterior glide of the humeral head.

When pronation is present, there is a longitudinal distraction of the radial head.

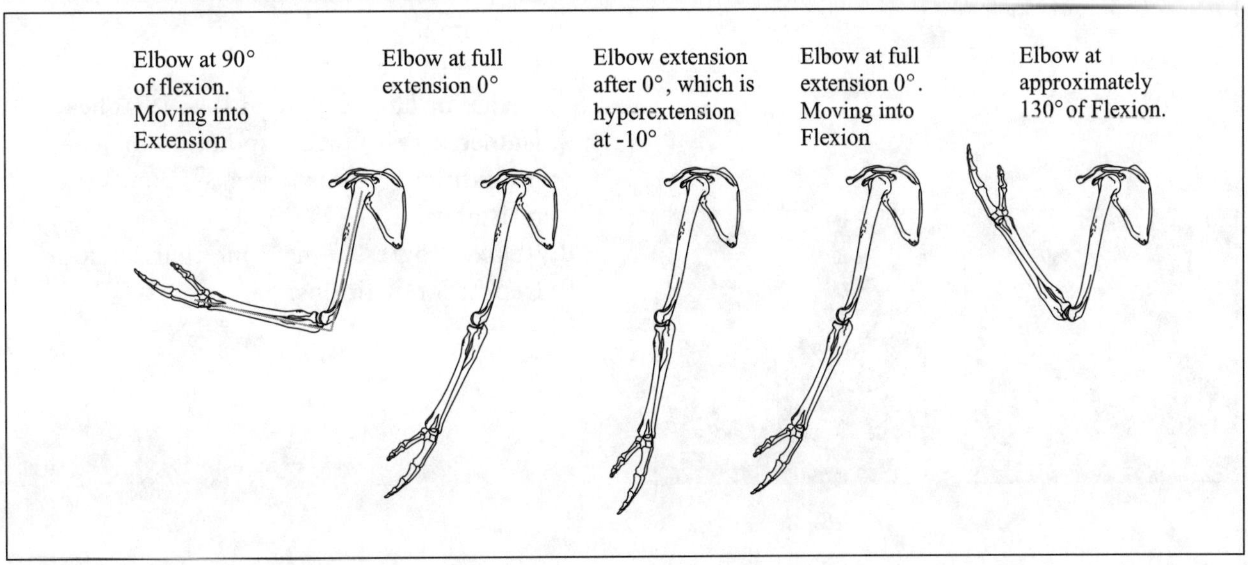

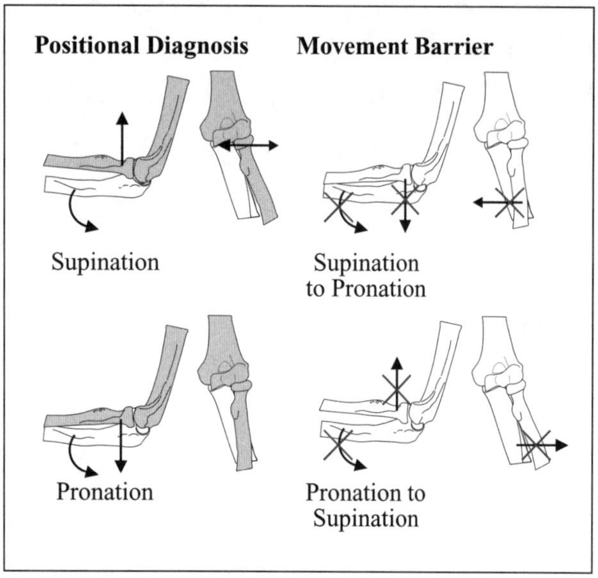

Movement Barrier

Palpate the muscle barrier at the radial head. There should be no medial or lateral movement of the radial head during extension. Palpate for medial and lateral motion of the radial head.

Flexed, pronated:

The radiohumeral joint is stuck flexed and pronated. The elbow cannot extend and supinate. The movement barrier is: extension, supination.

Flexed, supinated:

The radiohumeral joint is stuck flexed and supinated The elbow joint cannot extend and pronate. The movement barrier is: extension, pronation.

Treatment

1. In supine or in sitting, position the shoulder joint at 0 degree abduction, 0 degree rotation, 0 degree flexion. Position the forearm at 0 degree supination/pronation.
2. Flex the elbow to 100 percent flexion.
3. Passively extend the elbow from 100 percent flexion.
4. Palpate the radial head for medial and lateral glide of the radial head.
5. If there is lateral glide of the radial head, the positional diagnosis is: flexed, pronated. If there is medial glide of the radial head, the positional diagnosis is: flexed, supinated.
6. Extend to the interbarrier zone (before medial or lateral glide of the radial head).
7. Isometric resistance: supination or pronation; 5 grams force; 6 seconds resistance.
8. Relax. Progress to next interbarrier zone.
9. Repeat 3 repetitions.

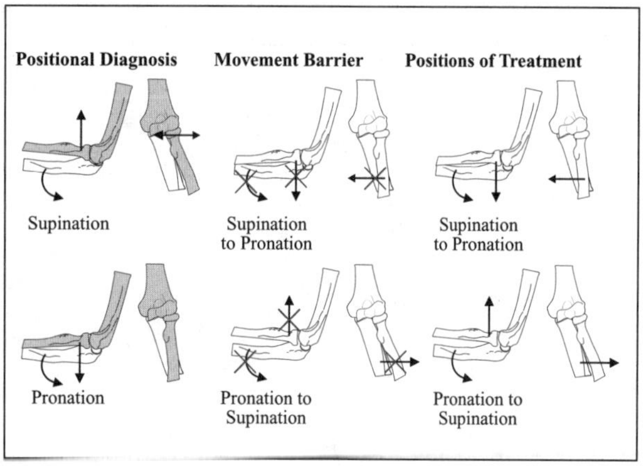

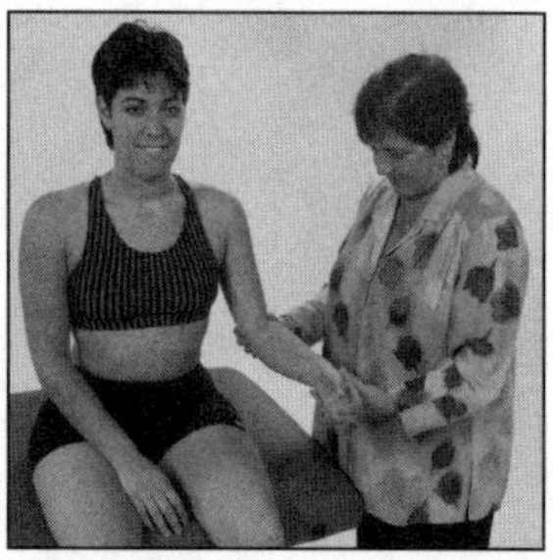

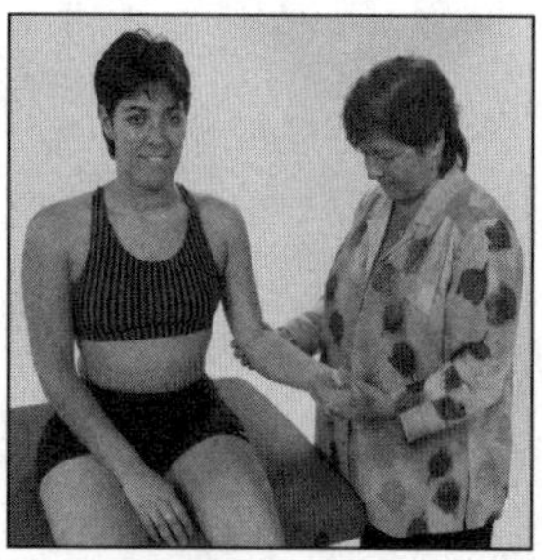

The Humeroulnar Joint

This joint is the primary flexion joint for the elbow. With this joint, there are extended dysfunctions that prevent flexion.

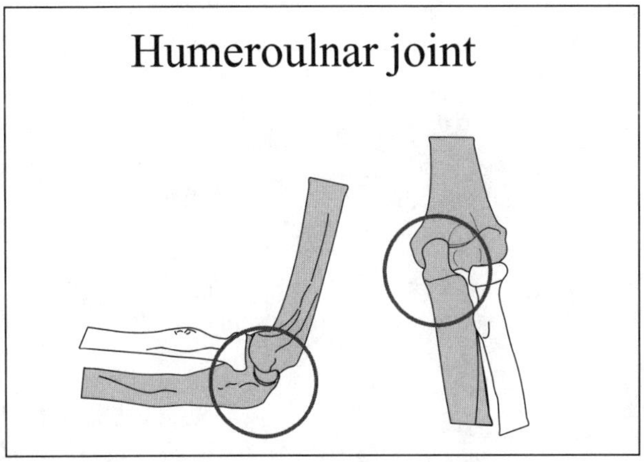

Positional Diagnosis

Extended, supinated:

> The humeroulnar joint is stuck extended and supinated, so that flexion and pronation are limited.

Extended, pronated:

> The humeroulnar joint is stuck extended and pronated, so that flexion and supination are limited.

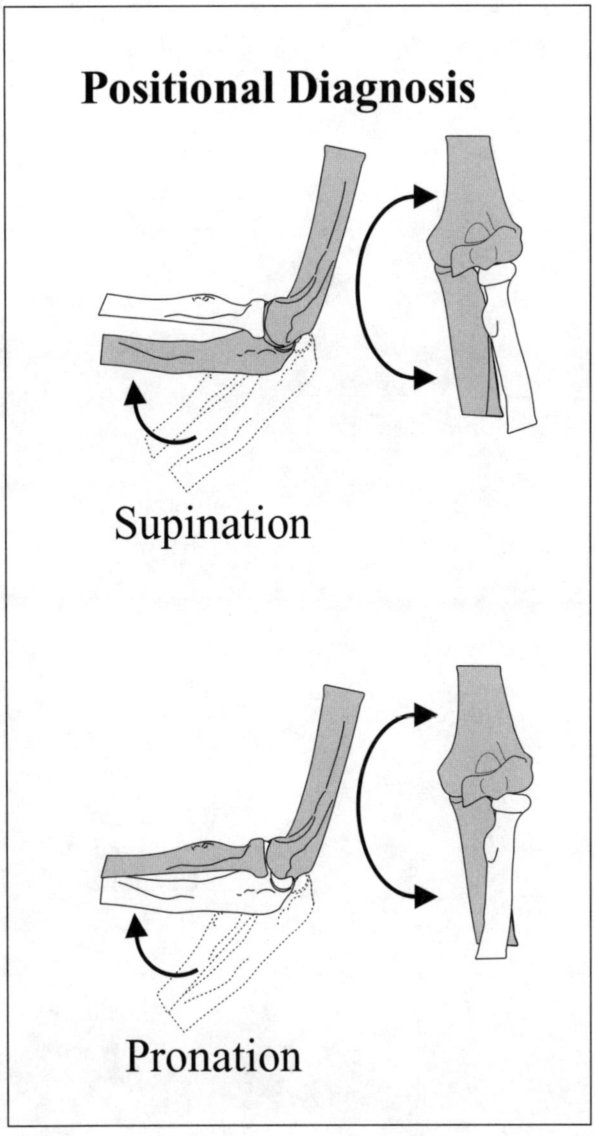

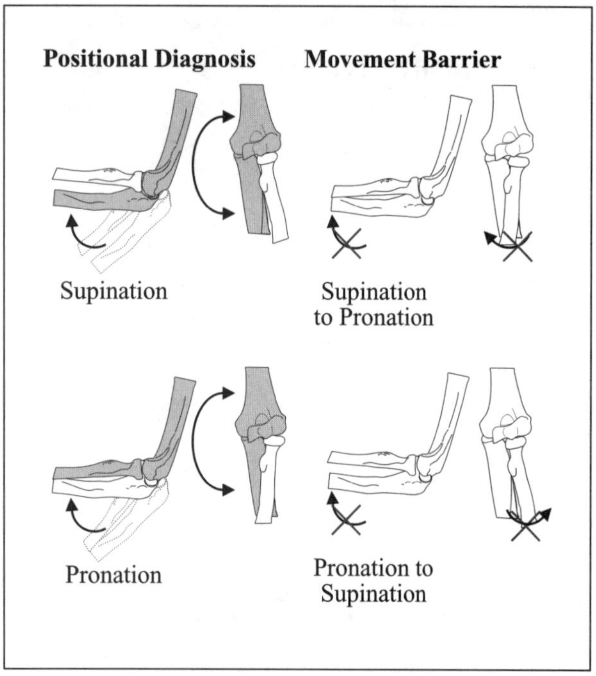

Movement Barrier

Palpate the muscle barrier at the cubital angle.

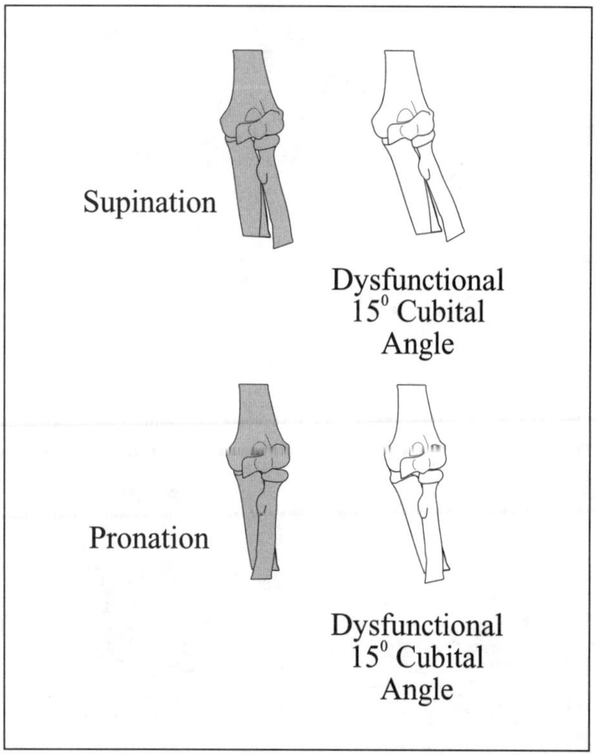

There should be no cubital angle until full extension. Full extension is 0 degree flexion. At 0 degree there is angulation with a medial gap that occurs during the locking mechanisms only.

Extended, supinated:
  The movement barrier is flexion and pronation.

Extended, pronated:
  The movement barrier is flexion and supination.

Treatment

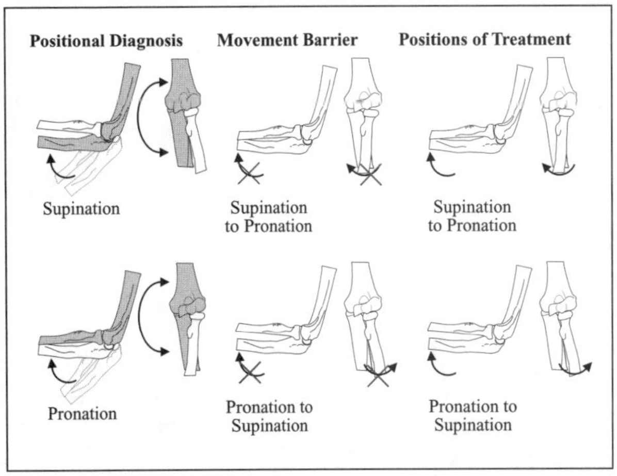

1. In supine or sitting. Position the shoulder at 0 degree abduction, 0 degree rotation, 0 degree flexion.
2. Position the forearm in 0 degree pronation/supination.
3. Position the elbow joint in full extension.
4. Passively flex the elbow joint.
5. Palpate the muscle barrier at the cubital angle.
6. As soon as the cubital angle increases, the barrier has been reached. Go to the interbarrier zone immediately before the cubital angle starts to appear. When the cubital angle appears with lateral glide of the proximal radial head, the positional diagnosis is: extended, pronated.

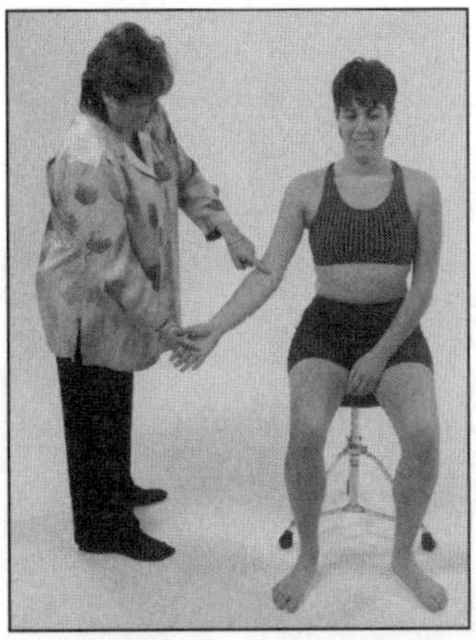
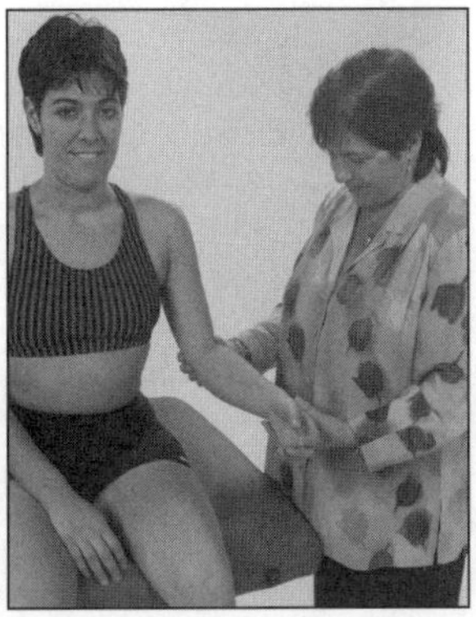

7. When the cubital angle appears with medial glide of the proximal radial head, the positional diagnosis is: extended, supinated.
8. Resistance: elbow flexion or extension; 6 seconds resistance, isometric, 5 grams force. Relax. Progress to the next interbarrier zone.
9. Repeat: 3 repetitions.
10. Reassess.

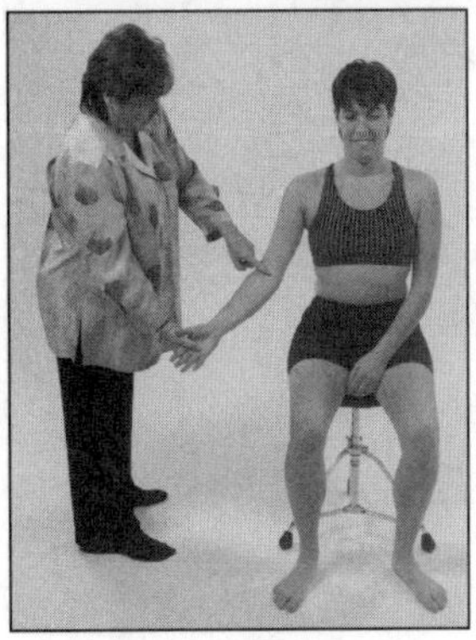
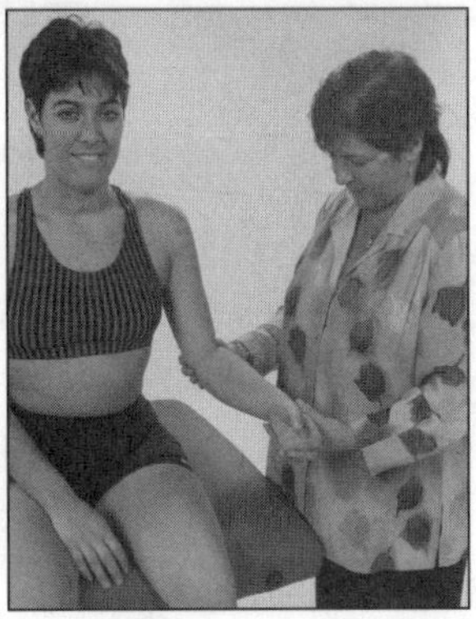

CHAPTER 33

THE WRIST JOINT

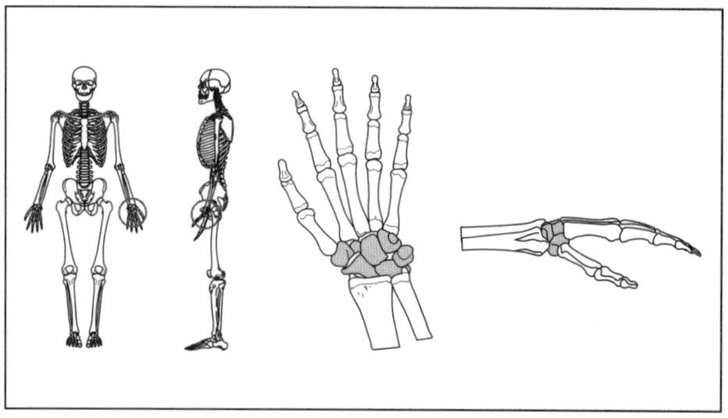

The wrist joint responds exceptionally well to Muscle Energy and "Beyond" Technique. Often, soft tissue restrictions of the wrist joint are really *double crush phenomena.* This means that a compromise of the proximal brachial plexus tissue (neurovascular tissue) will cause motion restrictions and soft tissue problems at the distal end of the arm, the wrist joint.

Assess soft tissue and joint restrictions at the neck and shoulder girdle region. Consider: spinal cord fibrosis of the cervical spine; cervical nerve root impingements, especially at C5; cervical plexus neural tissue tension; muscle spasm of the anterior and middle scalenes; costoclavicular joint dysfunction between clavicle and the first rib; muscle spasm of the pectoralis minor and fascial dysfunction of the clavipectoral fascia; tissue tension of the armpit with caudal subluxation of the humeral head (because of muscle spasm of the latissimus dorsi).

The radiocarpal joints are the primary joints of wrist extension.

The ulnarcarpal joints are the primary joints of wrist flexion.

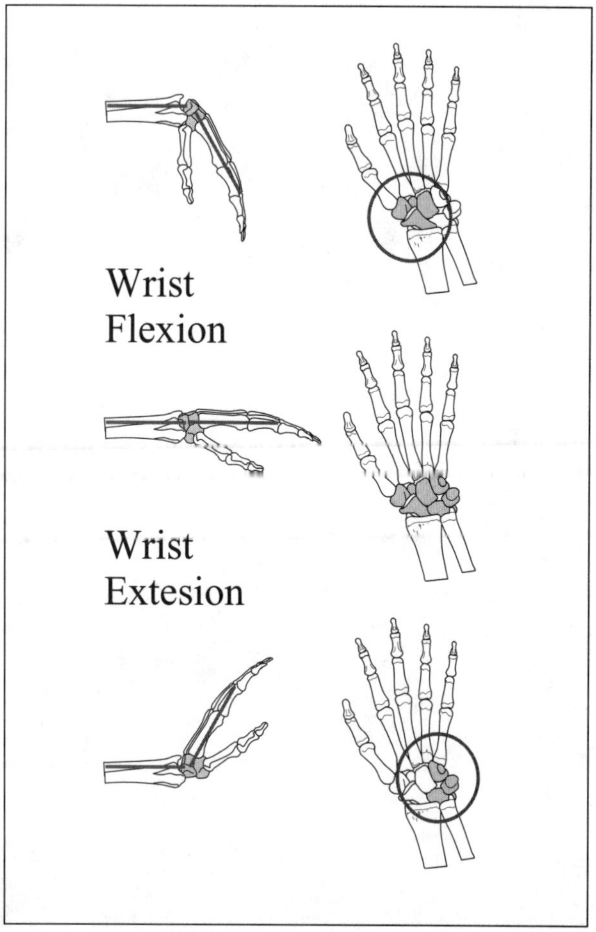

The proximal carpal row in articulation with the distal carpal row are the primary joints for radial and ulnar deviation.

The proximal carpal row glides medial for radial deviation.

The proximal carpal row glides lateral for ulnar deviation.

It is easy to attain excellent articular balance and good mobility with Muscle Energy and "Beyond" Technique for the radiocarpal joints and the ulnarcarpal joints.

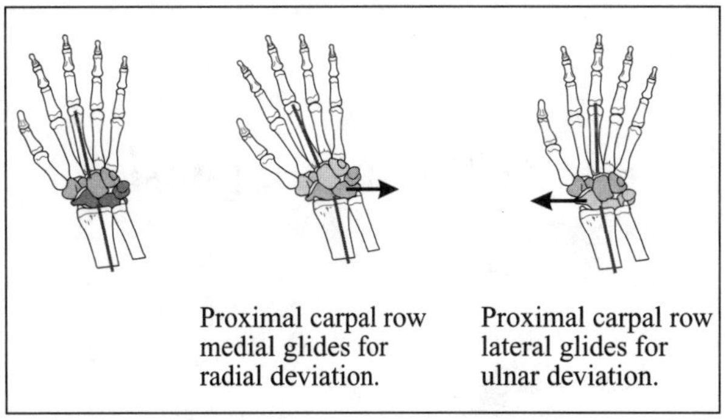

Direct Accessory Movements

1. During physiologic extension there is anterior glide of the proximal carpal row.
2. During physiologic flexion there is posterior glide of the proximal carpal row.
3. During physiologic radial deviation there is medial glide of the distal carpal row.
4. During physiologic ulnar deviation there is lateral glide of the distal carpal row.

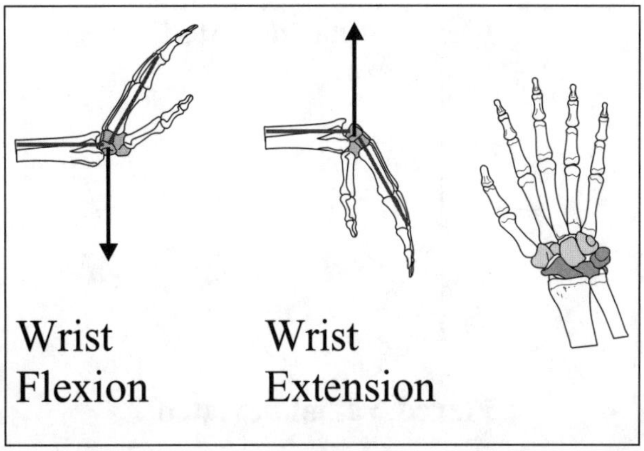

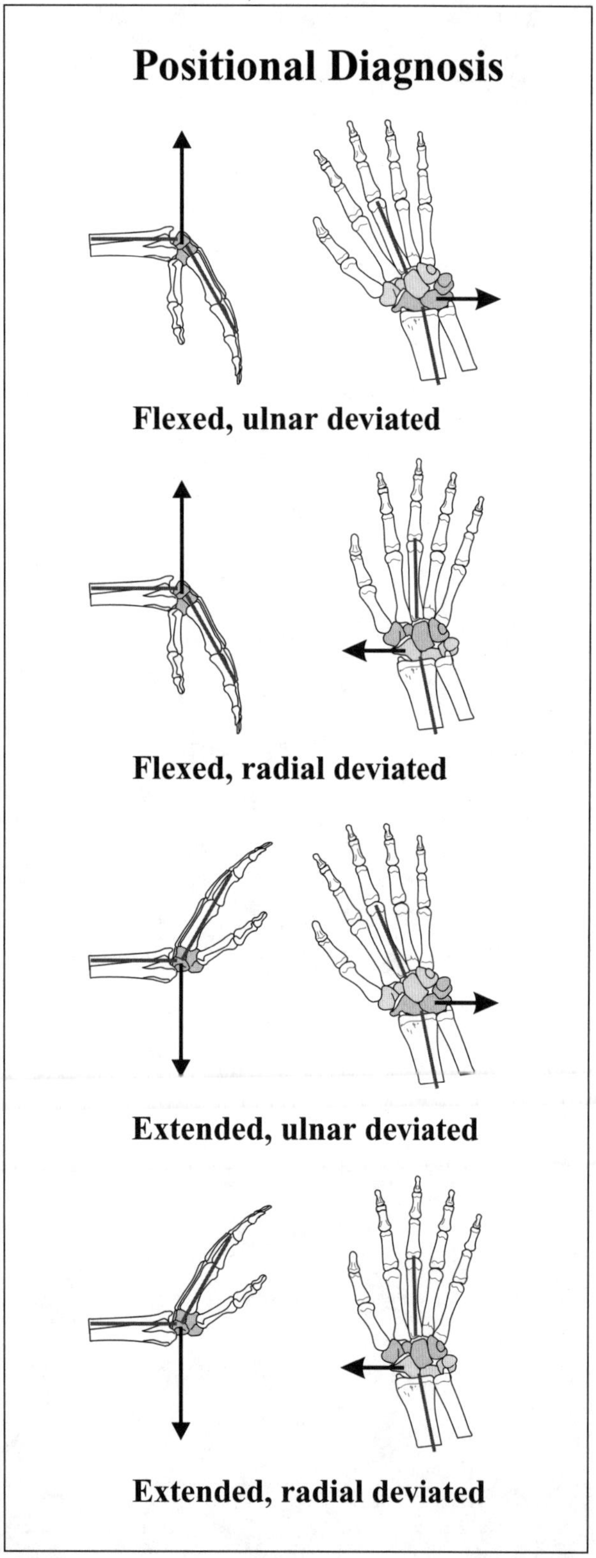

Positional Diagnosis

Flexed, ulnar deviated:
This occurs primarily at the radiocarpal joints.

Flexed, radial deviated:
This occurs primarily at the radiocarpal joints.

Extended, ulnar deviated:
This occurs primarily at the ulnarcarpal joints.

Extended, radial deviated:
This occurs primarily at the ulnarcarpal joints.

Movement Barrier

Palpate the movement barrier at the medial and lateral aspects of the proximal carpal row. Flexion and extension of the wrist joint occur without ulnar or radial deviation.

Flexed, ulnar deviated:
: The wrist is stuck flexed. The proximal carpal row is stuck in lateral glide. The movement barrier is extension and radial deviation.

Flexed, radial deviated:
: The wrist is stuck flexed. The proximal carpal row is stuck in medial glide. The movement barrier is extension and ulnar deviation.

Extended, ulnar deviated:
: The wrist is stuck extended. The proximal carpal row is stuck in lateral glide. The movement barrier is flexion and radial deviation.

Extended, radial deviated:
: The wrist is stuck extended. The proximal carpal row is stuck in medial glide. The movement barrier is flexion and ulnar deviation.

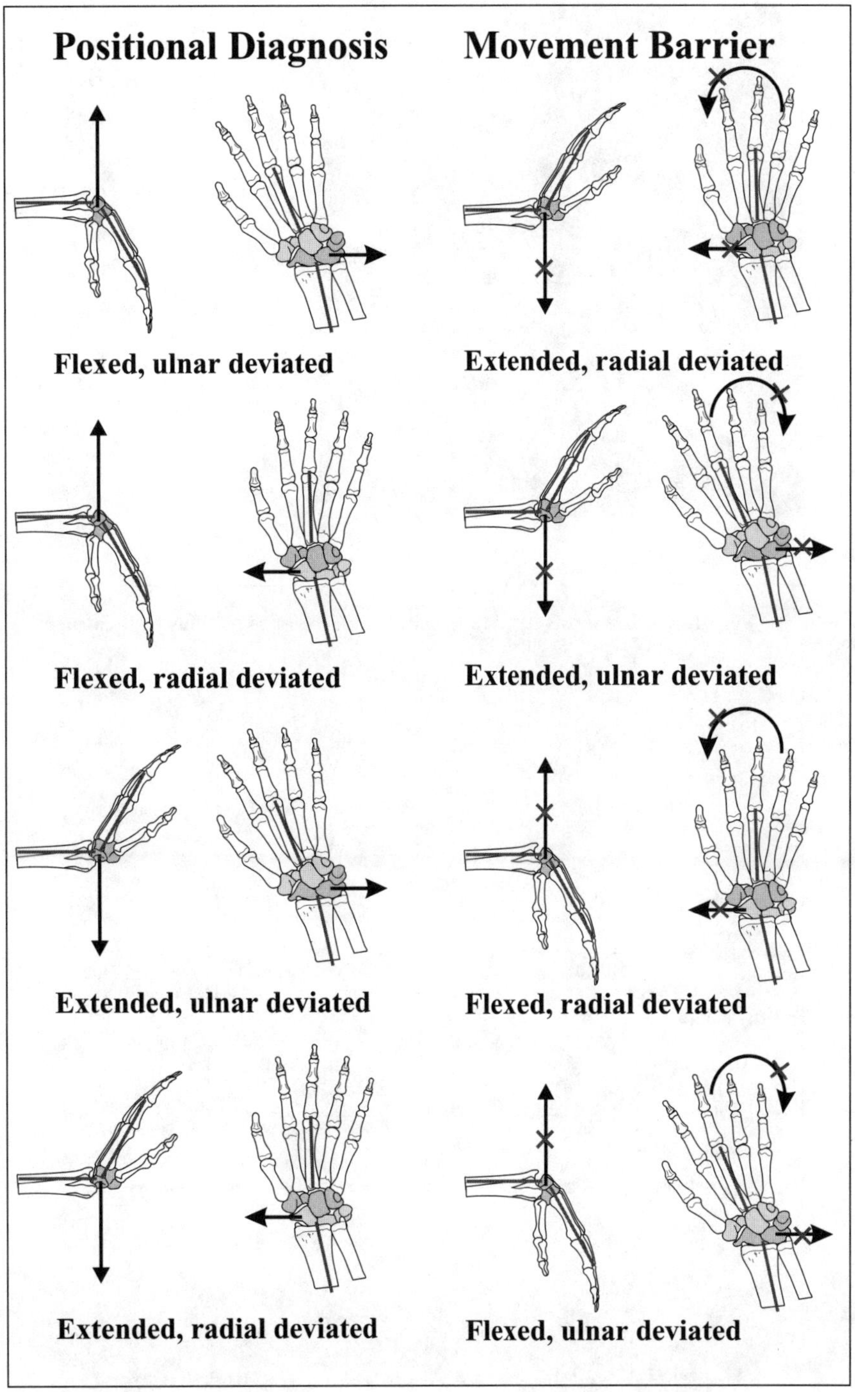

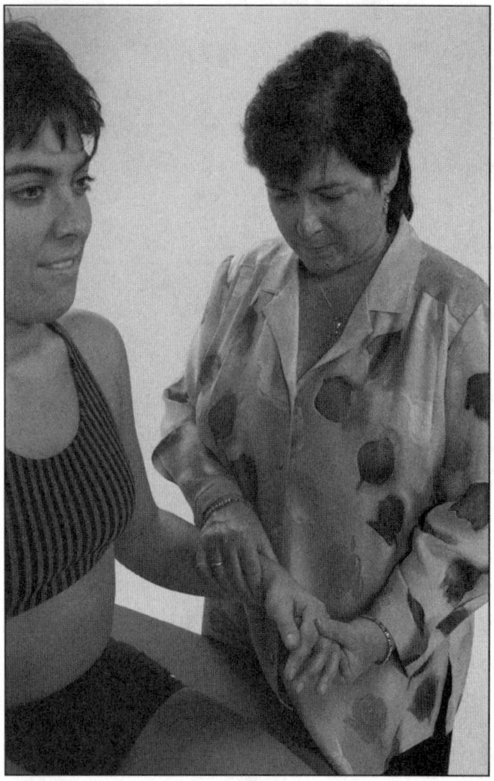

Treatment

1. Position in supine or sitting. Position the shoulder in 0 degree abduction, 0 degree rotation, 0 degree flexion.
2. Position the elbow joint at 90 degrees flexion.
3. Position the forearm in 0 degree supination/pronation.
4. Begin from 0 degree neutral wrist flexion/extension.
5. Palpate lateral aspects of the proximal carpal row to assess muscle barriers.

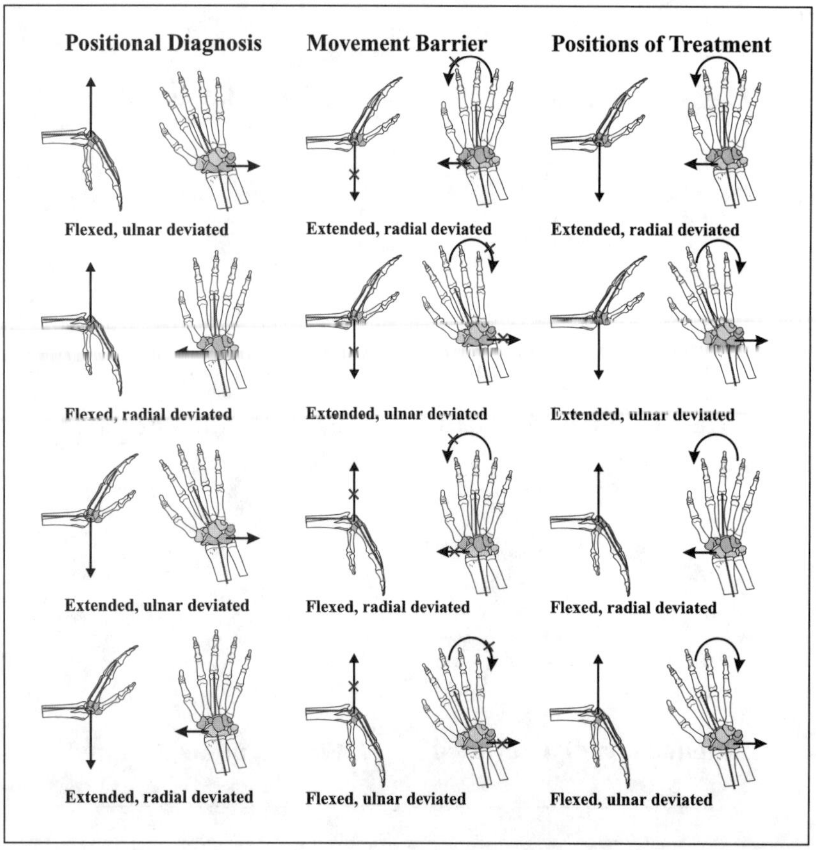

Flexed, ulnar deviated:

1. Extend the wrist.
2. Palpate lateral glide of the proximal carpal row.
3. Return to the interbarrier zone just before lateral glide of the proximal carpal row.
4. Perform radial deviation to the interbarrier zone.
5. Resist any wrist movement. Isometric resistance: 5 grams of force; 6 seconds resistance.
6. Relax. Progress to next interbarrier zone.
7. Repeat 3 repetitions.
8. Reassess.

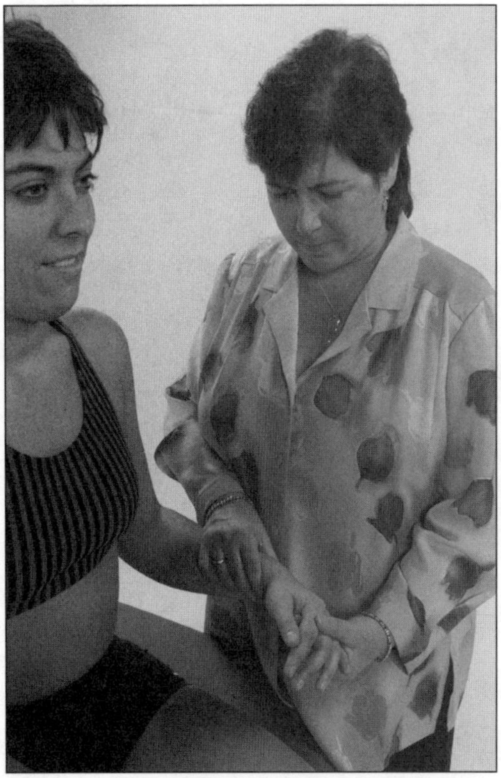

Flexed, radial deviated:

1. Extend the wrist.
2 Palpate the medial glide of the proximal carpal row.
3. Return to the interbarrier zone just before medial glide of the proximal carpal row.
4. Perform ulnar deviation to the interbarrier zone.
5. Resist any wrist movement. Isometric resistance: 5 grams of force; 6 seconds resistance.
6. Relax. Progress to next interbarrier zone.
7. Repeat 3 repetitions.
8. Reassess.

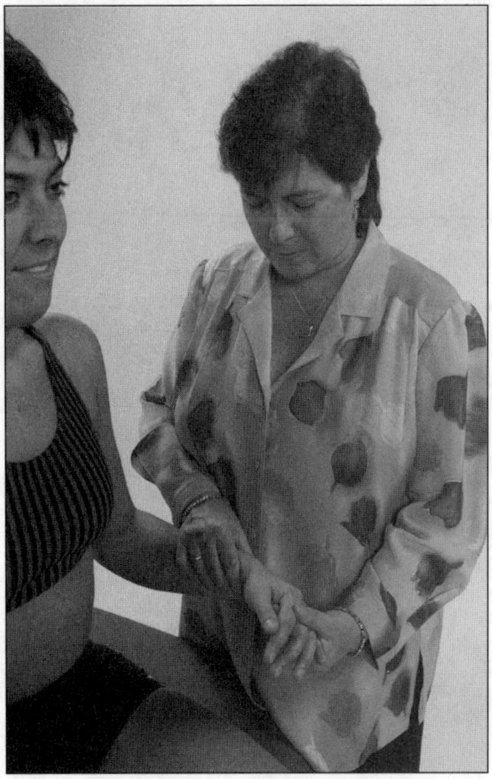

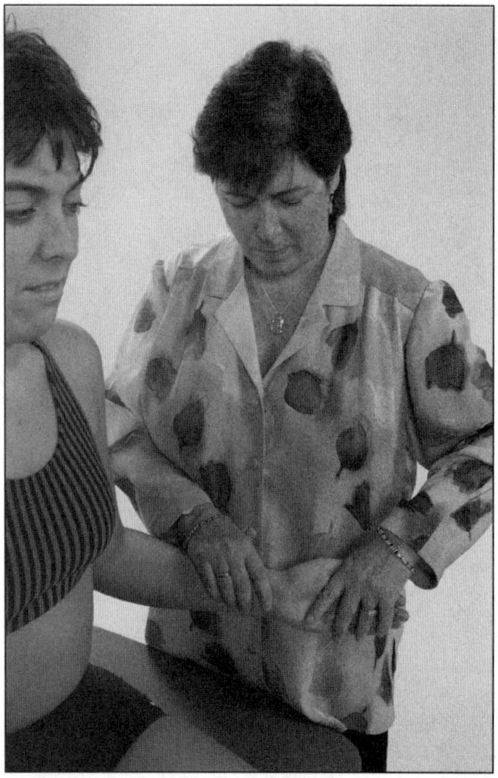

Extended, ulnar deviated:

1. Flex the wrist.
2 Palpate lateral glide of the proximal carpal row.
3. Return to the interbarrier zone just before lateral glide of the proximal carpal row.
4. Perform radial deviation to the interbarrier zone.
5. Resist any wrist movement. Isometric resistance: 5 grams of force; 6 seconds resistance.
6. Relax. Progress to next interbarrier zone.
7. Repeat 3 repetitions.
8. Reassess.

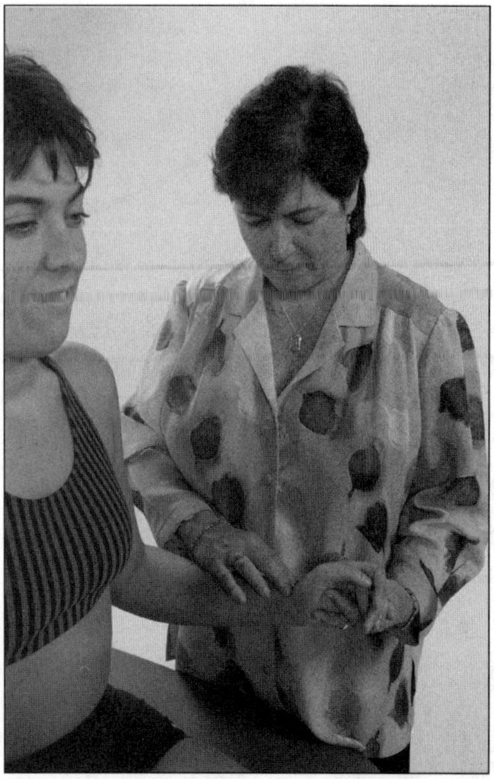

Extended, radial deviated:

1. Flex the wrist.
2. Palpate the medial glide of the proximal carpal row.
3. Return to the interbarrier zone just before medial glide of the proximal carpal row.
4. Perform ulnar deviation to the interbarrier zone.
5. Resist any wrist movement. Isometric resistance: 5 grams of force; 6 seconds resistance.
6. Relax. Progress to next interbarrier zone.
7. Repeat 3 repetitions.
8. Reassess.

REFERENCES

1 Adams, J. H., et al., "Diffuse Axonal Injury in Head-Injury—Definition, Diagnosis and Grading." *Histopathology,* 15, No. 1 (1989), 49–59.

2 Alexander, J. "The Nature of Contractures." *Clin Pros and Orthotics,* 8 (1985), 2–3.

3 Arbuckle, B. "Cranial Reinforcement from a Manipulative Standpoint: Articulations, Stress Bands, Buttresses." *JAOA,* 49 (1949), 188–94.

4 Arbuckle, B. "Subclinical Signs of Trauma." *JAOA,* 58 (1958), 160–66.

5 Arbuckle, B. "Through Cranial Base." JAOA, 48 (1949), 458–60.

6 Arcan, B. "A Fundamental Characteristic of the Human Body and Foot, the Foot-Ground Pressure Pattern." *J. Biomech,* 9 (1979), 453–57.

7 Arcan, et al. "FGP Assessment of Postural Disorders During the Process of Rehabilitation." *Scand. J. Rehab. Med.,* 9 (1977), 165–68.

8 Arnold, W., C.V., Ilberg. *Connections of the Cerebrospinal Fluid with the Lymphatic System of Head and Neck.* Frankfort, West Germany: Johann Wolfgang Goethe Universitat.

9 Ashby, P. "The Neurophysiology of Spasticity." *J. of Phys. Ther.,* 25, No. 4 (1973).

10 Bach-y-Rita, P. "A Conceptual Approach to Neural Recovery." Ed. Paul Bach-y-Rita. *Traumatic Brain Injury,* pp. 81–86.

11 Balliet, R. "Neurorehabilitation of Postacute Brain Injury: A Key to Optimizing Functional Motor Recovery." Ed. Paul Bach-y-Rita. *Traumatic Brain Injury,* pp. 119–133.

12 Barral, J.P. *Visceral Manipulation.* Seattle: Eastland Press, 1987.

13 Barral, J.P. *Visceral Manipulation II.* Seattle: Eastland Press, 1989.

14 Barral, J.P., P. Mercier. *Visceral Manipulation.* Seattle: Eastland Press, 1988.

15 Basmajian, J. *Muscles Alive.* Baltimore: Williams & Wilkins, 1977.

16 Basmajian, J. *Primary Anatomy.* Baltimore: Williams & Wilkins, 1970.

17 Becher, A. "Mobilizing the Thorax." *JAOA,* Jan. 1928, pp. 140–41.

18 Becker, D.P. *Textbook of Head Injury.* Philadelphia: W.B. Saunders, 1989.

19 Becker, D.P., et al. "Brain Cellular Injury and Recovery—Horizons for Improving Medical Therapies in Stroke and Trauma." *Western Journal of Medicine,* 148, No. 6 (1988), 670–84.

20 Becker, R.E. *The Body Electric.* New York: Morran, 1985.

21 Becker, R.E. "Cranial Therapy Revisited." *Osteopathic Annals,* 5 (1977), 11–40

22 Becker, R.E. "Craniosacral Trauma in the Adult." *Osteopathic Annals,* 4 (1976), 43–59.

23 Becker, R.E. "Diagnostic Touch: Its Principles and Application, Part I." *AAO Yearbook,* 63 (1963), 32–40.

24 Becker, R.E. "Diagnostic Touch: Its Principles and Application, Parts II and III." *AAO Yearbook,* 64 (1964), 153–60, 161–66.

25 Becker, R.E. "Diagnostic Touch: Its Principles and Application, Part IV." *AAO Yearbook,* 65, No. 2 (1965), 165–77.

26 Becker, R.E. "Meaning of Fascia and Fascial Continuity." *Osteopathic Annals,* 3 (1975), 8–32.

27 Becker, R.E. "Whiplash Injuries." *AAO Yearbook,* 1964, pp. 91–104.

28 Becker, R., G. Selder. *The Body Electric.* New York: Quill, 1985.

29 Beckwith, C.G. "Headache." *JAOA,* 48 (1988), 385–90.

30 Bell, W.E. *Clinical Management of Temporomandibular Disorders.* Chicago: Yearbook Medical Publishers, 1982.

31 Benson, H. *The Mind/Body Effect.* New York: Simon & Schuster, 1979.

32 Benson, H. *The Relaxation Response.* New York: Avon, 1975.

33 Bishop, B. "Neural Plasticity. Part 4. Lesion Induced Reorganization of the CNS. Recovery Phenomena." *J. of Phys. Ther.,* 62, No. 10 (1982).

34 Black, P., et al. *"An Interdisciplinary Perspective Study of Head Trauma in Children" Ed. C. R. Angle and E. A. Bering.* Physical Trauma as an Etiological Agent in Mental Retardation. Bethesda: Department of Health, Education and Welfare.

35 Bland, J. *Disorders of the Cervical Spine, Dx. and Medical Management.* W.B. Saunders, 1987.

36 Blumbergs, P. C., J. B. North, N. R. Jones. "Diffuse Axonal Injury in Head Trauma." *Journal of Neurology, Neurosurgery and Psychiatry,* 52, No. 7 (1989), 838–41.

37 Borysenko, J. *Minding the Body, Mending the Mind.* Reading: Addison Wesley Publishing, 1985.

38 Botte, M. J., W. H. Akeson, V. L. Nickel. "Spasticity and Contracture—Physiologic Aspects of Formation." *Clinical Orthopaedics and Related Research,* 233 (1988), 7–18.

39 Bourdillon, J., E. Day. *Spinal Manipulation.* Norwalk: Appleton and Lange, 1987.

40 Boyeson, M. G. "Neurotransmitter Aspects of Traumatic Brain Injury." Ed. Paul Bach-y-Rita. *Traumatic Brain Injury,* pp. 97–104.

41 Brunnstrom, S. *Clinical Kinesiology.* Philadelphia: F. A. Davis, 1979.

42 Buncher, C. R. "Statistics in Sports Injury Research." *Am J Sports Med,* 16(S) (1988), S57–S62.

43 Calliet, R. *Foot and Ankle Pain.* Philadelphia: F. A. Davis, 1978.

44 Calliet, R. *Hand Pain and Impairment.* Philadelphia: F. A. Davis, 1978

45 Calliet, R. *Knee Pain and Disability.* Philadelphia: F. A. Davis, 1979.

46 Calliet, R. *Low Back Pain Syndrome.* Philadelphia: F. A. Davis, 1976.

47 Calliet, R. *Neck and Arm Pain.* Philadelphia: F. A. Davis, 1978.

48 Calliet, R. *Shoulder Pain.* Philadelphia: F. A. Davis, 1977.

49 Calliet, R. *Soft Tissue Pain and Disability.* Philadelphia: F. A. Davis, 1978.

50 Cameron, J. *Principles of Physiological Measurement.* New York: Academic Press, 1st Ed., 1986, pp.179–80.

51 Carlon, R., B. Shield. *Healers on Healing.* Los Angeles: Jeremy Tarcher, Inc., 1989.

52 Carr, J. H., et al. *Movement Science: Foundations for Physical Therapy in Rehabilitation.* Aspen, MD, 1987.

53 Cash, J. *Neurology for Physiotherapists.* London: Faber and Faber, 1976.

54 Cathie, A. "Considerations of Fascia and Its Relation to Disease of the Musculoskeletal System." *American Academy of Osteopathy Yearbook,* 1974, pp. 85–88.

55 Cathie, A. "An Introduction to the Thoracoabdominal Junction." *AAO Yearbook,* 1960, pp. 29–32.

56 Caveness, W. F. "Post-Traumatic Sequelae." Ed. W. F. Caveness and A.E. Walker. *Head Injury.* Philadelphia: Lippincott.

57 Chusid, H. *Correlative Neuroanatomy and Functional Neurology.* Los Altos: Lane Medical, 18th Ed., 1982.

58 "Cranial Decompression." *LANCET,* 1, No. 8596 (1988), 1204.

59 Cyriax, J. *Textbook of Orthopedic Med-*

icine, Baltimore: Williams and Wilkins Co., Vol. 1, 6th Ed., 1976.

60 Cyriax, J., P. Cyriax. *Illustrated Manual of Orthopedic Medicine.* London: Butterworths, 1984.

61 Cyriax, J., G. Russel. *Textbook of Orthopedic Med.* Vol. II., Rx. by Manipulation, Massage and Injection. London: Baillier Lindall, 1977.

62 Daffner, R. H., et al. "Patterns of High-Speed Impact Injuries in Motor-Vehicle Occupants." *Journal of Trauma,* 28, No. 40 (1988), 498–501.

63 Devinsky, O., E. Feldman. *Examination of the Cranial and Peripheral Nerves.* New York: Churchill Livingstone, 1988.

64 Diller, L., Y. Ben-Yishay. "Assessment in Traumatic Brain Injury." Ed. Paul Bach-y-Rita. *Traumatic Brain Injury,* pp. 161–174.

65 Dorsey, L. *Beyond Illness. London: New Science Library, 1984.*

66 Dovesmith, E. "Growing Skull and Injured Child." *AAO Yearbook,* 1967, pp. 34–40.

67 Downs, J. R. "Treating the TMJ Dysfunction." *Osteopathic Physician,* 43 (1976), 106–13.

68 Drew, E. G. "Diagnosis of Acute Brain Injuries." *JAOA,* 36 (1937), 517–18.

69 Edeleng, J. *Manual Therapy for Chronic Headache.* Boston: Butterworths, 1988.

70 Eggleston, A. "Differential Diagnosis of the Cervical and Upper Thoracic Area." *AAO Yearbook,* 1961, pp. 19–27.

71 Elkind, A. H. "Headache and Facial-Pain Associated with Head Injury." *Otolaryngologic Clinics of North America,* 22, No. 6 (1989), 1251–71.

72 Erb, D. E., J. T. Povlishock. "Axonal Damage in Severe Traumatic Brain Injury—An Experimental Study in Cat." *Acta Neuropathologica,* 76, No. 4 (1988), 347–58.

73 Erlingheuser, R. F. "Circulation of the Cerebrospinal Fluid Through the Connective Tissue System." *AAO Yearbook,* 1959, pp.77–87.

74 Eyzaguirre, C., S. Fidone. *Physiology of the Nervous System.* Chicago: Yearbook Medical Publishers, 1975.

75 Feldenkrais, M. *Awareness Through Movement.* New York: Harper and Row, 1972.

76 Fox, E. L. *Sports Physiology.* Philadelphia: Saunders College Pub., 1984, pp. 207–212.

77 Frankel, J., M. Nordin. *Basic Biomechanics of the Skeletal System.* Philadelphia: Lea & Febiger, 1980.

78 Frost, H. M. *An Introduction to Biomechanics.* Springfield: Charles Thomas Publishers, 1971.

79 Frymann, V. M. "Core Link and the Three Diaphragms, Unit for Respiratory Function." *AAO Yearbook,* 1968, pp. 13–19.

80 Frymann, V. M. "Osteopathic Approach to Cardiac and Pulmonary Problems." *JAOA,* 77 (1978), 668–73.

81 Garland, D. E. "Clinical Observations on Fractures and Heterotopic Ossification in the Spinal Cord and Traumatic Brain Injured Populations." *Clinical Orthopaedics and Related Research,* 233 *(1988), 86–101.*

82 Gehin, A. *Atlas of Manipulative Techniques for the Cranium and Face.* Seattle: Eastland Press, 1988.

83 Gelb, H. *Clinical Management of the Head, Neck, and TMJ—Pain and Dysfunction.* Philadelphia: W.B. Saunders, 1985.

84 Gennarelli, T. A., et al. "Axonal Injury in the Optic-Nerve—A Model Simulating Diffuse Axonal Injury in the Brain." *Journal of Neurosurgery,* 71, No. 2 (1989), 244–53.

85 Gensemer, I. B., et al. "Behavioral Consequences of Trauma." *Journal of Trauma,* 28, No. 1 (1988), 44–49.

86 Gentry, L. R., J. C. Godersky, B. Thompson. "Imaging of Head Trauma—Review of the Distribution and Radiopathologic Features of Traumatic Lesions." *American Journal of Neuroradiology,* 9, No. 1 (1988), 101–10.

87 Gentry, L. R., J. C. Godersky, B.H. Thompson. "Traumatic Brain-Stem Injury—MR Imaging." *Radiology,* 171, No. 1 (1989), 177–87.

88 Gonza, E., I. Harrington. *Biomechanics of Musculoskeletal Injury.* Baltimore: Williams & Wilkins, 1982.

89 Gossman, Sahrman, Rose. "Review of Length Associated Changes in Muscles: Experimental Evidence and Clinical Implications." *J. of Phys. Ther.,* 62, No. 12 (1982).

90 Gower, D. J., K. S. Lee, J. M. McWhorter. "Role of Subtemporal Decompression in Severe Closed Head Injury." *Neurosurgery,* 23, No. 4 (1988), 417–22.

91 *Grant Atlas of Anatomy.* Baltimore: Williams & Wilkins, 1973.

92 Gray, H. *Gray's Anatomy.* New York: Crown Publishers, 1972.

93 Greenman, P. "Roentgen Findings in the Craniosacral Mechanism." *JAOA,* 70 (1970), 24.

94 Gribbin, J. *In Search of Schrodinger's Cat.* New York: Bantam.

95 Grieve, G. *Common Vertebral Joint Problems.* New York: Churchill Livingstone, 2nd Ed., 1988.

96 Grieve, G. *Mobilization of the Spine.* New York: Churchill Livingstone, 2nd Ed., 1986.

97 Haraka, J. H. "Dissection Offers Proof of Sutherland's Concept." *JAOA,* 82 (1982), 87.

98 Haraka, J. H. "An Osteopathically Integrated Approach to the Whiplash Complex." *JAOA,* 74 (1975), 941–55.

99 Hay, J. G. *The Biomechanics of Sports Techniques.* Englewood Cliffs, NJ: Prentice Hall, 1973, p. 236.

100 Hayes, R. L., et al. "Possible Protective Effect of Endogenous Opioids in Traumatic Brain Injury." *Journal of Neurosurgery,* 72, No. 2 (1990), 252–61.

101 Helfet, A., L. D. Gruebel. *Disorders of the Lumbar Spine.* JB Lippincott Co., 1978.

102 Helig, D. "The Importance of the Dorsolumbar Junction." *JAOA,* 55, No. 6 (1956), 167–170.

103 Hicks, C. *Practical Research Methods for Physiotherapy.* Edinburg: Churchill Livingstone, 1988.

104 Hinkel, R. *Tests of Significance.* Beverly Hills: Sage, 1976.

105 Ho, R. "Disorders of the Myofascia." *Osteopathic Medicine.* New York: McGraw Hill, Ch. 27.

106 Hoppenfeld, S. *Physical Examination of the Spine and Extremities.* Norwalk: Appelton-Century-Crofts, 1976.

107 Horwich, H. "Ocular Effects of Whiplash Injury." *Trans. Sect. Ophthalmology,* AMA, 1961, pp. 86–7.

108 Horwich, H., D. Kasner. "Effect of Whiplash Injuries on Ocular Functions." *SMJ,* 55 (1962), 69–71.

109 Iafrate P. R., K. L. Massey, L. Hendles. "Current Concepts in Clinical Therapeutics: Asthma." *Clin Pharm,* 5 (1986), 206–66.

110 Inman, Rolston, Todd. *The Joints of the Ankle.* Baltimore: Williams and Wilkins, 1976.

111 Inman, Rolston, Todd. *Human Walking.* Baltimore: Williams and Wilkins, 1976.

112 Jacobi, G., D. Klinter, G. Weiermann. "Clinic and Patho-Physiology of Intracranial-Pressure Increase." *Klinische Padiatrie,* 200, No. 5 (1988), 359–74.

113 Jones, L. *Strain and Counterstrain.* The American Academy of Osteopathy.

114 Kaltenborn. *Manual Therapy for the Extremities Joints.* Oslo, Norway: Olaf Nordis Bokhandel, 1974.

115 Kaltenborn, F. *Mobilization of the Spinal Column.* New Zealand University Press.

116 Kandel, E. R., J. H. Schwartz. *Principles of Neural Science.* New York: Elsevier Science Publishing, 2nd Ed., 1985, Ch. 37.
117 Kannus, P., M. Jarvinen. "Prediction of Torque Acceleration Energy and Power of Thigh Muscles from Peak Torque." *Med Sci Sport Ex.,* 21 (1989), 304–07.
118 Kapandji, I. A. *The Physiology of the Joints—The Upper Limb.* New York: Churchill Livingstone, 1974.
119 Kapandji, I. A. *The Physiology of the Joints—The Lower Limb.* New York: Churchill Livingstone, 1987.
120 Kapandji, I. A. *The Trunk and Vertebral Column. The Physiology of the Joints* Vol. III. New York: Churchill Livingston, 1979.
121 Kelikian, H. *Disorders of the Ankle.* Philadelphia: W.B. Saunders, 1985.
122 Kendall, H., F. Kendall, G. Wadworth. *Muscle Testing and Function.* Baltimore: Williams & Wilkins Co., 1971.
123 Kimmel, D. L. "Innervation of Spinal Dura Mater and Dura Mater of the Posterior Cranial Fossa." *Neurology,* 11 (1961), 800–09.
124 Kirby, R., Roberts, J. A. *Introductory Biomechanics.* Ithica, New York: Movement Publications, 1985, pp. 340–43.
125 Kissoon, N., J. Dreyer., M. Walia. "Pediatric Trauma—Differences in Patho-Physiology, Injury Patterns and Treatment Compared with Adult Trauma." *Canadian Medical Association Journal,* 142, No. 1 (1990), 27–34.
126 Knott and Voss. *Proprioceptive Neuromuscular Facilitation.* New York: Harper and Row, 1968.
127 Kolb, P., C. Denegar. "Traumatic Edema and the Lymphatic System." *Ath Train,* 18 (1983), 339–41.
128 Korr, I. M. "The Neural Basis of the Osteopathic Lesion." *JAOA,* 47 (1947), 191–98.
129 Korr, I. M. "Proprioceptors and Somatic Dysfunction." *JAOA,* 74 (1975), 638–50.
130 Korr, I. M., P. E. Thomas, H. M. Wright. "Symposium on the Functional Implications of Segmental Facilitation." *JAOA,* 54 (1955), 265–82.
131 Kraus, H. *Backache, Stress, Tension.* New York: Simon & Schuster, 1965.
132 Kruse, R., C. Kruse, R. Bolton. "Physical Therapy for the Patient with Peripheral Edema: Procedures for Management." *Phys Ther Rev,* 41 (1960), 29–33.
133 Kunz, K., B. Kunz. *The Hand Reflexology Workbook.* New York: Prentice Hall, 1985.
134 Kurz, I. *Introduction to Dr. Vodder's Manual Lymph Drainage,* Vol. II. Heidelberg: Haug Publishers, 1986.
135 LaFreniere, J. *LaFreniere Body Techniques.* U.S.A.: Masson Publishing, 1982.
136 Lance, J. W. "The Control of Muscle Tone, Reflexes, and Movement, Robert Wartenberg Lecture." *Neurology,* 30 (1980), 1303–13.
137 Larson, R. L. "Epiphyseal Injuries in the Adolescent Athlete." *Orth Clin North America,* 4 (1973), 839–51.
138 Larson, R. L., R. O. McMahon. "The Epiphysis and the Childhood Athlete. *JAMA,* 196 (1966), 607–12.
139 LaVine, R. *Neurophysiology: The Fundamental.* Lexington: Collamore, 1983.
140 Lay, E. M. "Osteopathic Management of Trigeminal Neuralgia." *JAOA,* 74 (1975), 373–89.
141 Lay, E. M., R. A. Cicora, M. Tettambel. "Recording of the Cranial Rhythmic Impulse." *JAOA,* 78 (1978), 149.
142 LeMay, M. "Asymmetries of the Skull and Handedness." *J. Neurol. Sci.,* 32 (1977), 243–53.
143 Leskell, L. "The Action Potential and Excitatory Effects of the Small Ventral Root Fibres to Skeletal Muscles." *Acta. Physiol. Scand.,* 10 (1945), 1–84.
144 Levitz, S. J., L. S. Whiteside, T. A. Fitzgerald. "Biomechanical Foot Therapy." *Clinical Pod Med Surg,* 5 (1988), 721–36.

145 Little, K. "Toward More Effective Manipulative Management of Chronic Myofascial Strain and Stress Syndromes." *Journal of American Osteopathic Association,* 68 (1969), 675–85.

146 Lobato, R. D., et al. "Post-traumatic Cerebral Hemispheric Swelling—Analysis of 55 Cases Studied with Computerized-Tomography." *Journal of Neurosurgery,* 68, No. 3 (1988), 417–23.

147 Lobato, R. D. "Management of Post-traumatic Brain Swelling—Response." *Journal of Neurosurgery,* 69, No. 6 (1988), 963.

148 Logan, G., W. McKinney. *Anatomic Kinesiology.* Dubuque, IA: W.C. Brown Co., 1983, p. 224.

149 Magoun, H. I. "Entrapment Neuropathy, Part I." *JAOA,* 67, No. 6 (1968), 643–52.

150 Magoun, H. I. "Entrapment Neuropathy, Part II." *JAOA,* 67, No. 7 (1968), 779–87.

151 Magoun, H. I. "Entrapment Neuropathy, Part III." *JAOA,* 67, No. 8 (1968), 889–99.

152 Magoun, H. I. *Osteopathy in the Cranial Field. Kirksville: Journal Printing Co.,* 3rd Ed., 1976.

153 Magoun, H. I. "Temporal Bone: Trouble Maker in the Head." *JAOA,* 73 (1974), 825–35.

154 Magoun, H. I. "Whiplash Injury: A Greater Lesion Complex." *JAOA,* 63 (1964), 524–35.

155 Maigne, R., W. T. Liberson., Eds. *A New Approach to Vertebral Manipulation.* Charles C. Thomas, 3rd Ed., 1979.

156 Maitland, G. *Vertebral Manipulation.* London: Butterworths.

157 Manter, J. "Movements of the Subtalar and Transverse Tarsal Joints." *The Anatomical Record,* 80, No. 4 (1941).

158 Marshall, C., G. Rossman. *Designing Qualitative Research.* Newbury Park: Sage, 1989.

159 Martin, M., et al. *Examination and Mobilization of the Extremities.* Part 1—Course Workbook.

160 Matiland, G. *Peripheral Manipulation. London: Butterworth & Co., Ltd., 1980.*

161 Mattis, S. "Mild Head Trauma." Ed. Paul Bach-y-Rita. *Traumatic Brain Injury,* pp. 41–50.

162 McKenzie, R. *The Lumbar Spine.* Spinal Publications, 1981.

163 Mehl, L. *Mind and Matter, A Healing Approach to Chronic Illness.* Berkeley, CA: Mind-Body Press, 1986.

164 Mennel, J. *Joint Pain.* Boston: Little, Brown & Co., 1964.

165 Micheli, L. J. "Overuse Injuries in Children's Sports: The Growth Factor." *Orthopedic Clinics of North America,* 14 (1983), 337–59.

166 Micheli, L. J. "Pediatric and Adolescent Sports Injuries: Recent Trends." *Clinics in Sports Medicine,* 18 (1987), 359–74.

167 Mitchell, F., P. Moran, N. Pruzzo. *An Evaluation and Treatment Manual of Osteopathic Muscle Energy Procedures.* Manchester, MO: Mitchell, Moran & Pruzzo Associates, 1979.

168 Netter, F. *The Nervous System* Vol. 1. West Caldwell: CIBA, 1983.

169 Nicosia, S. "The Minor Head Injury." *Axone,* 11, No. 1 (1989).

170 Norkin, C. C., D. J. White. *Measurement of Joint Motion: A Guide to Goniometry.* Philadelphia: F. A. Davis, 1985, pp. 90–93.

171 Noyes, F. R., T. N. Lindenfeld, M. T. Marshall. "What Determines an Athletic Injury (Definition)? Who Determines an Injury (Occurence)?" *Am J Sports Med,* 16, Suppl. 1 (1988), S5–S68.

172 O'Bryant, H. S. "Anaerobic Capacity: Programs for Basketball, Wrestling, Football." *National Strength and Conditioning Assn. Journal,* 5, No. 6 (1984), 41.

173 O'Donoghue, D. H. *Treatment of Injuries to Athletes.* Philadelphia: W.B. Saunders,

4th Ed., 1984.

174 Olsnes, B. T. "Neurobehavioral Findings in 80 Whiplash Patients with Long Lasting Symptoms." *Acta Neurologica Scandinavia,* 80, No. 6 (1989), 584–88.

175 Oxendine, J. B. *Psychology of Motor Learning.* New Jersey: Prentice-Hall, 2nd Ed., 1984.

176 Pang, L.Q. "Otological Aspects of Whiplash Injuries." *Laryngoscope,* 81 (1971), 1381–87.

177 Pepe, P. E. "Acute Post-Traumatic Respiratory Physiology and Insufficiency." *Surgical Clinics of North America,* 69, No. 1 (1989), 157–73.

178 Perry, J. "The Mechanics of Walking, a Clinical Interpretation." *J. of Phys. Ther.,* 47, No. 9.

179 Peterson, B., Ed. *Posture Balance and Imbalance.* Newark, NJ: American Academy of Osteopathy, 1983.

180 Peterson, C. A., H. A. Peterson. "Analysis of the Incidence of Injuries of the Epiphyseal Growth Plate." *Journal of Trauma,* 12 (1972), 275–81.

181 Podlas, H., K. L. Allen, E. A. Bunt. "Computed Tomography Studies of Human Brain Movements." *South African J. Surgery,* 22, No. 1 (1984), 57–63.

182 Polchaninoff, M. Gait. "Analysis Using a Portable, Microprocessor-Based Segmental Foot Force Measuring System." *IEEE Procedings of the Seventh Annual Symposium on Computer Application in Medical Care,* 1983.

183 Posner, M. I., S. E. Petersen. "The Attention System of the Human Brain." *Annual Review of Neuroscience,* 13 (1990), 25–42.

184 *Post Graduate Advances in Evaluation and Treatment of Low Back Dysfunction.* Forum Medicum, 1988.

185 *Post Graduate Advances in Physical Therapy.* Forum Medicum, 1989.

186 Povlishock, J. T. "Structural Aspects of Brain Injury." Ed. Paul Bach-y-Rita. *Traumatic Brain Injury,* pp. 87–96.

187 Quillin, W., K. Rouller. "Initial Management of Acute Sprains with Rapid Pulsed Pneumatic Compression and Cold." *JOSPT,* 4 (1982), 39–43.

188 Raney, R., H. Brashear. *Shands Handbook of Orthopedic Surgery.* Philadelphia: C.V., Mosby Co., 1971.

189 Rathbone, J. *Relaxation.* Philadelphia: Lea & Febiger, 1969.

190 Reid, C. C. "Cranial as Related to Eye, Ear, Nose & Throat." *JAOA,* 48 (1948), 428–31.

191 Reiser, D., D. Rosen. *Medicine as a Human Experience.* Rockville: Aspen Publications, 1984.

192 Retzlaff, E. W., D. K. Michael, R. M. Roppel. "Cranial Bone Mobility." *JAOA,* 74 (1975), 869–73.

193 Retzlaff, E. W., et al. "Nerve Fibers and Endings in Cranial Sutures." *JAOA,* Research Report, 77 (1978), 474–75.

194 Retzlaff, E. W., et al. "Possible Functional Significance of Cranial Bone Sutures." Report, 88th Session Am. Assn. of Anatomists, 1975.

195 Retzlaff, E. W., et al. "Structure of Cranial Bone Sutures." *JAOA,* Research Report, 75 (1976), 607–8.

196 Richardson, T., et al. "Improved Rebounding Performance through Strength Training." *National Strength and Conditioning Assoc. Journal,* 4, No. 6 (1983), 6–7, 70–71.

197 Riolo-Quinn, L. "Principles of Motor Learning." *Clinical Management in Physical Therapy,* 7 (1987), 9–11.

198 Roca, P. D. "Ocular Manifestations of Whiplash Injuries." *Ann. Ophthlmol,* 4 (1972), 63–73.

199 Rocobado, M. *Head, Neck and Temporomandibular Joint Dysfunctions.* Tacoma: Rocabado Institute, 1981.

200 Root, M. L., W. P. Orien, J. H. Weed.

Normal and Abnormal Function of the Foot. Los Angeles: Clinical Biomechanics, Vol. 2, 1977.

201 Rossi, P. H., H. E. Freeman. *A Systematic Approach,* 4th Ed., 1989.

202 Rossman, M. *Healing Yourself.* New York: Walker & Co., 1987.

203 Rovere, G. D., et al. "Retrospective Comparison of Taping and Ankle Stabilizers in Preventing Ankle Injuries." *Am J Sports Med,* 16 (1988), 228–33.

204 Runyon, R., A. Haber. *Fundamentals of Behavioral Statistics.* Reading: Addison-Wesley Publishing Co., 1977.

205 Sahrman, Norton. "The Relationship of Voluntary Movement to Spasticity in the Upper Motor Neuron Syndrome." *Annals of Neurology,* 2, No. 6 (1977).

206 Schneider, M. *Self Healing.* New York: Routledge & Kegan Paul, 1988.

207 Schooley, T. F. "Embryologic Development of the Central Nervous System." *Osteopathic Annals,* 4 (1976), 20–21.

208 Seitz, C. J., A. J. Goldfuss. "The Effect of Taping and Exercise on Passive Foot Inversion and Ankle Plantarflexion." *Athletic Training,* 19 (1984), 178–82.

209 Seminick, D. M., K. O. Adams. "The Vertical Jump: A Kinesiological Analysis with Recommendations for Strength and Conditioning Programming." *National Strength and Conditioning Association Journal,* 9, No. 3 (1987), 5–9.

210 Shankman, G. A. "Special Considerations in Conditioning the Young Athlete." *National Strength and Conditioning Association Journal,* 7, No. 3 (1985), 52–53.

211 Shealy, C. *The Pain Game.* Melbourne: Celestial Arts, 1976.

212 Siegel, B. *Love, Medicine and Miracles.* New York: Harper & Row, 1986.

213 Smith, F. F. *Inner Bridges—A Guide to Energy Movement and Body Structure.* Atlanta: Humanics, New Age, 1986.

214 Snyder, G. *Fascia—Applied Anatomy and Physiology,* pp. 65–75.

215 Stanitski, C. L. "Low Back Pain in Young Athletes." *The Physician and Sports Medicine,* 10 (1982), 77–90.

216 Sudnow, D. *Ways of the Hand.* New York: Harper & Row, 1978.

217 Sutherland, W. "The Cranial Bowl." *The Western Osteopath,* 1944.

218 Thie, J. *Touch for Health.* Marina del Rey, CA: DeVorss & Co., 1973.

219 Toglia, J. U., P. E. Rosenberg, M. L. Ronis. "Post-traumatic Dizziness, Vestibular, Audiologic and Medicolegal Aspects" *Arch. Otolaryngol,* 92 (1970), 485–92.

220 Trager, M. *Trager Mentastics—Movement as a Way to Agelessness.* New York: Station Hill Press, 1987.

221 Travell, J. "Lecture on Head Pain. Convocation of AAO." Colorado Springs, May 25, 1973.

222 Travell, J. "Temporomandibular Joint Pain Referred from the Muscles of the Head and Neck." *J. Prosthet. Dent.,* 10 (1960), 745.

223 Travell, J., D. G. Simons. *Myofascial Pain and Dysfunction: The Trigger Point Manual.* Baltimore: Williams and Wilkins Co., 1983.

224 Travell, J., S. H. Rinzler. "Myofascial Genesis of Pain." *Postgrad. Med.,* 2 (1952), 425.

225 Tursz, A. "Sports-Related Injuries in Children." *Am J Sports Med,* 14 (1986), 294–99.

226 Tweedle, C. D., H. Popiela, C.S. Thornton. "An Ultrastructural Investigation of the Trophic Role of the Nerve in the Development and Maintenance of Muscle." *JAOA,* 74 (1975), 880.

227 Upledger, J. E. "Craniosacral Function in Brain Dysfunction." *Osteopathic Annals,* 11 (1983), 318–24.

228 Upledger, J. E. *Craniosacral Therapy II—Beyond the Dura.* Seattle: Eastland Press, 1987.

229 Upledger J. E. "Diagnosis and Treatment of Temporoparietal Suture Head Pain." *Osteopathic Medicine*, 78 (1978), 19–26.

230 Upledger, J. E. "Relationship of Cranio-Sacral Examination Findings in Grade School Children with Developmental Problems." *JAOA*,77 (1978), 760–76.

231 Upledger, J. E. "Reproducability of Craniosacral Examination Findings: A Statistical Analysis." *JAOA*, 76 (1977), 890–99.

232 Upledger, J. E., Z. Karni. "Bioelectric and Strain Measurements During Cranial Manipulation." *JAOA*, 77 (1978), 476.

233 Upledger, J. E., Z. Karni, E. Retzlaff. "Mechanicoelectrically Recorded Physiological Patterns which Relate to Craniosacral Mechanism Phenomena." *JAOA*, 78 (1978), 297

234 Upledger, J. E., J. Vredevoogd. *Craniosacral Therapy*. Seattle: Eastland Press, 1983.

235 Wallace, R. B., W. R. Clark. "The Numerator, Denominator, and the Population-at-Risk." *Am J Sports Med*, 16, Supple. 1 (1988), S55–S56.

236 Warfel, J. *The Extremities*. Philadelphia: Lea & Febiger, 1978.

237 Warfel, J. *The Head, Neck, and Trunk*. Philadelphia: Lea & Febiger, 1973.

238 Watkins, R., J. Collis. *Lumbar Discectomy and Laminectomy*. Rockville: Aspen Publications, 1987.

239 Wellman, C. *Morals and Ethics*. Englewood Cliffs, NJ: Prentice-Hall, 1988.

240 "Where to Turn for Sports Injury Statistics." *Phys Sportsmed*, 16, No. 4 (1988), 183–88.

241 White, A., M. Panjabi. *Clinical Biomechanics of the Spine*. Philadelphia: Lippincott, 1978.

242 Wittlinger, G., H. Wittlinger. *Introduction to Dr. Vodder's Manual Lymph Drainage*. Heidelberg: Haug Publishers, 1981.

243 Wittlinger, H. *Introduction to Dr. Vodder's Manual Lymph Drainage*. Heidelberg: Haug Publishers, Vol. 1, 1984.

244 Woods, C. "Structural Normalization in Infants and Children." *JAOA*, 72 (1973), 903–08.

245 Woods, J. M., R. H. Woods. "Physical Finding Related to Psychiatric Disorder." *JAOA*, 60 (1961), 988–93.

246 Worthingham, C., L. Daniels. *Muscle Testing*. Philadelphia: W.B. Saunders Co., 1972.

247 Worthingham, C., L. Daniels. *Therapeutic Evaluation*. Philadelphia: W.B. Saunders Co., 2nd Ed., 1977.

248 Wuthers, R., J. Collis. *Lumbar Discectomy & Laminectomy*. Rockville, MD: Aspen Publications, 1987.

249 Wyke, B. "Neurological Mechanisms in the Experience of Pain." *Acupuncture & Electrotherapy Research*, 4, 27.

INDEX